Lecture Notes in Computer Science 16523

Founding Editors

Gerhard Goos
Juris Hartmanis

The series Lecture Notes in Computer Science (LNCS), including its subseries Lecture Notes in Artificial Intelligence (LNAI) and Lecture Notes in Bioinformatics (LNBI), has established itself as a medium for the publication of new developments in computer science and information technology research, teaching, and education.

LNCS enjoys close cooperation with the computer science R & D community, the series counts many renowned academics among its volume editors and paper authors, and collaborates with prestigious societies. Its mission is to serve this international community by providing an invaluable service, mainly focused on the publication of conference and workshop proceedings and postproceedings. LNCS commenced publication in 1973.

Penousal Machado · Juan J. Romero ·
Sérgio M. Rebelo

Editors

Artificial Intelligence in Music, Sound, Art and Design

15th International Conference, EvoMUSART 2026
Held as Part of EvoStar 2026
Toulouse, France, April 8–10, 2026
Proceedings

 Springer

Editors
Penousal Machado (ID)
University of Coimbra
Coimbra, Portugal

Juan J. Romero (ID)
University of Coruña
A Coruña, Spain

Sérgio M. Rebelo (ID)
University of Coimbra
Coimbra, Portugal

ISSN 0302-9743 ISSN 1611-3349 (electronic)
Lecture Notes in Computer Science
ISBN 978-3-032-24349-2 ISBN 978-3-032-24350-8 (eBook)
https://doi.org/10.1007/978-3-032-24350-8

This Springer imprint is published by the registered company Springer Nature Switzerland AG
The registered company address is: Gewerbestrasse 11, 6330 Cham, Switzerland

If disposing of this product, please recycle the paper.

Preface

This volume contains the proceedings of EvoMUSART 2026, the 15th International Conference on Artificial Intelligence in Music, Sound, Art and Design. The conference is part of Evo*, the leading event on bio-inspired AI in Europe, and was held in Toulouse, France, as a hybrid event, between Wednesday, 8th April, and Friday, 10th April 2026.

While the utilisation of Artificial Intelligence for artistic purposes can be traced back to the 1970s, the use of Artificial Intelligence for the development of artistic systems is a recent, trending, and significant area of research. There is a growing interest in the application of these techniques in fields such as visual art and music generation, analysis and interpretation; sound synthesis; architecture; video; poetry; design; and other creative tasks.

The main purpose of EvoMUSART 2026 was to bring together practitioners who are using Artificial Intelligence techniques for artistic tasks, providing the opportunity to promote, present, and discuss ongoing work in the area. As always, we strive to foster a fun, inclusive, welcoming, and constructive environment that encourages networking and contribution.

EvoMUSART has grown steadily since its first edition in 2003 in Essex, United Kingdom, when it was one of the Applications of Evolutionary Computing workshops. Since 2012, it has been a full conference as part of the Evo* co-located events. At the same time, under the Evo* umbrella, EvoApplications targeted research on the application of bio-inspired computing, EuroGP focused on the technique of genetic programming, and EvoCOP was dedicated to evolutionary computation in combinatorial optimisation. The proceedings for these co-located events are available in the LNCS series.

EvoMUSART 2026 received 58 submissions. The peer-review process was rigorous and double-blind. The Program Committee, listed below, was composed of 62 members from 18 countries. We selected 13 of these papers for long oral presentations (approximately a 22% acceptance rate), while 15 works were presented in short oral presentations and as posters, resulting in an overall acceptance rate of about 48%.

As always, the EvoMUSART proceedings cover a wide range of topics and application areas, including generative approaches to music, visual art and design. This volume of proceedings collects the accepted papers. As in previous years, the standard of submissions was high and good-quality papers had to be rejected. We thank all authors for submitting their work, including those whose work was not accepted for presentation on this occasion.

The work of reviewing is done voluntarily and generally with little official recognition from the institutions where reviewers are employed. Nevertheless, professional reviewing is essential to a healthy conference. Therefore, we particularly thank the members of the Program Committee for their hard work and professionalism in providing constructive and fair reviews.

Evo* 2026, which EvoMUSART 2026 was part of, would not be possible without the substantial contributions of a large group of people. We are grateful for the support provided by SPECIES, the Society for the Promotion of Evolutionary Computation in Europe and its Surroundings, for the coordination and financial administration.

We would like to thank Nuno Lourenço (University of Coimbra, Portugal) and Francisco Chicano (University of Málaga, Spain) for their dedicated work as Submission System Coordinator and EvoStar Web Administrator, respectively. We also thank the Evo* Graphic Identity Team, Sérgio M. Rebelo and Jéssica Parente (University of Coimbra, Portugal), for their dedication and excellence in graphic design. We are grateful to João Correia (University of Coimbra, Portugal) for his outstanding work managing publicity, and to Jessica Mégane (University of Coimbra, Portugal), Giorgia Nadizar (University of Trieste, Italy), and Karina Rebulli (University of Torino, Italy) for handling social media publications and publicity. We extend special gratitude to Dennis Wilson (ISAE-Supaero, University of Toulouse, France) and Sylvain Cussat-Blanc (Université Toulouse Capitole, France) for their role as local organisers.

We would like to acknowledge our invited keynote speakers, Simon Lucas (Queen Mary University of London, United Kingdom) and Guy Theraulaz (CNRS, France), as well as Jean-Baptiste Mouret (Inria, France), recipient of the 2025 Julian Francis Miller Award, for their thought-provoking and inspiring presentations. We also wish to express our sincere gratitude to the Steering Committee of EvoMUSART for their continued guidance, support, and supervision.

Finally, we would like to express our continued appreciation to Anna I. Esparcia-Alcázar, from SPECIES, Europe, whose considerable efforts in managing and coordinating Evo* helped build a unique, vibrant, and friendly atmosphere.

February 2026

Penousal Machado
Juan J. Romero
Sérgio M. Rebelo

Organization

Programme Chairs

Penousal Machado   University of Coimbra, Portugal
Juan J. Romero   University of A Coruña, Spain

Publication Chair

Sérgio M. Rebelo   University of Coimbra, Portugal

Program Committee

Michael Affenzeller	Upper Austria University of Applied Sciences, Austria
Wolfgang Banzhaf	Michigan State University, USA
Gilberto Bernardes	University of Porto, Portugal
Sebastian Berns	Queen Mary University of London, UK
Ying Bi	Victoria University of Wellington, New Zealand
Jean-Pierre Briot	Sorbonne University–CNRS, France
Inês Caetano	University of Lisbon, Portugal
Marcelo Caetano	University of California San Diego, USA
F. Amílcar Cardoso	University of Coimbra, Portugal
Miguel Carvalhais	Universidade do Porto, Portugal
Daniel Castro Silva	University of Porto, Portugal
Vic Ciesielski	RMIT University, Australia
Simon Colton	Queen Mary University of London, UK
João Correia	University of Coimbra, Portugal
João Miguel Cunha	University of Coimbra, Portugal
Roger Dannenberg	Carnegie Mellon University, USA
Alena Denisova	University of York, UK

Georgios Diapoulis	University of Gothenburg and Chalmers University of Technology, Sweden
Patrick Donnelly	Oregon State University-Cascades, USA
Luis F. Teixeira	Fraunhofer Portugal AICOS
Amalia Foka	University of Peloponnese, Greece
Jose Fornari	UNICAMP, Brazil
Björn Gambäck	Norwegian University of Science and Technology, Norway
Francesca Gasparini	University of Milano-Bicocca, Italy
John Gero	University of North Carolina at Charlotte, USA
Kazjon Grace	University of Sydney, Australia
Carlos Grilo	Polytechnic University of Leiria, Portugal
Erik Hemberg	Massachusetts Institute of Technology, USA
Andrew Horner	Hong Kong University of Science and Technology, China
Colin G. Johnson	University of Nottingham, UK
Anna Jordanous	University of Kent, UK
Maximos Kaliakatsos-Papakostas	Aristotle University of Thessaloniki, Greece
Man Hei Law	Hong Kong University of Science and Technology, China
António Leitão	Universidade de Lisboa, Portugal
Matthew Lewis	Ohio State University, USA
Carlos León	Complutense University of Madrid, Spain
Antonios Liapis	University of Malta, Malta
Alain Lioret	Université Paris VIII, France
Phil Lopes	Lusofona University, Portugal
Nuno Lourenço	University of Coimbra, Portugal
João Macedo	University of Coimbra, Portugal
Pedro Martins	University of Coimbra, Portugal
Tiago Martins	University of Coimbra, Portugal
Catarina Maçãs	University of Coimbra, Portugal
Jon McCormack	Monash University, Australia
Christos Mousas	Purdue University, USA
Mario Andrés Muñoz	University of Melbourne, Australia
Miguel Nicolau	University College Dublin, Ireland
Colm O'Riordan	University of Galway, Ireland
Fabian Ostermann	TU Dortmund University, Germany
Nereida Rodriguez-Fernandez	University of A Coruña, Spain
Jonathan Rowe	University of Birmingham, UK
Günter Rudolph	TU Dortmund University, Germany

Conor Ryan	University of Limerick, Ireland
Iria Santos	University of A Coruña, Spain
Paulo Urbano	University of Lisbon, Portugal
Igor Vatolkin	RWTH Aachen University, Germany
Olga Vechtomova	University of Waterloo, Canada
Sebastian von Mammen	Julius-Maximilians University Würzburg, Germany

Additional Reviewers

Roberto Gallotta
Konstantinos Soiledis

Contents

Long Talks

Addressing Dataset Scarcity in Music Emotion Recognition with LLMs 3
  Janita Aamir and Patrick J. Donnelly

Digital Artists' Perceptions of Generative AI in South Asia: Insights
from the South Asian Digital Art Archive (SADA) 21
  Najam-Ul Assar and Megan Smith

Algorithms for Collaborative Harmonization 34
  Eyal Briman, Eyal Leizerovich, and Nimrod Talmon

Asɛmpayɛtsia: An Afrocentric Framework for Computational Creativity
in Sound and Image 49
  Nana Amowee Dawson

The Art That Poses Back: Assessing AI Pastiches After Contemporary
Artworks 67
  *Anca Dinu, Andreiana Mihail, Andra-Maria Florescu,
  and Claudiu Creangă*

Generative Musical Exploration of Astronomical Catalogs 84
  Adrián García Riber

Generative Artificial Intelligence in Music: Towards an Aesthetics
of Co-creation 97
  Anthony Grégoire, Serge Lacasse, and Joanie Verviers De Blois

A Dataset for Automatic Vocal Mode Classification 113
  Reemt Hinrichs, Sonja Stephan, Alexander Lange, and Jörn Ostermann

A Novel Diffusion Model Based Approach for Sleep Music Generation 129
  Timo Hromadka, Kevin Monteiro, and Sam Nallaperuma-Herzberg

Prompt and Circumstances: Evaluating the Efficacy of Human Prompt
Inference in AI-Generated Art 145
  *Khoi Trinh, Scott Seidenberger, Joseph Spracklen,
  Raveen Wijewickrama, Bimal Viswanath, Murtuza Jadliwala,
  and Anindya Maiti*

Fluid Body: An Adaptive Embodied Sonification System for Cross-Cultural
Performance 161
   Yuting Xue and Yueshen Wu

EvoLiveDJ: A LLM-Based Agentic Framework for Interactive
Evolutionary Live Music Performance 176
   Kamer Ali Yuksel and Hassan Sawaf

AI Co-Artist: A LLM-Powered Framework for Interactive GLSL Shader
Animation Evolution 192
   Kamer Ali Yuksel and Hassan Sawaf

Short Talks

Music in the Age of Artificial Intelligence: Meaning and Creativity From
a Complex-Systems Perspective 209
   Güncel Gürsel Artıktay

Generative Artificial Intelligence, Musical Heritage and the Construction
of Peace Narratives: A Case Study in Mali 224
   *Nouhoum Coulibaly, Ousmane Ly, Michael Leventhal,
   and Ousmane Goro*

Crystallizing Semantics: Mapping the Journey of Word Meaning
in Language Models 236
   Himanshu Dwivedi

Probing for Advanced Music Theory Concepts in Generative Music Models ... 252
   Derek Kwan and Patrick J. Donnelly

End-To-End Song Structure Segmentation via Encoder–Decoder Network
Architecture and Hand-Crafted Features 271
   Phan Le Son, Nghi Nguyen, and Lam Pham

Life Beings: Living Quantum Art 287
   Alain Lioret, Kamal Hennou, and Florentin Eraud

Segmentation-Free Sound Hybridization for Creative Outcomes 303
   Pasquale Mainolfi and Carmine Emanuele Cella

Quantum Latent Spaces for Symbolic Music Generation 316
   Sanjay Majumder and Neal Anderson

Weather Sonification via a Latent Emotion Space: A Deep Learning
Approach 331
  Takeshi Matsumura and Gavin J. Pringle

Classifying Audio Timbre Without Audio Using Text-Only Training 347
  Peter McCabe and Patrick J. Donnelly

Decoding Emotions: Multimodal Integration of Deep Embeddings, Lyrics
and Music-Aware Cues 367
  Alessia Novacco, Francesca Gasparini, Giulia Rizzi, and Aurora Saibene

Mapping Artificial Neural Networks' Processing Data in Audiovisual
Artworks 383
  Tanguy Pocquet, Richard Allmendinger, and Ricardo Climent

LoopMatcher: Proof-of-Concept for AI-Assisted Music Loop Search 398
  *Subhrojyoti Roy Chaudhuri, Sai Pranav Madupu,
  Krishna Teja Guru Sai, and Vikram Jamwal*

Multi-objective Evolution of Diffusion Model Prompt Embeddings Using
CLIP-IQA 413
  Marcel Salvenmoser and Michael Affenzeller

EvoArtist: A Visual LLM-Driven Agentic AI Framework for Autonomous
Design Evolution 429
  Kamer Ali Yuksel and Hassan Sawaf

Author Index 443

Long Talks

Addressing Dataset Scarcity in Music Emotion Recognition with LLMs

Janita Aamir and Patrick J. Donnelly^(✉)

Oregon State University, Corvallis, OR 97330, USA
{aamirj,donnellp}@oregonstate.edu

Abstract. Music Emotion Recognition is an important task in the field of music information retrieval, with applications in music generation, song recommendation, and playlist generation. However, given copyright concerns and the expense of human subject studies, researchers lack large datasets annotated with emotions. This dataset scarcity significantly limits the ability of researchers to train machine learning models that recognize emotion directly from audio. To address this need, we introduce a novel approach that leverages language models to generate emotion annotations. Specifically, we task ChatGPT to provide (1) numeric estimations (2) set of emotion words, and (3) long-form descriptions that characterize the emotive qualities of a piece of music. We consider 22,968 songs across five public datasets to facilitate a comparison of our ChatGPT synthetic annotations against human and previous algorithmic annotations. Although indirect, these annotations have the potential to provide insight into the emotional content of music, opening new possibilities for research and applications in the area of music emotion recognition.

Keywords: Music Emotion Recognition · Language Models · ChatGPT · Dataset · Annotations · Circumplex Model · Valence · Arousal

1 Introduction

Music has the power to elicit strong emotional reactions in listeners. The ability to predict these emotions directly from music itself is a task of great interest to researchers in the field of music information retrieval. Music Emotion Recognition (MER) is the process of estimating the emotional content of a piece of music using computational methods, such as machine learning. If researchers could automatically predict common emotional reactions to music directly from the analysis of acoustic signals alone, this would open up new possibilities for music recommendation systems, mood-based playlist generation, and music therapy.

Prior to the rise of deep learning, researchers primarily relied on signal processing to extract features for predicting emotional content directly from audio signals. These include high-level features, such as tempo, pitch, and timbre, as well as low-level features, such as harmony, melody, and rhythm [23]. However, despite advances in machine learning techniques, researchers report limits in their ability to predict emotional content based on audio descriptors alone [22].

© The Author(s), under exclusive license to Springer Nature Switzerland AG 2026
P. Machado et al. (Eds.): EvoMUSART 2026, LNCS 16523, pp. 3–20, 2026.
https://doi.org/10.1007/978-3-032-24350-8_1

The single greatest challenge faced by researchers in MER is the scarcity of large, diverse, and reliable datasets. The task of annotating music with emotion labels is resource-intensive as it requires expensive and time-consuming human subject studies. Such studies take place in a controlled laboratory setting or online in a crowd-sourced task. Despite attempts to train annotators, inter-rater agreement on perceived emotions is often quite low [27, 32]. Furthermore, copyright concerns limit the ability of researchers to release large public datasets. Together these factors hinder the creation of datasets annotated with emotions.

Nevertheless, researchers have released a few annotated datasets for MER. However, these datasets are small, homogeneous, or copyrighted (see Sect. 3). Other researchers have explored folksonomy representations, such as using a proxy of LastFM[1] tags [18], as a means to estimate emotional responses to music from crowd-sourced text labels. Other approaches attempted to estimate emotion based on lyrics [5], social media discourse [2], and historical documents [16].

This paucity of large datasets labeled with emotion reactions to music hinders the ability of researchers to develop robust MER systems that generalize across different genres, languages, and cultures. State-of-the-art deep learning approaches require datasets that are orders of magnitude larger than any existing datasets. Because it is infeasible to curate human annotations on the scale required, researchers could potentially investigate creative computational approaches. The recent availability of large language models (LLMs), such as ChatGPT[2], has revolutionized many tasks. These models have been trained on a wide array of textual information including song titles, lyrics, music descriptions, tags, and reviews, and hold the potential to estimate the emotional content of songs.

We propose a new approach to creating datasets for the field of MER using text-based prompts to harvest valence and arousal values from a LLM. By leveraging existing LLMs, and bypassing the expense of manual annotation, there is potential to create large datasets necessary to train robust, accurate MER models. Although human annotations will always be preferred, synthetic estimations are able to produce large-scale datasets at a substantially lower cost.

Since the primary focus of this study is to explore the validity of our approach, we focus on songs that have previously existing human or algorithmic annotations. We obtain estimations on five commonly used datasets in MER: AMG1608 [6], DEAM [1], DEAP [17], Deezer [7], and PMEmo [33]. For each song within the five datasets, we use GPT-3.5-turbo to estimate annotations using three different approaches: (1) direct estimations of arousal and valence, (2) five descriptive words that represent the emotional content of the song, and (3) long-form textual descriptions that summarize the emotions conveyed. In the second approach, each word, or its synonym, is matched in a word list that contains the arousal and valence values [31]. In the third approach, the long-form text is filtered to remove stop words, and then each word from the paragraph,

---

or its synonym, is referenced from the same list. These individual values are averaged to produce an estimate of arousal and valence for that song. Through these three approaches, we generated arousal and valence estimates for a total of 22,968 songs.

This novel approach of annotating music for MER tasks unleashes creative possibilities to build large-scale datasets without extensive resources. By testing this method, we aim to lay the foundation for efficient and scalable MER dataset creation. While this study focuses on validating this approach, future work will explore diverse musical genres and cultures. With the release of larger datasets, the field of MER has potential for transformative growth and new applications.

2   Background and Related Work

Researchers in MER seek to correlate characteristics of music with both perceived and induced emotions. However, to do so effectively, researchers require knowledge of which emotions are associated with which music. The data creation process either requires listeners to manually annotate emotions in music or computational processes that synthetically estimate the common emotional reactions to music.

2.1   Types of Annotations

Datasets used in music emotion recognition tasks are most often annotated at one of two temporal levels, using either static or dynamic labels. Static annotations are given at the song level in which a single label serves as a proxy for the emotions of the entire song. Although common, static annotations fail to capture emotional variation within a song. As such, they are not particularly well suited for certain genres or longer pieces of music. Dynamic annotations are marked at fixed intervals to track emotional changes over time. While the use of dynamic labels may better capture the variable nature of emotion within a piece of music, collecting dynamic labels from human subjects presents a number of additional challenges compared to the simpler task of static annotations [1].

In addition to the temporal level of annotation, datasets in MER are labeled using one of two different concepts: categorical [12] or dimensional [26] models of emotion. Categorical labels reflect a finite set of basic emotion categories, such as anger, disgust, fear, happiness, sadness, and surprise [8]. In the widely used circumplex model of dimensional encoding [25], emotions are categorized in a two-dimensional space of *arousal* (calm to exciting) and *valence* (negative to positive). The arousal dimension attempts to measure wakefulness versus sleepiness while the valence dimension attempts to measure emotions along an axis of pleasure versus displeasure. Additionally, many studies consider a discretization of the arousal and valence space, considering only the four quadrants of the space.

2.2   Limitations and Criticisms

In many domains, datasets are prepared for supervised machine learning tasks by enlisting domain experts to provide annotations. However, these individuals may differ in opinion, leading to inconsistencies in annotations [29,32]. Furthermore, annotators in crowd-sourcing tasks experience fatigue which can negatively affect their efficiency and accuracy [34]. This noisy labeled data presents great challenges to researchers training machine learning models. The difficulty of curating high-quality annotations for MER tasks is compounded by the highly subjective nature of emotional reactions to music [19]. Emotions elicited from music often vary between individuals, given broad differences in personal and socio-cultural preferences. Although some characteristics of the music itself may correlate with perceived emotions, researchers also find differences in these perceived emotions depending on demographics or musical experience [27]. Furthermore, there are distinctions between perceived and induced emotions (see [10] for a review).

For these reasons, the MER task of annotating music with emotion has been widely criticized [11]. There exist many differences in the methodologies used to create datasets for MER. Although probabilistic approaches to map between categorical and dimensional labels have been proposed [30], researchers observed little generalizability between datasets when the method of creation differed [14].

3   Emotion Datasets

In this section we describe the five datasets we consider, review the original data collection methods, and discuss limitations and biases of each approach.

3.1   AMG1608

The AMG1608 dataset [6] consists of 1608 songs that have been annotated by 665 subjects (320 female; 345 male). The dataset provides 30-second audio clips for each song. The music is primarily contemporary Western, as categorized by the All Music Guide[3]. As part of their annotation collection process, the authors divide the 1608 songs into 134 different groups, attempting to provide a balanced mixture of songs with different emotions. Each subject is tasked to annotate a set of 13 songs in a group, which takes an average of 455 s to complete.

Despite the rigor of the study design, there exist some limitations in the data collection process. First, the authors only use a single genre of music, which likely introduces a cultural bias, limiting the generalization of the dataset to other genres. They also limit themselves to only 30-second audio previews of each song that may not fully represent the full range of emotions that could evolve over the course of the entire song. For their study, the authors recruited the majority of the subjects from Amazon Mechanical Turk, who were required to be residents of the United States with a high acceptance rate on that platform. This choice introduces a cultural and geographic bias, because the perception and interpretation of music may vary across different cultures and regions. Although the authors recruit 22 subjects from a campus setting, this small sample adds to potential bias because they likely share geographic and cultural characteristics.

Moreover, in a single session, subjects were tasked to annotate 13 songs within an annotation group. This number of consecutive annotations could increase the

---

[3] https://www.allmusic.com/ (last accessed on 2026/01/20).

risk of annotation fatigue and affect the annotation quality, especially towards the end of each session. Although the authors do not detail incentives provided to the subjects for task completion, the quality of annotations can be impacted significantly depending on the subjects' motivation and the value of the incentive.

3.2 DEAM

The DEAM dataset [1] consists of annotations for 1802 songs (58 full-length songs and 1744 45-second excerpts) collected over three years. The dataset features songs from a variety of Western popular music genres, such as rock, pop, electronic, country, and jazz. These songs were chosen because they are royalty-free, allowing the researchers to release excerpts of the songs with their dataset. However, the lack of demographic information from annotators is a significant oversight in their methodology. Demographic factors such as age, gender, country of origin, and music knowledge can significantly influence emotional responses to music. Without this, it is challenging to assess how representative the annotations are of a broader population or to understand the potential biases in the dataset.

Most of the audio in the dataset is 45-second music excerpts extracted from random moments in the song. Although this method aims for fairness, the selected excerpts might not accurately capture the emotional content of the songs because those 45 seconds may not be representative of the complete song. The authors also found that participants' preferences for certain songs influenced self-reported confidence in rating but they did not provide a clear analysis of how these factors might have influenced the actual annotations in their dataset.

3.3 DEAP

The DEAP dataset [17] is a multimodal dataset analyzing human affective states. The authors recorded the electroencephalogram and peripheral physiological signals of 32 (50% female; 50% male) participants. These participants were tasked to watch 40 music videos that were each one-minute long. These music videos were selected in a semi-automatic and manual approach in an attempt to reduce selection bias. In the semi-automatic approach, the authors made use of a list of stimuli values to retrieve songs with corresponding LastFM tags. They made sure that the songs they selected were equally spread out across the quadrants to reduce bias towards certain emotions. The authors also manually picked the songs that had the clear emotional content and were equally spread out on the arousal-valence space. The design choice of a one-minute duration for each music video was also selected by using a linear regression approach [28].

For each music video, participants provided ratings of arousal, valence, like/dislike, dominance, and familiarity. Frontal face videos were also recorded for 22 of the 32 participants. The authors highlight that during the collection process, participants were able to stop annotating, which helped reduce annotation fatigue, potentially leading to more reliable results. Overall, the authors attempted to conduct an annotation strategy that aimed to reduce bias. The inclusion of ratings such as like/dislike and familiarity could perhaps be beneficial to identify bias. However, the study lacks a demographic breakdown of participants and more information on the participant selection process, as the authors only provided the gender and age range. Although it may be possible that demographic statistics were intentionally withheld from publication, given the small sample size.

3.4   Deezer

The Deezer mood detection dataset [7] is a large and algorithmically annotated dataset. This dataset is one of the largest available, consisting of synthetic annotations for $18,644$ songs. The arousal and valence values were calculated by considering mood-related LastFM tags using the strategy proposed by Hu et al. [13]. The extracted tags are then referenced against a word list [31] to find the arousal and valence value of each tag. If several tags are present for one song, the authors take the average of the respective values. These arousal and valence values are then normalized to [0,1] to reduce the scale of the annotated values.

Although the audio and lyrics of the songs are not included in this dataset, the authors provide a mapping to the Million Song Dataset [3] so researchers can reference audio features and lyrics. The authors also do not provide any details on the average number of tags considered for each song. However, because there are only 2762 unique points across the 18,644 songs, it can be assumed that many songs share a small number of LastFM tags. Furthermore, the study did not provide details on their song selection process and the distribution of data across different genres of music. Even though human annotations are generally preferred in MER tasks, this dataset has received attention for its size and novelty.

3.5   PMEmo

The PMEmo dataset [33] contains annotations of 794 songs as well as electrodermal activity signals of the participants listening to the music. The authors recruited 457 subjects in total (236 female; 221 male), of which 366 were Chinese university students who are non-music majors, 44 were Chinese university students who are music majors, and 47 were identified as native English speakers. Each song in their dataset received a total of at least 10 emotion annotations including one by a music major and one by an English speaker. In the dataset, the authors mainly targeted trending pop songs on the Billboard Hot 100, the iTunes Top 100 Songs (USA), and the Top 40 Singles Chart (UK).

Because the majority of the annotators were Chinese students with non-music backgrounds, it is likely that there are age, academic, and cultural biases in the dataset. Although the dataset focuses on English pop songs, only 47 out of the 457 subjects (10.3%) were English speakers. This potentially introduces a significant linguistic bias, as many of the non-English speakers were unlikely to fully comprehend the nuanced emotions expressed through the English language lyrics. All these factors potentially affect the generalizability of the arousal and valence annotations collected for this dataset.

Additionally, the exclusive use of trending pop songs limits the diversity of the dataset. This focus, on English-language pop songs, might skew the balance of arousal and valence annotations towards one of four quadrants, which is problematic as it can overlook the broader spectrum of emotions. The authors even acknowledge that a large proportion of the songs received annotations that place them in the first quadrant, the quadrant that contains high energy songs.

The paper highlights that the facilitators created a relaxing environment for the subjects before collecting annotations. During this period, subjects listen to the song "In My Life" by Kevin Kern. This requirement to listen to one particular song is problematic and introduces bias that may affect the annotations for subsequent songs. The authors also report providing a 10-second break between each annotation. Although a break is likely a good idea, the authors provided no evidence that 10 seconds was sufficient for an emotional reset for the annotators.

While annotating, subjects were required to listen to each song twice in order to rate arousal and valence separately. This design choice could lead to fatigue or decreased attentiveness, potentially affecting the reliability of annotations. When collecting annotations for their dataset, the authors only consider the chorus of the song. This introduces yet another source of bias since the annotations might not reflect the full range of emotions expressed. In fact, the chorus is often more high-energy than other parts of a song. This choice likely influenced the dataset, skewing it towards capturing more intense emotional responses.

3.6   Summary and Implications

From the analysis of each data collection study, it is clear there is a lot of variation in the annotation approaches across the datasets. This variation includes the song selection process, setting, demographics, and methods of collection. These factors affect the consistency and the reliability of the emotion labels. For example, DEAP's controlled lab-based setting results in more consistent and reliable annotations but limits participation and dataset size. In contrast, the online methods used by PMEmo and DEAM increase the dataset and participant size but tend to reduce annotation reliability. It is not feasible to create large-scale datasets through human annotations alone and, given the inconsistencies in annotation collection studies, combining smaller datasets for MER model training becomes difficult as well. This highlights the need for a standardized and scalable approach to create reliable annotations across different types of music.

4   Dataset Creation Process

Our goal is to estimate the arousal and valence values of the songs that exist in published human-annotated datasets used by the MER field: AMG1608 (n = 1608 songs), DEAM (n = 1802), DEAP (n = 120), and PMEmo (n = 794). By estimating the emotions of the songs that exist in these datasets, we are able to compare the annotations estimated through LLMs to human-annotated baselines. We also compare to synthetic annotations of the Deezer dataset (n = 18,644). As one of the largest and most widely used datasets, it is an important

Table 1. Example prompt given to ChatGPT for the 5-word approach.

Example Prompt

Task:

Analyze the song {name} by {artist} to identify the emotional content conveyed in the song. You can look at lyrics, analyses, reviews, interviews, fan discussions, and more. Based on your analysis, provide five descriptive words that accurately capture the song's emotions.

Output:

Your output must be strictly in readable JSON format without any extra text:

```
{
  "words":
      [word1, word2, word3, word4, word5],
  "reasons":
      [reason for word1, reason for word2, reason for word3,
       reason for word4, reason for word5]
}
```

Requirements:

For the 'outputs' field, enter the words that accurately describe the emotional content of the song. No words should be repeated for the song.

For the 'reasons' field, provide reasoning for each word separately. The reason for each word should be in the same order as the word.

If you do not know the answer to a question, do not share false information. Instead, you should enter None for the 'words' and 'reasons' field.

baseline against which we compare our approach. In this study, we task GPT-3.5-turbo to obtain arousal and valence values using three different methods: (1) direct, numeric estimate, (2) list of five descriptive words, and a (3) long-form textual description.

We ask ChatGPT for the emotional content of songs, following the structured approach by Zhuo et al. [35]. An example prompt for one of the methods is provided in Table 1. The prompts for the other two methods are very similar.

We structured our prompts to minimize instances of hallucination in the responses provided by ChatGPT. Hallucination occurs when the text seems sensical, but the information provided is inaccurate or invented [15]. Following the approach of Zhuo et al. [35], we found reduced instances of hallucination in the responses provided by ChatGPT. The prompt uses specific task descriptions which reduces ambiguity and prevents the model from misinterpreting the objective. A structured output format, required by the prompt, also helps constrain the model for more consistency. The requirement of a clear JSON format helps prevent the language model from generating content that is disorganized or irrelevant. In the prompt, the language model is also instructed to respond None if unsure about a response, which prevented the model from fabricating

Table 2. Response provided for the Direct approach for the song 'Hello' by Adele.

Direct: ChatGPT Response for 'Hello' by Adele

{
    "valence": "0.4",
    "arousal": "0.6",
    "reasons": ["The valence estimate of 0.4 is based on the melancholic and reflective nature of the song's lyrics, which evoke feelings of sadness and longing.", "The arousal estimate of 0.6 is based on the powerful and emotionally charged vocal performance by Adele, as well as the dynamic range and intensity of the instrumentation."]
}

information. In the early stages of testing, we found that without this instruction, ChatGPT was more prone to providing responses that were inaccurate. We also instructed ChatGPT to provide reasoning with each response. This enabled the model to self-reflect on the responses and also helped us understand the decision making process of the LLM. The reasoning became especially useful when we were manually reviewing the responses. Lastly, we made sure to assign a task specific role to ChatGPT as it is a great way to provide context before prompting the language model. We primed the model to be an expert in music and the emotions conveyed in it.

We identified errors and potential hallucinations by reviewing responses, cross-referencing them, and re-prompting when necessary. Even after thorough review and re-prompts, we found that ChatGPT could not estimate the arousal and valence for a few songs. Nevertheless, the response rate was still very high.

4.1   Method 1: Direct Estimate

Initially, ChatGPT would not output numeric values, stating that it is not trained for such calculations. However, by clarifying the request as a "subjective estimate", ChatGPT then provided numeric values ranging from 0 to 1. This change encourages the LLM to interpret the emotional content of the songs with more flexibility by taking various factors into consideration such as lyrics, reviews, social media, and articles. Although ChatGPT's reasoning process is obscured by the model's algorithm, we hypothesize that subjective estimates may align more closely with the human annotation process since, like human annotators, the model is not calculating exact values but instead drawing on broader contextual cues. An example response soliciting a direct estimate is given in Table 2.

4.2   Method 2: Descriptive Words

In our second approach, we requested that ChatGPT describe each song with five descriptive words. We chose five words after testing ranges from 2–10 words. We observed that requesting fewer than five words led to insufficient information and more than five led to redundancy. We referenced each word in the response from ChatGPT against the corresponding arousal and valence scores using the extended Affective Norms for English Words (XANEW), which comprises 13,915 English lemmas [31], building upon the original list [4]. If a word was not listed, we used the NLTK package[4] to find a synonym for the word and matched it to the arousal and valence values. In an approach similar to other studies [18], we took the average of these scores to create a composite arousal and valence score for each song. These values were then normalized to [0–1] to maintain consistency with the direct estimates. An example response is given in Table 3.

Table 3. Response provided for the 5-word approach for the song 'Hello' by Adele.

5-words: ChatGPT Response for 'Hello' by Adele

{
   "words": ["nostalgic", "heartbreak", "longing", "regret", "resignation"],
   "reasons": ["The lyrics and Adele's vocal delivery evoke a sense of longing for the past and reminiscing about a lost love.", "The song is about a breakup and the emotions of heartbreak are evident in the lyrics and Adele's powerful vocals.", "The lyrics express a deep yearning for connection and a desire to reconcile with a past lover.", "There is a sense of regret and reflection on past mistakes in the lyrics, which contributes to the emotional content of the song.", "The overall tone of the song conveys a feeling of resignation and acceptance of the end of a relationship."]
}

4.3   Method 3: Long-Form Description

Next, we tasked ChatGPT to provide long-form textual descriptions of the songs that summarize their emotional content. These descriptions were no longer than 1024 tokens. From the response, we cleaned up the text and removed stop words. For the remaining words in the paragraph, we used a methodology similar to the second approach. Each word, or its synonym, was referenced against the XANEW list to find its corresponding arousal and valence scores. We then averaged the scores of all words or synonyms to obtain the arousal and valence values for each song. These values were also normalized to [0–1] to ensure consistency across all methodologies. An example response is given in Table 4.

---

[4] https://www.nltk.org/ (last accessed 2026/01/20).

Table 4. Response for the long-form approach for the song 'Hello' by Adele.

Long-form: ChatGPT Response for 'Hello' by Adele

{
    "paragraph": "The song 'Hello' by Adele conveys a deep sense of longing, heartbreak, and nostalgia. Adele's powerful and emotive vocals, combined with the poignant lyrics, evoke a strong emotional response from the listener. The lyrics speak of a past relationship and the desire to reconnect, highlighting the pain of lost love and the yearning for closure. The melancholic melody and soulful instrumentation further enhance the emotional intensity of the song.",
    "reasons": "The lyrics of the song express themes of longing, heartbreak, and nostalgia. Adele's vocals and delivery are known for their emotional depth, which adds to the overall sentiment conveyed in the song. The melody and instrumentation also contribute to the emotional intensity felt by the listener."
}

5   Statistical Analysis

We assessed ChatGPT's emotion estimation in two ways. First, we compared the results of the three methods (direct, 5-word, long-form). Second, we compared the estimated values to published human and algorithmic annotations. For both comparisons, we report Pearson's correlation coefficient (PCC). We also consider a discretization of the contiguous arousal and valence values into the four quadrants of the circumplex scale, namely low arousal/low valence, low arousal/high valence, high arousal/low valence, and high arousal/high valence. Through classification, we are able to see the percentage of arousal and valence values that are in the same quadrant. This way, if one set of arousal and valence values differ from another set, we can still assess the percentage of songs that reside in the same quadrant. This enables us to gain insights into high-level categorical similarities between emotions despite any numerical differences.

Lastly, we examined the five existing datasets to find any overlapping songs in either two, three, four, or all five of these datasets. For these overlapping songs, we calculated the PCC to evaluate the level of agreement between the original published annotations. Through this analysis, we were able to determine how well the original annotations align with each other. This provided insights into the consistency and reliability of emotion recognition across different datasets.

5.1   Comparison of Labeling Approaches

First, we compare the similarity of our annotations produced by the three methods of annotation. These results are shown in Table 5. When comparing these methods, we observe that the 5-words and long-form extraction approaches share a higher correlation for both valence and arousal when compared to direct/5-words and to direct/long-form approaches. For quadrant classification, we observe a similar trend; the classification accuracy between the 5-words and long-form approach is generally higher when either is compared to the direct estimations. Overall, these results show that the 5-word approach and long-form

Table 5. Comparison of the three approaches across five datasets: 5-words vs. Long-form (left), 5-words vs. Direct (middle), and Long-form vs. Direct (right). Results show PCC for arousal and valence, and accuracy for quadrant classification.

	5-words vs. Longform			5-words vs. Direct			Longform vs. Direct		
	Aro.	Val.	Quad.	Aro.	Val.	Quad.	Aro.	Val.	Quad.
AMG1608	0.685	0.772	59.3%	0.632	0.726	61.0%	0.600	0.700	54.8%
DEAM	0.571	0.623	55.1%	0.446	0.629	54.5%	0.447	0.610	47.0%
DEAP	0.712	0.843	69.2%	0.626	0.822	56.9%	0.541	0.796	56.9%
Deezer	0.626	0.707	54.8%	0.512	0.675	48.5%	0.459	0.684	44.7%
PMEmo	0.501	0.716	52.5%	0.410	0.621	54.7%	0.357	0.631	40.3%

Table 6. The mean ± std of arousal across the 5 different datasets and 4 methods.

Dataset	Direct	5-Words	Longform	Original
AMG1608	0.63 ± 0.15	0.57 ± 0.20	0.52 ± 0.18	0.55 ± 0.22
DEAM	0.60 ± 0.15	0.57 ± 0.19	0.49 ± 0.17	0.49 ± 0.20
DEAP	0.65 ± 0.16	0.50 ± 0.29	0.56 ± 0.25	0.49 ± 0.21
Deezer	0.61 ± 0.14	0.48 ± 0.19	0.47 ± 0.15	0.50 ± 0.19
PMEmo	0.67 ± 0.14	0.52 ± 0.16	0.45 ± 0.15	0.62 ± 0.18

method yield more similar results compared to the direct approach; this could be explained by the similar text-processing approach in the two methods. We also observe that the correlation for valence tends to be higher than that for arousal across all three comparisons in each of the five datasets. This likely indicates that valence, which measures negative to positive, is much easier to interpret than arousal, which measures calm to exciting. Diving deeper into specific datasets, we find that DEAP consistently has higher correlation across all three comparisons whereas, PMEmo and DEAM tend to have comparatively lower correlations. This trend may indicate that the emotions expressed in the musical pieces in the DEAP dataset are much more clearly defined, which leads to more consistent correlations across all the three methods tested. For each dataset and method, we provide the mean and standard deviation for arousal as Table 6 and for valence as Table 7.

Table 7. The mean ± std of valence across the 5 different datasets and 4 methods.

Dataset	Direct	5-Words	Longform	Original
AMG1608	0.56 ± 0.20	0.62 ± 0.23	0.50 ± 0.18	0.51 ± 0.17
DEAM	0.57 ± 0.18	0.69 ± 0.23	0.54 ± 0.18	0.49 ± 0.17
DEAP	0.54 ± 0.23	0.52 ± 0.30	0.50 ± 0.23	0.49 ± 0.25
Deezer	0.52 ± 0.18	0.63 ± 0.20	0.51 ± 0.16	0.56 ± 0.29
PMEmo	0.62 ± 0.18	0.66 ± 0.22	0.60 ± 0.18	0.60 ± 0.16

5.2   Comparison to Existing Annotations

Next, we compare the performance of our three LLM-based approaches with the human- (AMG1608, DEAM, DEAP, PMEmo) and algorithmically-annotated (Deezer) data. The goal of this comparison is to investigate the similarity of our LLM-derived annotations with the existing annotations widely used in MER research. The results of this statistical comparison are given in Table 8.

Table 8. Comparison of the original and ChatGPT annotations for Direct (left), 5-words (middle), and Long-form (right) approaches across five datasets. Results show PCC scores for arousal and valence, and accuracy for quadrant classification.

	Direct			5-words			Longform		
	Aro.	Val.	Quad.	Aro.	Val.	Quad.	Aro.	Val.	Quad.
AMG1608	0.582	0.635	53.6%	0.534	0.597	52.8%	0.532	0.579	54.7%
DEAM	0.185	0.201	32.5%	0.263	0.160	36.0%	0.240	0.101	35.1%
DEAP	0.560	0.688	49.1%	0.580	0.637	60.8%	0.496	0.637	55.0%
Deezer	0.208	0.446	35.6%	0.274	0.398	38.9%	0.284	0.394	41.2%
PMEmo	−0.091	0.044	44.0%	0.016	0.033	37.1%	−0.002	0.004	25.5%

We find that the LLM-generated annotations for the DEAP dataset generally achieved higher correlations to human annotations, followed by AMG1608. This may reflect the fact that both datasets were created using rigorous experimental designs employing many different human annotators. PMEmo, on the other hand, yielded the lowest agreement, perhaps explained by the less rigorous experimental design choices in that study. Given that most of the annotators were non-English speakers, this could have made it harder for our LLM-based approaches to align well with the original annotations. DEAM has the second lowest correlation values across all three approaches, which could be explained by its use of royalty-free music, which is typically less popular and it is possible that ChatGPT was not extensively trained on these songs. In this comparison, we also observe that correlations for valence are generally higher than arousal, which again suggests that the LLMs might find it easier to estimate valence than to estimate arousal. One noteworthy point is that, while the correlation values for PMEmo are quite low, the quadrant classifications are relatively higher, indicating some success in the model's ability to estimate the general emotion. Overall, we found modest agreement between our LLM estimations and the original data, but this agreement varies significantly across the different datasets with trends that are similar to those observed in Sect. 5.1.

5.3   Comparison Between Human Annotations

In human annotation studies, inter-rater reliability is often poor [27]. Because the extant MER datasets used different methodologies, we questioned the agreement between raters across these studies. We investigated the correlation between human annotations of the same song produced through different experimental designs. We examined the five existing datasets to find any songs that exist in more than one dataset. However, we found only 294 songs that overlapped in more than one datasets. The counts and correlations are shown in Table 9.

When examining the consistency of annotations across different datasets for the same song, we found the correlation values vary significantly. This likely results from differences in the experimental approach and underscores the subjective nature of the annotation task. This observation emphasizes the need for a consistent and scalable approach to annotate MER datasets. Although only 294 songs overlapped across the datasets, our LLM-derived annotations show higher correlations to the original annotations and between the three approaches. This suggests potential to achieve more consistency with LLM based annotations.

Table 9. PCC of original annotations for the songs ($n = 294$) that appear in more than one dataset. Correlations for arousal are given to the left of the diagonal and correlations for valence are given right of the diagonal. The parentheses contain the total count of the overlapping songs between any two datasets.

	AMG	DEAM	DEAP	Deezer	PmEmo
AMG		— (0)	-0.353 (3)	0.639 (246)	-0.254 (4)
DEAM	— (0)		— (0)	0.576 (3)	— (0)
DEAP	0.901 (3)	— (0)		0.705 (29)	— (0)
Deezer	0.571 (246)	-0.374 (3)	0.484 (29)		-0.014 (9)
PMEmo	0.257 (4)	— (0)	— (0)	-0.001 (9)	

6 Discussion

We present a novel approach to address challenges faced when creating annotated datasets for MER. We use an LLM to estimate arousal and valence using a direct numeric approach, a 5-word indirect estimation, and a long-form answer indirect estimation. This approach has potential to facilitate the generation of synthetic annotations on a large scale in a systematic manner. Our findings indicate that LLMs can contribute valuable insights into the emotional content of music.

6.1 Limitations and Future Work

Given the novelty of our approach, we acknowledge certain limitations which motivate our future work. One important concern is the tendency for LLMs to occasionally 'hallucinate'. To avoid this, we took steps to craft specific prompts and to identify hallucinations by manually inspecting the output and re-prompting the language models when appropriate. However, it is difficult to completely eliminate instances of hallucination. One solution to reduce hallucination could be to refine our prompt designs to include strategies such as Chain-of-Thought verification [20]. Another solution could necessitate re-prompting each song multiple times and choosing the most common response. However, this approach would limit scalability and increase financial costs. Another promising direction would be to explore how other language models perform with annotation estimation. While this study focuses on ChatGPT, we could compare other LLM models such as LLaMA or Claude. This comparison would be very beneficial to find an ensemble of models best suited for annotating music.

In this study, we focused on songs in existing datasets to facilitate comparison with human annotations. As future work, we plan to create a larger and genre-diverse dataset. We believe such a dataset is crucial for developing accurate, real-world emotion recognition systems, even if the annotations are indirect and imperfect estimations. Furthermore, incorporating other emotional classification scales, such as Plutchik's wheel of emotions [24], will provide researchers other avenues to explore the link between music and emotion.

6.2   Conclusion

This study presents a novel approach that leverages a LLM to estimate an average emotive reaction to pieces of music. This approach demonstrates the potential to generate emotion annotations which, although synthetic, have a moderate and positive correlation with emotions reported by humans. We emphasize that this relationship is comparable, if not stronger, to the level of correlation seen between humans annotating the same songs but in different settings. Ultimately, the true effectiveness of these annotations in practical applications, such as music recommendation systems [21] or playlist generation [9], is yet to be determined. Nevertheless, MER desperately needs large reliable datasets and this research invites further exploration of LLMs to estimate human reactions to music.

References

1. Aljanaki, A., Yang, Y.H., Soleymani, M.: Developing a benchmark for emotional analysis of music. PLoS ONE **12**(3), 1–22 (2017). https://doi.org/10.1371/journal.pone.0173392
2. Beery, A., Donnelly, P.J.: Learning affective responses to music from social media discourse. In: Abbas, M. (ed.) Practical Solutions for Diverse Real-World NLP Applications, pp. 93–119. Springer International Publishing, Cham (2024). https://doi.org/10.1007/978-3-031-44260-5_6
3. Bertin-Mahieux, T., Ellis, D.P.W., Whitman, B., Lamere, P.: The million song dataset. In: Klapuri, A., Leider, C. (eds.) Proceedings of the 12th International Conference on Music Information Retrieval (ISMIR), pp. 591–596. ISMIR, Miami (2011).https://doi.org/10.5281/zenodo.1415820

4. Bradley, M.M., Lang, P.J.: Affective norms for English words (ANEW): instruction manual and affective ratings. Tech. rep., Technical report C-1, the Center for Research in Psychophysiology (1999). https://api.semanticscholar.org/, CorpusID : 145474983, Accessed 10 Nov 2025

5. Çano, E., Morisio, M.: Moodylyrics: a sentiment annotated lyrics dataset. In: Proceedings of the 2017 International Conference on Intelligent Systems, Metaheuristics & Swarm Intelligence (ISMSI), pp. 118–124. Association for Computing Machinery, New York (2017). https://doi.org/10.1145/3059336.3059340

6. Chen, Y.A., Yang, Y.H., Wang, J.C., Chen, H.: The AMG1608 dataset for music emotion recognition. In: 2015 IEEE International Conference on Acoustics, Speech and Signal Processing (ICASSP), pp. 693–697. IEEE (2015). https://doi.org/10.1109/ICASSP.2015.7178058

7. Delbouys, R., Hennequin, R., Piccoli, F., Royo-Letelier, J., Moussallam, M.: Music mood detection based on audio and lyrics with deep neural net. In: Proceedings of the 19th International Society for Music Information Retrieval Conference (ISMIR), pp. 370–375. ISMIR (2018). https://doi.org/10.5281/zenodo.1492427

8. Ekman, P.: Expression and the nature of emotion. In: Scherer, K., Ekman, P. (eds.) Approaches to Emotion, vol. 3:19, pp. 319–344. Lawrence Erlbaum, Hillsdale, NJ (1984). https://doi.org/10.4324/9781315798806

9. Gaur, S., Donnelly, P.J.: Generating smooth mood-dynamic playlists with audio features and KNN. In: Johnson, C., Rebelo, S.M., Santos, I. (eds.) Proceedings of the 13th International Conference of Evolutionary and Biologically Inspired Music, Sound, Art and Design (EvoMUSART). vol. 14633, pp. 162–178. Springer (2024). https://doi.org/10.1007/978-3-031-56992-0_11

10. Gómez-Cañón, J.S., et al.: Music emotion recognition: toward new, robust standards in personalized and context-sensitive applications. IEEE Sign. Process. Magazine **38**(6), 106–114 (2021). https://doi.org/10.1109/MSP.2021.3106232

11. Gómez-Cañón, J.S., et al.: TROMPA-MER: an open dataset for personalized music emotion recognition. J. Intell. Inf. Syst. **60**(2), 549–570 (2023). https://doi.org/10.1007/s10844-022-00746-0

12. Hevner, K.: Experimental studies of the elements of expression in music. Am. J. Psychol. **48**(2), 246–268 (1936)

13. Hu, X., Downie, J.S., Ehmann, A.F.: Lyric text mining in music mood classification. In: Proceedings of the 10th International Conference on Music Information Retrieval (ISMIR), pp. 411–416 (2009). https://doi.org/10.5281/zenodo.1416790

14. Hult, S., Kreiberg, L.B., Brandt, S.S., Jónsson, B.T.: Analysis of the effect of dataset construction methodology on transferability of music emotion recognition models. In: Proceedings of the 2020 International Conference on Multimedia Retrieval, pp. 316–320 (2020). https://doi.org/10.1145/3372278.3390733

15. Ji, Z., et al.: Survey of hallucination in natural language generation. ACM Comput. Surv. **55**(12), 1–38 (2023). https://doi.org/10.1145/3571730

16. Khanal, P., Donnelly, P.J.: Emotionotes dataset: decoding emotions in classical music through concert program notes. In: International Conference on Computational Intelligence in Music, Sound, Art and Design (Part of EvoStar), pp. 339–356. Springer (2025). https://doi.org/10.1007/978-3-031-90167-6_23

17. Koelstra, S., et al.: DEAP: database for emotion analysis using physiological signals. IEEE Trans. Affect. Comput. **3**(1), 18–31 (2011). https://doi.org/10.1109/T-AFFC.2011.15

18. Laurier, C., Sordo, M., Serra, J., Herrera, P.: Music mood representations from social tags. In: Proceedings of the 10th International Conference on Music Information Retrieval (ISMIR), pp. 381–386 (2009)

19. Liebetrau, J., Schneider, S.: Music and emotions: a comparison of measurement methods. In: Audio Engineering Society Convention 134. Audio Engineering Society (2013). https://publica.fraunhofer.de/handle/publica/381224, Accessed 30 Jan 2026
20. Ling, Z., et al.: Deductive verification of chain-of-thought reasoning. In: Oh, A., Naumann, T., Globerson, A., Saenko, K., Hardt, M., Levine, S. (eds.) Proceedings of the 37th International Conference on Neural Information Processing Systems. NIPS '23, vol. 36, pp. 36407–36433. Curran Associates, Inc., Red Hook (2023). https://proceedings.neurips.cc/paper_files/paper/2023/file/72393bd47a35f5b3bee4c609e7bba733-Paper-Conference.pdf, Accessed 30 Jan 2026
21. Liu, Z., Xu, W., Zhang, W., Jiang, Q.: An emotion-based personalized music recommendation framework for emotion improvement. Inf. Process. Manag. **60**(3), 103256 (2023). https://doi.org/10.1016/j.ipm.2022.103256
22. Panda, R., Malheiro, R., Paiva, R.P.: Novel audio features for music emotion recognition. IEEE Trans. Affective Comput. **11**(4), 614–626 (2020). https://doi.org/10.1109/TAFFC.2018.2820691
23. Panda, R., Malheiro, R., Paiva, R.P.: Audio features for music emotion recognition: a survey. IEEE Trans. Affective Comput. **14**(1), 68–88 (2023). https://doi.org/10.1109/TAFFC.2020.3032373
24. Plutchik, R.: A general psychoevolutionary theory of emotion. In: Plutchik, R., Kellerman, H. (eds.) Theories of Emotion, pp. 3–33. Academic Press (1980). https://doi.org/10.1016/B978-0-12-558701-3.50007-7
25. Russell, J.A.: A circumplex model of affect. J. Personality Soc. Psychol. **39**(6), 1161–1178 (1980). https://doi.org/10.1037/h0077714
26. Russell, J.A., Weiss, A., Mendelsohn, G.A.: Affect grid: a single-item scale of pleasure and arousal. J. Personality Soc. Psychol. **57**(3), 493 (1989). https://doi.org/10.1037/0022-3514.57.3.493
27. Schedl, M., Gómez, E., Trent, E.S., Tkalčič, M., Eghbal-Zadeh, H., Martorell, A.: On the interrelation between listener characteristics and the perception of emotions in classical orchestra music. IEEE Trans. Affect. Comput. **9**(4), 507–525 (2018). https://doi.org/10.1109/TAFFC.2017.2663421
28. Soleymani, M., Kierkels, J.J., Chanel, G., Pun, T.: A Bayesian framework for video affective representation. In: 2009 3rd International Conference on Affective Computing and Intelligent Interaction and Workshops, pp. 1–7. IEEE (2009). https://doi.org/10.1109/ACII.2009.5349563
29. Sylolypavan, A., Sleeman, D., Wu, H., Sim, M.: The impact of inconsistent human annotations on ai driven clinical decision making. NPJ Digit. Med. **6**(1), 26 (2023). https://doi.org/10.1038/s41746-023-00773-3
30. Wang, J.C., Yang, Y.H., Chang, K., Wang, H.M., Jeng, S.K.: Exploring the relationship between categorical and dimensional emotion semantics of music. In: Proceedings of the Second International ACM Workshop on Music Information Retrieval with User-centered and Multimodal Strategies, pp. 63–68 (2012). https://doi.org/10.1145/2390848.2390865
31. Warriner, A.B., Kuperman, V., Brysbaert, M.: Norms of valence, arousal, and dominance for 13,915 English lemmas. Behav. Res. Methods **45**, 1191–1207 (2013). https://doi.org/10.3758/s13428-012-0314-x
32. Watcharasupat, K.N., Ding, Y., Ma, T.A., Seshadri, P., Lerch, A.: Uncertainty estimation in the real world: a study on music emotion recognition. In: European Conference on Information Retrieval, pp. 218–232. Springer (2025). https://doi.org/10.1007/978-3-031-88711-6_14

33. Zhang, K., Zhang, H., Li, S., Yang, C., Sun, L.: The PMEmo dataset for music emotion recognition. In: Proceedings of the 2018 International Conference on Multimedia Retrieval, pp. 135–142. ACM, Yokohama Japan (2018). https://doi.org/10.1145/3206025.3206037
34. Zhang, Y., Ding, X., Gu, N.: Understanding fatigue and its impact in crowdsourcing. In: 22nd International Conference on Computer Supported Cooperative Work in Design (CSCWD), pp. 57–62. IEEE (2018). https://doi.org/10.1109/CSCWD.2018.8465305
35. Zhuo, L., et al.: Lyricwhiz: robust multilingual zero-shot lyrics transcription by whispering to ChatGPT. In: Proceedings of the 24th International Society for Music Information Retrieval Conference, pp. 343–351 (2023). https://doi.org/10.5281/zenodo.10265295

Digital Artists' Perceptions of Generative AI in South Asia: Insights from the South Asian Digital Art Archive (SADA)

Najam-Ul Assar[✉] and Megan Smith[✉]

University of British Columbia, 3505 Spectrum Ct, Kelowna, BC V1V 2Y7, Canada
najam@lahoredigitalfestival.com, megan.smith@ubc.ca

Abstract. This paper examines how contemporary South Asian digital artists perceive, engage with, and resist generative artificial intelligence (AI) in creative practice. Drawing on thirty-six qualitative interviews conducted for the South Asian Digital Art Archive as part of the study approved by the University of British Columbia ethics board (Study ID: H25–00591) analyses artists' reflections on authorship, ethics, posthumanism, data coloniality, and infrastructural inequality. Rather than presenting a unified stance, the interviews reveal heterogeneous, often ambivalent positions: some artists approach AI as a conditional collaborator, while others articulate a refusal grounded in ecological, political, or epistemic concerns. Using reflexive thematic analysis informed by postcolonial digital humanities scholarship, the paper identifies six overlapping conceptual clusters that situate AI within the lived conditions of uneven access, historical marginalization, and cultural specificity. The findings suggest that South Asian engagements with generative AI are shaped less by narratives of technological optimism or fear than by situated ethical practices that emphasize care, locality, and embodied forms of creativity. By foregrounding voices from the Global South, the paper contributes qualitative insight to global debates on AI and art and demonstrates the value of decolonial methodologies for studying emerging technologies.

Keywords: Postcolonial AI · Digital Art Archive · AI Ethics

1 Introduction

In recent years, the rapid proliferation of generative artificial intelligence (AI) enhanced by tools such as Midjourney, Stable Diffusion, and ChatGPT, have generated sustained debate across creative industries, the humanities, and artistic research. These debates often centre on questions of authorship, automation, labour, and creativity, and they are frequently framed through narratives of technological promise or existential risk. Much of this discourse has emerged from Euro-American institutional contexts, shaped by specific histories of industrialization, platform capitalism, and access to computational infrastructure. At the same time, critical scholarship produced within these contexts has rigorously challenged techno-optimistic narratives by foregrounding AI's environmental

© The Author(s), under exclusive license to Springer Nature Switzerland AG 2026
P. Machado et al. (Eds.): EvoMUSART 2026, LNCS 16523, pp. 21–33, 2026.
https://doi.org/10.1007/978-3-032-24350-8_2

costs, labour exploitation, and embedded biases (Crawford, 2021; Noble, 2018). This paper does not treat such "Western" perspectives as homogeneous. Rather, it distinguishes between globally dominant AI imaginaries circulated through platforms, corporations, and policy regimes, and situated engagements with technology that emerge under conditions of historical marginalization, infrastructural precarity, and linguistic exclusion. For practitioners and scholars in the Global South, AI is not only a technological artifact but also an epistemic apparatus: a system of power that reproduces colonial hierarchies through biased datasets, extractive infrastructures, and uneven access to computation (Couldry & Mejias, 2019; Birhane, 2021). In South Asia, artists encounter generative AI through layered experiences of digital coloniality, material constraint, and cultural misrepresentation. The region's creative ecosystems that are rooted in hybrid traditions of craft, performance, and technological improvisation engage AI neither as novelty nor inevitability, but as a contested terrain in which questions of agency, ethics, ecology, and representation intersect. These engagements offer critical perspectives for rethinking creative AI beyond universalizing frameworks, revealing how artists navigate, resist, and reimagine algorithmic systems from within historically uneven technological landscapes.

This study draws on research conducted as part of a graduate thesis at the University of British Columbia, Okanagan Campus. Its empirical foundation is the South Asian Digital Art Archive (SADA), an open, multilingual research archive documenting contemporary digital arts practices across South Asia. SADA operates at the intersection of archival ethnography and digital-humanities praxis, foregrounding how artists theorize technology through lived experience rather than through imported vocabularies of innovation. Between March and August 2025, thirty-six semi-structured interviews were conducted with artists working across video, sound, installation, moving image, animation, electronic literature, and computational media. Participants include both locally embedded practitioners and diaspora artists based in Europe and North America, enabling comparative insight into how access, mobility, and infrastructure shape artistic engagements with AI. Each interview followed a semi-structured format that situated generative AI within broader discussions of artistic practice and digital ethics. Rather than foregrounding AI as an isolated topic, participants were first asked to reflect on their creative processes and the role of digital technologies within them, followed by questions concerning ethical challenges encountered when working with digital platforms. Generative AI was subsequently examined in relation to these reflections, allowing artists to situate AI within their existing material, ethical, and conceptual frameworks. This approach avoided prescriptive or technologically deterministic framing and enabled comparative analysis while preserving participants' own vocabularies and priorities. Through this dialogical structure, artists articulated a wide spectrum of positions toward generative AI, ranging from pragmatic experimentation and cautious adoption to explicit refusal grounded in ecological, political, or epistemic concerns. Taken together, the interviews constitute a qualitative dataset that captures a moment of transition in how generative AI is being received, negotiated, and contested within South Asian creative communities.

While much existing literature on AI art focuses on Western creative economies and technological centres, relatively few studies examine how artists outside the Global North situate AI within their socio-political and infrastructural contexts (Elgammal et al., 2017; McCormack et al., 2019). The SADA corpus helps address this gap by articulating perspectives shaped by postcolonial histories, multilingual digital environments, and uneven technological access. Artists consistently describe AI not as a neutral medium but as a historically situated phenomenon entangled with extraction, privilege, and resistance. In doing so, they contribute to what Risam terms the postcolonial digital humanities: an approach that reorients digital knowledge production toward histories of power, colonialism, and inequity (Risam, 2018). This paper treats SADA not only as a repository of interviews but also as a methodological intervention. Following Caswell (2021), archives are understood as living infrastructures of memory and power rather than neutral containers of data. Through dialogical interviewing, collaborative transcription, and reflexive thematic analysis, artists are positioned as co-producers of knowledge rather than as sources of extractable data. Analysis of the thirty-six interviews identifies six recurring, overlapping conceptual clusters: AI as collaborator; ethical anxiety and authorship; refusal and resistance; posthuman and metaphysical engagement; decolonial data and representation; and access, infrastructure, and inequality. These themes are elaborated in the sections that follow to demonstrate how regional specificities shape the ethical and aesthetic vocabularies through which artists engage with generative AI. Ultimately, this paper argues that South Asian digital artists perceive generative AI not as a discrete technology but as a social condition; one embedded in longer histories of colonialism, development, and global inequality. Their responses reframe dominant debates on AI and creativity away from technological determinism and toward cultural negotiation, care, and situated ethics. By centering voices from the Global South, the study contributes to a more plural understanding of digital creativity and underscores the need for ethical, ecological, and context-sensitive approaches to AI in artistic practice.

2 Methodology

This study is grounded in qualitative and interpretive research traditions within the digital humanities and art research, informed by postcolonial and decolonial approaches to knowledge production. Methodologically, it treats SADA not only as a source of empirical material but also as a research infrastructure shaped by ethical, political, and epistemic commitments. The study therefore adopts a reflexive, hybrid methodology combining semi-structured interviews, thematic analysis, and critical reflexivity (Braun & Clarke, 2006; Pink et al., 2016). This approach enables close engagement with artists' discourse while acknowledging the researcher's positionality as a curator and interlocutor embedded within South Asian digital-arts networks. All research activities were conducted as part of a graduate thesis at the University of British Columbia, Okanagan Campus, under institutional ethics approval.

2.1 Research Design and Data Collection

Between June and October 2025, thirty-six (n = 36) in-depth, semi-structured interviews were conducted with digital artists across South Asia, including India, Pakistan, Bangladesh, Sri Lanka, Nepal, Afghanistan, Bhutan, and Maldives, as well as artists from the South Asian diaspora. Participants were selected through purposive sampling to reflect diversity in geography, artistic medium, and career stage within the South Asian digital-arts ecosystem, complemented by an open call that extended participation beyond established networks and frequently cited practitioners (Patton, 2002). The corpus includes artists working across sound art, moving image, installation, animation, electronic literature, immersive environments, and computational practices. Approximately half of the participants identify as women or non-binary, and a significant portion are diaspora artists currently based in Europe or North America, enabling comparative insight into how mobility, access, and infrastructure shape artistic engagements with digital technologies. Interviews were conducted online via Zoom and lasted 60–90 min. Rather than foregrounding generative AI as an isolated topic, the interviews were structured around broader conversations on artistic practice, ethics, archives, and postcolonial context. Participants were invited to reflect on how personal and collective memory shape their work, their relationship to South Asian artistic traditions, and the evolution of their digital practices. Subsequent questions addressed creative processes, the role of digital technologies, and ethical challenges encountered when working with digital platforms. Artists were also asked to discuss the significance and accessibility of digital archives, the influence of colonial and postcolonial histories on their practice, and the ways in which language, symbolism, and cultural references negotiate between local specificity and global circulation. Social and environmental concerns such as systemic inequality, ecological crisis, and socio-political engagement were examined alongside questions of technological access, infrastructural limitation, and future imaginaries of digital preservation. Generative AI was explored within this wider ethical and material framework, allowing artists to situate AI within their existing practices rather than treating it as a standalone or prescriptive topic.

2.2 Analytic Framework

Data analysis followed reflexive thematic analysis as outlined by Braun and Clarke (2006, 2019). The process was iterative and interpretive rather than linear, reflecting the dialogical nature of the interviews and the complexity of artistic meaning-making. Analysis proceeded in three stages:

Initial Coding: Each transcript was reviewed line by line, and statements directly addressing *AI*, *automation*, *data ethics*, or *creative authorship* were extracted. This step produced an initial edited corpus of approximately 12,200 words focused on and around AI-related content.

Categorical Clustering: Excerpts were grouped under six emergent categories derived inductively from the data: (1) AI as Collaborator; (2) Ethical Anxiety and Authorship; (3) Refusal and Resistance; (4) Posthuman and Metaphysical Engagement; (5) Decolonial Data and Representation; and (6) Access, Infrastructure, and Inequality.

Interpretive Synthesis: The final phase involved synthesizing thematic patterns into conceptual arguments about the epistemological and ethical positioning of South Asian artists vis-à-vis generative AI. Attention was paid not only to what artists *said* but also to how they articulated emotion, uncertainty, and ambivalence, treating these affective registers as data (Ahmed, 2014).

2.3   Reflexivity and Positionality

The researcher's positionality is central to the methodological design. All interviews were conducted by Najam-Ul Assar, who serves as both interviewer and curator of the South Asian Digital Art Archive (SADA). In this dual role, the author occupies a liminal position between scholar and practitioner, necessitating ongoing reflexivity regarding power, interpretation, and representation (England, 1994; Rose, 1997). Many participants were professional peers or collaborators within South Asian digital-arts networks, and interviews often took the form of extended conversations rather than extractive data collection. This proximity facilitated trust and openness, particularly when discussing sensitive topics such as censorship, precarity, and ethical dilemmas surrounding AI. To mitigate potential bias, interpretive triangulation was employed. Emerging themes were discussed with independent scholars in digital humanities and art–technology research, and selected quotes were returned to participants for contextual verification prior to publication. Reflexivity also extended to archival practice. Following Caswell (2021) and Mbembe (2002), SADA is understood as a living, political archive rather than a neutral repository. Decisions concerning transcription, metadata, and categorization were made collaboratively, ensuring that the archive itself embodied the ethical values articulated by participants.

2.4   Decolonial Research Ethics

The research adheres to decolonial ethics, prioritizing consent, reciprocity, and care (Smith, 2012). Participants retain ownership of their interview transcripts and may withdraw or redact material at any stage. Quotes used in this paper were selected and edited collaboratively to ensure accuracy and mutual agreement on meaning. While SADA is conceived as an open and evolving archive, full interview transcripts are not publicly released at this stage. This decision reflects consent agreements, concerns around miscontextualisation, and a deliberate resistance to extractive open-data paradigms that disproportionately affect artists from historically marginalized contexts. Openness is therefore treated as relational and negotiated rather than automatic. To further resist data extractives, metadata collection is intentionally minimal, and limited to non-intrusive descriptors such as name, country, artistic medium, and year of interview (Sayers, 2018). Biometric, demographic, or behavioural analytics are excluded. This approach aligns with principles of minimal and ethical computing and reflects the artists' own critiques of data-intensive AI systems.

2.5   Limitations and Scope

This study prioritizes analytical depth over statistical generalizability. While thirty-six interviews provide a rich qualitative corpus, the findings do not claim to represent all South Asian artists' perspectives on generative AI. Participants were self-selecting and already engaged in digital or new-media practices, which may have privileged voices with some degree of technological access. Future research could extend SADA to include artisans, craft practitioners, or non-digitally trained creators whose engagements with AI may differ substantially. Nevertheless, the corpus spans multiple countries, artistic media, and positionalities, offering, to date, qualitative accounts of how generative AI is being negotiated within South Asian digital art. By framing methodology as both practice and ethics, the study demonstrates that researching AI in postcolonial contexts requires not only analytical rigour but also attentiveness to infrastructure, consent, and power.

3 Findings and Analysis

The analysis of thirty-six artist interviews conducted under the South Asian Digital Art Archive (SADA) reveals a complex and often ambivalent landscape of attitudes toward generative artificial intelligence (AI). The findings resist any singular narrative of acceptance or rejection. Instead, they articulate plural, situated positions shaped by ethical reflection, material constraint, and creative agency. Through iterative coding and reflexive thematic analysis, six recurring and overlapping thematic categories emerged: (1) AI as collaborator; (2) ethical anxiety and authorship; (3) refusal and resistance; (4) posthuman and metaphysical engagement; (5) decolonial data and representation; and (6) access, infrastructure, and inequality. Numerical descriptors are used here to indicate approximate thematic distribution rather than statistical significance, reflecting the overlapping and interpretive nature of qualitative analysis.

3.1 AI as Collaborator

Approximately twenty-five interviewees described AI as a conditional collaborator or assistive tool rather than an autonomous creative agent. Artists in this group emphasized pragmatic uses of AI, such as subtitling, translation, prototyping, or pre-visualization, while maintaining clear boundaries around authorship and creative control. An Indian artist framed this position explicitly: "I welcome the ways AI can contribute to workflow," describing AI as "an intelligent collaborator" and "a good intern—useful, efficient, but very much under guidance." In her account, AI is useful in the production contexts e.g., "AI-based light meters" and using AI "to shorten subtitles, which then become something I actively work on." Similarly, another Indian artist described AI as most valuable at the planning and prototyping stage: "AI becomes a practical way to quickly prototype something visual that helps communicate intent," especially when a design document needs early visualization. She also noted a functional use case in post-production and scene generation: "I used AI to animate those images, and it worked surprisingly well," though she emphasized she does not rely on it for "visual aesthetics" as a primary aim. In the same pragmatic register, a British artist working on a project in Afghanistan highlighted AI translation as a tool that can support cross-linguistic work, while emphasizing limits and the necessity of human mediation: "we did use AI translators and systems in the beginning," but "it was really important that we had cultural consultants," since context and idiom can be "lost in translation quite easily." Across these accounts, AI is positioned as assistive and bounded. It is useful for workflows and interim steps but not treated as an autonomous author.

3.2 Ethical Anxiety and Authorship

Ethical concern emerged as one of the most pervasive themes across the corpus. Around thirty-six interviewees expressed explicit anxieties about authorship, consent, and data provenance. Artists frequently expressed unease about unconsented data scraping, stylistic appropriation, and the opacity of training datasets. A Pakistani artist articulated a sharp ethical objection to style replication: "it was so unethical," describing how quickly a living artist's visual language could be copied. In a different register, a British-Indian artist connected authorship anxiety to the erasure of labour: when people asked her whether her self-made 3D characters were AI, she "felt offended," because she is "primarily self-taught" and the assumption collapses years of learning into automation.

Data provenance and corporate extraction also surfaced strongly. A Sri Lankan artist drew a hard line around attribution and ownership, rejecting hidden use: "I don't believe in… having ChatGPT… write a sentence and then say, oh, look at this sentence I wrote. I think that's completely absurd." He connected this to training data: large language models are "trained on stolen fiction," stressing "disclose what exactly it is you're doing," and arguing that if one uses such systems, "you should hold no copyright over what you've created."

For several interviewees, ethical anxiety intersected with ecological concerns. A Nepalese artist stated plainly: "My opposition to AI is primarily environmental," adding that "even with uploading things to the cloud I'm against it," and linking institutional use of AI imagery to both ecological cost and the bypassing of paid artistic labour. Likewise, A Bangladeshi artist flagged extractive infrastructures, saying "they use a lot of resources… the data mining", forming for him as a central ethical struggle in his limited, necessity-driven experiments. Such reflections align with critiques of AI's hidden material and environmental costs (Crawford, 2021) and are often intensified by participants' awareness of colonial histories of cultural and resource extraction.

3.3 Refusal and Resistance

Around fifteen interviewees articulated an explicit stance of refusal toward generative AI. Importantly, refusal was framed not as technophobia but as an ethical, ecological, or process-based practice. An Indian artist described a refusal rooted in protecting thinking and the creative struggle: "I don't really want to use these tools," arguing that AI can make it "very tempting to outsource thinking itself," which leads to "a loss." He uses ChatGPT only in a limited informational way, "more like I would use Google or translation tools", while insisting that "making, thinking, and imagining have to come from the person." A Maldivian artist points out that AI removes "struggle," erodes "craftsmanship," and reduces "desire to think and challenge yourself," which can bolster refusal-as-process. But she isn't refusing outright, and is instead experiencing more critical ambivalence.

A refusal grounded in process and pleasure appears in a Pakistani artist, who emphasized that AI "take[s] away from the creative act… that process," including the "eureka moment" of making intellectual connections. For her, the issue is not moral superiority but a commitment to the lived experience of discovery: "I find a lot of joy in in the process." These positions resonate with Ahmed's (2014) articulation of refusal as an affective and political stance against dominant progress narratives. Refusal, however, was rarely absolute. Several interviewees combined critique with tactical or experimental engagement, using AI strategically while remaining critically aware of its epistemological limits. As one artist explained, "To hack a system, you must first dance inside it." This ambivalence reflects what Bhabha (1994) describes as mimicry; partial adoption used to expose and destabilize dominant technological logics.

3.4 Posthuman and Metaphysical Engagement

A smaller yet conceptually significant subset engaged generative AI through philosophical frameworks that re-situate "AI" beyond a narrow tool-discourse. An Indian artist explicitly moved away from extractive prompt logic, describing skepticism toward "simply inputting a prompt to extract an image or video," and instead framing AI as a collaborator with "agency," influenced by "animist traditions common across South Asia." In this view, AI is approached "conversationally": "The system is not commanded but engaged," with room for indeterminacy and response.

A different philosophical articulation appears in a non-binary Indian artist, who does not use generative AI in practice because it does not align with how they conceptualize world-making and plurality. They foregrounded the tension between contradictory South Asian epistemologies and a model that is structured to always answer: "Coming from South Asia… we live with contradiction as a fundamental condition," and asked what happens when "a system designed to always provide a response encounters contradictory worldviews." Here, the critique is not only ethics, but ontology and interface: the model may be multidimensional, but interaction is "question/answer," producing hierarchy rather than dialogue. These reflections align with decolonial posthumanist frameworks (Ferrando, 2019) and situate AI within long-standing South Asian ontologies of distributed agency, memory, and relational existence.

3.5 Decolonial Data and Representation

Concerns about dataset coloniality and representational bias were raised by around half of the interviewees. Artists repeatedly described AI systems as privileging Western aesthetics, languages, and cultural references. A Pakistani artist connected bias to the absence of robust archives: "AI systems are only as good as the data they are trained on," and "when something hasn't been archived, it effectively doesn't exist for AI systems." She therefore argued for building "strong, democratic, and culturally specific archives," and when using AI preferred "my own data or carefully curated datasets" as an ethical responsibility. Another Pakistani artist linked representational bias to who builds AI: "Most dominant AI models come from Western… institutions," and "this bias isn't limited to images; it extends to political and ideological positions as well." He emphasized a counter-strategy: "I would like to train my own data sets" if using AI seriously in visual work. Bias was also empirically observed through prompt experiments.

An Indian artist reported that when she tested prompts (e.g., poems about "Indian women" and "Indian men"), she encountered "a strong pattern of bias," where identities were "reduced to narrow stereotypes." Cultural misrecognition emerged in relation to Bhutan as well. A Bhutanese artist observed that AI-generated characters often "miss out… details" in traditional attire and that "it still it misses some details," framing this as a dataset and representation gap rather than a purely aesthetic complaint. These demands articulate a decolonial claim to data sovereignty and frame AI as a contemporary extension of colonial knowledge extraction (Noble, 2018; Chakrabarty, 2000).

3.6 Access, Infrastructure, and Inequality

Material access emerged as the most widely shared concern across the dataset. Nearly all interviewees referenced constraints such as cost, limited training, unstable electricity, and restricted bandwidth. A Pakistani artist described the financial barrier directly: "Software is expensive, and not everyone can afford it," recalling how, in Pakistan, many relied on pirated software to work; she also cited subscription pricing as prohibitive. Infrastructure constraints intersected with production time and risk. A Nepalese artist contrasted earlier rendering conditions adding that "it took me two days to render that" and experienced anxiety about "load settings" during that time, however, with AI's speed, they required "five minutes" to reach something comparable, underscoring how technical conditions shape who can experiment and how.

Institutional access and educational ecosystems also mattered. A Sri Lankan artist emphasized that the challenge was not only access to software but also "the lack of critical frameworks around how and why to use these tools," noting that "very few teachers or artists locally" were engaging digital media thoughtfully, forcing self-teaching and informal learning networks. Finally, inequality was also framed as environmental and institutional responsibility. A Nepalese artist argued that large institutions that can pay artists but choose AI compound both ecological and labour injustice: "they can afford to pay," yet using AI becomes "an environmental burden and an ethical shortcut," often producing imagery that feels "exoticized" and detached from local visual culture. This infrastructural imagination positions scarcity not only as a limitation but also as a generative condition of practice (Larkin, 2013).

Taken together, the findings reveal a set of plural, situated ethics rather than a unified position on generative AI. Across the thirty-six interviews, artists foreground responsibility over novelty, locality over abstraction, and relationality over individual authorship. AI is understood not as a discrete technology but as a social condition entangled with ecology, labour, language, and infrastructure. These perspectives challenge universalizing narratives of machine creativity and underscore the necessity of culturally embedded and historically informed approaches to AI ethics. As one participant succinctly concluded, "AI opens possibilities for storytelling, but access decides who gets to imagine."

4 Conclusion: Toward Plural Ethics of Technological Creativity

The South Asian Digital Art Archive (SADA) corpus demonstrates that contemporary artists working across South Asia engage with generative AI neither as passive adopters nor as uniformly oppositional actors. Instead, they approach algorithmic systems as critical interlocutors, situating AI within historically informed, ethically attentive, and materially grounded frameworks. Across the thirty-six interviews analyzed, artists expressed ambivalence toward generative AI. This is an ambivalence that is neither contradictory nor productive. AI appears simultaneously as a tool for experimentation and a source of ethical concern; as a means of extending creative practice and as a reminder of structural inequality. Rather than resolving these tensions, artists often deliberately inhabit them, using ambivalence itself as a mode of critique and reflection. Taken together, the findings support an understanding of plural ethics of technological creativity: an ethical orientation that is situated, relational, and context-dependent rather than universal or prescriptive. These ethics emerge inductively from artists' reflections rather than from external normative frameworks. While grounded in South Asian contexts, they contribute to broader discussions in AI studies and the digital humanities by demonstrating how ethical reasoning around technology is shaped by infrastructure, history, and cultural practice.

4.1 Access, Infrastructure, and Inequality

A central insight of this study is that artistic engagement with AI is always situated. Artists' practices cannot be separated from the socio-technical infrastructures that condition access to computation, software, training, and institutional support. Across the SADA corpus, participants repeatedly linked their engagement with digital tools to material realities such as unstable electricity, limited bandwidth, high costs of proprietary tools, and uneven educational infrastructures. These conditions shape not only what forms of AI experimentation are possible but also how AI is conceptualized and evaluated. This finding complicates universalist narratives that frame AI as a globally uniform creative medium. Instead, AI emerges as an assemblage of tools, bodies, institutions, and environments whose meanings are produced through use. For example, practices of creative problem solving and improvisation, often described through the idiom of jugaad, illustrate how artists work creatively within constraints, repurposing available resources rather than following linear models of technological progress. In this sense, AI is not an autonomous agent but a situated practice, acquiring significance only through culturally specific forms of engagement, aligning with Suchman's (2007) account of situated action.

4.2 Care, Refusal, and Ethical Deliberation

A second insight concerns the ethical framing of refusal. For a group of interviewees, refusal of generative AI does not indicate rejection of technology per se, but a deliberate ethical stance grounded in care, responsibility, and ecological awareness. Artists frequently described refusal as a way to preserve embodied forms of making, relational authorship, and attentiveness to environmental impact. These positions draw on feminist and decolonial traditions that understand care not as withdrawal but as an active mode of resistance within extractive systems (Tronto, 1993; Puig de la Bellacasa, 2017).

4.3 Plural Posthuman Imaginaries

A smaller set of interviewees articulated engagements with AI through spiritual, metaphysical, or relational frameworks. These perspectives resonate with decolonial posthumanist scholarship (Ferrando, 2019) and demonstrate how non-Western ontologies can inform alternative understandings of human–machine relations. Rather than displacing human agency, these imaginaries recast creativity as co-emergent and relational. While not representative of all participants, such reflections underscore the plurality of philosophical resources artists draw upon when making sense of AI, challenging the assumption that posthuman thought is rooted solely in Euro-American theory.

4.4 Decolonial Data Futures

Across multiple themes, artists articulated a sustained critique of data coloniality. AI systems were frequently described as reproducing Western-centric aesthetics, linguistic hierarchies, and cultural assumptions. These critiques align with broader work in critical data studies that highlights how data infrastructures reflect existing power relations (Noble, 2018; D'Ignazio & Klein, 2020). Importantly, participants' concerns extended beyond representational bias to questions of ownership, consent, and benefit. Calls for local datasets, transparent training processes, and community-controlled models reflect an emerging decolonial approach to data governance.

4.5 Implications and Scope

The findings of this study have implications for AI scholarship, arts policy, and digital-humanities research. First, they suggest that ethical frameworks for AI must account for infrastructural inequality and cultural specificity rather than assuming uniform conditions of access. Second, they highlight the value of artistic practice as a site of ethical knowledge production. Artists' reflections that are grounded in material experience, ecological awareness, and cultural memory can offer insights that complement technical and policy-oriented approaches to AI governance. At the same time, the study does not claim generalizability beyond its qualitative scope. The SADA corpus represents a diverse but self-selecting group of artists already engaged with digital media. The aim is not to speak for all South Asian artists, but to provide analytically transferable insights into how generative AI is being negotiated within specific postcolonial contexts. Future research could extend this work by engaging non-digital practitioners, comparative regional studies, or mixed methods approaches.

4.6 The Future of Ethical Technological Creativity

Looking forward, the SADA corpus gestures toward a redefinition of what it means to create ethically in a world increasingly mediated by algorithms. The challenge is not simply to regulate AI but to cultivate technological cultures of care and ecosystems that balance innovation with responsibility. South Asian artists, through their practices of improvisation, refusal, and relational imagination, model such cultures in action. Their work demonstrates that technological progress need not be opposed to ethical consciousness; rather, ethics can be a generative force within creative innovation. By drawing on plural epistemologies such as to be feminist, ecological, spiritual, and decolonial, these artists expand the vocabulary of AI art and offer new paradigms for cross-cultural dialogue. Ultimately, SADA's value lies not only in its archival contribution but in its theoretical proposition: that artistic reflection is itself a mode of ethical computation. When artists reimagine AI through care, locality, and plurality, they rewrite the algorithmic futures that will shape collective life.

Acknowledgments. This research was conducted as part of a graduate thesis at the University of British Columbia, Okanagan Campus (Study ID: H25-00591). The authors would like to thank all participating artists for their time, generosity, and willingness to share their practices and reflections. Any errors or omissions remain the responsibility of the authors.

Disclosure of Interests. The authors declare that they have no competing interests that are relevant to the content of this article.

References

Ahmed, S.: Willful Subjects. Duke University Press, f'Durham (2014)

Bhabha, H.K.: The Location of Culture. Routledge, London (1994)

Birhane, A.: Algorithmic injustice: a relational ethics approach. Patterns **2**(2), 100205 (2021). https://doi.org/10.1016/j.patter.2021.10020. Accessed 30 Jan 2026

Braun, V., Clarke, V.: Using thematic analysis in psychology. Qual. Res. Psychol. **3**(2), 77–101 (2006). https://doi.org/10.1191/1478088706qp063oa. Accessed 30 Jan 2026

Braun, V., Clarke, V.: Reflecting on reflexive thematic analysis. Qual. Res. Sport Exer. Health **11**(4), 589–597 (2019). https://doi.org/10.1080/2159676X.2019.1628806. Accessed 30 Ja 2026

Caswell, M.: Urgent Archives: Enacting Liberatory Memory Work. Routledge, London (2021)

Chakrabarty, D.: Provincializing Europe: Postcolonial Thought and Historical Difference. Princeton University Press, Princeton (2000)

Couldry, N., Mejias, U.A.: The Costs of Connection: How Data Is Colonizing Human Life and Appropriating It for Capitalism. Stanford University Press, Stanford (2019)

Crawford, K.: Atlas of AI: Power, Politics, and the Planetary Costs of Artificial Intelligence. Yale University Press, New Haven (2021)

D'Ignazio, C., Klein, L.F.: Data Feminism. MIT Press, Cambridge (2020)

Elgammal, A., Liu, B., Elhoseiny, M., Mazzone, M.: CAN: Creative adversarial networks—Generating "art" by learning about styles and deviating from style norms. In: Proceedings of the Eighth International Conference on Computational Creativity (ICCC 2017). arXiv:1706.07068 (2017)

England, K.V.L.: Getting personal: reflexivity, positionality, and feminist research. Professional Geograph. **46**(1), 80–89 (1994). https://doi.org/10.1111/j.0033-0124.1994.00080.x. Accessed 30 Jan 2026

Ferrando, F.: Philosophical Posthumanism. Bloomsbury, London (2019)

Larkin, B.: The politics and poetics of infrastructure. Ann. Rev. Anthropol. **42**, 327–343 (2013). https://doi.org/10.1146/annurev-anthro-092412-155522. Accessed 30 Jan 2026

Mbembe, A.: The power of the archive and its limits. In: Hamilton, C., Harris, V., Taylor, J., Pickover, M., Reid, G., Saleh, R. (eds.) Refiguring the Archive, pp. 19–27. Springer, Dordrecht (2002). https://doi.org/10.1007/978-94-010-0570-8_2

McCormack, J., Gifford, T., Hutchings, P., Brown, A.R.: Autonomy, authenticity, authorship and intention in computer generated art. In: Proceedings of the Tenth International Conference on Computational Creativity (ICCC 2019), pp. 37–44 (2019)

Noble, S.U.: Algorithms of Oppression: How Search Engines Reinforce Racism. NYU Press, New York (2018)

Patton, M.Q.: Qualitative Research & Evaluation Methods, 3rd edn. Sage, Thousand Oaks (2002)

Pink, S., Horst, H., Postill, J., Hjorth, L., Lewis, T., Tacchi, J.: Digital Ethnography: Principles and Practice. Sage, London (2016)

Puig de la Bellacasa, M.: Matters of Care: Speculative Ethics in More than Human Worlds. University of Minnesota Press, Minneapolis (2017)

Risam, R.: New Digital Worlds: Postcolonial Digital Humanities in Theory, Praxis, and Pedagogy. Northwestern University Press, Evanston (2018)

Rose, G.: Situating knowledges: positionality, reflexivities and other tactics. Progress Human Geograp. **21**(3), 305–320 (1997). https://doi.org/10.1191/030913297673302122. Accessed 30 Jan 2026

Sayers, J. (ed.): The Routledge Companion to Media Studies and Digital Humanities. Routledge, London (2018)

Smith, L.T.: Decolonizing Methodologies: Research and Indigenous Peoples, 2nd edn. Zed Books, London (2012)

Suchman, L.: Human-Machine Reconfigurations: Plans and Situated Actions, 2nd edn. Cambridge University Press, Cambridge (2007)

Tronto, J.C.: Moral Boundaries: A Political Argument for an Ethic of Care. Routledge, New York (1993)

Algorithms for Collaborative Harmonization

Eyal Briman[✉], Eyal Leizerovich, and Nimrod Talmon

Ben-Gurion University of the Negev, Beer-Sheva, Israel
`briman@post.bgu.ac.il`

Abstract. We study collaborative aggregation of structured sequences, using musical harmonization as a concrete and well-defined testbed. While our motivating application is music, the broader goal of this work is to understand how collective inputs over structured strings can be aggregated algorithmically in a principled way–an issue that also arises in text aggregation and related domains. Concretely, given multiple harmonization suggestions for a fixed melody, we design and analyze aggregation algorithms that aim to balance two competing objectives: (i) representing the agents' suggestions faithfully, and (ii) producing a sequence that is internally coherent according to a deliberately simple transition-based notion of harmony. We introduce a formal model, propose several aggregation rules inspired by social choice, and study their computational properties. Our experimental evaluation shows that relatively simple methods, in particular plurality- and Kemeny-based rules, perform consistently well under both representation and coherence criteria.

Keywords: Aggregation · Structured sequences · Harmonization · Social choice · Algorithms

1 Introduction

Many collaborative settings require aggregating multiple individuals' inputs into a single collective output that is both representative and internally coherent. Classical social choice typically studies outcomes such as a single alternative or a ranking; by contrast, many contemporary applications involve aggregating *structured objects*, such as strings or sequences, where both local and global constraints matter. This paper studies aggregation in such structured domains from a social choice and algorithmic perspective.

We use *musical harmonization* as a concrete and well-defined testbed for structured aggregation. Harmonization naturally yields sequences of symbols (chords) drawn from a fixed alphabet, making it amenable to formal modeling and algorithmic analysis. Importantly, our goal is *not* to provide a detailed computational model of harmony. Rather, we deliberately adopt a simplified musical abstraction in order to isolate aggregation questions: how to combine multiple proposed sequences into a single sequence that (i) represents the agents' inputs and (ii) satisfies a notion of internal coherence.

P. Machado et al. (Eds.): EvoMUSART 2026, LNCS 16523, pp. 34–48, 2026.
https://doi.org/10.1007/978-3-032-24350-8_3

In our model, each agent provides a candidate chord sequence of fixed length. Representation is quantified by comparing the selected chord at each position to the agents' proposed chords using a pitch-set–based distance (Jaccard distance). Coherence is captured via a lightweight transition model that assigns weights to consecutive chord pairs (a 2-gram model). These modeling choices abstract away from musically important aspects such as functional harmony, voice-leading, tonal context, and melodic constraints. Consequently, our results should be interpreted as statements about aggregation methods within this controlled framework rather than as claims about musically optimal harmonizations.

Within this setting, we develop a formal model, propose several aggregation rules inspired by social choice (including plurality- and Kemeny-based methods), and analyze their computational properties. We also introduce 2-gram–aware variants that explicitly trade off representation and coherence. Our experimental results indicate that relatively simple aggregation rules perform robustly across the considered measures. While this finding is encouraging from a computational and practical standpoint, it also raises broader questions about when and why simple methods remain competitive in structured aggregation problems.

Contributions. Our main contributions are: (i) a formal social-choice-inspired model for aggregating chord sequences as a representative testbed for structured sequence aggregation; (ii) a collection of aggregation algorithms balancing representation and coherence, including 2-gram–based variants; (iii) computational complexity results identifying which objectives admit polynomial-time solutions and which are intractable; and (iv) an experimental evaluation comparing the proposed methods under our representation and coherence criteria.

Paper Structure. After reviewing related work, we introduce the musical preliminaries in Sect. 2. We then formalize the aggregation framework in Sect. 3, present and analyze aggregation algorithms in Sect. 4, and study their computational complexity in Sect. 5. Finally, we report and discuss experimental results in Sect. 6.

Related Work. We mention related work from both the social choice literature and chord sequence generation. First, we mention the work on general aggregation in metric spaces [2,18], that also includes suggestions on how to perform text aggregation. Another model that is relevant is that of *multiple attribute list aggregation* [1], in which a sequence of elements (not necessarily text characters) are the output of the social choice instance. We also wish to mention judgment aggregation [7], which corresponds to a very general social choice setting and to which some of our algorithms are related, and get inspiration from the general work on aggregating preferences under constraints in formatted ways [9] Next we mention works on decision-theoretic planning techniques into automatic harmony generation and chords progression generation based on stochastic processes [5,10,17]. Our work makes assumptions for simplicity needs that differ from the classic chord progression problems, as we are interested in aggregation of chords preferences of agents, with respecting the probability of a chord progression to appear (based on pre-trained 2-gram model), but with no explicit use of the melody itself – we assume the user is responsible of giving a valid harmony to a melody, and we offer an aggregation method.

2   Musical Preliminaries

While our focus is primarily on social choice, we briefly introduce the musical concepts used in the paper. Music is often described in terms of three components: *rhythm, melody,* and *harmony* [8,13]. We outline these notions in an abstract (symbolic) form to keep the preliminaries accessible to readers without a music-theory background.

Rhythm. Rhythm is the organization of sound and silence in time. It is typically structured into beats grouped into bars (measures), where the time signature specifies how many beats occur in each bar and which note value constitutes one beat.

Melody. A melody can be viewed as a sequence of notes played over time. Abstractly, it is a string over a finite alphabet of pitch symbols (e.g., C, D, E, ...). For example, the opening of "Twinkle, Twinkle, Little Star" can be written as the sequence C–C–G–G–A–A–G.

Harmony. Harmony concerns notes sounding simultaneously, forming *chords,* and is the central musical component in our study of collaborative harmonization. In the same symbolic framework, a chord is a small set of pitch symbols played together (e.g., the C major chord contains $\{C, E, G\}$). A harmonization is then a sequence of such chords accompanying a melody. Different chord choices and progressions can change the perceived character of the same melody; for instance, accompanying a melody with C major ($\{C, E, G\}$) yields a different feel than accompanying it with A minor ($\{A, C, E\}$).

2.1   A Basic Grammar of Harmony

Our main motivation for studying the collaborative harmonization problem is that harmony can be naturally represented as text (where each chords corresponds to its character) and, more importantly, it has some general structured grammar that is more basic than general text documents. In what follows we provide a simplified view of the grammar of harmony that is useful for our purposes.

In this paper, we consider a finite set of chords we address as "alphabet of chords" as shown in the supplementary material (this is not a complete list of all chords used in western music, though, but it is sufficient for the purposes of our study). Correspondingly, an harmonic sequence is simply a sequence of characters from that alphabet of chords.

Chord Similarity – A Spatial (Grammatic) Aspect. Given the chord alphabet (presented in the supplementary material), it is natural to observe that some chords are acoustically more similar than others. In analogy to language—where some words are closer in meaning than others (e.g., *bicycle* and *bike* versus *apple* and *spaceship*)—we would like to quantify how "close" two chords are.

Accordingly, we introduce a distance metric that measures chord dissimilarity and will be used to assess how well an aggregated harmonization represents the agents' suggestions.

Below we discuss two options.

- **Jaccard Distance**: Recalling that each chord (internally) contains few notes, this metric measures *dissimilarity* by comparing the overlap between the note sets of two chords (treated as sets of 4 notes) to the size of their union. It accounts for overlapping notes, yielding a distance that ranges from 0 (identical chords) to 1 (no common notes).

Concretely, the Jaccard distance between two chords A and B is

$$d_{\text{Jaccard}}(A, B) = 1 - \frac{|A \cap B|}{|A \cup B|} \,, \tag{1}$$

where $|A \cap B|$ denotes the number of common notes between chords A and B, and $|A \cup B|$ represents the total number of unique notes in both chords.

Example 1. CMaj7 $= \{C, E, G, B\}$ and FMaj7 $= \{F, A, C, E\}$ have two common notes, namely C and E, and six unique notes in their union, namely $\{A, B, C, E, F, G\}$. Thus, their Jaccard distance is

$$d_{\text{Jaccard}}(\text{CMaj7}, \text{FMaj7}) = 1 - \frac{2}{6} = \frac{2}{3}.$$

- **Tonal Distance**: Besides the simple note-counting process that underpins the Jaccard distance, there are other psychoacoustic features that affect chord similarity. The tonal distance relates to the *acoustic-functional relationship* between chords: it assesses the harmonic function of chords, rather than focusing solely on the pitch content [11,12].

As the concept of *acoustic-functional relationship* between chords is challenging to model directly, we have chosen the Jaccard distance as our metric for this work although it is a less common "chord-distance" in practice, as we are interested in the aggregation process itself. We speak about this decision more in Sect. 8 and recommend investigating other distance metrics for future work.

Chord Progression – Temporal Grammatical Aspect. Besides the similarity (and interchangeability) of chords, there is also importance to chord progression – how chords in a sequence relate to each other. Essentially, this is somehow similar to the fact that certain **words** are more likely to come after other words (e.g., the word *dog* is somehow likely to come after the word *barking* but perhaps less likely to come after the word *fruitful*; this concept is known as the n-gram model in natural language processing [3]).

Concretely, in this paper, for the calculation of transition probabilities used in the context of chord progressions, we took a data-based approach. In particular, we considered a data set consisting of 1,410 Jazz songs (iReal Pro). To extract the chord symbols from this dataset, we used ireal-reader tool. After acquiring the chord symbols, our next step involved converting each chord into a simplified representation using the alphabet shown in the supplementary material. For the purpose of aggregation, we filtered 1015 songs from the data set (in particular, those that are precisely 32 bars in length, each encompassing a maximum of two chords per bar; technically, when encountering instances where only a single chord was present in a bar, we replicated it to ensure uniformity and consistency throughout the data set). Then, we used the adjusted data set to compute the probabilities of two chords being successive in an harmonic sequence.

As a result, we have a complete, directed graph with weighted arcs that represent the probability of one chord appearing after another, defined as $G = (T, w)$. In this context, each vertex $t \in T$ corresponds to a chord from "chords alphabet" as shown in the supplementary material. The set of edges, E encompasses all possible arcs, where the weight $w_{(u,t)}$ on the arc $(u, t) \in E$ signifies the degree of smoothness associated with transitioning from chord u to chord t, with $0 \leq w_{(u,t)} \leq 1$.

3   Formal Model

We describe our formal model and discuss how to evaluate the quality of different winning harmonizations.

A Formal Model of Collaborative Harmonization. With respect to a certain "harmonic alphabet A (containing the set of m possible chords), an *instance* of our model contains the following ingredients:

- A given *size k* of the harmonic sequence to produce.
- A set $V = \{v_1, \ldots, v_n\}$ of n agents (i.e. n sets of harmonic sequences); each agent suggests its ideal harmonic sequence, denoted by v_i, where $v_i \in A^k$.
- A fixed *chord transition graph G* over A, with edge weights $w_{(u,t)} \in (0, 1]$ encoding the likelihood (or smoothness) of transitioning from chord u to chord t, as described in Sect. 2.

It is thus convenient to denote an instance of the model by (k, V). Given an instance (k, V) of the model, a *solution W* corresponds to an aggregated harmonic sequence; formally, $W \in A^k$.

An *aggregation method* (i.e., a voting rule) for the setting of collaborative harmonization takes an instance (k, V) as its input and outputs a solution W.

Remark 1. Note that, perhaps surprisingly, the melody itself is not part of instances of our model, as we address only the harmonization; however, refer to Sect. 8 to a more elaborate discussion regarding this point.

Solution Quality. What is missing from the section above is a discussion on how to evaluate the quality of solutions (i.e., aggregated harmonic sequences) of instances of our model. Informally speaking, we are looking for aggregation methods whose output strike a balance between these two aspects: (1) first, a solution shall respect the suggestions of the agents – i.e., the agents ideal harmonic sequences shall be taken into account by the aggregation algorithm; and (2) second, a solution should be musically appealing on its own.

There are several ways to formally capture these two desires. Below we describe our approach, which builds on (1) chord similarity for the first aspect; and (2) chord progression probabilities for the second aspect.

3.1   Agent Satisfaction

Our approach to the consideration of the correspondence between the ideal harmonic sequences of the agents and a possible winning harmonization is to define *agent satisfaction* (this follows the utility-based approach in social choice, such as used, e.g., for participatory budgeting [16]). Concretely, consider some agent v_i and a possible winning harmonization W; intuitively, the more similar v_i is to W (with respect to the metric used for chord similarity) the more satisfied v_i shall be. Formally, we define as follows:

– Let d be the Jaccard distance (refer to Section Similarity).
– We then define, for agent i voting as v_i and a possible winning harmonization W, the satisfaction of agent i from W to be (assume that $|v_i| = |W| = k$):

$$sat(i, W) := \sum_{j \in [k]} d(v_i[j], W[j]) \ .$$

Using the concept of agent satisfaction, one may, e.g., consider the goal of maximizing the social welfare (i.e., maximizing the sum of agent satisfaction $\Sigma_{i \in [n]} sat(i, W)$). In this paper we largely take this utilitarian approach.

3.2   Musical Coherence

Besides corresponding to the ideal harmonic sequences of the agents, a solution of a collaborative harmonization instance shall also be musically coherent. Our approach towards the assessment of the musical quality of a potential winning harmonization builds on the 2-gram approach described in Sect. 2. While, indeed, musical coherence is more involved than our simple 2-gram approach (such as using an n-gram), our approach nevertheless captures the basics of chord progression: acoustically, a smoother transition sequence between chords contributes to a smoother "flow".

Remark 2. While it makes sense to consider Musical Coherence of the whole harmonization as a whole, a natural simplification is to consider pairwise transitions.

Formally, we define *Musical Coherence* as follows:

- We start with the *chord transition graph* G, as described in Sect. 2. Each directed edge (u, t) is assigned a weight $w_{(u,t)} > 0$ that represents the smoothness of transitioning from chord u to chord t.
- Next, for a candidate winning harmonization $W = (W[1], \ldots, W[k])$, we measure coherence by aggregating the smoothness of all consecutive transitions. Since smoothness values are naturally combined multiplicatively (each transition contributes a factor), we apply a logarithm to obtain an equivalent additive objective that is numerically stable and easier to optimize and interpret. Formally,

$$\text{Coherence}(W) := \sum_{j=1}^{k-1} \log\big(w_{(W[j], W[j+1])}\big) \,. \tag{2}$$

Equivalently, maximizing $\text{Coherence}(W)$ is the same as maximizing the product $\prod_{j=1}^{k-1} w_{(W[j], W[j+1])}$, because log is strictly increasing.

This definition quantitatively evaluates a harmonization by rewarding sequences whose consecutive chord transitions are smooth according to the transition graph.

3.3 Example and Discussion

Consider the following collaborative harmonization scenario involving a sequence of 4 chords and a community of 3 agents. The ideal harmonic sequence of the three agents are as follows:

- Agent 1: Cmaj7, Dm7, Db7, Cmaj7.
- Agent 2: Am7, Dm7, E7, Am7.
- Agent 3: Cmaj7, Fmaj7, G7, Am7.

Table 1. Chords and Notes they Contain.

Chord	Notes
Cmaj7	C,E,G,B
Am7	A,C,E,G
Dm7	D,F,A,C
Db7	Db,F,Ab,B
Fmaj7	F,A,C,E
E7	E,Ab,B,D
G7	G,B,D,F

Suppose we have a potential solution, as follows:

$W = (\text{Cmaj7}, \text{Dm7}, \text{E7}, \text{Am7})$

Agent satisfaction (based on Jaccard distance of common notes of two chords) is calculated using table 1 as follows:

$$\text{Agent 1: } (0 + 0 + \frac{2}{3} + 0.4) = 1.0667$$

$$\text{Agent 2: } (0.4 + 0 + 0 + 0) = 0.4$$

$$\text{Agent 3: } (0 + 0.4 + \frac{2}{3} + 0) = 1.0667$$

Thus the total satisfaction is 2.53334.

Using the transition graph (2-gram) probabilities calculated as explained in Sect. 2 we have $w_{(\text{Cmaj7},\text{Dm7})} = 0.0252903$, $w_{(\text{Dm7},\text{E7})} = 0.00706179$, and $w_{(\text{E7},\text{Am7})} = 0.0608952$. Therefore,

$$\text{Coherence}(W) = \log(0.0252903) + \log(0.00706179) + \log(0.0608952) \approx -11.429$$

4   Approaches and Algorithms

In this section, we develop and discuss algorithms for solving instances of collaborative harmonization. For convenience, we will use the following notation:

- The preferences of the agents are denoted using a matrix $\mathcal{B}$ of size $n \times k$ (for n agents and k being the length of the harmonic sequence). Each element $b_{i,j}$ in this matrix represents the chord selected by agent i in position j.
- Given such an instance, a solution is denoted by $W \in A^k$.

Plurality. Plurality performs a majority vote independently at each position $j \in [k]$, selecting a most frequent chord among $\{b_{1,j}, \ldots, b_{n,j}\}$. For any $W \in A^k$, define

$$M(W) = \sum_{i=1}^{n} \sum_{j=1}^{k} \mathbb{I}(b_{i,j} = W[j]), \tag{3}$$

and output $W \in \arg\max_{W \in A^k} M(W)$. Ties at any position are broken by a fixed deterministic total order $\prec$ over A (e.g., lexicographic chord names).

Kemeny Rule. We adapt the Kemeny principle by minimizing, at each position, the total Jaccard distance to the agents' chords. For any $W \in A^k$, define

$$K(W) = \sum_{j=1}^{k} \sum_{i=1}^{n} d(b_{i,j}, W[j]). \tag{4}$$

Since $K(W)$ decomposes across positions, we compute W by choosing for each j: $W[j] \in \arg\min_{c \in A} \sum_{i=1}^{n} d(b_{i,j}, c)$, breaking ties using the same order $\prec$.

PAV. Next, we are interested in proportionality [15]; and, to this end, we adapt the PAV voting rule to our setting. Concretely, we consider the *proportional Borda count* rule: this rule utilizes the harmonic series to assign scores to candidates or preferences based on their rankings, aiming for proportional representation according to their ranking [6]. In our context, the objective is to create a chord sequence that effectively represents the diverse preferences of the agents, proportionally. To accomplish this, we employ an objective functions that is largely similar to the proportional Borda count, albeit where the Borda score is replaced by our Jaccard metric. Formally, the objective function is defined as follows:

$$P(W) = \left(\sum_{i=1}^{n} \sum_{j=1}^{k} \frac{1}{j} \cdot (s\,(U\,(B_i, W))\,[j]) \right). \tag{5}$$

In this equation:

- U is a utility function that takes two k-sized vectors and returns a k-sized vector of utilities. We set $U(B_i, W)[j] = 1 - d(b_{i,j}, W[j])$.
- B_i represents a k-sized vector of chords, representing the ith row in matrix B.
- s denotes a sorting function that takes a k-sized vector and arranges it in descending order.

The output of the algorithm is $\arg\max_W (P(W))$. As this problem is naturally NP-Hard (see Sect. 5), for its computation (in our computer-based simulations), we utilized a heuristic approach a local search algorithm of simulated annealing.

Clustered-Kemeny. We introduce Clustered-Kemeny, an algorithm based on the Kemeny voting rule, adapted for the natural division of musical pieces into sections. The goal is to identify clusters of individuals with similar chord preferences, enabling the partitioning of the musical piece into sections and matching each cluster to its most representative section.

Assuming a given partition of the harmonic sequence's length k into $x \leq n$ continuous sections, we formulate a linear program for optimizing voter clustering and section matching. The partition, represented by vector Z, must satisfy specific conditions.

The problem is divided into two nested sub-problems: (1) Given a partition Z, find the optimal clustering of agents into sections; and (2) Find the optimal Z from all possible partitions into at most X sections, returning the optimal agent clustering. The objective is to maximize total satisfaction by minimizing the distance of selected chord solution W and their respective sections within the clustering. Formally:

$$\min_{Z \in \text{partitions}} \min_{a_i \in Z, W} \sum_{z \in Z} \sum_{i=1}^{n} \sum_{j=1}^{k} P(j, z, Z) \cdot Q(i, z, Z) \cdot (d(b_{i,j}, W[j]))$$

where: (1) $Q(i, z, Z)$ indicates whether agent i is in section z (1 if true, otherwise a specified value less than 1); (2) $P(j, z, Z)$ is 1 if position j belongs to section

z, 0 otherwise; and (3) For the last section: $P(j, x, Z)$ is 1 if $Z[x] \leq j \leq k$, 0 otherwise. These collectively define the optimization problem for the Clustered-Kemeny algorithm.

4.1   2-Gram-Based Approaches

Note that, above, we have only relied on the Jaccard distance for quantifying the chord similarity (between the ideal harmonic sequence of each agent and some proposed solution). Next we consider also the musical coherence (or harmonic flow) of a candidate harmonization denoted by Coherence(W) as described in Sect. 3.2.

These 2-Grams Based Algorithms expand upon the previous approaches, considering transition probabilities to enhance predictive accuracy while accommodating the collaborative nature of chord sequences. Below we formally describe the details of taking into account such transition probabilities in the computation of the specific aggregation algorithms described above.

Plurality with 2-gram. The objective is given by $\arg\max_W \big(x_M \cdot M(W) + (1 - x_M) \cdot$ Coherence$(W)\big)$, where $x_M \in [0, 1]$ is a constant.

Kemeny with 2-gram. The objective is determined by $\arg\min_W \big(x_K \cdot K(W) - (1 - x_K) \cdot$ Coherence$(W)\big)$, where $x_K \in [0, 1]$ is a constant.

PAV with 2-gram. The objective is determined by $\arg\min_W \big(x_P \cdot P(W) - (1 - x_P) \cdot$ Coherence$(W)\big)$, where $x_P \in [0, 1]$ is a constant.

Clustered-Kemeny with 2-gram. The objective is determined by $\arg\min_W \big(x_{KC} \cdot KC(W) - (1 - x_{KC}) \cdot$ Coherence$(W)\big)$, where $x_{KC} \in [0, 1]$ is a constant.

5   Computational Complexity

We present the computational complexity of the aggregation goals defined above in Table 2. Additionally, we provide two sketches of hardness proofs. The complete proofs, along with all other missing proofs of the complexity results, are included in the supplementary material.

Theorem 1. *PAV and PAV with 2-gram are NP-hard.*

Proof. We prove the NP-hardness of the Proportional algorithm by noting that it includes Proportional Approval Voting (PAV) as a special case, where $U[i] \in \{0, 1\}$. This implies Proportional is NP-hard [14]. By setting $x_p = 1$, which reduces to the Proportional algorithm, we establish that Proportional with 2-gram is also NP-hard.

Theorem 2. *Clustered-Kemeny and Clustered-Kemeny with 2-gram are NP-hard.*

Table 2. Various problems of collaborative harmonization and their computational complexity.

Problem	Complexity
Plurality	Poly-time
Kemeny	Poly-time
Proportional	NP-hard
Plurality with 2-gram	Poly-time
Kemeny with 2-gram	Poly-time
Proportional with 2-gram	NP-hard
Clustered-Kemeny	NP-hard
Clustered-Kemeny with 2-gram	NP-hard

Proof. We reduce the k-median problem, known to be NP-Hard [4], to Clustered-Kemeny. Given n strings $S = \{s_1, s_2, \ldots, s_n\}$ of length ℓ, a distance metric d, and a threshold t, the k-string median problem seeks k median strings $M = \{m_1, m_2, \ldots, m_k\}$ such that:

$$\sum_{i=1}^{n} \min_{1 \leq j \leq k} d(s_i, m_j) \leq t.$$

We construct a Clustered-Kemeny instance by replicating each input string k times, transforming it into an agent representation (e.g.,"abc" becomes "abcab-cabc"). Partitions and lengths are defined as $Z = \{L, L \cdot 2, L \cdot 3, \ldots, L \cdot k\}$. Setting $Q(i, z, Z) = 0$ for each agent i, we show that the k-median problem is a yes instance if and only if Clustered-Kemeny is a yes instance, establishing its NP-hardness.

For Clustered-Kemeny with 2-gram, setting $x_{KC} = 1$ aligns it with the hardness proof of Proportional with 2-gram, thus proving it NP-hard as well.

Remark 3. Plurality is computable polynomial-time since it performs a majority vote independently (from linearity) at each position $j \in [k]$ (with fixed tie-breaking). Kemeny is polynomial-time as well, because its objective $K(W) = \sum_{j=1}^{k} \sum_{i=1}^{n} d(b_{i,j}, W[j])$ decomposes across positions: for each j we select $W[j] \in \arg\min_{c \in A} \sum_{i=1}^{n} d(b_{i,j}, c)$ (again with fixed tie-breaking).

In both cases of Plurality-2gram and Kemeny-2gram, the combined objective can be written as a sum of (i) *unary* terms that depend only on the chosen chord at each position and (ii) *pairwise* terms that depend only on consecutive chords:

$$\sum_{j=1}^{k} f_j(W[j]) + \sum_{j=1}^{k-1} g_j(W[j], W[j+1]),$$

where f_j is the contribution of the base rule at position j (e.g., Plurality counts or Kemeny distances), and $g_j(c, c') = \log(w(c, c'))$ is the 2-gram coherence contribution. Therefore, the optimal sequence can be computed by a standard dynamic program (Viterbi-style) over k layers and $|A|$ states per layer:

$$\mathrm{DP}[1,c] = f_1(c), \qquad \mathrm{DP}[j,c] = f_j(c) + \max_{c' \in A}\big(\mathrm{DP}[j-1,c'] + g_{j-1}(c',c)\big),$$

(and analogously with min if the base objective is formulated as minimization). This yields $O(k|A|^2)$ time, plus polynomial preprocessing to compute the values $f_j(c)$.

6 Computer-Based Simulations

Next, we report on computer-based simulations to evaluate the quality of the proposed algorithms. Given the NP-hard nature of the problems, we used a heuristic approach with a simulated annealing solution. The local search was initiated with the Plurality algorithm, and each search had 1000 iterations.

Due to the scarcity of relevant data for collaborative harmony composition, we adopted a semi-artificial approach. We used real harmonizations of songs and introduced random perturbations to simulate ideal harmonies by different agents.

For the dataset, we used 8, 16, and 32 agents, each contributing an ideal chord progression. Each agent's progression had 8, 16, and 32 variations, respectively, of the original progression of songs from a processed dataset of 1015 jazz songs (see Sect. 2.1). Variations were introduced by random chord swaps within ranges (0, 1), (1, 2), (2, 3), and (3, 4). The new chord was selected based on Jaccard distances, ensuring musical coherence.

For the 2-gram-based algorithms, we fine-tuned the weights through several rounds of testing, arriving at optimal weights: $X_M = 0.5$, $X_K = 0.9$, $X_P = 1 - 2e^{-4}$, and $X_{KC} = 0.9$.

Evaluation metrics were based on three measures for any solution W:

- **Song Similarity Measure**: Assesses the proximity of the aggregated chord progression to the original sequence. Lower distances indicate higher adherence to the original progression:

$$\sum_{j=1}^{m} d(W[j], \text{original song}[j]) \ .$$

- **Cluster Coherence**: Evaluates the coherence of chords within each 16-bar section of the aggregated chord progression. Lower cluster distances indicate smoother transitions:

$$\frac{1}{(m-16)\cdot n}\sum_{i=1}^{n}\sum_{j=1}^{m-16}\sum_{k=j}^{j+16} d(W[k], b_{i,k}) \ .$$

- **Musical Coherence**: Measures the appropriateness of the aggregated chord progression within the context of the 2-gram. Higher scores indicate more harmonious sequences, as detailed in the supplementary material.

7   Results and Discussion

The full results shown in the supplementary material the average scores of all 1015 instances, each scaled by a factor of 100. These scores are associated with various parameters: a specific measurement, an algorithm used, the number of chord swaps employed to generate different variations, and the number of agents (representing the variations).

We also see a trend: an increase in the number of chord swaps (from 0 to 1) leads to a significant decline in musical coherence. This decline occurs because employing the Jaccard distance for chord swaps often generates chord progressions or variations that are less probable within the data set. Consequently, these less probable variations adversely affect the outcomes produced by the aggregation algorithms.

Further, direct analysis of the results are given in the supplementary material, and the source code is available at GitHub repository.

7.1   Insights and Implications

Here are some of the main insights we derive from the results of the simulations:

- Robustness to Errors- Plurality-based algorithms exhibit robustness to errors, maintaining better proximity to original sequences and higher Musical Coherence across varying error ranges and agent counts.
- Balancing Chord Adherence and Coherence- Kemeny methods strike a balance between chord adherence and musical coherence, particularly effective when integrated with 2-gram.
- Scalability Concerns- Plurality-based algorithms demonstrate scalability and consistent performance across different agent counts. In contrast, Clustered-Kemeny might face challenges in larger-scale settings, potentially indicating scalability issues.
- Clustering - It is intriguing to note that despite the anticipated ability of Clustered-Kemeny to generate coherent and adherent sections, it appears to have fallen short in achieving this objective. One plausible explanation could be attributed to the creation of numerous small clusters, each smaller than 16 bars, coupled with only a few agents assigned to each cluster. This scenario led to the formation of multiple 16-bar clusters that lacked coherence, consequently resulting in an inadequate evaluation of the measure's effectiveness.

7.2   Summary

Essentially, we observe that, while Kemeny + 2-gram doesn't consistently yield the best results across all combinations of measures, number of agents, and number of errors affecting the variations of a song, it reliably positions itself in a solid middle ground. It may not excel in every scenario, but it consistently maintains a respectable performance, offering a balance among different evaluation measures when considering diverse variations and conditions.

8   Conclusions and Future Work

This work studied collaborative aggregation of structured sequences from a social choice and algorithmic perspective, using musical harmonization as a deliberately simplified and well-structured testbed. Our goal was not to model harmony in depth, but to aggregate multiple proposed sequences into a single outcome balancing representation and internal coherence. Within this controlled framework, we introduced a formal model and several aggregation algorithms, and found that relatively simple methods often perform robustly under the considered measures. Several limitations suggest natural directions for future work, as we discuss next.

Our notion of similarity and coherence relies on pitch-set overlap (Jaccard distance) and local transition probabilities (2-grams), which enable controlled comparisons but do not capture functional harmony, voice-leading, or tonal context; exploring more musically grounded measures is therefore important. Moreover, since the same constructs appear in the objectives, data generation, and evaluation metrics, our empirical results should be interpreted comparatively within this framework rather than as claims of musical optimality, and robustness under alternative measures remains open. We also abstract away from melody, a substantial simplification. Melody could be incorporated by restricting the feasible chord alphabet at each position or by adding melody-conditioned penalties to the objectives. Another limitation is the vertical nature of our algorithms, which compare chords only at identical positions; addressing equivalences across non-aligned positions may improve contextual aggregation. Algorithmically, the strong performance of plurality- and Kemeny-based rules suggests that local aggregation can be effective in structured domains, though a deeper theoretical explanation of when and why this occurs remains to be developed. A natural heuristic not explored here is to apply Plurality at each position and break ties by maximizing coherence with the previously chosen chord. Finally, external validation against human musical judgment would strengthen the connection to musical practice. More broadly, the chord alphabet and transition model induce a grammar-like structure, suggesting follow-on work that treats harmonization as a generative grammar and applies grammar-based evolutionary or search methods optimized using our objectives. Beyond music, our results support the broader view that exploiting structured sequence representations can facilitate principled aggregation in domains such as text and other symbolic data.

References

1. Briman, E., Talmon, N.: Multiple attribute list aggregation and an application to democratic playlist editing. In: Malvone, V., Murano, A. (eds.) Multi-Agent Systems - 20th European Conference, EUMAS 2023, Naples, September 14-15, 2023, Proceedings. Lecture Notes in Computer Science, vol. 14282, pp. 1–16. Springer (2023). https://doi.org/10.1007/978-3-031-43264-4_1
2. Bulteau, L., Shahaf, G., Shapiro, E., Talmon, N.: Aggregation over metric spaces: proposing and voting in elections, budgeting, and legislation. J. Artif. Intell. Res. **70**, 1413–1439 (2021). https://doi.org/10.1613/JAIR.1.12388

3. Cavnar, W.B., Trenkle, J.M., et al.: N-gram-based text categorization. In: Proceedings of SDAIR-94, 3rd Annual Symposium on Document Analysis and Information Retrieval. vol. 161175, p. 14. Las Vegas, NV (1994)
4. Charikar, M., Guha, S., Tardos, É., Shmoys, D.B.: A constant-factor approximation algorithm for the k-median problem. J. Comput. Syst. Sci. **65**(1), 129–149 (2002). https://doi.org/10.1006/JCSS.2002.1882
5. Clement, B.J.: Learning harmonic progression using markov models (1998)
6. Dey, P., Talmon, N., van Handel, O.: Proportional representation in vote streams. In: Larson, K., Winikoff, M., Das, S., Durfee, E.H. (eds.) Proceedings of the 16th Conference on Autonomous Agents and MultiAgent Systems, AAMAS 2017, São Paulo, BMay 8-12, 2017, pp. 15–23. ACM (2017). http://dl.acm.org/citation.cfm?id=3091134
7. Grossi, D., Pigozzi, G.: Judgment aggregation: a primer. Synthesis Lectures on Artificial Intelligence and Machine Learning, Morgan & Claypool Publishers (2014). https://doi.org/10.2200/S00559ED1V01Y201312AIM027
8. Levine, M.: The jazz theory book. " O'Reilly Media, Inc." (2011)
9. Li, M., Vo, Q.B., Kowalczyk, R.: Aggregating multi-valued CP-nets: a CSP-based approach. J. Heuristics **21**(1), 107–140 (2015). https://doi.org/10.1007/S10732-014-9276-8
10. Paiement, J., Eck, D., Bengio, S.: A probabilistic model for chord progressions. In: ISMIR 2005, 6th International Conference on Music Information Retrieval, pp. 312–319, London, 11-15 September 2005, Proceedings (2005). http://ismir2005.ismir.net/proceedings/1091.pdf
11. Persichetti, V.: Twentieth century harmony: creative aspects and practice (1961)
12. Rohrmeier, M.: Towards a generative syntax of tonal harmony. J. Math. Music **5**(1), 35–53 (2011)
13. Rohrmeier, M.: The syntax of jazz harmony: diatonic tonality, phrase structure, and form. Music Theory Anal. (MTA) **7**(1), 1–63 (2020)
14. Skowron, P., Faliszewski, P., Lang, J.: Finding a collective set of items: from proportional multirepresentation to group recommendation. Artif. Intell. **241**, 191–216 (2016). https://doi.org/10.1016/J.ARTINT.2016.09.003
15. Skowron, P., Lackner, M., Brill, M., Peters, D., Elkind, E.: Proportional rankings. In: Sierra, C. (ed.) Proceedings of the Twenty-Sixth International Joint Conference on Artificial Intelligence, pp. 409–415, IJCAI 2017, Melbourne, August 19-25, 2017. ijcai.org (2017). https://doi.org/10.24963/IJCAI.2017/58
16. Talmon, N., Faliszewski, P.: A framework for approval-based budgeting methods. In: The Thirty-Third AAAI Conference on Artificial Intelligence, AAAI 2019, The Thirty-First Innovative Applications of Artificial Intelligence Conference, IAAI 2019, The Ninth AAAI Symposium on Educational Advances in Artificial Intelligence, EAAI 2019, pp. 2181–2188, Honolulu, Hawaii, January 27 - February 1, 2019. AAAI Press (2019). https://doi.org/10.1609/AAAI.V33I01.33012181
17. Yi, L., Goldsmith, J.: Automatic generation of four-part harmony. In: Laskey, K.B., Mahoney, S.M., Goldsmith, J. (eds.) Proceedings of the Fifth UAI Bayesian Modeling Applications Workshop (UAI-AW 2007), Vancouver, British Columbia, July 19, 2007. CEUR Workshop Proceedings, vol. 268. CEUR-WS.org (2007). https://ceur-ws.org/Vol-268/paper10.pdf
18. Zvi, G.B., Leizerovich, E., Talmon, N.: Iterative deliberation via metric aggregation. In: Fotakis, D., Insua, D.R. (eds.) Algorithmic Decision Theory - 7th International Conference, ADT 2021, Toulouse, France, November 3-5, 2021, Proceedings. Lecture Notes in Computer Science, vol. 13023, pp. 162–176. Springer (2021). https://doi.org/10.1007/978-3-030-87756-9_11

Asɛmpayɛtsia: An Afrocentric Framework for Computational Creativity in Sound and Image

Nana Amowee Dawson[✉] [iD]

Department of Music and Dance, Faculty of Arts, University of Cape Coast, Cape Coast, Ghana
amowee@gmail.com

Abstract. This paper introduces *Asɛmpayɛtsia*, an Afrocentric compositional framework derived from Ghanaian-Akan-Mfantse folklore (*Kodzi*), as a new model for computational creativity. Developed through artistic research in music and audiovisual composition, the framework proposes a Triadic Process of Artistic Translation—cultural excavation, compositional translation, and audiovisual reinscription—as a generative system through which intangible heritage can inform artificial intelligence–based creative processes. Inspired by UNESCO's 2003 Convention on the Safeguarding of Intangible Cultural Heritage, the Fourth Industrial Revolution, and the African Union's Agenda 2063 (Aspiration 5), *Asɛmpayɛtsia* redefines African cultural forms as dynamic creative algorithms rather than archival artifacts. The paper maps the logic of oral reasoning, cyclical temporality, and communal improvisation in Akan epistemology, or artistic reasoning, onto computational principles of iteration, recursion, and emergence. By reframing indigenous artistic cognition as a computational aesthetic model, this work contributes to current discourses on Decolonial AI and Cultural Computation, offering an alternative to Eurocentric paradigms of generative art and music. Ultimately, *Asɛmpayɛtsia* positions artistic research as a method for designing AI systems that think *with*—rather than *about*—African heritage, advancing new possibilities for intercultural creativity and algorithmic imagination.

Keywords: Computational creativity · Afrocentric aesthetics · Artistic research · Generative sound and image · Cultural computation · Decolonial AI · *Asɛmpayɛtsia* framework

1 Introduction

Advancements in artificial intelligence (AI) and generative systems are transforming composition, sound design, and visual media. However, the theoretical and aesthetic foundations of many computational-creative systems remain shaped by Eurocentric ideas about authorship, time, and cultural knowledge. Scholars warn that this dominance risks "algorithmic colonization," whereby externally developed technological solutions are imposed without regard for local values or creative epistemologies (Lugonzo, 2025; Arora, 2024). At the same time, critics argue that technological progress without social

© The Author(s), under exclusive license to Springer Nature Switzerland AG 2026
P. Machado et al. (Eds.): EvoMUSART 2026, LNCS 16523, pp. 49–66, 2026.
https://doi.org/10.1007/978-3-032-24350-8_4

and historical reflection exacerbates bias, resource exploitation, and opaque value judgments embedded in system design (Arora, 2024; Bartlett, Pfeffekorn, & Sunde, 2025). These concerns have intensified calls for historically grounded, pluralistic, and ethically oriented approaches to AI.

Within artistic research, practitioners advocate methods that generate knowledge through creative practice rather than merely about artworks. Practice-led inquiry is widely recognized as a legitimate scholarly mode, providing conceptual and methodological tools for framing creative processes as generative systems (Smith & Dean, 2009; Skains, 2018; Benschop et al., 2025; Borgdorff, 2012, 2013; Candy & Edmonds, 2018; Kara, 2020; Leavy, 2020). This orientation is particularly relevant for culturally rooted computational-creativity models, as it positions artistic practice as both method and evidence. Drawing on doctoral artistic research that transformed Ghanaian-Akan-Mfantse folklore (*Kodzi*) into a visual-programmatic form (Dawson, 2023), this paper introduces *Asɛmpayɛtsia*: an Afrocentric compositional framework organized through a Triadic Process of Artistic Translation—cultural excavation, compositional translation, and audiovisual reinscription. Rather than treating heritage as static data, *Asɛmpayɛtsia* proposes a computational aesthetic grounded in oral reasoning, cyclical temporality, improvisational mutation, and communal performance.

The contribution is twofold. First, the paper presents *Asɛmpayɛtsia* as an aesthetic-epistemic model that reframes indigenous compositional intelligence as a resource for computational creativity. Second, it outlines how the triadic process maps onto computational principles—iteration, recursion, fitness-as-resonance, and distributed co-creation—thereby suggesting pathways for designing AI tools that think with, rather than about, African heritage. This orientation resonates with UNESCO's emphasis on the living transmission of intangible heritage (UNESCO, 2003) and the African Union's Agenda 2063, which situates cultural renewal at the center of continental development (African Union, 2015).

To situate this intervention, the paper engages three strands of literature: (1) computational-creativity research defining creative systems (Eglash et al., 2024); (2) decolonial and critical studies of AI that challenge Western technical dominance (Lugonzo, 2025; Arora, 2024; Lewis et al., 2025); and (3) artistic-research approaches foregrounding practice-led scholarship (Smith & Dean, 2009; Skains, 2018; Hamilton & Hansen, 2024). Rather than replicating existing algorithmic paradigms, the analysis shows how Afrocentric creative logic can recalibrate system goals, evaluation criteria ("what counts as 'fitness' or 'quality'"), and human–machine relations.

The paper addresses the following research questions:

1. How can indigenous artistic logics—particularly those embedded in Akan-*Mfantse* oral and performative practices—be translated into computationally actionable design principles?
2. In what ways does the *Asɛmpayɛtsia* triadic process challenge Eurocentric assumptions in generative-art practices?
3. What are the ethical and decolonial implications of operationalizing cultural resonance and communal authorship in AI tools?

This paper does not present an implemented AI system or empirical evaluation. Instead, it offers a theoretically grounded, artistically derived framework to inform future design of computational systems within decolonial and culturally situated AI research. Following this introduction, Sect. 2 reviews relevant literature on computational creativity, decolonial AI critique, and artistic-research methods. Section 3 details the *Asɛmpayɛtsia* framework and its triadic components, supported by diagrams. Section 4 maps the model to computational primitives and prototyping pathways. Section 5 discusses ethical and decolonial implications, and Sect. 6 concludes with future directions, including postdoctoral work toward a functioning prototype.

2 Background and Context

2.1 Computational Creativity and Generative Systems

Computational creativity investigates the capacity of machines to perform tasks traditionally associated with human creativity, encompassing generative art, algorithmic composition, and AI-assisted design. Recent studies suggest that generative systems can enhance creative productivity by supporting ideation and iterative exploration, particularly in visual domains (Jordanous, 2019; Zhou & Lee, 2024). At the same time, ongoing debates question whether such systems exhibit deeper forms of creativity involving emotional engagement, intention, and cultural understanding. Boden's foundational distinction between combinational, exploratory, and transformational creativity remains influential in framing these discussions (Boden, 1998). Subsequent advances—including deep neural networks, transformer architectures, diffusion models, and hybrid symbolic–subsymbolic systems—have greatly expanded machine-generated expression across sound and image (McCormack et al., 2019; Cook et al., 2019; Veale & Cardoso, 2019; Veale & Pérez y Pérez, 2020; Carnovalini & Rodà, 2020; Cibotaru, 2025).

The central concern of this paper, however, is not whether AI systems can produce outputs that appear transformational, but how creativity is framed, evaluated, and situated within dominant paradigms. Many contemporary approaches prioritize optimization-driven objectives, statistical fitness, or novelty-based evaluation while abstracting away cultural context and relational meaning. From the perspective of *Asɛmpayɛtsia*, transformational creativity emerges instead through ethical engagement with cultural memory, communal processes, and situated interpretation. The framework, therefore, does not reject current AI capabilities but proposes a complementary orientation in which computational creativity is guided by cultural resonance, relational accountability, and dialogic emergence rather than optimization alone.

2.2 Decolonial Critique in AI

From a decolonial perspective, scholars have criticized the integration of AI technologies for embedding Western-centric epistemologies into technical systems and institutional practices. Correa Lucero and Martens (2025), for example, analyze how AI may reproduce colonial matrices of domination in Latin American contexts by marginalizing indigenous knowledge systems. In African settings, the imposition of externally developed AI without engagement with local epistemologies has similarly been described as "algorithmic colonization," in which technological infrastructures override indigenous creative logics. Such critiques underscore the need for culturally situated and ethically aligned AI systems that respond to the communities in which they operate.

2.3   Artistic Research Methodologies

Artistic research methodologies, particularly practice-led and practice-based approaches, provide important frameworks for integrating creative practice with scholarly inquiry. These methods treat the act of making as a mode of knowledge production, enabling the exploration of complex concepts through artistic processes. Hamilton and Hansen (2024) emphasize the pedagogical and critical potential of practice-led research, while Skains (2018) highlights its capacity to address questions that exceed conventional methodological boundaries. Such approaches are especially pertinent for AI systems engaging with indigenous cultural practices. By foregrounding practice-led inquiry, researchers can ensure that technological design and deployment remain responsive to specific cultural contexts rather than abstracted from them.

3   The *Asɛmpayɛtsia* Framework

3.1   Origins in the *Kodzi* (Akan Folklore)

The *Asɛmpayɛtsia* framework developed from my doctoral research on the revival of Ghanaian-Akan-Mfantse *Kodzi*—a collection of oral narratives that serve both as folklore and as moral philosophy (Gyekye, 1995; Dawson, 2023). The *Kodzi* tradition reflects an indigenous understanding of translation among words, sounds, and gestures, in which the act of performance itself becomes a way of theorizing (Yankah, 1995; Agawu, 2003). In this oral cosmology, knowledge is shared through condensed proverbs, layered metaphors, and rhythmic storytelling—principles that resemble algorithmic recursion and self-similarity (Mbiti, 1990; Anku, 2000). *Asɛmpayɛtsia* (literally "concise are words of worth") was conceived as a decolonial compositional model that reclaims these logics as valid epistemic and creative approaches. Instead of treating African traditions as "source material," it positions Akan thought as the driving force behind the compositional structure. This approach aligns with Ramose's (1999) Ubuntu epistemology, which holds that knowledge is relational, cyclical, and value-driven. It also echoes Dei's ElderCrit—a framework for engaging indigenous wisdom as theory rather than folklore (Dei, 2009, 2020; Dei & Adjei, 2024; Adjei & Dei, 2025).

Through this grounding, *Asɛmpayɛtsia* transforms folklore into compositional infrastructure, functioning simultaneously as theory, method, and artistic intervention.

Akan Rhythmic and Temporal Logics as Computational Inspiration

A central source of the *Asɛmpayɛtsia* framework is the rhythmic and temporal logics of Akan musical practice, which differ fundamentally from dominant Western compositional paradigms. In many Akan performance contexts, rhythm is organized not through linear metric hierarchy or harmonic progression but through cyclical time, polyrhythmic interaction, and call-and-response structures that unfold through communal participation. Musical meaning emerges through repetition with variation rather than through thematic development or goal-directed resolution. For example, Akan drumming practices often operate through interlocking rhythmic layers, in which no single pattern functions as the definitive temporal anchor. Instead, stability emerges relationally, through the interaction of multiple rhythmic voices over time. This contrasts with Western metrical systems, which typically rely on a fixed pulse, hierarchical accentuation, and linear formal development. From a computational perspective, this suggests models based on recursive iteration, relational timing, and emergent synchronization, rather than strictly sequential or rule-based generation. Similarly, the call-and-response principle in Akan vocal and instrumental music functions as a dynamic feedback mechanism. Each musical utterance is shaped in real time by its social and performative context, allowing improvisation to operate within culturally defined constraints. In *Asɛmpayɛtsia*, this logic informs the use of feedback loops and stochastic recombination, where generative processes remain responsive to contextual input rather than optimizing toward a predetermined aesthetic goal. By grounding computational inspiration in these specific rhythmic and temporal practices, *Asɛmpayɛtsia* foregrounds African musical reasoning not only as cultural content but as a source of algorithmic logic. This vignette illustrates how indigenous performance structures can inform alternative models of computational creativity grounded in relationality, cyclicality, and communal emergence, rather than in linear optimization.

3.2   The Triadic Process of Artistic Translation

The core logic of *Asɛmpayɛtsia* unfolds through what I call the Triadic Process of Artistic Translation—three interconnected movements that show how cultural meaning transforms into sound and visual art: Cultural Excavation, Compositional Translation, and Audiovisual Reinscription. Each stage mirrors a cycle of oral transmission in which knowledge is continuously reinterpreted, renewed, and technologized.

Cultural Excavation
This phase involves retrieving ancestral ways of knowing, symbols, and sound memory. It draws on *Sankɔfa*—the Akan principle of returning to reclaim what has been forgotten or suppressed. In practice, this includes decoding proverbs, reconstructing rhythmic patterns, and engaging with elders' philosophical commentaries. Dei (2020) and Eshun (1998) both stress the importance of speculative listening as a way of cultural retrieval—listening not only to what was said but also to what the ancestors meant through rhythm, silence, and gesture. In this way, cultural excavation functions like dataset curation in machine learning—an interpretive and ethical process that defines the system's knowledge boundaries. Each selected cultural fragment becomes a "data point" that carries emotional and historical significance.

Compositional Translation
Here, the excavated materials are reinterpreted through creative transformation. Traditional rhythms, narratives, and tonal logics are restructured into new musical or audiovisual forms. This stage reflects the processes of feature extraction and transformation in

generative systems—finding patterns and reassembling them through iterative improvisation (Euba, 1989; Nzewi, 1997, 2007). Instead of algorithmic optimization, however, *Asɛmpayɛtsia* favors cultural resonance as its measure of success. The translation process remains cyclical and non-linear, embracing indeterminacy, improvisation, and community feedback as part of its compositional logic (Kisiedu & Lewis, 2023; Voegelin, 2019).

Audiovisual Reinscription

The final stage introduces technological mediation, in which digital tools function as extensions of oral imagination. Using techniques such as audio sampling, generative animation, and AI-assisted recomposition, the sonic archive is re-envisioned in new sensory realms. This stage treats technology as an ancestral collaborator rather than a neutral tool (Haraway, 2016; Bhagwati, 2020). In practice, this can manifest as AI-generated visuals that respond to rhythmic cycles or as machine-learning models trained to simulate call-and-response dynamics. The goal is not automation but amplification—to extend indigenous compositional logics into posthuman contexts (Machado & Cardoso, 1998; Kisiedu & Lewis, 2023).

3.3   Structural and Aesthetic Logic

Asɛmpayɛtsia works through a repeating three-part cycle, where each stage loops back into the next. Visually, it is pictured as:

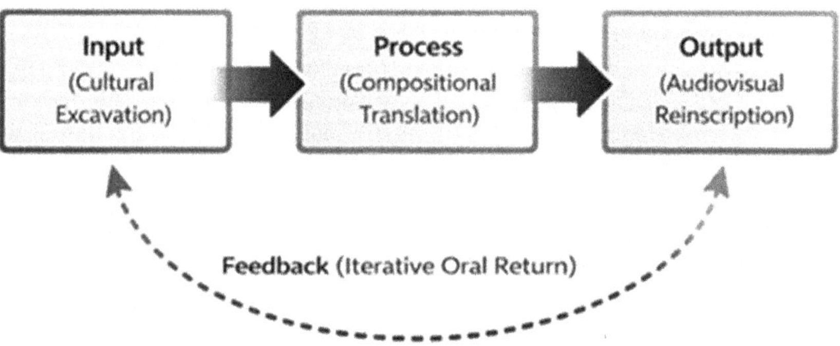

Fig. 1. *Asɛmpayɛtsia*'s triadic process

Each iteration enables mutation (variation through improvisation), iteration (retelling through performance), and fitness (judged by cultural or emotional resonance rather than statistical optimization). In this way, *Asɛmpayɛtsia* parallels 'evolutionary computation,' but with aesthetic evaluation rooted in communal sensibility instead of numeric fitness (Machado & Cardoso, 1998; Whitelaw, 2004). Improvisation, repetition, and feedback thus shape the model's computational aesthetics, reflecting how oral traditions develop through collective participation and memory (Chernoff, 1981; Agawu, 2003). This perspective presents African oral art as proto-algorithmic—an ancient yet living form of data processing rooted in rhythm, sociality, and improvisation, as further illustrated in Fig. 2.

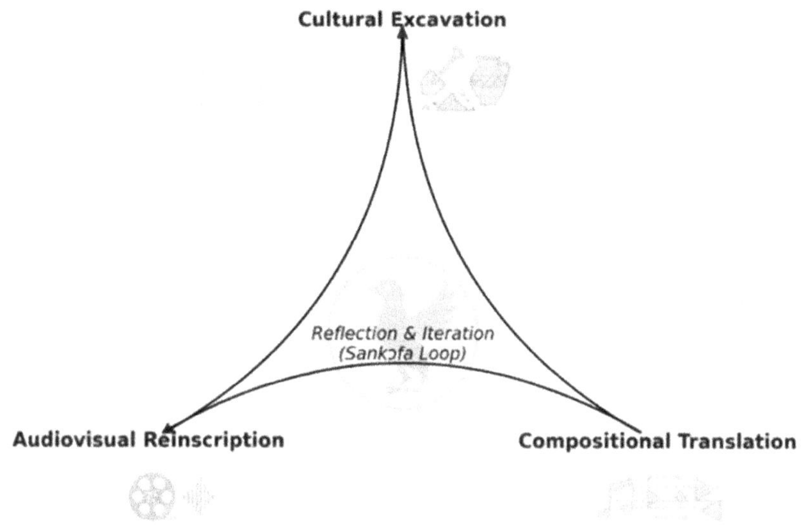

Fig. 2. Structural and Aesthetic Logic of the *Asɛmpayɛtsia* Framework.

3.4 Implementation Thought Experiment

This thought experiment is not proposed as a functioning algorithm but as a conceptual scaffold for translating indigenous epistemologies into computational design logic. While Asɛmpayɛtsia currently functions as a philosophical and compositional model, it can also inspire a conceptual framework for AI-assisted artistic systems. Below is a speculative workflow representing how the triadic process might be encoded:

Algorithm *Asɛmpayɛtsia*_AI:

Input: Cultural Archive (recorded folktales, proverbs, rhythms, oral performances).

Step 1: Cultural Excavation.

- Perform feature mapping: Identify recurring metaphors, rhythms, tonal motifs
- Encode cultural semantics as weighted vectors (e.g., "truth", "community", "ritual")

Step 2: Compositional Translation.

- Generate rhythmic and melodic recombinations using stochastic improvisation
- Evaluate each output using "cultural resonance metrics" (derived from community feedback loops)

Step 3: Audiovisual Reinscription.

- Translate sonic features into visual motion data

– Train image/video generation models to reflect rhythmic density and metaphorical tone

  Output: Audiovisual composition with embedded ancestral logic.
  End Algorithm.

This thought experiment explains how AI can support indigenous creativity by integrating ethical and cultural assessment functions—what could be called decolonial computation. This approach aligns with recent studies on cultural AI and critical machine creativity (McCormack et al., 2019; Herremans et al., 2017; Herremans & Chew, 2018; Carnovalini & Rodà, 2020), which promote culturally responsive algorithms that honor local ways of knowing. Figure 3 summarizes the *Asɛmpayɛtsia* framework, illustrating how the Triadic Process of Artistic Translation connects to computational mapping, decolonial implications, and future development pathways.

4   From Framework to Computational Model

4.1   Mapping Cultural Logic to Computational Principles

The *Asɛmpayɛtsia* framework acts as an interpretive bridge between Akan oral epistemologies and algorithmic architectures. In oral tradition, meaning is created through metaphor, rhythm, and communal performance rather than through fixed representations. This requires a shift from representation-based computing to process-based computation (Veale & Cardoso, 2019). The idea of oral reasoning can relate to symbolic processing, but in *Asɛmpayɛtsia*, this processing remains dynamic, with symbols changing in importance depending on context—reflecting relational-materialist views of creativity in AI that ask "where is creativity?" rather than "what is creativity?" (Celis et al., 2025). Similarly, the call-and-response performance style common in African musical traditions becomes a feedback loop in computational terms: each generation prompts a response, whether from humans, machines, or communities, creating a dialogic cycle of co-creation (Lee, Pasquier, & Yuri, 2025). The collective aspect of African performance—where creativity is shared among participants—aligns with the concept of distributed co-creativity in computational systems (Glăveanu, 2014). This shifts agency from a single human author or a lone AI system to a networked system of actors, artifacts, and practices. Building on these parallels between oral and computational reasoning, Table 1 outlines how key features of Akan creative logic can be conceptually mapped onto computational principles. This mapping is not a one-to-one translation but a philosophical approximation that shows how indigenous epistemologies might influence algorithmic thinking. It presents oral knowledge as processual, relational, and performative—qualities that can enhance computational creativity models grounded in co-agency and contextual adaptation.

Table 1. Mapping Cultural Logic to Computational Principles

Cultural Logic	Computational Principle	Interpretive Note
Oral Reasoning	Symbolic Processing	Meaning emerges dynamically through metaphor and rhythm
Call-and-Response	Feedback Loop	Recursive interactive generation rather than one-off output
Communal Performance	Distributed Co-Creativity	Shared authorship across human and machine agents
Proverb/Metaphor	Weighted Semantic Vectors	Cultural concepts encoded with relational weights
Improvisation	Stochastic Recombination	Creative variation guided by cultural resonance metrics

4.2 Prototyping Pathways

The next stage of *Asεmpayεtsia* envisions computational prototyping, translating the artistic logic into a design for a generative system. The modules correspond with the framework's phases.

- **Cultural Excavation:** analyzing oral archives (folktales, rhythms, spoken word) and mapping tonal, metaphorical, and performative patterns.
- **Compositional translation:** creating recombinations of these features using stochastically guided algorithms, assessed by cultural-resonance metrics rather than solely aesthetic optimization.
- **Audiovisual Reinscription:** mapping sonic features to visual motion or image generation models, embedding the relational logic of rhythm and community listening rather than linear multimedia production.

In postdoctoral development, this could be implemented through hybrid systems (e.g., symbolic + neural networks) where the "fitness function" is replaced by resonance feedback loops involving community participants (Machado & Cardoso, 2022; Machado, Romero, & Greenfield, 2021). This method embodies a participatory design philosophy that prioritizes relationality over novelty. In terms of tools, such a system could draw on existing musical AI frameworks, such as DDSP (Engel et al., 2020) and Magenta (Roberts et al., 2018), in combination with generative visual models (e.g., Stable Diffusion; Rombach et al., 2022) and custom feedback interfaces that capture performer–community responses, thereby conceptually closing the loop from creation to evaluation to iteration.

4.3 Comparison with Existing Models

While many creative-AI paradigms (GANs, diffusion models, evolutionary art) emphasize novelty, optimization, and statistical fitness, *Asεmpayεtsia* diverges by prioritizing cultural resonance, relational authorship, and processual co-creation. For example, evolutionary art often places a human or algorithmic "fitness function" at the center of aesthetic optimization (Machado & Cardoso, 1998). In contrast, *Asɛmpayɛtsia* positions the algorithm as a participant in a network of cultural agents. Philosophically, this aligns with relational-materialist theories of creativity (Celis et al., 2025), and ethically, it responds to critiques of generative art's bias, appropriation, and cultural flattening (Srinivasan & Uchino, 2020; Ivanona, 2025).

In other words, in contrasting optimization-driven models with the *Asɛmpayɛtsia* framework, the distinction should not be understood as a rejection of optimization or formal evaluation. Instead, it concerns the role optimization plays within the creative system. In many dominant approaches to computational creativity—such as evolutionary art, GAN-based aesthetics, or diffusion-based generation—optimization functions as the primary creative driver, with aesthetic value defined through statistical fitness, loss minimization, or novelty metrics. While these approaches have proven effective at producing complex, stylistically convincing artifacts, they often treat creativity as a problem of convergence within an abstract search space, essentially detached from cultural context. By contrast, *Asɛmpayɛtsia* positions optimization as a secondary or supportive mechanism embedded within a broader, relational, and culturally situated process. The framework prioritizes dialogic interaction, communal feedback, and contextual responsiveness as central creative forces, with optimization serving these values rather than determining them. Evaluation, in this model, is therefore not abandoned but reconfigured: instead of relying solely on numeric fitness functions, *Asɛmpayɛtsia* allows hybrid assessment strategies that combine measurable system properties (e.g., structural coherence, rhythmic consistency, responsiveness) with participatory and interpretive evaluation informed by cultural resonance and communal listening practices. This shift does not preclude comparing systems; rather, it reframes what is compared. Instead of asking whether a system produces more novel or statistically optimized outputs, the *Asɛmpayɛtsia* framework invites evaluation of how systems sustain relational coherence, ethical engagement with cultural material, and adaptive responsiveness in situated creative contexts. In this sense, the framework proposes not an alternative to computational rigor but an expanded notion of rigor—one that recognizes creativity as an emergent, relational process rather than a purely optimization-driven outcome. In summary:

- Optimization-driven model → human + machine as opponent/opponent or producer/consumer
- *Asɛmpayɛtsia* model → human + machine + culture as co-creators
- This shift challenges assumptions about autonomy in AI, authorship in art, and the role of heritage in system design.

5 Discussion: Decolonial Implications and Creative Potentials

5.1 Repositioning Creativity Through an African-Centered Epistemic Base

The *Asɛmpayɛtsia* framework redefines creativity as a relational, ethical, and community-based act rather than an individualistic, innovation-driven process. This perspective stems from Akan philosophies of knowledge, where wisdom (*nyansa*) is closely linked to moral and social responsibility (Wiredu, 1997; Gyekye, 1997). In this worldview, creativity does not come from competition or novelty but from the ability to listen, remember, and rearticulate ancestral knowledge within changing contexts. *Asɛmpayɛtsia* thus functions as a decolonial creative epistemology—a model that affirms African knowledge systems as theoretical rather than merely cultural (Dei, 2020; Ndlovu-Gatsheni, 2018). In contrast to Western modernist paradigms that emphasize originality and authorship, *Asɛmpayɛtsia* views the composer or artist as a mediator between memory and the future. The focus on brevity and metaphor as knowledge tools reflects the Akan principle that "truth hides in the folds of proverb." Therefore, the creative act becomes an ethical act—what Mignolo and Walsh (2018) describe as epistemic delinking: breaking away from colonial dependencies in meaning-making. By grounding creative production in indigenous logic, *Asɛmpayɛtsia* challenges the assumed universality of Western aesthetic systems and reclaims space for multiple understandings of art, sound, and technology.

5.2   Relevance to AI Ethics and Human–Machine Collaboration

In the field of AI ethics, debates have mainly focused on accountability, bias, and transparency (Jobin, Ienca, & Vayena, 2019). However, these frameworks often remain anthropocentric and Western in their moral assumptions. *Asɛmpayɛtsia* offers an alternative ethical perspective grounded in relational accountability (Wilson, 2008), in which the value of a system is judged not only by its performance but also by its ability to foster harmony among human, nonhuman, and ancestral actors. By embedding Akan ethical principles—listening, reciprocity, and communal responsibility—into computational design, *Asɛmpayɛtsia* redefines human–machine collaboration as co-listening rather than control. This aligns with emerging paradigms of "caring AI" and responsible relational design (Floridi, Cowls, Beltrametti, et al., 2018; Crawford, 2021). A potential approach to managing the risks and opportunities AI may present in this work is to implement ISO/IEC 42001:2023 within a structured, balanced framework, supported by an appropriate governance structure (International Organization for Standardization [ISO], 2023). The concept of "artificial" conveys a simulated version of a creation; this is no exception to the study, as this feature of AI necessitates thorough risk and management systems to address the inherent precariousness associated with its use. In this case, with respect to the Decolonial Implications (Ethical Design) aspect of this project, the AI policy will serve as the basis for future directions for this creation. Instead of machines optimizing outputs, they engage in cycles of response and reflection. This shift questions the power imbalance between humans and algorithms by incorporating cultural protocols of respect and consent into the design process. In practice, such a model could inspire participatory AI systems for creative industries where communities co-define evaluation metrics ("cultural resonance" rather than "user engagement") and decide which sonic or visual materials can be ethically reinterpreted. This decentralizes authorship and promotes a decolonial ethic of shared agency—where the machine acts as a conduit for cultural continuity rather than a site of extraction.

5.3   Implications for Artists, Composers, and AI Developers

For composers and artists, *Asɛmpayɛtsia* creates a methodological link between artistic intuition and cultural philosophy. It validates indigenous forms of theorization—such as song, proverb, and ritual—as sources of compositional logic. This challenges the long-standing academic hierarchy that separates "theory" from "practice" or "art" from "philosophy." In this way, the framework can serve as an educational tool for re-Africanizing curricula in composition, media arts, and sound studies (Nzewi, 2007; Agawu, 2016, 2023). For AI developers, the model proposes a new design approach: instead of coding creativity as random exploration or optimization, they can encode it as dialogue and relationships. This requires systems that consider contextual ethics, culturally specific datasets, and participatory validation. Such methods align with recent efforts for intercultural machine learning—AI models that are sensitive to cultural patterns and epistemic diversity (Russo, Schliesser, & Wagemans, 2023).

5.4   Interdisciplinary Value and Future Potentials

Asɛmpayɛtsia shows how artistic research can connect with philosophy, anthropology, and computational design to create new knowledge systems. It broadens decolonial theory beyond critique into technological practice, illustrating that decolonization is not just about rejecting Western models but also about building alternative ways of knowing and making. The framework's triadic logic—Cultural Excavation, Compositional Translation, Audiovisual Reinscription—can steer cross-disciplinary collaborations among musicians, animators, and computer scientists. It also points toward AI systems capable of promoting cultural sustainability, where machine learning serves to preserve living heritage. The ethical considerations extend into environmental and social realms. By framing creativity as a communal act of care, *Asɛmpayɛtsia* challenges extractive technological paradigms and models a more ecological relationship between humans and machines (Parikka, 2023).

The root of decolonial approaches within a deeply colonial social and artistic order raises a question that often lingers in my mind: Is AI a Western invention within this so-called "global village"? And if so, could its use in decolonial contexts be a case of "fishing with fish"? Possibly so. In many ways, AI appears to be an academic and technological innovation that I can comfortably classify as Western. Most laboratories developing such models are located in the West, are funded by Western institutions, and are shaped by Western epistemologies. At a recent symposium, Dei (2025) reminded us of a crucial distinction—that between "universities in Africa" and "African universities." His words struck me deeply. As long as we continue to operate within the framework of universities in Africa, rather than truly African universities, the shadow of Aristotelian logic and Eurocentric paradigms will continue to define our intellectual spaces. I have yet to come across an entirely African laboratory dedicated to Artificial Intelligence research rooted in indigenous epistemologies—and, truthfully, I suspect the reverse is also rare. Still, within academic culture, the value of replicability persists. We are trained to test, adapt, and reapply theories across contexts, even when those contexts are not our own. This brings me to a subtle tension—perhaps even a discomfort—about advancing a framework like *Asɛmpayɛtsia* within the global AI discourse. The world may indeed be a "global village," but this village is not neutral; uneven access, epistemic hierarchies, and cultural exclusions structure it. Yet, it is precisely within these imbalances that my work finds its urgency—to reimagine technology not as an imported construct but as a space where African heritage can think, breathe, and create on its own terms.

Ultimately, *Asɛmpayɛtsia* is not just a framework for artistic creation—it is a decolonial approach to creativity that honors ancestral connections while shaping future diversity. It turns the AI lab into a listening space, the composer's studio into a philosophical platform, and coding into an ethical act.

6 Conclusion and Future Work

The *Asɛmpayɛtsia* framework reimagines creativity as an epistemic act of return as a unit of specific consideration, i.e., an artistic and computational process rooted in African moral philosophy, orality, and sonic reasoning. By translating the Akan proverb "concise are words of worth" into compositional logic, it suggests a computational aesthetic that emphasizes cultural resonance, ethical listening, and ancestral continuity over algorithmic optimization. This contribution challenges existing paradigms in artificial intelligence and creative practice by affirming that African thought systems are not only sources of content but also systems of reasoning that can inform design, computation, and theory.

Asɛmpayɛtsia serves as both a conceptual and methodological model, providing a framework for artistic research and a template for decolonial system design. Its three-part process—Cultural Excavation, Compositional Translation, and Audiovisual Reinscription—offers a means of encoding relational values, ethical protocols, and sonic memory into AI systems. In doing so, it encourages a shift from extractive approaches to computational creativity toward generative practices that honor communal knowledge, contextual specificity, and interdependence. The framework thus redefines AI not as an autonomous creator but as a co-listener—a technological participant in the ongoing dialogue between heritage and innovation. While the framework has not yet been instantiated as a working AI system, its value lies in establishing an epistemic and aesthetic foundation upon which such systems can be responsibly developed.

Moving forward, this research will develop into a postdoctoral program that creates a working prototype of the *Asɛmpayɛtsia*_AI system. This prototype will examine how algorithms can interpret and reassemble cultural archives—such as recorded folktales, proverbs, and rhythms—into audiovisual pieces that preserve symbolic and emotional connections to their sources. The project will also evaluate community-informed "cultural resonance metrics" as alternatives to existing aesthetic assessment models in computational art. Concurrent creative outputs will include generative soundscapes, algorithmic video projects, and live AI-assisted performances, each reflecting the framework's ethical and artistic principles.

Ultimately, *Asɛmpayɛtsia* contributes to a growing movement of Decolonial AI—a reorientation that challenges technological universality and restores cultural diversity to the computational imagination. It shows that technology can act as a space of cultural continuity, where ancestral philosophies influence the design of future systems. In this reimagined landscape, to compose, to code, and to create are no longer separate actions—they are unified acts of remembering, translating, and becoming.

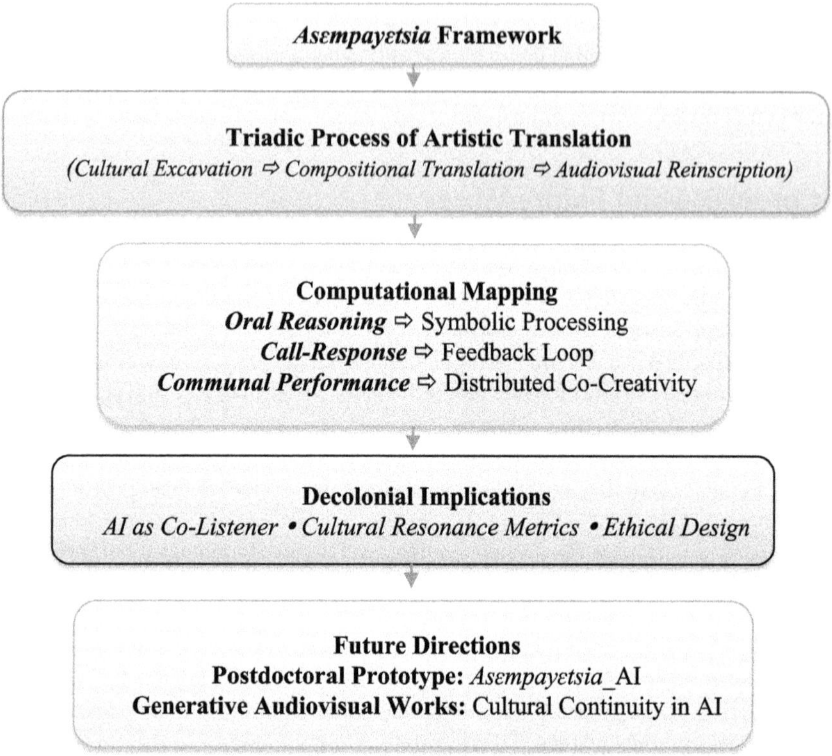

Fig. 3. *Asɛmpayɛtsia* Framework Summary

References

Adjei, P.B., Dei, G.J.S.: Indigenous African Elders Critical Teachings (ElderCrits) as a methodology. J. Critical Res. Methodologies **1**(1), 6–28 (2025)

African Union.: Agenda 2063: The Africa We Want – Framework Document (2015). https://au.int/sites/default/files/documents/33126-doc-framework_document_book.pdf. Accessed 28 Jan 2026

Agawu, K.: African imagination in music. Oxford University Press (2016)

Agawu, K.: Representing African Music: Postcolonial Notes, Queries. Routledge, Positions (2003)

Agawu, K.: Tonality as a colonizing force in Africa. In On African Music: Techniques, Influences, Scholarship (pp. 53–74). Oxford University Press (2023) https://doi.org/10.1093/oso/9780197664063.003.0003

Anku, W.: Circles and Time: A Theory of Structural Organization of African Music. Ethnomusicology **44**(2), 265–296 (2000). https://mtosmt.org/retrofit/mto.00.6.1/mto.00.6.1.anku.php#:~:text=Willie%20Anku&text=ABSTRACT:%20Much%20of%20African%20music,of%20performance%20and%20analytical%20applications. Accessed 28 Jan 2026

Arora, P.: Creative data justice: a decolonial and indigenous approach. J. Cultural Analyt. **9**(1), 1–19 (2024). https://doi.org/10.1080/1369118X.2024.2420041

Bartlett, V., Pfefferkorn, J., Sunde, E. K.: Decentering Ethics: AI Art as Method. Open Humanities Press (2025). https://openhumanitiespress.org/books/download/Bartlett-Pfefferkorn-Sunde_2025_Decentring_Ethics.pdf. Accessed 28 Jan 2026

Benschop, R., Peters, P., Lemmens, B.: Artistic researching: expositions as matters of concern. In: Schwab, M., Borgdorff, H. (eds.) The Exposition of Artistic Research: Publishing Art in Academia (Chapter 4). Cambridge University Press (2025). https://doi.org/10.1017/978940 0600928.004

Bhagwati, S.: New music: towards a diversity of practices. OnCurating, **47**, 32–39 (2020). https://www.on-curating.org/issue-47-reader/new-music-towards-a-diversity-of-practices.html. Accessed 28 Jan 2026

Boden, M.A.: Creativity and artificial intelligence. Artif. Intell. **103**(1–2), 347–356 (1998). https://doi.org/10.1016/s0004-3702(98)00055-1

Borgdorff, H.: The Conflict of the Faculties: Perspectives on Artistic Research and Academia. Leiden University Press (2012)

Borgdorff, H.: The Exposition of Artistic Research: Publishing Art in Academia. Leiden University Press (2013)

Candy, L., Edmonds, E.: Practice-based research in the creative arts: foundations and futures from the front line. Leonardo **51**(1), 63–69 (2018). https://doi.org/10.1162/LEON_a_01471

Carnovalini, F., Rodà, A.: Computational creativity and music generation systems: an introduction to the state of the art. Frontiers in Artificial Intelligence, 3, Article 14 (2020). https://doi.org/10.3389/frai.2020.00014

Carnovalini, F., Rodà, A.: Computational creativity and music generation systems: an introduction to the state of the art. Front. Artif. Intell. **3**, 14 (2020). https://doi.org/10.3389/frai.2020.00014

Celis Bueno, C., Chow, P.S., Popowicz, A.: Not "what", but "where is creativity?": towards a relational-materialist approach to generative AI. AI & Soc **40**, 339–351 (2025). https://doi.org/10.1007/s00146-024-01921-3

Chernoff, J.M.: African rhythm and African sensibility: Aesthetics and social action in African musical idioms (revised ed.). University of Chicago Press (1981)

Cibotaru, V.: Is there computational creativity? AI & Soc. (2025). https://doi.org/10.1007/s00146-025-02708-w

Cook, M., Colton, S., Pease, A., Llano, M.T.: Framing in computational creativity: A survey and taxonomy. In: Grace, K., Cook, M., Ventura, D., Maher, M.L. (eds.) Proceedings of the 10th International Conference on Computational Creativity (ICCC 2019), pp. 156–163. Association for Computational Creativity (ACC) (2019)

Correa Lucero, H., Martens, C.: Colonial structures in AI: a Latin American decolonial literature review of structural implications for marginalized communities in the Global South. AI Soc. (2025). https://doi.org/10.1007/s00146-025-02547-9

Crawford, K.: Atlas of AI: Power, Politics, and the Planetary Costs of Artificial Intelligence. Yale University Press (2021)

Dawson, N.A.: "Pɛrpl Greyps": Kodzi in 3re Beatz (Doctoral thesis). University of Cape Coast (2023). https://ir.ucc.edu.gh/xmlui/handle/123456789/12237. Accessed 28 Jan 2026

Dei, G.J.S., Adjei, P.B.: ElderCrits as a building framework for Black-Indigenous peoples' decolonizing solidarities. Canadian Soc. Work Rev. **39**, 39–66 (2024). https://doi.org/10.7202/111 4540ar

Dei, G.J.S.: Elders' Cultural Knowledges and African Indigeneity. In: Abidogun, J., Falola, T. (eds.) The Palgrave Handbook of African Education and Indigenous Knowledge. Palgrave Macmillan, Cham (2020) https://doi.org/10.1007/978-3-030-38277-3_14

Dei, G.J.S: Dondology and the African Cultural Library: Rethinking issues in African education and anti-colonial praxis [Keynote presentation]. The First Annual Symposium for Dondology: Repositioning Dondology as an African-centered Field of Study, Balme Library Conference Hall, University of Ghana, Legon (2025)

Dei, G.S.: The Indigenous as a site of decolonizing knowledge for conventional development and the link with education. In: Langdon, J. (ed.) Indigenous Knowledges, Development and

Education, pp. 15–36 (Chapter 2). Sense/Brill (2009) https://doi.org/10.1163/9789087906993_003

Eglash, R., Robinson, K.P., Bennett, A., Robert, L., & Garvin, M.: Computational reparations as generative justice: Decolonial transitions to unalienated circular value flow. Big Data Soc. **11**(1) (2024) https://doi.org/10.1177/20539517231221732

Engel, J., Hantrakul, L., Gu, C., Roberts, A.: DDSP: Differentiable digital signal processing. arXiv preprint (2020). https://doi.org/10.48550/arXiv.2001.04643. Accessed 28 Jan 2026

Eshun, K.: More Brilliant Than the Sun: Adventures in Sonic Fiction. Quartet Books (1998)

Euba, A.: Essays on Music in Africa 1: Intercultural Perspectives. Bayreuth African Studies (1989)

Floridi, L., Cowls, J., Beltrametti, M., et al.: AI4People—an ethical framework for a good AI society: opportunities, risks, principles, and recommendations. Mind. Mach. **28**, 689–707 (2018). https://doi.org/10.1007/s11023-018-9482-5

Glăveanu, V.P.: Distributed creativity: thinking outside the box of the creative individual. Springer (2024). https://doi.org/10.1007/978-3-319-05434-6

Gyekye, K.: Tradition and Modernity: Philosophical Reflections on the African Experience. Oxford University Press (1997)

Gyekye, K.: An Essay on African Philosophical Thought: The Akan Conceptual Scheme. Temple University Press (1995)

Hamilton, D., Hansen, L.: An artful becoming: the case for a practice-led research approach to open educational practice research. Teach. High. Educ. **29**(7), 1757–1774 (2024). https://doi.org/10.1080/13562517.2024.2336159

Haraway, D.J.: Staying with the Trouble: Making Kin in the Chthulucene. Duke University Press (2016). https://doi.org/10.2307/j.ctv11cw25q

Herremans, D., Chew, E.: MorpheuS: generating structured music with constrained patterns and tension. IEEE Trans. Affect. Comput. **10**(4), 510–523 (2018). https://doi.org/10.1109/TAFFC.2017.2737984

Herremans, D., Chuan, C.-H., Chew, E.: A functional taxonomy of music generation systems. ACM Comput. Surv. **50**(5), Article 69 (2017) https://doi.org/10.1145/3108242

International Organization for Standardization & International Electrotechnical Commission.: ISO/IEC 42001:2023 Information technology — Artificial intelligence — Management system (1st ed.). ISO (2023). https://www.iso.org/standard/42001?utm. Accessed 28 Jan 2026

Ivanova, M.: AI, art and morality.: AI Ethics **5**, 4269–4278 (2025) https://doi.org/10.1007/s43681-025-00735-3

Jobin, A., Ienca, M., Vayena, E.: The global landscape of AI ethics guidelines. Nature Machine Intell. **1**(9), 389–399 (2019). https://doi.org/10.1038/s42256-019-0088-2

Jordanous, A.: Evaluating evaluation: assessing progress and practices in computational creativity research. In: Veale, T., Cardoso, F. (eds.) Computational Creativity. Computational Synthesis and Creative Systems. Springer, Cham (2019). https://doi.org/10.1007/978-3-319-43610-4_10

Kara, H.: Creative Research Methods: A Practical Guide (2nd ed.). Policy Press (2020)

Kisiedu, H., Lewis, G.E. (eds.): Composing while Black: Afrodiasporic new music today / Afrodiasporische Neue Musik Heute (bilingual ed.). Wolke Verlag (2023)

Leavy, P.: Method meets art: Arts-based research practice (3rd ed.). Guilford Publications (2020)

Lee, K.J.M., Pasquier, P., Yuri, J.: Revival: Collaborative artistic creation through human-AI interactions in musical creativity. arXiv preprint (2025)

Lewis, J.E., Whaanga, H., Yolgörmez, C.: Abundant intelligences: placing AI within Indigenous knowledge frameworks. AI & Soc. **40**, 2141–2157 (2025). https://doi.org/10.1007/s00146-024-02099-4

Lugonzo, T.: From Colonial Bias to Relational Intelligence: Decolonizing AI with Indigenous and African Epistemologies. Liberated Arts Journal (2025). https://ojs.lib.uwo.ca/index.php/lajur/article/download/22436/17676. Accessed 28 Jan 2026

Machado, P., Cardoso, A.: Computing aesthetics. In F. M. de Oliveira (ed.) Proceedings of the XIV Brazilian Symposium on Artificial Intelligence (SBIA'98). Lecture Notes in Computer Science, pp. 219–228. Springer (1998) https://doi.org/10.1007/10692710_23

Machado, P., Romero, J., Greenfield, G.: Artificial intelligence and the arts: computational creativity, artistic behavior, and tools for creatives. Springer (2021). https://doi.org/10.1007/978-3-030-59475-6

Mbiti, J.S.: African Religions and Philosophy. Heinemann (1990)

McCormack, J., Gifford, T., Hutchings, P.: Autonomy, Authenticity, Authorship and Intention in Computer Generated Art. In: Ekárt, A., Liapis, A., Castro Pena, M.L. (eds.) Computational Intelligence in Music, Sound, Art and Design — EvoMUSART 2019. Lecture Notes in Computer Science, vol. 11453, pp. 35–50. Springer, Cham (2019). https://doi.org/10.1007/978-3-030-16667-0_3

Mignolo, W.D., Walsh, C.E.: On Decoloniality: Concepts, Analytics. Duke University Press, Praxis (2018)

Ndlovu-Gatsheni, S.J.: Epistemic Freedom in Africa: Deprovincialization and Decolonization. Routledge (2018)

Nzewi, M.: A Contemporary Study of Musical Arts: Illuminations, reflections and explorations. African Minds (2007)

Nzewi, M.: African Music: Theoretical Content and Creative Continuum. Institut für Ethnomusikologie (1997)

Parikka, J.: Operational Images: From the Visual to the Invisual. University of Minnesota Press (2023)

Ramose, M.B.: African Philosophy Through Ubuntu. Mond Books (1999)

Roberts, A., Engel, J., Raffel, C., Hawthorne, C., Eck, D.: A hierarchical latent vector model for learning long-term structure in music. arXiv preprint (2018). https://doi.org/10.48550/arXiv.1803.05428

Rombach, R., Blattmann, A., Lorenz, D., Esser, P., Ommer, B.: High-resolution image synthesis with latent diffusion models. In: Proceedings of the 2022 IEEE/CVF Conference on Computer Vision and Pattern Recognition (CVPR), pp. 10674–10685 (2022). https://doi.org/10.1109/CVPR52688.2022.01042

Russo, F., Schliesser, E., Wagemans, J.: Connecting ethics and epistemology of AI. AI & Soc. **39**, 1585–1603 (2024). https://doi.org/10.1007/s00146-022-01617-6

Skains, R.L.: Creative practice as research: discourse on methodology. Media Practice Educ. **19**(1), 82–97 (2018). https://doi.org/10.1080/14682753.2017.1362175

Smith, H., Dean, R.T.: Practice-led research, research-led practice – towards the iterative cyclic web. In: Smith, H., Dean, R.T. (eds.) Practice-led Research, Research-led Practice in the Creative Arts, pp. 1–38. Edinburgh University Press (2009). https://doi.org/10.1515/9780748636303-002

Srinivasan, R., Uchino, K.: Biases in generative art - a causal look from the lens of art history. arXiv preprint (2020)

UNESCO.: Convention for the Safeguarding of Intangible Cultural Heritage. https://ich.unesco.org/en/convention (2003). Accessed 28 Jan 2026

Veale, T., Cardoso, A. (eds.): Computational Creativity: The Philosophy and Engineering of Autonomously Creative Systems. Springer (2019). https://doi.org/10.1007/978-3-319-43610-4

Veale, T., Cardoso, F. Amílcar (eds.): Computational Creativity: The Philosophy and Engineering of Autonomously Creative Systems. Computational Synthesis and Creative Systems, Springer, Cham (2019). https://doi.org/10.1007/978-3-319-43610-4

An Introduction to the Field of Computational Creativity: Veale, T., Pérez y Pérez, R. *Leaps and Bounds*. N. Gener. Comput. **38**, 551–563 (2020). https://doi.org/10.1007/s00354-020-00116-w

Voegelin, S.: The Political Possibility of Sound. Bloomsbury (2019)

Whitelaw, M.: Metacreation: Art and artificial life. MIT Press (2004). https://mitpress.mit.edu/9780262232340/metacreation/?utm. Accessed 28 Jan 2026

Wilson, S.: Research Is Ceremony: Indigenous Research Methods. Fernwood Publishing (2008)

Wiredu, K.: Cultural Universals and Particulars: An African Perspective. Indiana University Press (1997)

Yankah, K.: Speaking for the Chief: Okyeame and the Politics of Akan Royal Oratory. Indiana University Press (1995)

Zhou, E., Lee, D.: Generative artificial intelligence, human creativity, and art. PNAS Nexus, **3**(3), Article pgae052 (2024) https://doi.org/10.1093/pnasnexus/pgae052

The Art That Poses Back: Assessing AI Pastiches After Contemporary Artworks

Anca Dinu[(✉)] [iD], Andreiana Mihail [iD], Andra-Maria Florescu [iD],
and Claudiu Creangă [iD]

University of Bucharest, ISDS, 90 Panduri Road, 050107 Bucharest, Romania
`anca.dinu@lls.unibuc.ro`

Abstract. This study explores artificial visual creativity, focusing on ChatGPT's ability to generate new images intentionally pastiching original artworks such as paintings, drawings, sculptures and installations. The process involved twelve artists from Romania, Bulgaria, France, Austria, and the United Kingdom, each invited to contribute with three of their artworks and to grade and comment on the AI-generated versions. The analysis combines human evaluation with computational methods aimed at detecting visual and stylistic similarities or divergences between the original works and their AI-produced renditions. The results point to a significant gap between color and texture-based similarity and compositional, conceptual, and perceptual one. Consequently, we advocate for the use of a "style transfer dashboard" of complementary metrics to evaluate the similarity between pastiches and originals, rather than using a single style metric. The artists' comments revealed limitations of ChatGPT's pastiches after contemporary artworks, which were perceived by the authors of the originals as lacking dimensionality, context, and intentional sense, and seeming more of a paraphrase or an approximate quotation rather than as a valuable, emotion-evoking artwork.

Keywords: generative contemporary art · artificial visual creativity · pastiche · direct human experts evaluation · style transfer

1 Introduction

The rapid technological development of Large Language Models (LLMs) has extended their creative potential and ability to imitate art across various modalities, from literary works to visual art generation [2,8,36]. This expansion raises important questions regarding the nature of stylistic imitation in computational art creativity and the understanding of artificial artistic creation. [6,21].

Central to this discussion is the concept of *pastiche*, which originates from Italian *pasticcio*, meaning a blending of meat and pasta turned into a pie. This etymology suggests that a pastiche creates something new from available and recognizable elements, without introducing a new substance [10]. Moreover, before

---

All authors contributed equally to this work.

© The Author(s), under exclusive license to Springer Nature Switzerland AG 2026
P. Machado et al. (Eds.): EvoMUSART 2026, LNCS 16523, pp. 67–83, 2026.
https://doi.org/10.1007/978-3-032-24350-8_5

postmodernist theories, the term had a negative connotation equivalent to a lack of creativity [30]. However, the pastiche is now seen as an acknowledgment of previous works across a wide range of domains [23]. In art, the pastiche represents an example of eclecticism that usually pays homage to the original work of art, going beyond mere imitation, by emulating its style and content [13]. In short, a pastiche intentionally refers to an original by paying tribute to its motifs, genre, and time period instead of parodying or mocking it [19].

There are numerous examples throughout the history of art of artists who have created pastiches or commentaries on famous masterpieces. From the Renaissance, when Giorgio Vasari imitated the styles of Raphael and Michelangelo, and Caravaggio's followers produced similar works as homage to their master, to the nineteenth-century Pre-Raphaelites, and the twentieth-century innovators such as Picasso and Braque. Marcel Duchamp's *L.H.O.O.Q.* (1919) [31], the mustached *Mona Lisa*, stands as an iconic example of conceptual pastiche. Andy Warhol and Salvador Dalí both reinterpreted Leonardo da Vinci's *Last Supper*, while Postmodernism introduced a new wave of artistic pastiche through Sherrie Levine, Cindy Sherman, Jeff Koons, Damien Hirst, Glenn Brown, or Richard Prince. Contemporary Romanian artists have also explored this lineage: Ion Grigorescu's works after Adolf Wölfli, and Ciprian Mureșan's re-creations featuring artists such as Andrea Mantegna and Maurizio Cattelan, continue this dialogue between imitation, reflection, and originality.

When modern artists develop recognizable visual signatures, whether through thematic content or compositional structure, these elements can become subject to being pastiched both by human creators and nowadays by generative AI systems. Such an example can be represented by the Ghibli trend in which users employed ChatGPT to turn their photographs in the same style as the Japanese animation studio Ghibli, that sparked debates[1].

In this article, we examine ChatGPT's capacity to produce pastiches after contemporary artworks provided by twelve artists from various countries. First, we quantitatively measure the distance between the original works and the pastiches, by embedding the features of the artworks in a common vectorial space where we compute the cosine distance between the vectors. Second, we turn to a qualitative analysis with the help of the artists themselves, who were asked to grade and comment on the artificially created art that supposed to pastiche their own.

Therefore, the novelty of this study is twofold. First, the automatic evaluation of the similarity of the pastiches to the original artworks employed five state-of-the-art (SOTA) vision models to extract various features capturing different aspects of artistic style. Second, the artists were actively involved, not only contributing with the original artworks but also participating in the evaluation and commenting on the AI-generated creative products.

The rest of the article is structured as follows. In Sect. 2 we summarize related work. We continue with the presentation of the dataset in Sect. 3. In Sect. 4,

---

[1] https://www.forbes.com/sites/danidiplacido/2025/03/27/the-ai-generated-studio-ghibli-trend-explained/, last accessed 2026/01/30.

we describe the experimental setup, while in Sect. 5 we showcase the results. Section 6 is dedicated to results analysis and Sect. 7 to some empirical observations. We conclude in Sect. 8.

2   Related Work

In the last few years, computer vision and AI research have increasingly focused on identifying and modeling artistic style, often by using large datasets of both digitized original artworks [5,39], and artificially generated pieces [1].

Early exploration of computational style imitation emerged with the advent of neural style transfer [11], which optimized a model to reproduce the content of an artwork while adopting the visual style of another through iterative optimization of convolutional neural activations. Previous research developed Generative Adversarial Networks (GANs) [12], making it possible for end-to-end learning of artistic distributions and eliminating the need for explicit optimization [41]. Additional research and development of diffusion models [16,29] and vision-language models [28] have further advanced the field by enabling text-based generation that creates images in various artistic styles using natural language prompts. Thus, text-to-image systems such as DALL-E[2], Midjourney[3], and Stable Diffusion[4] were designed to replicate visual artistic style [34].

Following the rapid development of visual AI generation, various studies have analyzed its abilities and limitations [1]. Even if AI systems produce high-quality generations, they still make common mistakes [20]. Studies show that these models often fail to combine objects with varying attributes and relations [37], struggle with basic syntax like negation or word order [22,25], and misrepresent numbers or text in images [3,4]. Most of the research on image generation focuses on investigating the impact of GenAI, particularly when comparing AI-generated images with original human work in creative and industry contexts [7,40]. [9,35] showed how LLMs prefer certain artistic styles, while [33] experimented with evaluation metrics for assessing AI-generated art. [27] developed a novel image style transfer method by enabling an LLM to handle multiple styles efficiently. The results showed that it managed to generate good visual outputs and worked faster than traditional style methods. Conducting a quantitative analysis of human perceptions and preferences for generative art, [15] discovered that even if humans could distinguish human and AI-generated artworks, there was a preference for AI-generated work, which led to a more in-depth discussion on the future of art and its value to society.

The majority of the research on consumer reaction to AI-generated visuals in marketing is mixed. While [38] found that AI-assisted artists often receive positive reactions, [17] noticed that AI art can be devalued even when it is indistinguishable from human-made work. [24] analyzed more than 3 million text prompts for diffusion models and discovered that user behavior is mostly

---

[2] https://openai.com/index/dall-e/, last accessed 2026/01/30.
[3] https://www.midjourney.com/home, last accessed 2026/01/30.
[4] https://github.com/CompVis/stable-diffusion/, last accessed 2026/01/30.

recreational for personal use, rather than generating works with novel artistic value.

3   Data

We invited twelve contemporary artists working across various mediums, including drawing, painting, sculpture, and installation, to each submit three images of their artworks, preferably executed in different styles or techniques. Using the same prompt for all, we then asked ChatGPT to generate two new works inspired by each original. This process resulted in a dataset of 108 images: 36 original artworks and 72 AI-generated pastiches. The artists included were: Adi Matei, Ciprian Mureșan, Ion Grigorescu, Iulia Uță, Karine Fauchard, Lazar Lyutakov, Marius Tănăsescu, Mathias Poeschl, Oana Năstăsache, Philip Patkowitsch, Răzvan Botiș, and Tom Chamberlain. The prompt used reads: *"Make/create something which is in the spirit, technique and style of the artist. But different composition and concept. Do not copy, improve, explain, or translate the original work. Do not question the aesthetic behind it. The original artist should be recognized in the new work. Do not make derivatives but new and different art work."*

4   Experimental Setup

We employ five SOTA computer vision models to extract high-dimensional embeddings capturing different aspects of artistic style. All images were preprocessed to RGB format and normalized according to each model's specifications.

The AdaIN-Style model [18] extracts 1920-dimensional style statistics by computing channel-wise mean and standard deviation from four layers of a VGG19 [32] encoder ($relu1_1$, $relu2_1$, $relu3_1$, $relu4_1$), isolating pure texture and color patterns independent of spatial composition.

The ResNet50-Style model produces 2048-dimensional embeddings from the pre-logit layer of a ResNet-50 [14] network, capturing mid-level features relevant to artistic style classification.

For semantic understanding, we use CLIP-ViT-L [28] (openai/clip-vit-large-patch14) which generates 768-dimensional vision-language aligned embeddings.

DINOv2 [26] (facebook/dinov2-large) provides 1024-dimensional self-supervised visual features from the CLS token output, capturing fine-grained visual patterns.

Finally, VGG19 [32] extracts 4096-dimensional perceptual features that encode high-level visual representations correlating with human perception, allowing the model to capture semantic and structural information rather than low-level pixel details.

For each artwork group (original and its pastiches), we compute three pairwise cosine distances, that measure angular similarity in the embedding space:

1. original to pastiche 1;
2. original to pastiche 2;
3. pastiche 1 to pastiche 2;

5   Results

Our analysis quantified the similarity between original artworks and their pastiches using five distinct feature embedding models. The results reveal a multifaceted view of style, where different models capture complementary aspects of artistic similarity.

5.1   Overall Model Comparison

The five models produced rather different average distances, indicating that each measures some other characteristics when it comes to similarity. Table 1 summarizes the average distances between the originals and the pastiches (Org→Pst1, Org→Pst2) and between the two pastiches themselves (Pst1↔Pst2), also visualized in Fig. 1.

Table 1. Average Cosine Distances by Model. Lower values (the lowest in bold) indicate higher similarity. AdaIN-Style is the most forgiving, while VGG19 is the strictest.

Model	Org→Pst1	Org→Pst2	Average	Pst1↔Pst2	Feature type
AdaIN-Style	**0.055**	**0.071**	**0.063**	**0.068**	texture, color
CLIP-ViT-L	0.194	0.200	0.197	0.173	conceptual
ResNet50-Style	0.427	0.481	0.454	0.402	artistic style
DINOv2	0.441	0.484	0.463	0.396	fine-grained visual
VGG19	0.648	0.701	0.674	0.662	perceptual

A clear hierarchy emerges:

1. **AdaIN-Style** registered the lowest average distance (0.063). As this model captures only channel-wise feature statistics (mean and standard deviation) and discards all spatial information, this low distance suggests the pastiches are highly successful at replicating the originals' texture and color palettes.
2. **CLIP-ViT-L** reported the next-lowest distance (0.197), indicating a high degree of semantic or conceptual consistency between originals and pastiches.
3. **ResNet50-Style** (0.454) produced larger distances, indicating that, while concepts or textures might align, specific artistic style features are less similar between originals and generated pastiches.
4. **DINOv2** (0.463) also showed moderate distances, suggesting that fine-grained visual details differ more significantly.
5. **VGG19** returned the highest average distance (0.674), indicating rather dissimilar perceptual features.

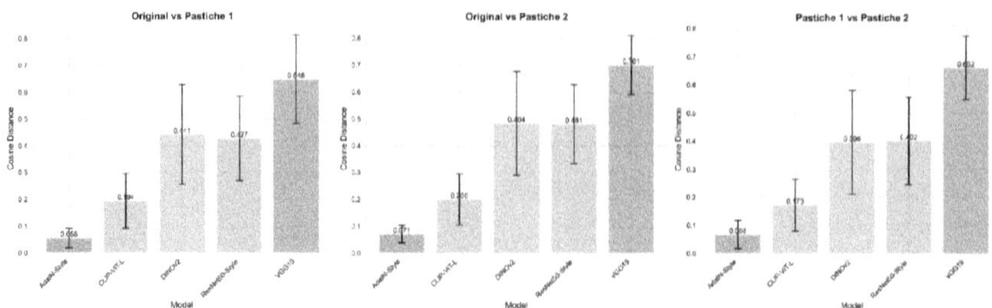

Fig. 1. Bar charts comparing average cosine distances for Orig→Past1, Orig→Past2, and Past1↔Past2 comparisons across all five models.

5.2   Model Discrimination and Consistency

Beyond average distance, the models' discrimination power (variance) and consistency reveal their underlying characteristics. The distribution of these distances is shown in Fig. 2, with statistical summaries in Table 2.

DINOv2 exhibited the highest variance (0.0349), making it the most discriminative model. This aligns with its design to capture fine-grained visual features, allowing it to detect subtle differences between artworks. Conversely, it also showed the lowest consistency (0.131 average difference), indicating that the two pastiches generated for the same original work often varied significantly in their visual execution.

AdaIN-Style was the antithesis, with 25 times less variance (0.0014). This extremely low variance suggests it views most pastiches as texturally similar to their originals. Its high consistency (0.034) reinforces that both pastiches successfully captured the same target texture statistics.

CLIP-ViT-L demonstrated a "best of both worlds" consistency, with low variance (0.0103) and the second-highest pastiche consistency (0.046). This suggests that at a semantic level, both pastiches were equally and consistently close to the original's concept.

Table 2. Model Discrimination (Variance) and Pastiche Consistency. Discrimination measures the spread of all measurements, while Consistency measures the average difference between the two pastiche distances for the same artwork.

Model	Variance	Discrimination	Pastiches (Avg. Diff.)	Consistency
DINOv2	0.0349	Most Discriminative	0.131	Most Variable
VGG19	0.0274	High Discrimination	0.111	Moderate
ResNet50-Style	0.0249	Balanced	0.115	Moderate
CLIP-ViT-L	0.0103	Consistent	0.046	Very Consistent
AdaIN-Style	0.0014	Least Discriminative	0.034	Most Consistent

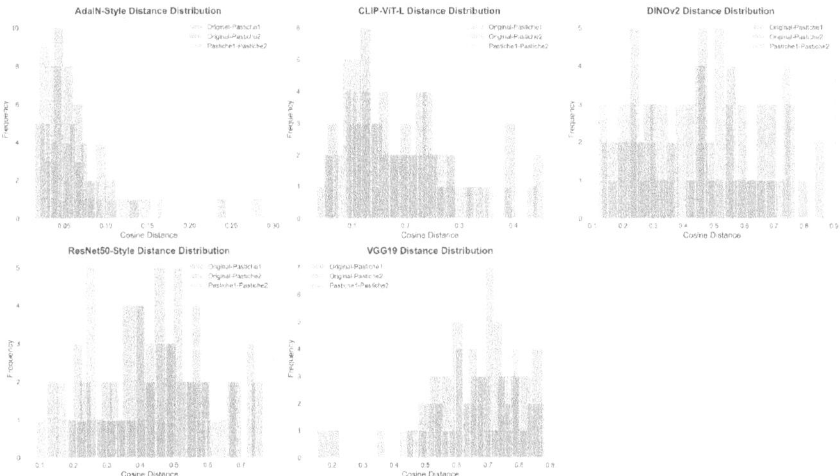

Fig. 2. Distance distributions for each of the five models. Note the tight, low-distance grouping of AdaIN-Style versus the wide, high-distance spread of DINOv2 and VGG19, illustrating their respective discrimination power.

5.3 Model Agreement and Correlation

A final component of the results is understanding whether the models agree with each other. We computed pairwise correlations to see if models that rank one pastiche as very similar to the original (low distance) also rank others similarly. The scatter plots in Fig. 3 visualize this agreement for all model pairs.

All pairs show a moderate, positive correlation (r values between 0.538 and 0.604). This indicates that while the models generally agree, their rankings are far from identical, and each model captures unique information that the others do not. The highest agreement is between DINOv2 and VGG19 ($r = 0.604$), suggesting a strong link between fine-grained visual features and classic perceptual metrics.

6 Discussion

The quantitative results from our 5-model analysis provide a new lens through which to interpret artistic style, pastiche quality, and the nature of visual similarity. We discuss the primary implications of these findings.

6.1 The Multi-Dimensional Nature of Artistic Style

We find that "style" is not a monolithic, singular concept. Instead, it is a multi-dimensional property, and our five models effectively capture distinct facets

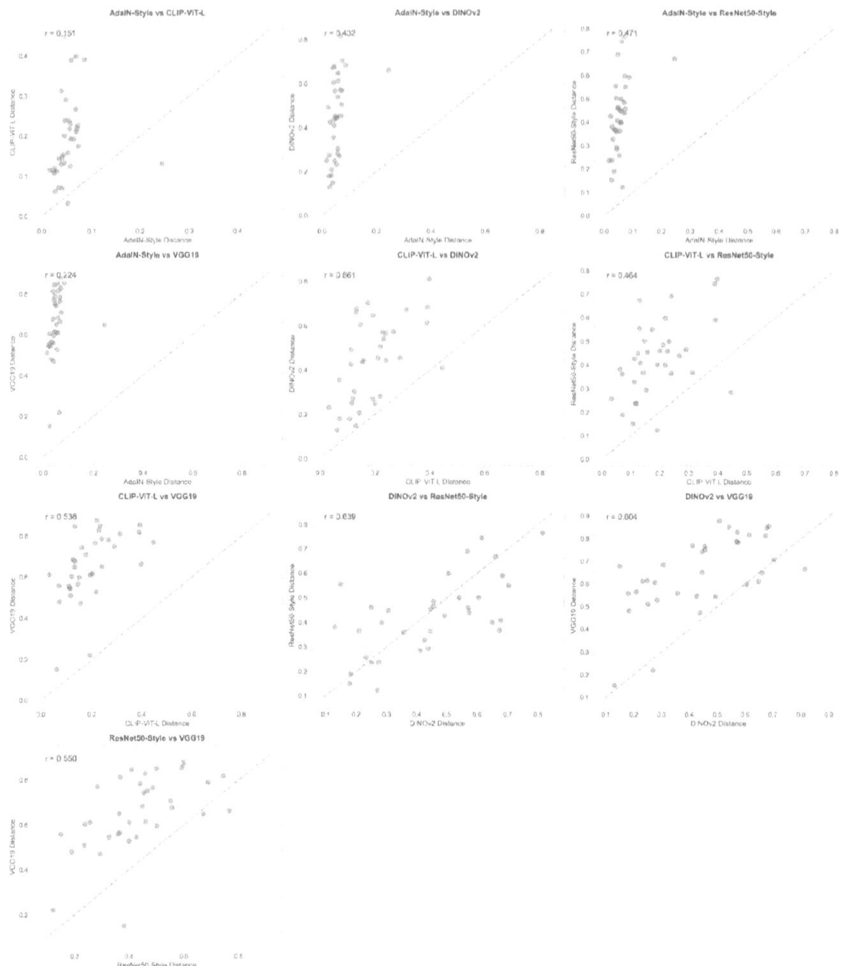

Fig. 3. Pairwise correlation scatter plots between models. Each plot shows the distance from one model (x-axis) against another (y-axis) for the same artwork pairs. A tight diagonal line indicates high agreement (redundant information), while a diffuse cloud indicates low agreement (complementary information).

of this property. The 11-fold difference in average distance between AdaIN-Style (0.063) and VGG19 (0.674) is a stark quantitative measure of this multi-dimensionality (see Fig. 1 and 2).

This framework allows us to dissect similarity: two artworks can be (1) texturally similar (low AdaIN distance) but (2) compositionally different (high DINOv2 distance), and (3) semantically aligned (low CLIP distance). This implies that any computational evaluation of style must first define which dimension of style is being measured. Our models provide a vocabulary for this:

1. **AdaIN-Style:** The statistical dimension (texture, color).
2. **ResNet50-Style:** The categorical dimension (artistic movement/class).
3. **CLIP-ViT-L:** The semantic dimension (concept, theme, intent).
4. **DINOv2:** The structural/detailed dimension (composition, fine features).
5. **VGG19:** The classical perceptual dimension.

6.2   The Compositional Gap: Texture Versus Structure

The most significant finding from our analysis is the profound gap between texture-based similarity and all other similarity types. The extremely low average distance (0.063) and variance (0.0014) of the AdaIN-Style model demonstrate that the pastiches were overwhelmingly successful at matching the pure statistics of texture and color. This is logical, as the AdaIN method itself is foundational to style transfer techniques that optimize for these exact statistics (e.g., Gram matrices).

However, the high distances from DINOv2 (0.463) and VGG19 (0.674) reveal what we term the "Compositional Gap". Despite matching texture, the pastiches largely failed to replicate the originals' spatial relationships, compositional structure, and fine-grained visual details.

This finding is critical for the field of neural style transfer (NST). It suggests that methods relying primarily on feature statistics are solving only part of the problem. While they create texturally plausible images, they miss the important structural and compositional elements that are clearly detected by models like DINOv2. The high discrimination power of DINOv2 (0.0349 variance) makes it an ideal tool for measuring and potentially optimizing for this compositional gap in future work. Two visual examples of this phenomenon are in Fig. 4, which displays a work of Oana Năstăsache ($DINO = 0.269$, $AdaIN = 0.070$), where texture is preserved but structure is lost, and, in Fig. 5, which illustrates a work by Ciprian Mureșan, where the AI captures the composition but deviates in texture ($DINO = 0.584$, $AdaIN = 0.034$).

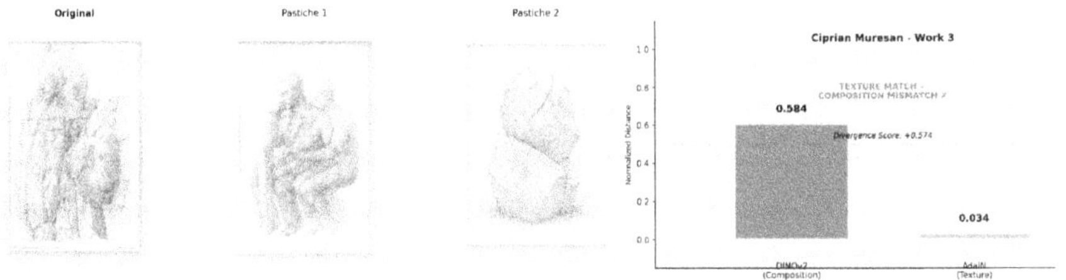

Fig. 4. Visual example of the "Compositional Gap" (High DINO, Low AdaIN). This case illustrates pastiches that successfully mimic the texture and color palette of the original (low AdaIN distance) but fail to capture the structural composition (high DINO distance).

6.3 Quantifying Style Reproducibility and Individual Characteristics

This framework can be applied by art historians to quantify stylistic consistency. For example, one could measure the average distance between all works within an artist's portfolio. A low intra-artist distance would suggest a highly consistent style, while a high intra-artist distance would suggest an artist who varied their style significantly.

Furthermore, the model-specific rankings provide a "style fingerprint." An artist who ranked as the hardest to match by a specific model, such as the AdaIN-Style model, can be understood as having a signature defined by his unique texture and color palettes, while another who ranked as the hardest to match by another model, say the VGG19 model, can be interpreted as having a very pronounced personal perceptual signature.

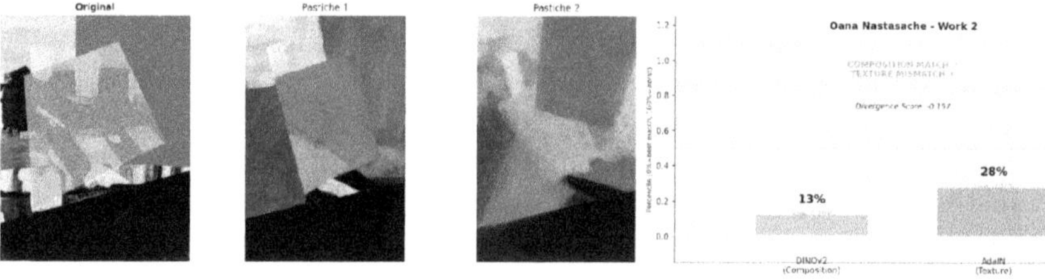

Fig. 5. Visual example of Structural Alignment (Low DINO, High AdaIN). Representing the inverse of the "Compositional Gap," this case shows where the AI successfully replicates the spatial composition and geometric blocking (low DINO distance) but deviates in texture or color statistics.

7 The Artists' Evaluation of the Artificially Generated Pastiches

We showed the artists the artificially generated pastiches after their three artworks and asked them to grade and comment on them by the following questions:

1. To what extent do you recognize your personal artistic language and the coherence of your visual style in this new work? (1 = not at all, 10 = completely)
2. How does this new work inspire you or what thoughts does it provoke? (open answer)
3. To what extent do you consider that the work generated by ChatGPT has aesthetic or artistic value? (1 = not at all, 10 = very high)

The grades vary widely by artist, as can be seen in Fig. 6. The mean was very low, 3.58 for recognizing their own style in the pastiches and 4.83 for their aesthetic value. The average grade for style similarity of 3.583 translates into a distance of 0.642 $(1 - 3.583/10)$, which perfectly aligns the human judgment with the VGG19 model's average cosine distance of 0.648. This shows that perceptual features play an important role in judging the style resemblance.

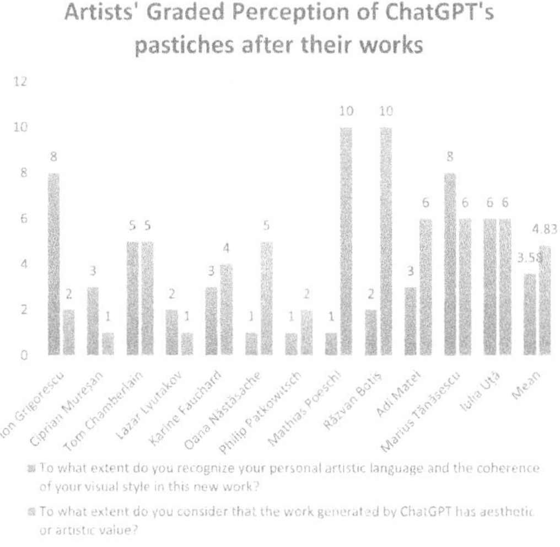

Fig. 6. Artists' Graded Perception of ChatGPT's pastiches after their works.

We did not have any means to automatically assess the quality of the pastiches, so we relied on the artists' opinion on that matter, reflected in their answers to the third question. The low 4.83 score out of 10 indicates that artificially generated pastiches are still far behind human artistry.

The artists' comments on the generated pastiches, obtained as their responses to the question "How does this new work inspire you, or what thoughts does it provoke?", reveal the essential limitation of AI in the field of artistic creation. They highlight a void of context and meaning, a lack of dimensionality and intentional sense, and an accent on imitation rather than originality, often accompanied by a form of controlled hallucination. The AI-generated work tends to function as a paraphrase or an approximate quotation rather than as a valuable, emotion-evoking artwork. In contemporary conceptual art the visual component is inseparable from its theoretical structure, from its ideology; together they articulate the work's meaning and significance. In the case of ChatGPT the theoretic appendage was missing and it worked solely with the visual material.

For example, artist Ion Grigorescu (RO) notes that ChatGPT did not understand the conceptual intention behind his painting, *Măriuca*, in Fig. 7a, that the

(a) Ion Grigorescu's *Măriuca*   (b) Pastiche 1 after *Măriuca*

Fig. 7. Ion Grigorescu's artwork *Măriuca* versus its Pastiche 1.

work should not be visually consummated. The painting it is not about the perfect plasticity, but about a specific idea and the emotion behind it. He observed that the AI model produced only the bed cover in that spirit, the rest of the pastiches's composition being visually excessive, as it can be seen in Figure 7b. The artist wonders: "Who taught AI to "paint"? Its works look like the Munich Academy of Art from 1900, drawings on different planes, somberly colored.".

Another example of such AI limitations is the simplistic and obvious pastiche after Tom Chamberlain's (UK) drawing *Dimmer* illustrated in Fig. 8a. This artist's drawings and paintings are built through repeated marks (e.g. erased, redrawn, softened, and reapplied) technique until the surface becomes almost ethereal, which Chat GPT did not understand and obviously could not replicate, as can be observed in Fig. 8b. The dimensionality is gone, and, as Chamberlain said, the pastiche needs more "hand", an organically constructed surface: "They make me worry about my work looking like something. I mean like cliche or formalism [...] They don't really give me much to think with, rather things to think against. [...] These images are like paraphrase. Without wanting to sound reactionary, they make me want more hand, more touch."

It was unexpected to note that ChatGPT, after multiple pastiches generated from different artworks, made a new pastiche after Philip Patkowitsch's piece in which it included a version of Ion Grigorescu's *Măriuca*, as it can be seen in Fig. 9.

(a) Tom Chamberlain's *Dimmer*

(b) Pastiche 1 after *Dimmer*

Fig. 8. Tom Chamberlain's artwork *Dimmer* versus its Pastiche 1.

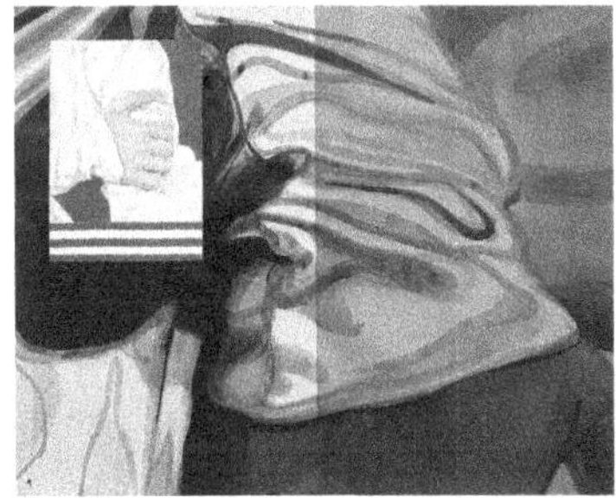

(a) Philip Patkowitsch's *Untitled*

(b) Pastiche 2 after *Untitled*

Fig. 9. Philip Patkowitsch's artwork *Untitled* versus its Pastiche 2.

8   Conclusions

Our multi-model framework has practical implications across several domains.

For AI researchers, this study shows that evaluating pastiche or style transfer quality with a single metric (e.g., LPIPS, FID) is insufficient. We propose using a "style dashboard" of complementary metrics, like the ones produced by the following models: AdaIN-Style for texture validation, CLIP-ViT-L for semantic and conceptual alignment, ResNet50-Style for style features, DINOv2 for compositional fidelity, and VGG19 for similarity of perceptual features.

For visual arts, these tools can enhance traditional expertise by adding automated measurements and objective analysis. One could quantify the influence of one artist on another by measuring the distance between their works or track an artist's stylistic evolution over time by plotting their works in these high-dimensional feature spaces.

Moreover, aesthetically, AI models do not reach the level of human creation: they lack meaning, context, and dimensionality, which explains the constant predicament that divides art theorists, oscillating between the enthusiasm provoked by a convincing copy, the disappointment generated by the absence of originality, and the disillusionment of seeing a dry simulacrum.

9   Limitations and Future Work

First, we used only one model to generate artworks and only twelve artists. In future work, we plan to also experiment with other models and compare them and invite more artists to participate with artworks. Second, some of the works were tridimensional, in particular sculptures and installations, and the photos cannot reveal the depth and the real structure, so the model might have had issues in perceiving these features, which might impacted their pastiche quality. Finally, we designed the prompt ourselves, which might bias the pastiche results. In future work, we plan to also invite artists to participate in the writing of the prompt instructions.

Acknowledgments. We are grateful to all the artists who agreed to let us use their works and for their insightful feedback: Adi Matei, Ciprian Mureșan, Ion Grigorescu, Iulia Uță, Karine Fauchard, Lazar Lyutakov, Marius Tănăsescu, Mathias Poeschl, Oana Năstăsache, Philip Patkowitsch, Răzvan Botiș, and Tom Chamberlain. This research is supported by:

– the project "Romanian Hub for Artificial Intelligence - HRIA", Smart Growth, Digitization and Financial Instruments Program, 2021-2027, MySMIS no. 351416;

– a grant of the Ministry of Research, Innovation and Digitization, CNCS - UEFIS-CDI, project SIROLA, number PN-IV-P1-PCE-2023-1701, within PNCDI IV;

– a grant of the Ministry of Education and Research, CCCDI - UEFISCDI, project number PN-IV-P6-6.1-CoEx-2024-0042, within PNCDI IV

Ethical Statement. There are no ethical issues with the publication of our work. We have respected all licenses and agreements of the software used, as well as the works of the artists who agreed to lend their artworks for this research.

References

1. Asperti, A., George, F., Marras, T., Stricescu, R.C., Zanotti, F.: A critical assessment of modern generative models' ability to replicate artistic styles. Big Data Cogn. Comput. **9**(9) (2025). https://doi.org/10.3390/bdcc9090231, https://www.mdpi.com/2504-2289/9/9/231

2. Avlonitou, C., Papadaki, E.: Ai: An active and innovative tool for artistic creation. Arts **14**(3) (2025). https://doi.org/10.3390/arts14030052, https://www.mdpi.com/2076-0752/14/3/52

3. Borji, A.: A categorical archive of chatgpt failures (2023). https://arxiv.org/abs/2302.03494, Accessed 30 Jan 2026

4. Borji, A.: Qualitative failures of image generation models and their application in detecting deepfakes. Image Vision Comput. **137**, 104771 (2023b). https://doi.org/10.1016/j.imavis.2023.104771

5. Chung, J., Hyun, S., Heo, J.: Style injection in diffusion: a training-free approach for adapting large-scale diffusion models for style transfer. In: Proceedings of the IEEE/CVF Conference on Computer Vision and Pattern Recognition (CVPR 2024), pp. 8795–8805. Seattle (2024)

6. Colton, S., Wiggins, G.A.: Computational creativity: the final frontier? In: Proceedings of the 20th European Conference on Artificial Intelligence (ECAI 2012), pp. 21–26 (2012). https://doi.org/10.3233/978-1-61499-098-7-21

7. Cunningham, C.V., Radvansky, G.A., Brockmole, J.R.: Human creativity versus artificial intelligence: source attribution, observer attitudes, and eye movements while viewing visual art. Front. Psychol. **16** (2025). https://doi.org/10.3389/fpsyg.2025.1509974

8. Dinu, A., Florescu, A.M., Dinu, L.: Analyzing large language models' pastiche ability: a case study on a 20th century Romanian author. In: Hämäläinen, M., Öhman, E., Bizzoni, Y., Miyagawa, S., Alnajjar, K. (eds.) Proceedings of the 5th International Conference on Natural Language Processing for Digital Humanities, pp. 20–32. Association for Computational Linguistics, Albuquerque (2025). https://doi.org/10.18653/v1/2025.nlp4dh-1.3, https://aclanthology.org/2025.nlp4dh-1.3/

9. Du, Z., Zeng, A., Dong, Y., Tang, J.: Understanding emergent abilities of language models from the loss perspective. In: Globerson, A., et al., (eds.) Advances in Neural Information Processing Systems. vol. 37, pp. 53138–53167. Curran Associates, Inc. (2024)

10. Dyer, R.: Pastiche. Routledge, Film studie/Media studies (2007)

11. Gatys, L.A., Ecker, A.S., Bethge, M.: Image style transfer using convolutional neural networks. In: 2016 IEEE Conference on Computer Vision and Pattern Recognition (CVPR), pp. 2414–2423 (2016). https://doi.org/10.1109/CVPR.2016.265

12. Goodfellow, I.J., et al.: Generative adversarial nets. Adv. Neural Inf. Process. Syst. **27** (2014)

13. Greene, R., Cushman, S., Cavanagh, C., Ramazani, J., Rouzer, P.: The Princeton Encyclopedia of Poetry and Poetics, 4th edn. Princeton University Press, Princeton Reference (2012)

14. He, K., Zhang, X., Ren, S., Sun, J.: Deep residual learning for image recognition. In: Proceedings of the IEEE Conference on Computer Vision and Pattern Recognition (CVPR), pp. 770–778 (2016)

15. van Hees, J., Grootswagers, T., Quek, G.L., Varlet, M.: Human perception of art in the age of artificial intelligence. Front. Psychol., **15** (2025). https://doi.org/10.3389/fpsyg.2024.1497469

16. Ho, J., Jain, A., Abbeel, P.: Denoising diffusion probabilistic models. In: Larochelle, H., Ranzato, M., Hadsell, R., Balcan, M., Lin, H. (eds.) Advances in Neural Information Processing Systems. vol. 33, pp. 6840–6851. Curran Associates, Inc. (2020)

17. Horton, C.B., Jr., White, M.W., Iyengar, S.S.: Bias against AI art can enhance perceptions of human creativity. Sci. Rep. **13**(1), 19001 (2023). https://doi.org/10.1038/s41598-023-45202-3

18. Huang, X., Belongie, S.: Arbitrary style transfer in real-time with adaptive instance normalization. In: Proceedings of the IEEE International Conference on Computer Vision (ICCV) (2017)

19. Hutcheon, L.: A Theory of Parody: The Teachings of Twentieth-Century Art Forms. University of Illinois Press (2000)

20. Ismayilzada, M., Paul, D., Bosselut, A., van der Plas, L.: Creativity in AI: progresses and challenges (2024). https://arxiv.org/abs/2410.17218, Accessed 30 Jan 2026

21. Khatiwada, P., Washington, J., Walsh, T., Hamed, A.S., Bhatta, L.: The ethical implications of AI in creative industries: a focus on AI-generated art (2025). https://arxiv.org/abs/2507.05549, Accessed 30 Jan 2026

22. Leivada, E., Murphy, E., Marcus, G.: Dall·e 2 fails to reliably capture common syntactic processes. Soc. Sci. Hum. Open **8**(1), 100648 (2023). https://doi.org/10.1016/j.ssaho.2023.100648

23. McArthur, T., McArthur, T., McArthur, R.: The Oxford Companion to the English Language. Oxford University Press, Oxford Companions Series (1996)

24. McCormack, J., Llano, M.T., Krol, S.J., Rajcic, N.: No longer trending on artstation: prompt analysis of generative AI art. In: Johnson, C., Rebelo, S.M., Santos, I. (eds.) Artificial Intelligence in Music, Sound, Art and Design, pp. 279–295. Springer Nature Switzerland, Cham (2024)

25. Murphy, E., de Villiers, J., Morales, S.L.: A comparative investigation of compositional syntax and semantics in dall·e and young children. Soc. Sci. Hum. Open **11**, 101332 (2025). https://doi.org/10.1016/j.ssaho.2025.101332

26. Oquab, M., et al.: Dinov2: learning robust visual features without supervision. arXiv preprint arXiv:2304.07193 (2023), Accessed 30 Jan 2026

27. Pan, B., Ke, Y.: Efficient artistic image style transfer with large language model (LLM): a new perspective. In: 2023 8th International Conference on Communication and Electronics Systems (ICCES), pp. 1729–1732 (2023). https://doi.org/10.1109/ICCES57224.2023.10192799

28. Radford, A., et al.: Learning transferable visual models from natural language supervision. In: Proceedings of the 38th International Conference on Machine Learning (ICML). Proceedings of Machine Learning Research, vol. 139, pp. 8748–8763. PMLR (2021)

29. Rombach, R., Blattmann, A., Lorenz, D., Esser, P., Ommer, B.: High-resolution image synthesis with latent diffusion models. In: 2022 IEEE/CVF Conference on Computer Vision and Pattern Recognition (CVPR), pp. 10674–10685. IEEE Computer Society, Los Alamitos (2022). https://doi.org/10.1109/CVPR52688.2022.01042

30. Rose, M.A.: Post-modern pastiche. British J. Aesthetics **31**(1), 26–38 (1991). https://doi.org/10.1093/bjaesthetics/31.1.26

31. Schwarz, A.: The Complete Works of Marcel Duchamp. Delano Greenidge Editions (2000)

32. Simonyan, K., Zisserman, A.: Very deep convolutional networks for large-scale image recognition. arXiv preprint arXiv:1409.1556 (2014), Accessed 30 Jan 2026

33. Wang, B., Zhu, Y., Chen, L., Liu, J., Sun, L., Childs, P.: A study of the evaluation metrics for generative images containing combinational creativity. AI Edam **37**, e11 (2023). https://doi.org/10.1017/S0890060423000069

34. Wasielewski, A.: The reification of style in ai image generation. Hertziana Studies in Art History (2024). https://doi.org/10.48431/hsah.0302

35. Wei, J., et al.: Emergent abilities of large language models. Trans. Mach. Learn. Res. (2022). https://openreview.net/forum?id=yzkSU5zdwD, Accessed 30 Jan 2026

36. Yin, S., et al.: A survey on multimodal large language models. National Sci. Rev. **11**(12), nwae403 (2024). https://doi.org/10.1093/nsr/nwae403

37. Zarei, A., et al.: Mitigating compositional failures in text-to-image models with causal text embedding refinement. In: 2025 IEEE International Conference on Pervasive Computing and Communications Workshops and other Affiliated Events (PerCom Workshops), pp. 74–79 (2025). https://doi.org/10.1109/PerComWorkshops65533.2025.00044
38. Zhang, X., Zhou, M., Lee, G.M.: AI voice in online video platforms: a multimodal perspective on content creation and consumption (2024). https://ssrn.com/abstract=4676705, Accessed 30 Jan 2026
39. Zhang, Z., et al.: Artbank: artistic style transfer with pre-trained diffusion model and implicit style prompt bank. In: Proceedings of the Thirty-Eighth AAAI Conference on Artificial Intelligence (AAAI 2024), pp. 7396–7404. Vancouver (2024)
40. Zhong, X., Zhu, J., Liu, W., Hu, C., Deng, Y., Wu, Z.: An overview of image generation of industrial surface defects. Sensors **23**(19) (2023). https://doi.org/10.3390/s23198160
41. Zhu, J.Y., Park, T., Isola, P., Efros, A.A.: Unpaired image-to-image translation using cycle-consistent adversarial networks. In: 2017 IEEE International Conference on Computer Vision (ICCV), pp. 2242–2251 (2017)

Generative Musical Exploration
of Astronomical Catalogs

Adrián García Riber[(✉)] [ID]

Universitat Pompeu Fabra, School of Engineering, 08018 Barcelona, Spain
adrian.garcia@upf.edu

Abstract. This article proposes the application of generative models
for the autonomous musical exploration of astronomical data to be used
both in creative applications and scientific outreach. It is focused on the
catalog of variable sources observed by the Optical Monitoring Camera (OMC) onboard the International Gamma-Ray Astrophysics Laboratory (INTEGRAL), which is hosted by the Spanish Virtual Observatory (SVO). The work describes a methodology, based on Box Least
Squares (BLS) periodograms and Long Short-Term Memory (LSTM)
networks, for the exploration of astronomical catalogs through sound and
music. The proposal explores the representation of light flux time-series
through musical notes calculated from the main periodic signal detected
in the represented light curves. The resulting notes are cross-matched
with a generative composition created from a selection of musical pieces
by Wolfgang Amadeus Mozart. The autonomous composition *Mozart's
sky* is presented as a proof of concept that provides real sky examples of
the creative and informative possibilities behind this approach.

Keywords: Sonification · Music · Astronomy · Astrophysics ·
Computer Science · Deep learning

1   Introduction

The expanding use of auditory displays to complement visualization [1–3] in
astronomical data provides innovative perspectives in the intersections between
Sound, Music and Astronomy [4–8]. In the last years, the potential of Sonification [9–11] for scientific analysis [12,13], accessibility and inclusion [14,15],
education, public engagement and outreach [16,17], has been demonstrated by
research teams like Trayford et al. (2023–2024), using STRAUSS to explore different applications of sonification that range from scientific audifications [18]
to planetarium shows [19]. The study of Arkand et al.(2024), also found positive feedback for sonification regarding enjoyability, learning, enhancement of
the experience, trust or accuracy in the representation and accessibility, both
from Blind or Low Vision (BLV) and no-BLV participants [20]. The current
development of digital technology has also opened up the possibility of introducing sonification within science education environments, allowing Fovino et

P. Machado et al. (Eds.): EvoMUSART 2026, LNCS 16523, pp. 84–96, 2026.
https://doi.org/10.1007/978-3-032-24350-8_6

al.(2023) to evaluate the effectiveness of the motion-sensing sonification tool *Edukoi* [21]. Focused on inclusion across different musical cultures, and aligned with the efforts of Garcia-Benito (2025) to draw parallelisms between historical music practices and contemporary sound design [22], Harrison et al. (2025) explored the collaborative adaptation of the show "Audio Universe: Tour of the Solar System", to create a relatable experience for Caribbean audiences of a show originally designed with Euro-American musical influences [23].

Framed in this inspiring context, and focused in the intersections of sonification and music composition [24], this article introduces the use of generative models for the autonomous musical exploration of astronomical catalogs, providing examples which blend astrophysical data analysis with classical music-based generative compositions. The work explores the light curves captured by the Optical Monitoring Camera (OMC) onboard the International Gamma-Ray Astrophysics Laboratory (INTEGRAL) mission [25] to provide real sky examples of commonly used time-series analysis. Additionally, it presents the autonomous composition *Mozart's sky* as a proof of concept for the creative and informative potential of the proposal. The approach is designed to be used in scientific outreach activities and creative applications. It enhances the aesthetic quality of the representations to make them more accessible and engaging, while bridging the fields of Data science, Astronomy and Music. Additionally, an Open Source simplified version of the prototype is provided to allow the exploration of the catalog with any user selected MIDI file.

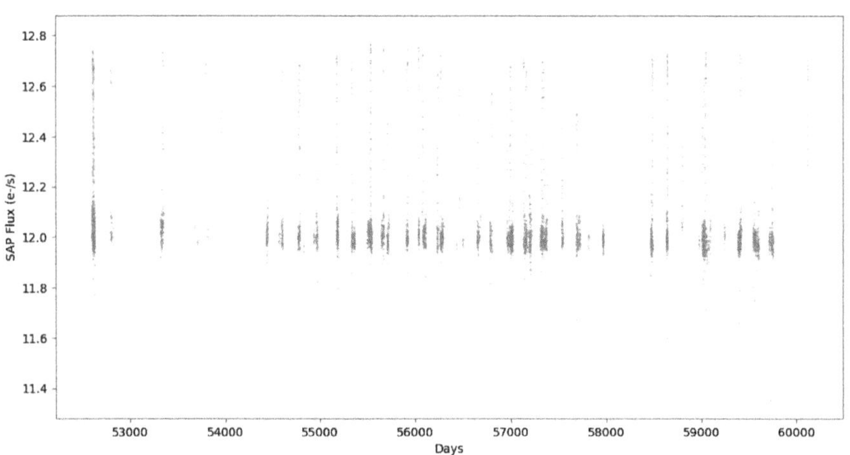

Fig. 1. OMC light curve from the Eclipsing binary of Algol type star V1011 Cyg, OMC ID: 2677000065. RA(2000): 298.8125 DEC(2000): 34.2083 (SVO). SAP flux vs time.

2 Light Flux Time-Series

The INTEGRAL-OMC archive includes light curves from more than 5,263 variable sources with a median magnitude ranging from 7.10 to 16.27. The observations are recorded from a high orbit, allowing continuous tracking during several weeks, something that is not possible from ground-based observatories [26]. The catalog provides data in the V-band from October 2002 to February 2010. The time base for the observations varies from 0.69 days to 2,815.19 days.

Light curves are two-dimensional graphical representations of the variations in the brightness flux observed in a research object over time. The analysis of these time-series, mainly focused on the detection and evaluation of periodicity, is commonly used for the classification of binary systems, variable stars, and supernovae, as well as for the discovery of extrasolar planets based on the transit detection method (the passage of an object between a light source and the observer). Figure 1 shows an example of this type of time series extracted from the OMC-VAR catalog.

The first INTEGRAL-OMC catalog is described in Garzón et al. (2018), a study providing statistical methods to detect variability and periodicity for the classification of variable objects. The catalog includes 2,631 pulsating stars, 1,132 eclipsing binaries, and various other types such as X-ray binaries and cataclysmic variables [27]. Table 1 briefly enumerates the main characteristics of these types of objects, additional information can be found in Trümper and Hasinger (2008) [28]. All OMC-VAR light curves are accessible through Virtual Observatory (VO) protocols hosted by the Spanish Virtual Observatory (SVO)[1]. The archive browser, descriptive papers [29–33], and project documentation can be found in the OMC Archive webpage[2].

Table 1. Main types of variable objects that can be found in the INTEGRAL-OMC catalog. Representative examples include name and ID reference.

Variable object	Physical system	Cause of variability	Name/OMC ID
Pulsating star	Single variable star	Radial pulsations	UZ Cen
		Non-radial pulsations	IOMC 8976000074
Eclipsing binary	Two orbiting stars	Mutual eclipses	V1011 Cyg
			IOMC 2677000065
X-ray binary	Star and	Variable accretion with	V615 Cas
	compact object	X-ray emission	IOMC 4047000079
Cataclysmic variable	White dwarf and	Accretion outbursts	V1223 Sgr
	low mass object	Thermonuclear events	IOMC 7409000033

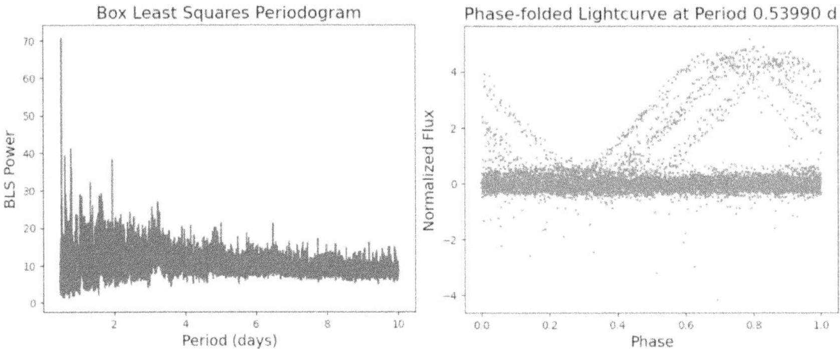

Fig. 2. Periodogram analysis of the light curve from V1011 Cyg, OMC ID: 2677000065. Left: Box Least Squares (BLS) power vs period. Right: Phased folded light curve at best fit (0.54 days), normalized flux vs phase.

3   Box-Least Squares (BLS) Periodogram Analysis

The Lomb-Scargle periodogram analysis is one of the standard statistical methods used to search for periodicity in irregularly sampled data when Fourier analysis cannot be applied [34]. It is primarily used in the search for variable stars and in the analysis of radial velocity data. The periodogram shows an estimation of the power of the Fourier transform as a function of the period, from which we can read the periods detected in the signal. It essentially involves fitting a sinusoidal model to the data for each frequency, obtaining greater power at the periods for which the model fits better the analyzed signal.

All the mathematical and historical details related to this method can be found in VanderPlas (2018), which is also a good reference for complementary techniques for periodic analysis of discrete observations, such as Bayesian approaches or the Fourier, Phase-folding, and BLS methods [35]. This latter type of analysis fits the input time series to periodic box-shaped functions, which represent better the behavior of a light curve during a transit than sines and cosines, providing useful information in the search of exoplanets and in binary star characterization [36].

Figure 2 shows the distribution of the fit periods found in V1011 Cyg and its the phase folded light curve. This process involves folding the observational time data over the found period so that each measurement is assigned a phase value between 0 and 1, representing its position within that periodic cycle.

Phase folding allows visual and quantitative analysis by overlaying multiple cycles on top of each other, which enhances the signal-to-noise ratio of periodic features. For example, in exoplanet transit detection or variable star analysis, phase folding reveals the characteristic dip of flux when individual scattered observations are hard to interpret. Approaches like the fast BLS (*fBLS*) allow the detection of small rocky transiting planets with periods shorter than one day at low computational costs [37].

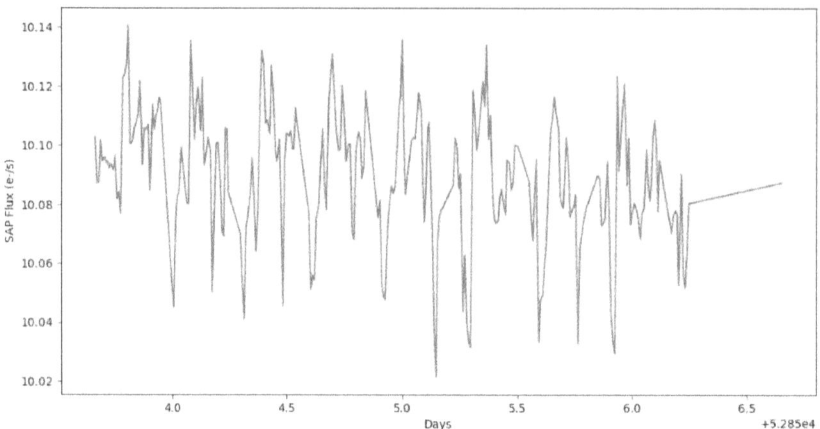

Fig. 3. Zoomed light curve from the Variable star BD-01 339. Best fit period 0.98 days, OMC ID: 4691000019. RA(2000): 36.9082 DEC(2000): -0.4115 (SVO). SAP flux vs time.

Astropy library[3] allows the implementation of the standard and generalized Lomb-Scargle and BLS periodograms, appropriately handling magnitudes with their attached units. *Astropy*'s astrostatistics module[4] provides numerous functions, algorithms, and statistical tools that even allow the normalization of the flux (sigma-clipping). All the documentation, functionality, common tools and statistical processes needed to perform Astronomy and Astrophysics in Python are available at astropy:docs[5].

Using these tools we calculated the BLS periodograms of 1,263 OMC light curves obtaining values ranging from 0.5 to 9.8 days, with 54.85 percent of the periods under one day. The results illustrate the abundance of sources with fast variability such as δ Scuti stars, AGNs binaries, X-ray binaries, gamma-ray burst, eclipsing binaries or pulsating variables, which commonly have periods from a couple of hours to less than a day. As an example, Fig. 3 shows in detail the fast flux variations of BD-01 339.

Aimed at obtaining the envelope of the complete light curve, we additionally used *Astropy*'s 1D polynomial fitting[6] with the Levenberg-Marquardt algorithm to compute the curve with the magnitude error of the observation. Figure 4 shows the the light curve with the polynomial fitting for the Pulsating variable star V2131 Cyg.

---

[3] Astropy: https://www.astropy.org/index.html.

[4] Astrostatistics tools: https://docs.astropy.org/en/stable/stats/index.html.

[5] Astropy docs: https://docs.astropy.org/en/stable/index.html.

[6] Astropy fitting: https://docs.astropy.org/en/stable/modeling/fitting.html.

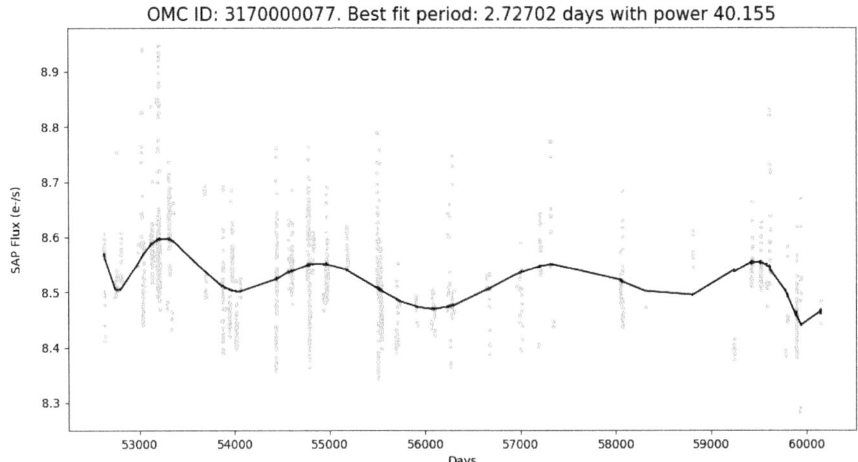

Fig. 4. Light curve with polynomial fitting from the Pulsating variable star V2131 Cyg, OMC ID: 3170000077. RA(2000): 311.2435 DEC(2000): 40.4117 (SVO). SAP flux vs time.

4   Sound Generation from BLS Periodograms

The results of a BLS periodogram analysis can be represented by sound through the correspondence between time and frequency domains (f=1/t), or with any alternative strategy such as direct time parsing through OSC or MIDI messages [38].

The following video shows a periodogram-driven score generated from 100 light curves of the Kepler Exoplanet Object of Interest (KOI) catalog[7] from the Mikulsky Archive for Space Telescopes (MAST) of the Space Telescope Science Institute (STScI)[8]. The proposal provides a musical representation of the main period of each light curve translated into a musical note. Each best fit period was interpreted as the period of the fundamental frequency for each note (multiplied by a factor of 100 to adjust it to the audible range). The results were approximated to the fundamental frequency of the nearest note over a chromatic scale of 12 notes .

Video 1. https://vimeo.com/678882792

The amplitudes of the notes were obtained from the power of the best fit periods, scaled logarithmically to match the 128 velocity stages of the MIDI standard [39]. The durations of the notes were mapped to the mean power of the periodogram. The durations were grouped in four categories ranging from *eighth* to *whole* musical figures to generate the final score. Figure 5 provides the light curve of Kepler-4 star as a high quality representative sample from the KOI catalog used to generate the score.

---

[7] KOI catalog: https://data.nasa.gov/dataset/kepler-objects-of-interest-koi.

[8] MAST Archive: https://archive.stsci.edu/missions-and-data.

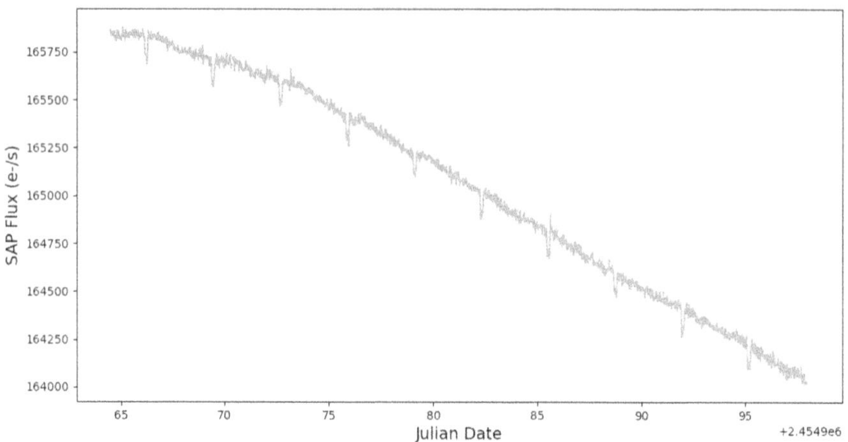

Fig. 5. Kepler light curve showing planetary transits with relative depth of $0.87.10^{-3}$, orbital period of 3.213 days and duration of about 3.95 h around the sunlike star Kepler-4, KIC: 11853905 [40]. RA(2000): 285.6156 DEC(2000): 50.1358 (STScI). SAP flux vs time.

5 Musical Exploration of the INTEGRAL-OMC Catalog

Inspired by the potential of the approach described in previous section and as a continuation of the work with stellar spectra by Riber and Serradilla (2023) [41], Fig. 6 illustrates the pipeline of an autonomous music-driven system for the exploration of the INTEGRAL-OMC Archive. The proposal consists on two parallel blocks, a BLS analysis and a LSTM network, which converge into a Pitch Class Set Theory algorithm to create a final multimodal (auditory and visual) exploration of the catalog.

The first block calculates the BLS periodograms of the light curves, extracts the best fit periods, multiplies them by 100 to bring the frequencies to the audible range, and scales the values from 0 to 127 to convert them into MIDI notes. The following video shows the score (Score 1), corresponding to the sequential playback of the periodogram results.

Video 2. https://vimeo.com/1137544349

The second block generates the musical sequence (Score 2) using a LSTM network trained with the seven sonatas by Wolfgang Amadeus Mozart listed in Table 2, which were transcribed into MIDI by Bernd Krueger[9]. For this purpose we used the *music generation notebook* [42] by Jha (2024) [43]. The musical pieces used for the training are open source and the MIDI files are licensed under Creative Commons Germany License[10].

---

[9] Classical Piano MIDI page: http://piano-midi.de/mozart.htm.
[10] http://piano-midi.de/copy.htm.

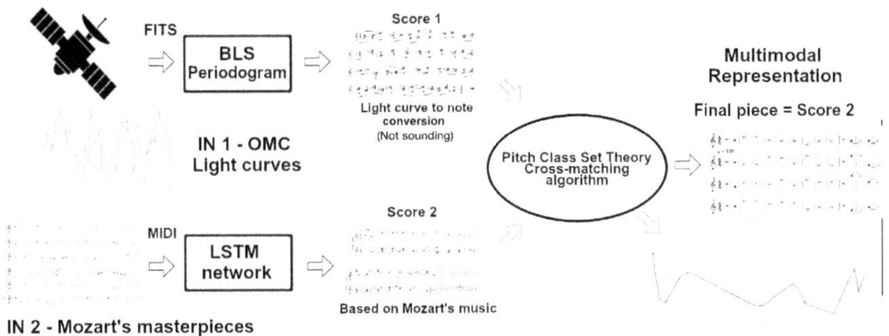

Fig. 6. Generative system based on the Pitch-Class Set Theory for the musical exploration of the INTEGRAL OMC Archive. Continuation of the work on stellar spectra by Riber and Serradilla (2023) [41].

Regarding the technical details, the network uses a multi-layer LSTM structure with 2 LSTM layers fed by a linear Input Layer. It includes batch normalization after the encoding and dropout for stability. The linear Output Layer maps the predictions to pitch classes, providing the final sequence of notes. The training process consisted in 100 epochs that produced an MSE loss $= 0.092$ for the training set.

Table 2. Wolfgang Amadeus Mozart's Sonatas used in the training of the LSTM network [44].

Year	Sonata	Catalog
1777	Sonata No. 8 D major	KV 311
1783	Sonata No. 10 C major	KV 330
1783	Sonata No. 11 A major (Alla Turca)	KV 331
1783	Sonata No. 12 F major	KV 332
1783	Sonata No. 13 Bb major	KV 333
1788	Sonata No. 16 C major (Sonata facile)	KV 545
1789	Sonata No. 17 Bb major	KV 570

Finally, the algorithm cross-matches the notes corresponding to the best fit periods of the light curves with the root note of the chords generated by the LSTM network, using the Pitch Class Set Theory (implemented with *music21* library [45]) to provide the final multimodal representation. This approach allows the generation of groups of all registers corresponding to a note with octave equivalence (A4 - A2 - A8), without distinguishing enharmonic equivalents (like G4 sharp and A4 flat). Translating those classes into numbers, with 0 corresponding to C and 11 corresponding to B, each generated chord is reduced to a single code that allows their comparison for finding positive matches between samples. Figure 7 provides the last bars of the piece *Mozart's Sky*, which is presented in the following video as a proof of concept for this proposal of music-driven exploration of astronomical data.

Video 3. https://vimeo.com/1137472800

Fig. 7. Last bars of the generative piece *Mozart's sky*.

An additional Jupyter notebook[11] based on the RtMidi[12] library is provided to allow the exploration of the INTEGRAL-OMC archive with any user selected MIDI file. The following video shows an example of the light curves returned by one of the most famous note sequences in science fiction soundtracks.

Video 4. https://vimeo.com/1153768425

6   Conclusion and Prospective

This work explores the potential use of generative models to create non-supervised auditory representations of astronomical data. It is framed within the context of the SVO open science initiatives aimed at bringing science closer to society, with a special focus on the development of tools that could contribute towards an horizontal access to astronomical data for people with and without visual impairment. Through the exploration of the time-series from the INTEGRAL-OMC archive, the proposal offers a proof of concept for the conversion of light flux variations into sound and music. The results are expected to be useful in creative applications and scientific outreach. Near future work includes a user study based on listening tests through an online questionnaire aimed at confirming the potential of the approach to provide a horizontal experience for sighted, blind or low vision audiences, and to enhance the accessibility, engagement, and understanding of astronomical information through sound.

The MIDI protocol can be useful to facilitate the musification of analytical results (such as those obtained from periodograms), the mappings of light flux variations in virtual musical instruments, and the generation of automatic musical scores. The proposal is expected to be used in scientific dissemination, science

---

[11] https://github.com/AuditoryVO/INTEGRAL-OMC-Exploration.
[12] https://caml.music.mcgill.ca/gary/rtmidi/.

engagement applications and science outreach, as well as in creative applications, concerts and conferences.

Focusing on the accuracy of the approach, the frequency quantization intrinsic to each MIDI musical note increases the uncertainty of the results. Nevertheless, this type of conversion provides an overall auditory representation correlated with the flux distribution of the represented light curve, which can be useful in preliminary representations for the exploration of massive data archives and live musical performances. Although it has not been implemented in this work, the use of different master tunings and the reproduction of microtonal sections could be included in future developments with the addition of Controller Change (CC) and System Exclusive (SysEx) MIDI messages within the playback loop.

Regarding the aesthetics of the representations [46], the inclusion of simple complements such as reverberation can emphasize musicality, bringing the results closer to the sonorities used in the creation of original soundtracks for films and television or in electro-acoustic music. We find that this approach can reduce the listener fatigue and can improve the acceptance of the auditory representations, as previously suggested by Roads (1996) [47], Kramer (2000) [48], Risset (2003) [49], and Barrass and Vickers (2011) [50].

The results of BLS periodogram analysis can be represented through sine waves and musical notes to provide useful information that can be horizontally understood by people with and without visual impairment. This method, used for detecting periodical signals in time series, suggests a natural relationship with the periodical sounds of the representations. Nevertheless, in our study we mapped the best fit periods directly to MIDI notes, making them potentially easier to understand. As proved with the piece *Mozart's Sky*, the sonification and musification frameworks can be creatively applied to BLS analysis, transforming detected periodic box-shaped variations into musical pitches, to enrich interdisciplinary bridges between data science, astronomical catalogs, and musical expression.

Acknowledgments. This article is based on data from the OMC Archive at CAB (INTA-CSIC), pre-processed by ISDC and further processed by the OMC Team at CAB. The OMC Archive is part of the Spanish Virtual Observatory project. Both are funded by MCIN/AEI/10.13039/501100011033 through grants PID2020-112949GB-I00 and PID2019-107061GB-C61, respectively. It also uses data collected with the TESS and Kepler missions, obtained from the MAST data archive at the Space Telescope Science Institute (STScI). Funding for the TESS and Kepler missions are provided by the NASA Explorer Program and by the NASA Science Mission Directorate. STScI is operated by the Association of Universities for Research in Astronomy, Inc., under NASA contract NAS 5âĂŞ26555.

This work made use of Astropy: (https://www.astropy.org/team.html.) a community-developed core Python package and an ecosystem of tools and resources for Astronomy [51].

Disclosure of Interests. The authors have no competing interests to declare that are relevant to the content of this article.

References

1. Morrissey, W., Zajicek, M.: Can sound output enhance graphical computer interfaces?. Contem. Ergonom., 172-176 (2000)
2. Brewster, S.: Nonspeech auditory output. In: The Human-computer Interaction Handbook, pp. 273–290. CRC Press (2007)
3. Enge, K., et al.: Open your ears and take a look: a state-of-the- art report on the integration of sonification and visualization. Comput. Graph. Forun. **43**(3), e15144 (2024). https://doi.org/10.1111/cgf.15114
4. Lovett, J.M.: The Sound Culture of Space Science. (Doctoral Thesis) University of Toronto, ProQuest Dissertations & Theses, 2021.28259760. Canada (2021)
5. Pesic, P.: Music and the making of modern science. MIT Press (2022)
6. Wilson, S.: Information arts: intersections of art, science, and technology. MIT Press (2003). https://doi.org/10.7551/mitpress/3765.001.0001
7. Gómez, A.H.: Stellar per-formations at the edge of the collapse. (Doctoral Thesis), University of Huddersfield, UK
8. Bigg, C., Vanhoutte, K.: Spectacular astronomy. Early Popular Visual Cult. **15**(2), 115–124 (2017). https://doi.org/10.1080/17460654.2017.1319037
9. Kramer, G., et al.: Sonification report: status of the field and research agenda (2010). https://digitalcommons.unl.edu/psychfacpub/444/
10. Hermann, T., Hunt, A., Neuhoff, J.G.: The sonification handbook (Vol. 1). Berlin: Logos Verlag (2011). https://sonification.de/handbook/
11. Worrall, D.: Sonification design. Springer Cham, pp. 1571–5035 (2019). https://doi.org/10.1007/978-3-030-01497-1
12. Bornmann, L.: The sound of science: data sonification has emerged as possible alternative to data visualization. EMBO Rep. **25**(9), 3743–3747 (2024). https://doi.org/10.1038/s44319-024-00230-6
13. García Riber, A., García-Benito, R., Serradilla, F.: Interactive multimodal integral field spectroscopy. RAS Techn. Instr. **3**(1), 748–758 (2024). https://doi.org/10.1093/rasti/rzae049
14. Diaz-Merced, W.L., et al.: Sonification of astronomical data. Proc. Int. Astron. Union **7**(S285), 133–136 (2011). https://doi.org/10.1017/S1743921312000440
15. Casado, J., et al.: Sonification as a tool for data analysis: usability and compliance evaluation study. Int. J. Sociotechnol. Knowl. Dev. (IJSKD) **14**(1), 1–27 (2022). https://doi.org/10.4018/IJSKD.299048
16. Zanella, A., Harrison, C.M., Lenzi, S., Cooke, J., Damsma, P., Fleming, S.W.: Sonification and sound design for astronomy research, education and public engagement. Nat. Astronomy **6**(11), 1241–1248 (2022). https://doi.org/10.1038/s41550-022-01721-z
17. Misdariis, N., Özcan, E., Grassi, M., Pauletto, S., Barrass, S., Bresin, R., Susini, P.: Sound experts' perspectives on astronomy sonification projects. Nat. Astronomy **6**(11), 1249–1255 (2022). https://doi.org/10.1038/s41550-022-01821-w
18. Trayford, J.W., Harrison, C.M., Hinz, R.C., Kavanagh Blatt, M., Dougherty, S., Girdhar, A.: Inspecting spectra with sound: proof-of-concept and extension to datacubes. RAS Techn. Instr. **2**(1), 387–392 (2023). https://doi.org/10.1093/rasti/rzad021
19. Trayford, J. W., Youles, S., Bonne, N., Harrison, C. M.: Ear to the sky: astronomical sonification for accessible outreach, education and research with STRAUSS. In: Revista Mexicana de Astronomia y Astrofisica Conference Series, vol. 57, p. 42 (2024). https://researchportal.port.ac.uk/en/publications/ear-to-the-sky-astronomical-sonification-for-accessible-outreach-/

20. Arcand, K.K., et al.: A universe of sound: processing NASA data into sonifications to explore participant response. Front. Commun. **9**, 1288896 (2024). https://doi.org/10.3389/fcomm.2024.1288896

21. Fovino, L.G.N., et al.: Evaluating the effectiveness of sonification in science education using Edukoi. Pers. Ubiquit. Comput. **28**(5), 693–711 (2024). https://doi.org/10.1007/s00779-024-01809-5

22. García-Benito, R.: Beyond universality: cultural diversity in music and its implications for sound design and sonification. arXiv preprint (2025). https://doi.org/10.48550/arXiv.2506.14877

23. Harrison, C. M., et al.: A case study of translating sonifications across musical cultures for an educational application. arXiv preprint (2025). https://doi.org/10.48550/arXiv.2506.08096

24. Mermikides, M.: Hidden music: the composer's guide to sonification. Cambridge University Press (2025). https://doi.org/10.1017/9781009258555

25. Winkler, C., et al.: The INTEGRAL mission. Astron. Astrophy. **411**(1), L1–L6 (2003). https://doi.org/10.1051/0004-6361:20031288

26. Gutiérrez, R., Solano, E., Domingo, A., García, J.: The optical monitoring camera data server: contents and functionalities. In: Astronomical Data Analysis Software and Systems (ADASS) XIII, vol. 314, p. 153 (2004)

27. Alfonso-Garzón, J., Domingo, A., Mas-Hesse, J.M., Giménez, A.: The first INTEGRAL-OMC catalogue of optically variable sources. Astron. Astrophys. **548**, A79 (2012). https://doi.org/10.1051/0004-6361/201220095

28. Trümper, J. E., Hasinger, G. (Eds.): The Universe in X-rays. Springer Science & Business Media (2008)

29. Zejda, M., Domingo, A.: Reference frame and time standard used in INTEGRAL/OMC datasets. Inf. Bull. Vari. Stars **5996**, 1–2 (2011)

30. Mas-Hesse, J.M., et al.: OMC: an optical monitoring camera for INTEGRAL-instrument description and performance. Astron. Astrophy. **411**(1), L261–L268 (2003). https://doi.org/10.1051/0004-6361:20031418

31. Mazy, E., et al.: Optical design of the Optical Monitoring Camera (OMC) of INTEGRAL. Astron. Astrophy. **411**(1), L269–L273 (2003). https://doi.org/10.1051/0004-6361:20031480

32. Walton, D.M., et al.: The CCD and readout electronics for the OMC instrument on Integral. Astron. Astrophy. **411**(1), L275–L279 (2003). https://doi.org/10.1051/0004-6361:20031453

33. Domingo, A., et al.: The input catalogue for the OMC camera onboard INTEGRAL. Astron. Astrophys. **411**(1), L281–L289 (2003). https://doi.org/10.1051/0004-6361:20031400

34. Ivezić, Ž., Connolly, A.J., VanderPlas, J.T., Gray, A.: Statistics, data mining, and machine learning in Astronomy: a practical Python guide for the analysis of survey data, vol. 8. Princeton University Press (2020)

35. VanderPlas, J.T.: Understanding the lomb–scargle periodogram. Astrophys. J. Suppl. Ser. **236**(1), 16 (2018). https://doi.org/10.3847/1538-4365/aab766

36. Kovács, G., Zucker, S., Mazeh, T.: A box-fitting algorithm in the search for periodic transits. Astron. Astrophys. **391**(1), 369–377 (2002). https://doi.org/10.1051/0004-6361:20020802

37. Shahaf, S., Zackay, B., Mazeh, T., Faigler, S., Ivashtenko, O.: FBLS–a fast-folding BLS algorithm. Mon. Not. R. Astron. Soc. **513**(2), 2732–2746 (2022). https://doi.org/10.1093/mnras/stac960

38. Riber, A.G.: Deep Learning Auditory Virtual Observatory. (Doctoral Thesis), E.T.S.I. y Sistemas de Telecomunicación (UPM) (2025). https://doi.org/10.20868/UPM.thesis.90828
39. Anderton, C.: The MIDI protocol. In: Audio Engineering Society Conference: 5th International Conference: Music and Digital Technology. Audio Engineering Society (1987)
40. Borucki, W.J., et al.: Kepler-4b: a hot neptune-like planet of a G0 star near main-sequence turnoff. Astrophysical J. Lett. **713**(2), L126 (2010). https://doi.org/10.1088/2041-8205/713/2/L126
41. Riber, A. G., Serradilla, F.: AI-rmonies of the Spheres. In: International Conference on Computational Intelligence in Music, Sound, Art and Design (Part of EvoStar), pp. 132-147. Cham: Springer Nature Switzerland (2023). https://doi.org/10.1007/978-3-031-29956-8_9
42. Mastering PyTorch https://github.com/arj7192/MasteringPyTorchV2/tree/main, Accessed 17 Jan 2026
43. Jha, A. R.: Mastering PyTorch: create and deploy deep learning models from CNNs to multimodal models, LLMs, and beyond. Packt Publishing Ltd. (2024)
44. Classical Piano MIDI Page. http://piano-midi.de/mozart.htm, Accessed 17 Jan 2026
45. Cuthbert, M. S., Ariza, C.: music21: a toolkit for computer-aided musicology and symbolic music data (2010). http://hdl.handle.net/1721.1/84963
46. Grond, F., Hermann, T.: Aesthetic strategies in sonification. AI Soc. **27**, 213–222 (2012). https://doi.org/10.1007/s00146-011-0341-7
47. Roads, C.: The computer music tutorial. MIT Press (1996)
48. Kramer, G.: Auditory display: sonification, audification and auditory interfaces. Addison-Wesley Longman Publishing Co., Inc (2000)
49. Risset, J.C.: The perception of musical sound. Comput. Music J. **27**(1), 1–12 (2003). https://muse.jhu.edu/article/41090
50. Barrass, S.: The aesthetic turn in sonification towards a social and cultural medium. AI Soc. **27**(2), 177–181 (2012). https://doi.org/10.1007/s00146-011-0335-5
51. Price-Whelan, A.M., et al.: Astropy collaboration the astropy project: sustaining and growing a community-oriented open-source project and the latest major release (v5. 0) of the core package. Astrophys. J. **935**(2), 167 (2022)

Generative Artificial Intelligence in Music: Towards an Aesthetics of Co-creation

Anthony Grégoire⬤, Serge Lacasse⁽⊠⁾⬤, and Joanie Verviers De Blois⬤

Université Laval, Quebec City, QC, Canada
{anthony.gregoire.2,joanie.verviers-deblois.1}@ulaval.ca
serge.lacasse@mus.ulaval.ca

Abstract. The rapid rise of generative artificial intelligence (GAI) is reshaping the paradigms of artistic creation. In music, this transformation is particularly striking: new systems capable of composing, performing, and producing original works challenge traditional boundaries between author, instrument, and machine. Platforms such as SUNO and UDIO, capable of generating complete songs with vocals, instrumentation, and mixing from simple text prompts, represent an unprecedented technical and aesthetic revolution. This paper presents a research-creation project exploring the aesthetic and practical potentials of co-creation between humans and AI systems in recording studio settings. Based on ethnographic observation of music creators working with GAI tools, we analyze how AI transforms collaborative dynamics, creative processes, and the very notion of musical authorship. Our findings reveal that rather than replacing creators, AI acts primarily as an inspiration tool and collaborative catalyst, while exposing fundamental questions about creativity, agency, and artistic value in the age of algorithmic intelligence.

Keywords: Generative AI · Music creation · Research-creation · Co-creation · Computational aesthetics · Recording studio · Human-machine collaboration

1 Introduction

The rapid rise of generative artificial intelligence (GAI) is reshaping the paradigms of artistic creation today. In the musical field, this transformation is particularly striking: new systems capable of composing, performing, and producing original works challenge traditional boundaries between author, instrument, and machine. The emergence of platforms such as SUNO and UDIO, capable of generating complete songs with vocals, instrumentation, and mixing

---

In accordance with our commitment to place generative AI at the heart of the creative process as a collaborative agent, this text was written in part using ChatGPT before being read, revised, and reworked by the authors. The analyses are the product of the team's reflections and exchanges and were not produced by artificial intelligence.

© The Author(s), under exclusive license to Springer Nature Switzerland AG 2026
P. Machado et al. (Eds.): EvoMUSART 2026, LNCS 16523, pp. 97–112, 2026.
https://doi.org/10.1007/978-3-032-24350-8_7

from a simple text description (prompt), represents an unprecedented technical and aesthetic revolution. These seemingly autonomous technologies question the place of human gesture in the creative process: if the machine can compose, what becomes of artistic subjectivity?

Far from conceiving this mutation as a threat to musical art, our project, *Generative AI in Music: Creation and Production* (funded by the International Observatory on the Societal Impacts of Artificial Intelligence), explores instead the aesthetic and practical potentials of co-creation between humans and AI systems. Situated within a research-creation approach [1], this study aims to understand how the integration of GAI can transform collaborative work in recording studios. We seek to determine how AI tools can enrich composition, production, and recording while modifying modes of interaction between creators and technologies. This questioning arises particularly with the recent emergence of new tools aimed at professional creators, notably SUNO Studio (since September 25, 2025) and AI Studio by Moises (since August 20, 2025).

Our approach rests on a dual observation. On one hand, generative AI raises legitimate concerns related to ethics, copyright, or loss of authenticity [2–4]. On the other hand, it opens new creative perspectives that go beyond simple automation of technical tasks [5]. Between these two poles—fear and fascination—an unprecedented field of experimentation unfolds: that of an aesthetics of human-machine collaboration, founded on exploration of shared creativity. In this perspective, and contrary to the rather pessimistic vision of certain authors [6], we consider AI not as a passive instrument but as a creative partner, capable of stimulating, diverting, and reorienting human aesthetic processes [7]. This posture is notably based on a posthumanist onto-epistemology, which aims among other things to decenter the human by integrating it into an ensemble composed of human and non-human entities [8]. This posture thus proposes a *relational* framework that can be established between these entities:

This produces [...] a new understanding of the human, not as an autonomous agent endowed with transcendental consciousness, but rather an immanent—embodied and embedded relational—entity that thinks with and through multiple connections to others, both human and nonhuman, organic and inorganic others [9, p. xii].

This means for us that creativity can no longer be envisioned as an exclusively human quality but as an emergent property, here, of a socio-technical system. In the studio context, each interaction between musician, sound engineer, and machine produces a field of distributed agencies. The introduction of tools like Synthesizer V, Audimee, LANDR, or AIVA, for example, adds an additional layer to this network: AI proposes, suggests, sometimes improvises, but always through frameworks imposed by humans. Analysis of this co-agency constitutes a central contribution of the project.

The central question running through this project is therefore: to what extent can generative AI act as a true collaborator within the musical creation and production process? In other words, how can we redefine the notion of creativity when the act of creation results from a dialogue between humans and algorithmic systems? More precisely, in the context of the present project, how does this potential expanded vision of musical creativity manifest itself (or not) among the group of authors and composers who participated in the project? Thus, the objective of this research is twofold: 1) evaluate AI's effective contribution to the creative process, and 2) rethink aesthetic and epistemic categories based on this collaboration.

The problematic can therefore be formulated as follows: How does collaboration between humans and GAI systems reconfigure creative dynamics, power relations, and decision-making modes in the musical creation process? This interrogation exists within a dual tension. On one side, GAI's promise is that of a democratization of creative means: anyone can now generate a song, text, or voice without prior skills. On the other, this same automation raises the fear of aesthetic standardization and loss of singularity. These two poles form the dialectic at the heart of our problematic, which also unfolds in five sub-axes:

1. **Aesthetic:** To what extent do productions resulting from human-AI collaboration generate new sonic or expressive forms?
2. **Ethical and epistemic:** How are notions of author, intention, and responsibility redistributed in a framework where the machine actively participates?
3. **Methodological:** How to document and analyze this collaboration in a way that is both rigorous and reflexive?
4. **Axiological:** What value is accorded to musical productions created with AI assistance?
5. **Ethical:** To what extent do humans, collaborating with the machine, consider an AI-supported creative approach equitable?

This project, conducted at the Laboratoire de recherche-création en musique et multimédia (LARCEM) and the Laboratoire audionumérique de recherche et de création (LARC) at Université Laval, is based on the production and ethnographic analysis of songs created using GAI tools at all stages of the process: writing, composition, arrangement, and production. The study relies on a collaborative and reflexive method, combining participant observation, interviews, and multimodal analyses, to document interactions between human actors and algorithmic devices.

The choice we made of an ethnographic approach, combined with research-creation, allows us to partially respond to these questions. By observing and recording the work process, it becomes possible to describe the fabric of musical creation in its most concrete dimension: collective decisions, trials, rejections, adjustments, emotions, discussions around proposals generated by AI. This material allows us to better understand not only what AI produces but how it transforms the ways of making and thinking about music.

2   Computational Aesthetics and Research-Creation

The articulation between musical creation and AI is not new. However, the emergence of large-scale generative models based on deep learning and audio synthesis has disrupted the very nature of this relationship. If the computer was long perceived as a calculation or processing tool, it now becomes a cognitive and aesthetic actor in its own right, capable of imitating, interpreting, and proposing new forms. It is in this context that the work of Emanuele Arielli and Lev Manovich [5,10], which resonates with posthumanist thought, takes on their full meaning.

According to Arielli, AI allows us to go beyond the instrumental conception of technology to enter a logic of "artificial aesthetics," where the machine becomes co-producer of sensible objects. In his model, four modes of aesthetic interaction between humans and AI can be distinguished:

1. analysis of existing aesthetic objects;
2. analysis of subjects who admire them (pattern recognition);
3. generation of aesthetic objects;
4. generation of subjects themselves (pattern generation), whose tastes and expectations can be modeled.

This theoretical framework offers a powerful analytical grid for understanding creative dynamics in an algorithmic environment. The project *Generative AI in Music: Creation and Production* primarily relies on the first three levels of this model. GAI intervenes as an analysis tool, production agent, and vector for reflection on reception. By enabling the generation of lyrics, melodies, or timbres, but also analyzing results or evaluating them collectively, AI acts simultaneously on the structure, content, and perception of the musical work.

On the methodological and epistemological level, our approach falls within the tradition of research-creation [1,11], where artistic production constitutes a mode of knowledge in itself. Unlike a simple technological case study, this involves an integrated experimental device: the creative process becomes both the object and site of research. GAI tools are mobilized not to replace creators but to shift the center of gravity of creativity toward a relational and collective space.

Collaboration thus occupies a central place. As Wilsmore and Johnson [12] have shown, contemporary musical production rests on varied forms of co-production: integrative collaboration (between long-term partners), familiar collaboration (between members of the same community), and complementary collaboration (between distinct expertises). By introducing AI into this system, we introduce a fourth modality: algorithmic collaboration, where the machine becomes an interlocutor whose proposals must be interpreted, evaluated, and negotiated by humans.

Finally, our theoretical framework also draws from the sociology of studio practices [13]. This work has shown that creativity in the studio results from a series of complex interactions between technologies, musicians, and producers, rather than from an individual act of inspiration. The introduction of GAI exacerbates this dynamic by making visible and expressible the mediations that structure the creative act, while redefining boundaries between spontaneity and programming, intention and suggestion, reflection and explicitation. Thus, the project's conceptual framework articulates three complementary dimensions:

1. a computational aesthetics that conceives AI as an aesthetic actor [5,8,10];
2. a research-creation methodology [1,11] that makes the process a space for reflexive experimentation;
3. an analysis of collaboration in the studio [12,13] that repositions creation within a collective and technological ecosystem.

3   An Ethnographic Research-Creation Device

The methodological device implemented within our project aims to reconcile the requirements of empirical research and those of artistic creation through a structure in two creation phases:

- Weeks 1–2: experimentation with GAI tools and first impressions (exploratory phase)
- Weeks 3–5: design and production of first songs
- Week 6: preliminary analysis of data from the first phase
- Weeks 7–9: design and production of other songs (optimized phase)
- Weeks 10–11: presentation of latest creations, final analysis and synthesis of results

Moreover, the project is based on an articulation between two teams: 1) a so-called creation team, responsible for the composition and production of two songs; and 2) a so-called ethnographic observation team composed of an ethnomusicologist and a Master's student in musicology, charged with documenting, analyzing, and interpreting the creative process. The first team, composed of musicians, Master's and PhD students in composition and musicology, and researchers in music, was given the mandate to design musical works using a series of AI tools, generative or not. The mobilized software covers the entire spectrum of the musical process

- ChatGPT-4o and (Anthropic) Claude for generating lyric ideas
- SUNO and UDIO for generating melodic motifs from texts (prompts)
- Synthesizer V and Audimee for creating virtual vocal tracks
- AIVA for orchestration
- Ozone for mastering

Members of this first team took the necessary time to experiment with these AI tools to better understand their functioning. They subsequently grouped into two teams to create one or two original pieces. One of the creators, more experienced with creation and vocal synthesis software, joined both creation sub-teams to offer support in the final rendering of creations.

The second "ethnographic" team had the objective of observing the entire creative process by adopting a reflexive and non-intrusive posture. Creative activities were not recorded to avoid disturbing the natural flow of the creators' work, but individual meetings were held at the beginning of the experimentation phase, then after the first creation phase and before the presentation of the second series of creations. During these interviews, the creators were able to present their exploratory and experimental approach and their creation process, while emphasizing the use of GAI: obstacles, facilities, discoveries, frustrations, and progress. All these interviews were recorded and transcribed alongside continuous ethnographic note-taking (decisions, discussions, hesitations, reactions to proposals generated by AI, and negotiations between team members). These qualitative notes are completed by semi-directed interviews conducted before and after song production: participants express their expectations, their perceptions of AI's role, as well as transformations experienced in their relationship to creation and collaboration. Team meetings during creation phases were also recorded, without the observers being present.

This iterative model allowed integrating the reflexive dimension at the very heart of the process: the first song acts as an experimental prototype, while the second consolidates learning from the first. A plenary meeting was therefore organized to share the work done after each of the two creation phases, where creators and observers could share questions and reflections throughout musical listening sessions in the studio. Creators were thus able to put their experience of (co-)creation with GAI into words and develop this experience during successive interviews, but they were also led to produce a summary document of their AI-assisted creation process.

3.1   Data Analysis and Instrumentation

The collected data (audio, video, verbal, and textual) were manually processed through a combination of tools and techniques, in parallel, from an in-depth and iterative reading of all ethnographic notes and recorded exchanges with participants. These are therefore essentially discourse analyses and ethnographic observations. Each interview was first examined by the researchers individually to identify major discursive axes and differentiated impressions (actors' representations, lexical choices, uses and justifications) and thematic regularities. An analytical grid was then developed to classify segments according to issues raised, modes of justification, identity positionings, or legitimation strategies. Coding was performed directly on texts using highlighting and notes allowing identification of recurrences, oppositions, and implicit elements, before being shared among observation team members. This work helped identify transversal trends between reports and better understand how participants represented and structured meaning production around (co-)creation issues.

The project's structure in two creation phases also allowed for iterative reflections: during the first phase, creators had to create a song using only AI, while intervening as little as possible; whereas during the second, they could use it at their discretion. After the first two interview series, we noticed some changes in attitudes and perceptions among creators, following their experimentation or simply due to the fact that they were working in teams. Thus, we concluded that it would be better to separate analysis phases into three parts: a first personal experimentation phase, a second for writing the first song, and a third for writing the second song. During each of these phases, creators' opinions and expectations toward GAI modulated, which led them to question AI's contribution as a collaborator.

From our first interviews, we found several similarities but also certain differences regarding the perception and expectations creators had toward GAI, which led us to establish typical profiles that we illustrated on a diagram showing the level of reluctance/acceptance (X axis) and difficulty/ease (Y axis) for each participant (see below). In this diagram, we can see that the five creators are located at different places on the axis, according to their experiences and impressions. Some were, from the beginning, very open to working with AI and seeing what it could bring them, while others were more closed to its use. Still others were nuanced, with discourse oscillating from one side to the other.

With the second interviews, we observed certain changes in perception toward GAI use, which led us to produce a second diagram. Two major elements emerge from these interviews: the impact of collaboration (for example teamwork as a remedy for a first negative experience) and questions about the very nature of this collaboration (collaboration vs co-creation, and sometimes doubt that there is even collaboration). However, these changes being still very close to the first profiles, we incorporated these second changes within the first diagram.

The third and final phase of the project allowed the five creators to use GAI at their discretion: they could therefore use it to varying degrees and at the intensity of their choice. Our last interviews highlighted new reflections allowing us to produce a second diagram for each profile. We thus paired the first and last graphs to explain each participant's psychological journey following their creation work with AI.

Psychological Profiles. All observations used for generating these analyses come from notes taken during individual interviews and plenary feedback sessions. To move from qualitative observations to a position on two axes (X = Resistance→Acceptance; Y = Difficulty→Ease), we proceeded by thematic coding followed by heuristic translation into 0–10 values, the steps of which were as follows:

1. **Marker extraction:** For each participant, we identified lexical and behavioral indices in our notes and transcriptions (e.g., words or expressions such as "not comfortable," "surprised," "appropriation," "work tool"), as well as frequency of tool use, weekly practical work hours with these tools, and intention to mention or not AI use in the creative process and publication of a work.
2. **Continuum construction:**
   - X axis: 0 = strong resistance (ethical posture or refusal of AI), 10 = active acceptance (frequent use, appropriation, willingness to integrate)
   - Y axis: 0 = great difficulty (does not grasp the tool, technical or aesthetic frustration), 10 = ease and confidence in use
3. **Numerical attribution:** These markers were qualitatively transformed into numerical values. Main conventions are as follows:
   - Low X (1–3): explicit references to non-disclosure or moral refusal

- Medium-high X (6–9): declarations of "AI as work tool," increase in hours of use, progressive appropriation
- Low Y: comments focused on aesthetic frustration (e.g., "ugly results")
- High Y: indications of rapid learning, "very rapid progression," or phrase elements like "mastery"

These choices reflect a constructed and explicated interpretation (not a raw measurement) useful for visualizing individual and heuristic trends. Thus, five distinct profiles appeared in the creative team, one for each person, showing that if none of them were used to GAI in music composition, each one developed different affinities at different levels with GAI in their compositional process.

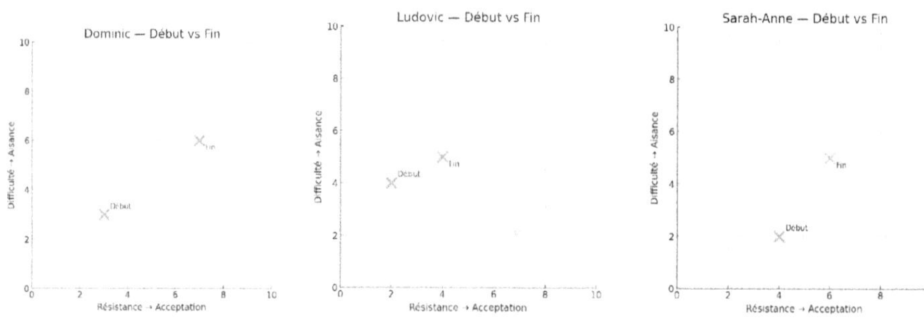

Fig. 1. Dominic's trajectory.

Fig. 2. Ludovic's trajectory.

Fig. 3. Sarah-Anne's trajectory.

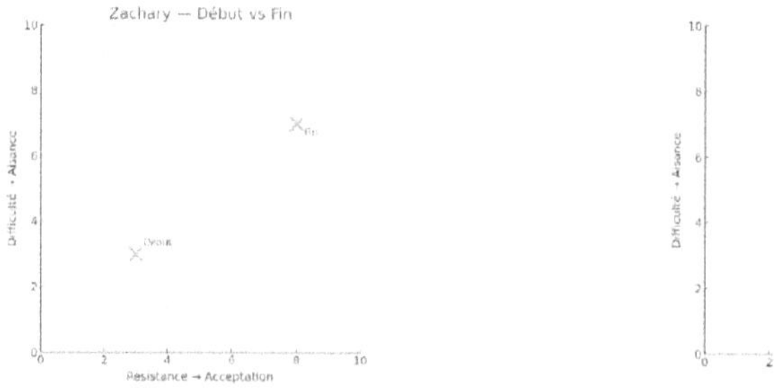

Fig. 4. Zachary's trajectory.

Fig. 5. Patrick's trajectory.

Profile 1: Dominic presents a start marked by curiosity mixed with pragmatic doubt (see Fig. 1). He experiments with several tools (Suno, AIVA), but his initial relationship is utopian/exploratory: he expects AI to understand his artistic intentions but quickly realizes it does not grasp the nuances of his discourse (through his prompts), especially linguistic ones. His evolution passes through partial disillusionment, followed by instrumental acceptance: AI becomes a tool for creating demos, not a creative partner. At project's end, AI is for him an efficiency accelerator, but he would not publish an AI-generated piece even if his group encourages more frequent GAI use for producing new original creations the group could reappropriate.

Profile 2: Ludovic presents the most ideological and ethical posture of the creator group (see Fig. 2). He distinguishes generative AI from assistance tools (e.g., Synth V), preferring controllable tools over AI that "erases" creative intention. Moreover, his quite marked awareness of AI use's ecological impact is a constant brake on broader and more frequent use. His resistance is thus principled, not technological, as evidenced by recurring phrases such as "Not comfortable publishing a 100% AI piece" or "Importance of human action in collaboration." At the end, he recognizes AI's value for inspiration or post-production but will not integrate it into his artistic identity.

Profile 3: Sarah-Anne initially has cautious but anxious openness, with a desire for efficiency rather than listening to the tool (see Fig. 3). She quickly encounters frustration and a feeling of creative impoverishment: "Ugly results without pleasure," "Not the impression of thinking as a creator but as a technician." Throughout the project, she develops critical lucidity: AI does not replace compositional skills, and creativity returns in human manipulation of generated product.

Profile 4: Zachary presents the most dramatic evolution (see Fig. 4): from defensive skepticism to reflexive appropriation, as evidenced by certain passages in ethnographic notes: "AI is not yet a mirror of what he wants." But he discovers a form of inspiration and mutual learning dynamic with Claude (Anthropic) where very rapid progression is seen in his work with AI, even if the desired level of technical detail forces him to take much more time working with AI.

Profile 5: Patrick is the most socio-affective of the group (see Fig. 5). His evolution is structured around the feeling of collaboration: "ChatGPT becomes your friend," or "AI stimulates ideas." He discovers that AI can become an extension of human brainstorming. At the end, he also highlights a kind of recognition of AI's emotional limits but presents strong adoption for its efficiency.

Methodological Limits and Recommendations. The main limitation we encountered concerns the conversion of our qualitative observations into numerical scores since such a process necessarily contains a share of interpretation. It is therefore important to understand these positions as heuristic markers rather than as psychometric measures in the strict sense. To strengthen this type of analysis, it would be desirable to triangulate data with standardized questionnaires (scales of technological acceptance, perceived efficiency). Repeating the same codification using a second coder would also be useful to evaluate intercoder reliability and better objectify observed trends. From a usage perspective, the produced graphs would finally gain from being mobilized as exploratory tools in writing as illustration of psychological trajectories. They should however be completed by interview excerpts, in the form of direct quotations, to reinforce descriptive validity and restore the richness of qualitative material.

4 Results

4.1 AI as Inspiration Tool Rather Than Creation Engine

The project's initial protocol postulated that AI could support musical ideation and facilitate certain creation stages (lyrics, melodies, arrangements). In the field, this is verified but only in a precise direction: AI triggers ideas, it does not develop them. In this sense, Patrick speaks of ChatGPT as a "nest of suggestions" in a brainstorming role rather than final writing. Sarah-Anne also reinforces this idea by mentioning that it is not the result that interests her but rather the ideas and inspirations emerging from GAI. For his part, Zachary adds that he considers AI as a *mirror*: it forces him to clarify his formulations and "obliges him to put in writing what he would like." Dominic completes this idea by discussing a phenomenon of "discovery" and "adaptation/introspective process" facing AI's limits. In light of these results, a direct contradiction with the project's hypothesis emerges. The initial objective supposed that AI could contribute to the *generation* of innovative musical content, but observations show that the most significant contributions do not relate to generation but to ideation. In terms of Arielli's typology [10], AI mainly functioned in *pattern recognition* mode (reacting to human ideas) but not in *pattern generation* mode (proposing true aesthetic directions).

4.2 Saving Time... If One Does Not Seek to "Create"

The project's objectives foresaw "optimizing time management in the studio," notably in mixing or mastering. This was indeed confirmed, at a certain level, by our interviews with participants. According to Dominic, AI is a work tool that can be good for facilitating the task for someone in a bind, it is an "efficiency and time gain." Patrick's experience reinforces this point since he judges that AI saved him enormous time since he was not alone and it offered him ideas on a silver platter. Normally, he would have set it aside after a few hours, but AI stimulated him by bringing him to work longer. During our interviews, it was also brought to our attention by certain participants that AI also saves time on less mastered points. Sarah-Anne mentions that AI could be useful for working on a style she masters less well or for bringing her beyond her own universe. On Patrick's side, AI would help him with mastering, a compositional aspect with which he is not familiar. Ludovic, for his part, would use AI for points where its use is indispensable for him to arrive at what he wishes.

However, as soon as one moves to music—notably with Suno or Udio—the time saved is canceled by correction time. Sarah-Anne mentions it was difficult to find herself in AI results since it generated up to 70–80 extracts for a single track. To find her way, she made a journal to note the different steps she was taking to keep track. Moreover, several participants mentioned that results were random or outside aesthetic intention. Sarah-Anne finds that AI's aesthetic choice is "quite bad" melodically, harmonically, but also at the level of chosen structure. Moreover, it sometimes happened that she asked AI to generate a piece in a precise style and the result unfortunately did not correspond to these desires. It was also mentioned by several participants that it was tedious work of fragmentation, pruning, and reassembly of proposals. Working with AI, Sarah-Anne did not have the impression of thinking as a creator but rather as a technician since the machine does everything in her place. Patrick and Ludovic rather have the impression of having been AI's artistic directors, the song does not belong to them since they only guided it so the result makes sense.

These observations therefore bring a paradoxical result: AI saves time *on peripheral work* but loses it as soon as one seeks to produce a serious musical work. The time taken trying to understand how it works and how to address it productively, the same time would have been taken for a composition without AI. This reality strongly nuances the project's hypothesis according to which AI could "improve efficiency" of musical production in the studio.

4.3   Musical Creation Generated by AI Is Still Basic

The project starts from the idea that these new tools could transform musical creation in the studio. However, participants detected certain limits that make AI a still basic tool. Ludovic, who considers himself as a "soundscape composer", notes that models "are based on standards" and prevent going out of these standards. He also mentions that GAI tools were not made by musicians for musicians.[1] For his part, Zachary underlines that AI does not understand intention nor emotional context of texts: it sometimes goes too far into images or not enough, there is no middle ground. This perception joins one of the project's passages: "control over the result remains very limited." In other words, AI generates *content* on demand but does not understand the *creation* process. What it generates is *sufficient for someone who does not know how to compose* but is frustrating for a creator who cannot let their creativity go as they wish. This all contradicts the hypothesis of a balanced AI/human collaboration: participants

---

[1] At the time this research was conducted (spring-summer 2025), the professional-oriented versions SUNO Studio and Moises AI Studio were not yet available.

do not see AI as a collaborator in Banerji's sense [7] but rather as a rough intern capable of proposing ideas but incapable of developing them.

4.4 Poor Sound Quality Reduces Threat to Creators

On generative platforms:

- imprecise artificial voices (diction problems in French)
- questionable rhythmic placements
- "ugly" or "without pleasure" sound textures

This limitation constitutes a major result: AI's so-called "threat" collapses as soon as one listens. Dominic and Ludovic refuse to publish a 100% AI piece. Sarah-Anne speaks of an "unsatisfactory" result even after correction. Patrick and Zachary mention that at the beginning they were reluctant to embark on this project because they did not perceive AI well; but at the end of the process, they finally feel less threatened by it. Ludovic also mentions this idea that creators are not in danger. Thus, contrary to the concern evoked in the description: "these platforms [...] seem the most *threatening*," observations rather show that AI is less aesthetically threatening than expected. It can impress *on first listening* but collapses *on critical re-listening*—where musicians' expertise lies.

5 Discussion: Towards an Ecology of Distributed Creation

One of the first findings from this project concerns the mutation of AI's role in the musical creation process: increasingly, the machine acts as an implicit co-author. GAI tools do not simply generate content: they influence decision-making, work temporality, and the group's affective dynamic. This hybridization generates a form of distributed creativity, where the original idea emerges less from an individual than from a network of human and non-human interactions. However, if the project primarily aimed to test AI's concrete contribution to musical creation, ethnographic data show that it:

- inspires the creator but does not create;
- accelerates creative work but only in what is peripheral;
- generates creative content but does not compose;
- impresses through its possibilities but does not convince.

Thus, far from replacing creators, AI seems to us to expose what makes their value. It acts as a mirror, and this mirror is multiple: 1) it reflects the person's degree of ease in platform manipulation, but 2) also reflects this person's ability to clearly explain their ideas, while highlighting 3) machine limits that take humans as a reference example through their designers. Biases are therefore diverse regarding AI use in musical creation, and these, when not ideological in nature, immediately refer creators to a discursive distancing of AI-assisted creation in an opposition between co-construction and collaboration. In this sense, if the definition of what constitutes an "author" and a "collaborator" can be seen in developing and writing ideas, collaborating with each other in making significant intellectual or artistic contributions, we can say that essentially composers and AI in our study are not working together on the same artistic effort. The key difference lies in the primary role: the composer conceptualizes and executes the core idea, whereas the GAI proposes variations of what has been asked by the composer in an attempt to "know" better what that person wants to create, without sharing easy accountability for the finished work.

5.1   Redefinition of Author and Subjectivity

In observed practices, the notion of author shifts toward a collective and augmented subjectivity, reflecting Ferrando's notion of "decentering the human in relation to the nonhuman" [8, p. 54]. AI does not replace human gesture: it multiplies it, by proposing variations, amplifying intuitions, or prompting unexpected bifurcations. This process evokes Arielli's *pattern generation* logic [10]: the machine learns aesthetic regularities and proposes new combinations that humans interpret and select. However, this opening is accompanied by tension: partial delegation of creative choice to the algorithm weakens perception of intention. Participants report a form of strangeness facing certain musical proposals generated by AI—"beautiful" sounds but whose genesis they cannot explain, for example. This feeling, which Lacasse [14] designates as "artistic AIpocalypse," translates both fascination and anxiety linked to the emergence of non-human creativity.

5.2   AI as Collaboration Catalyst

Another striking result from our observations is GAI's capacity to stimulate human collaboration. Generative tools become mediators in aesthetic discussion: team members debate, justify, criticize, or adjust proposals from the machine. In this sense, AI acts as a dialogic third party, a trigger for collective reflexivity. This phenomenon joins the analyses of Arsenault [11] and Wilsmore and Johnson [12] on musical co-production: the best collaborations do not rest on fusion of intentions but on negotiation of differences. The introduction of an algorithmic entity accentuates this dialogic logic: it imposes an additional mediation space, obliging humans to make explicit their aesthetic criteria and implicit values.

5.3   Temporality and Efficiency

On the practical level, AI contributes to production time optimization, notably at the post-production stage. Tools such as Ozone considerably accelerate certain technical tasks, allowing the team to concentrate more on the artistic and creative dimension of the approach. However, this efficiency raises a paradox:

as operations become automated, "saved" time is often reinvested in reflection, successive listening sessions, and discussion around aesthetics and authorship. In other words, AI does not replace human work; it shifts its intensity of involvement toward the critical and interpretive dimension of the process.

5.4 Performativity of Interaction

Finally, observations show that the human-AI relationship possesses a highly performative dimension. Each interaction with the algorithm produces a mirror effect: creators discover themselves through their reactions to the machine. This active reflexivity joins the conception of a relational aesthetics in Arielli's sense [10]: the work no longer resides in the finished object but in the network of interactions that gave it birth. Thus, the co-creation experience reveals less a delegation of creativity than an extension of the field of sensibility. The machine thus becomes the partner of a subjectivation experience: by negotiating with it, the human explores new forms of self and relationship to the sonic world. "Collaboration" therefore does not implicitly refer to the sounding of ideas but rather to an explicit formulation of discourse on these ideas whose sounding then imposes a negotiation of taste, intentions, and values.

6   Towards an Aesthetics of Distributed Intelligences

This project has allowed highlighting the complexity of relationships that today unite musicians, technologies, and studio knowledge. Far from announcing humanity's disappearance, generative AI invites reconfiguring musical creation as an ecology of distributed intelligence, where each actor—human, machine, interface—participates in meaning construction.

On the theoretical level, the research confirms the relevance of Arielli's proposed model: GAI acts simultaneously as an analysis tool, creation agent, and cognitive mirror. On the methodological level, it illustrates the fruitfulness of an ethnographic research-creation approach, capable of grasping aesthetic dynamics *in situ*. On the aesthetic level, it highlights the emergence of a new poetics of mediation, where artistic value is measured less by individual originality than by the relational quality of the process.

By situating the machine at the heart of the musical creation process, this approach opens a field of ethical and political reflection that is already being addressed by rights management organizations. In October 2025, ASCAP, BMI, and SOCAN announced alignment on AI registration policies, establishing that works created with AI assistance can be registered for copyright protection provided there is meaningful human creative contribution [15]. This landmark decision recognizes AI as a legitimate tool in the creative process while maintaining human authorship as central. However, critical questions remain: how to ensure that these collaborations remain respectful of rights, consents, and sensibilities of all creators involved? How to preserve artists' autonomy while welcoming algorithmic creativity? How to define and measure "meaningful human creative contribution" in practice? These questions, far from being resolved, outline the contours of a new era of musical production, marked by co-evolution of humanities, technologies, and legal frameworks.

Ultimately, AI is not the end of music but the emergence of a sonic posthumanism. A distributed, reflexive (post)humanism, conscious of the mediations that traverse it. By placing GAI at the heart of musical practice, this project does not seek to produce an aesthetics of substitution but an aesthetics of relationship: one where creativity becomes a common language between humans and machines.

Acknowledgments. This study was funded by the Observatoire international des impacts sociétaux de l'intelligence artificielle (OBVIA). We thank all the musicians and creators who participated in this research-creation project.

Disclosure of Interests. The authors have no competing interests to declare that are relevant to the content of this article.

References

1. Stévance, S., Lacasse, S.: Research-Creation in Music and the Arts: Towards an Interdiscipline. Routledge, London (2018)
2. Smith, G.W., Leymarie, F.F.: The machine as artist: an introduction. Arts **6**(2), 5 (2017)
3. Azzaria, G.: Intelligence artificielle et droit d'auteur: L'hypothèse d'un domaine public par défaut. Les cahiers de la propriété intellectuelle **30**(3), 925–946 (2018)
4. Gervais, D.J.: The machine as author. Vanderbilt Law School Facul. Publ. **105**, 2053–2106 (2020)
5. Manovich, L., Arielli, E.: Artificial Aesthetics: Generative AI, Art, and Visual Media (2021–2024)
6. White, C.C.: The AI Music Problem: Why Machine Learning Conflicts with Musical Creativity. Routledge, New York (2025)
7. Banerji, R.: De-instrumentalizing HCI: social psychology, rapport formation, and interactions with artificial social agents. In: Filimowicz, M., Tzankova, V. (eds.) New Directions in Third Wave Human-Computer Interaction, Volume 1: Technologies, pp. 43–66. Springer, Cham (2018)
8. Ferrando, F.: Philosophical Posthumanism. Bloomsbury, London (2019)
9. Braidotti, R.: Foreword. In: Ferrando, F.: Philosophical Posthumanism, pp. xi–xvi. Bloomsbury, London (2019)
10. Arielli, E.: Chapter 1. In: Manovich, L., Arielli, E.: Artificial Aesthetics: Generative AI, Art, and Visual Media (2021–2024)
11. Arsenault, S.: Écrire une chanson avec des jeunes: Analyse herméneutique et réflexive du processus de co-création. Presses de l'Université Laval, Quebec City (2023)
12. Wilsmore, R., Johnson, C.: Coproduction: Collaboration in Music Production. Routledge, London (2022)
13. Thompson, P.: Creativity and collaboration in the recording studio: an empirical study. PhD thesis, University of Liverpool (2015)

14. Lacasse, S.: L'IApocalypse artistique annoncée: La réponse de l'esthétique artificielle en musique. Presented at OBVIA Seminar L'IA générative et la création de contenus culturels et médiatiques (2024)
15. SOCAN, ASCAP, BMI: ASCAP, BMI and SOCAN Announce Alignment on AI Registration Policies (2025). https://www.socan.com/ascap-bmi-and-socan-announce-alignment-on-ai-registration-policies/. Accessed 15 Nov 2025

A Dataset for Automatic Vocal Mode Classification

Reemt Hinrichs[(✉)] [iD], Sonja Stephan, Alexander Lange [iD], and Jörn Ostermann

Institut für Informationsverarbeitung, Leibniz University Hannover,
Hannover, Germany
`hinrichs@tnt.uni-hannover.de`

Abstract. The Complete Vocal Technique (CVT) is a school of singing developed in the past decades by Cathrin Sadolin et al.. CVT groups the use of the voice into so called vocal modes, namely Neutral, Curbing, Overdrive and Edge. Knowledge of the desired vocal mode can be helpful for singing students. Automatic classification of vocal modes can thus be important for technology-assisted singing teaching. Previously, automatic classification of vocal modes has been attempted without major success, potentially due to a lack of data. Therefore, we recorded a novel vocal mode dataset consisting of sustained vowels recorded from four singers, three of which professional singers with more than five years of CVT-experience. The dataset covers the entire vocal range of the subjects, totaling 3,752 unique samples. By using four microphones, thereby offering a natural data augmentation, the dataset consists of more than 13,000 samples combined. An annotation was created using three CVT-experienced annotators, each providing an individual annotation. The merged annotation as well as the three individual annotations come with the published dataset. Additionally, we provide some baseline classification results. The best balanced accuracy across a 5-fold cross validation of 81.3 % was achieved with a ResNet18. The dataset can be downloaded under https://zenodo.org/records/14276415.

Keywords: complete vocal technique · vocal modes · artificial intelligence · convolutional neural network · singing voice dataset

1 Introduction

Singing voice analysis and more specifically vocal technique analysis is commonly employed by singing teachers as well as singing students, either as a first step in assisting a student to improve their voice (or performance) or to assess how to achieve a certain sound [15]. Due to limited time and budget, time spent with a singing teacher is usually a scarce resource. Additionally, singing teachers are attention constrained, i.e. likely to miss or forget certain details when giving their feedback to a student. These two points could be improved on through automatic singing voice analysis. Automatic singing voice analysis aims to algorithmically replicate some of the analyses performed by singing teachers, either

P. Machado et al. (Eds.): EvoMUSART 2026, LNCS 16523, pp. 113–128, 2026.
https://doi.org/10.1007/978-3-032-24350-8_8

for entertainment purposes (usually simple pitch estimation) or to assist singing students and teachers in their work. Thanks to the explosion of deep learning in the past decade, the field has seen some growth in the recent years. A good introduction is given by Humphrey et al. [10]. Terminology in singing pedagogy can be confusing [9,14]. A common example being support or belting, which can be difficult to grasp for early voice students. Different vocal pedagogy "schools" exist, especially in the context of contemporary commercial music, which provide different ways to think and speak about singing and come with different ideas on how to approach the development of ones singing voice. Examples are Estill Voice Training [7], Speech Level Singing [12] but also the Complete Vocal Technique (CVT) developed by Cathrin Sadolin [16]. CVT is a system that attempts to describe the entire range of sounds the human voice can healthily produce as well as providing rules and advice on how to actually achieve these sounds. Vocal modes, introduced by CVT and which will be described in Sect. 2.1, are a useful way of thinking about certain characteristics of the voice. These vocal modes ought to refine for instance the vocal category belting [14].

Aim and Contribution. While some advances have been made in automatic singing voice analysis, the field is still impeded by a lack of publically available datasets compared to, e.g., computer vision. A well known dataset for singing voice analysis is VocalSet [19] which consists of arpeggios sang in different styles by professional singers. However, the voice traits recorded for the VocalSet cannot (or hardly) be used to analyze the singing voice with respect to CVT terminology. For the example of CVT, which provides useful categories of certain qualities of the singing voice, no dedicated datasets have been published to date. Automatic classification of vocal modes has been previously attempted [3,18] but so far with unsatisfying results and without publishing a dedicated dataset. This considerably hinders the application of machine learning algorithms. Because the automatic classification of vocal modes can be a helpful tool for singers [18], in this work we present a novel dataset. The dataset is made up of samples recorded from four singers covering their entire vocal range for all four vocal modes defined by CVT, namely Neutral, Curbing, Overdrive and Edge. Besides aiding research on vocal mode classification, the dataset can be helpful to similar singing voice analysis tasks such as distinguishing belting from non-belting.

2   Method and Materials

2.1   The Complete Vocal Technique

The Complete Vocal Technique (CVT) is a school of voice pedagogy developed by Cathrin Sadoline [16] that aims to provide research-backed descriptions and instructions on how to create any sound of the human voice found in music healthily [6,13]. Some of its elements are found under different names in other voice pedagogies. Its three basic principles are support, "necessary twang", and avoiding protrusion of the jaw [16].

Vocal Modes CVT categorizes the human voice into four distinct classes - the vocal modes - namely Neutral, Curbing, Overdrive, and Edge. The vocal modes are determined by the properties or qualities "hold", twang, and "metal". These three qualities can in principle be quantified, e.g., one can say that there is more or less metal/hold/twang in a voice, albeit usually a rather coarse quantification is used. The term "hold" corresponds to the bodily feeling of holding back the sound, resulting in a restrained sounding voice, which can be heard, for example, in moaning. "Metal" refers to a sharper, brighter trait of certain sounds typically heard in shout-like sounds. "Twang" is a term commonly used in different schools of voice pedagogy and refers to a duck-like sound color [1]. Neutral is the vocal mode without any metal. The two other properties can be present without altering the vocal mode. Curbing is a reduced metallic mode, i.e., the voice contains some metal and a certain amount of hold. It does not use twang. Overdrive is a full metallic mode in contrast to the reduced metallic mode Curbing. Edge is a full metallic mode which, in contrast to Overdrive, adds twang to yield a very sharp, bright sound often found in, e.g., hard rock belting [4]. However, because of the so called necessary twang, which is a certain amount of generally required minimal twang to produce a sound, and CVT's idea of "density", which aims to describe more or less full sounding vocal modes (e.g. a reduced sounding/reduced-density Overdrive) [11], the distinctions between the vocal modes are gradual, similar to a color ray, which has regions of uncertainty between clearly distinct colors. Consequently, the distinction between the vocal modes becomes increasingly unclear as one deviates from the so called "center" of the mode. Roughly speaking, the "center" of a mode is the archetypal way of singing/producing a vocal mode. Except for Overdrive, which for males is limited from above by C5 and for females by D5, CVT does not impose limits (regarding healthy production) on the vocal range for the vocal modes. Specifically, there is no lower bound for the vocal range for any of the vocal modes. CVT imposes vowel restrictions for the different vocal modes, that is, except for Neutral, in general only a certain subset of all possible vowels can be used for a vocal mode. These subsets are summarized in Table 1.

Table 1. List of vowels for each vocal mode according to the Complete Vocal Technique (CVT). The chosen list of vowels followed the official CVT book [16]. Note that the most recent CVT book (and the corresponding smartphone app) slightly altered the Curbing vowels.

Curbing	Edge	Neutral	Overdrive
'I' (as in 'sit'), 'O' (as in 'woman'), 'UH' (as in 'hungry')	'I' (as in 'sit'), 'EH' (as in 'bed'), 'A' (as in 'cat'), 'OE' (as in 'herb')	'EE' (as in 'see' [i]), 'OO' (as in 'food' [u]), 'AH' (as in 'father'), 'I' (as in 'sit'), 'EH' (as in 'bed' ϵ), 'A' (as in 'cat' [ae]), 'O' (as in 'woman'), 'OH' (as in 'so' [o])	'EH' (as in 'bed'), 'OH' (as in 'so')

2.2   Dataset

The only research known to the authors attempting to automatically classify the four vocal modes of CVT was performed by the Complete Vocal Institute (CVI) [2,18]. However, no dataset was published and as such no dedicated dataset for vocal mode classification exists. Therefore a novel data set was recorded in this work. For this, four vocalists, labeled s1, s2, s3, and s4, were recorded at the Institut für Informationsverarbeitung, two males and two females. The subjects s1, s3 and s4 have at least 15 years of singing experience each, s2 has about five years of experience. The age range covered 37 to about 48 years. All subjects gave their informed consent to publication of the recorded data. The initial plan was to record six subjects. However, it proved to be impossible to find two additional subjects. Except for s2, who is the main author of this manuscript, all were professional singers with university level training and experience as vocal coaches or musical singers. s2 was a hobby singer experienced in CVT through his vocal lessons. One Behringer B-5 condenser microphone, one Stage Line ECM-40 condenser microphone, one IPhone, and one MotorolaOne smartphone were used for recording. The first two were used to record at high quality with slightly different frequency responses, the latter two to capture the audio in a more realistic scenario. The smartphones' purpose was to provide a natural augmentation of the recordings, which can be crucial to achieve generalization for machine learning algorithms. The basic setup was meant to approximate the approach described in [2]. The condenser microphone recordings were sampled at 48 kHz, but were later downsampled to 44.1 kHz to match the IPhones sampling rate. The MotorolaOne recordings, recorded natively at 16 kHz, were also resampled to 44.1 kHz. Sampling bit depth was 16 bits in all cases. The condenser microphones were recorded using a Focusrite 18i20 audiointerface and Cubase 9.5. Each subject sang major arpeggios across their entire vocal range, starting at the highest note they felt comfortable with in the respective vocal mode. The major arpeggios were gradually reduced by one half step such that the entire vocal range of the singers was captured without gaps. This way, five such iterations per vowel were performed, each starting a semitone lower. CVT imposes restrictions on the allowed vowels for each vocal mode as shown in Table 1. Only in Neutral all vowels can be used. The number of allowed vowels is considerably reduced in the other modes. To maintain at least an approximately similar number of samples per mode, each vowel was repeated twice for Curbing and Edge, and three times for Overdrive.   To steer the pronunciation of the vowels, each subject was provided with a list of vowels per vocal mode together with a (German) key word containing the respective vowel. Before singing the respective arpeggios, the subjects were presented corresponding major chords played by the experiment conductor, s2. During annotation, a few recorded samples were found to have improper vocal oscillation. These samples were removed/excluded from the main dataset, but are provided in the online repository to assist possible future research on, e.g., singing mistakes. After data polishing, the final

Table 2. Comparison of the nominal and the corresponding annotated vocal mode. For Curbing and Overdrive, a larger number of samples was annotated as Neutral.

Nominal \ Annotated	Neutral	Curbing	Overdrive	Edge
Neutral	1580	20	4	1
Curbing	101	471	20	29
Overdrive	104	22	462	71
Edge	13	31	71	723

annotated dataset is made up of 1605 unique nominal productions[1] for Neutral, 642 unique nominal productions for Curbing, 664 unique nominal productions for Overdrive and 841 unique nominal productions for Edge. As such, the entire dataset is made up of 3752 unique productions. As mentioned, most of these productions were recorded using in total four microphones, resulting in a total of 13335 samples. The nominal vocal modes are the vocal modes the subjects were asked to produce. The nominal mode is not necessarily identical with the annotated mode. For example, the production of a vocal mode can fail, and the resulting recording/sound might be perceived by a listener/annotator as belonging to a different mode than intended. For example, near the lower end of a singer's range, maintaining the desired vocal mode can be challenging (with the exception of Neutral). While previous works such as [18] do not make this distinction, it is crucial for developing classification algorithms, which are supposed to map audio samples to the appropriate vocal mode. Such a classification has to be based on the actual sound of the samples (as perceived by a listener). Each sample of the dataset consists of a single note, in almost all cases without onset and offset. The recorded audio data was manually cut for the Behringer recordings, and the well-known locally normalized cross-correlation method [20] was used to automatically extract corresponding samples for the three other data sources/microphones. While this automatic extraction was not manually checked for correctness in its entirety, performed isolated comparisons between the automatically and manually extracted samples showed a 100 % agreement. Furthermore, a vast number of automatically extracted samples were listened to without finding a single obviously incorrectly extracted sample. The dataset, including the annotation, is available at https://zenodo.org/records/14276415, published under a Creative Commons Attribution Non-Commercial 4.0 International license.

Annotation. Annotation was done by s1 (the vocal teacher of s2 and second author of this manuscript); s3 and an additionally invited annotator, because s4 declined the invitation to annotate the dataset. The invited third annotator

---

[1] In this work, the term "production" is used to refer to the original, unique vowels sang/produced by the subjects. Because several microphones were used to capture each production, the individual productions correspond to more than one sample.

was a voice student familiar with CVT through university course work as well as additional choir work. Peak-normalized samples were passed to the annotators in randomized, consecutively numbered sequences, with each annotator receiving a different order. Thus the annotators operated on files labeled '0001.wav', '0002.wav' etc. Annotation was performed without the presence of the main author during the private time of the respective annotators. Random repetitions of previous samples, about 50 per annotator, were used to assess intra-rater reliability. The three individual annotations are included in the dataset to be hopefully combined with additional annotations in the future. For usage in classification tasks, however, a unique, merged/final annotation had to be constructed from the individual annotations. Because each annotator annotated/rated about 50 samples twice, there are about 50 samples per individual annotation where the respective annotator gave two ratings, which were not necessarily identical. A single sample occurred for which two annotators each gave two ratings. This sample, nominal Neutral, was rated by all annotators and all ratings as Neutral. Considering this, the three individual annotations received from the individual annotators were combined/merged in the following way: For each sample, all ratings of the annotators were aggregated, with the consequence, that up to four[2] ratings had to be considered. The fourth rating was due to random repetitions as mentioned in the previous paragraph. If the annotators reached a majority consensus for this aggregated annotation, that is, one of the four vocal modes was annotated by the majority (4:0, 3:1 or 2:1) of the aggregated annotations, this vocal mode became the annotated, ground truth vocal mode of the sample. For 367 samples, no majority was found this way. For these 367 samples, the nominal mode was used as a tie-breaker, but only if the nominal mode agreed with a majority (2:2 or 1:1:1) of the annotators. This allowed to resolve all but 29 of these 367 samples. These final 29 samples were deemed undecidable and marked accordingly in the final, merged annotation. As such, these 29 samples were left out in the classification tasks reported in this work. This merged annotation, consisting of 3723 samples, alongside the three individual annotations, is included in the published dataset.

Annotation Metrics. The raw agreement, i. e., without considering class imbalances, between the annotated and the nominal modes was 86.9 %. The balanced agreement between the annotated and the nominal modes, which considers class imbalances, was 82.7 %. The latter first computes the agreement separately for each class and then averages across the individual agreements. These values are in agreement, both quantitatively and qualitatively, with previous results [3, 18] - [18] reports a balanced agreement of 80.7 % - which showed that even professionals can disagree to some extent regarding the vocal mode of a vocal segment. This issue will be addressed in the discussion. A correspondence table of the nominal and annotated vocal modes is given in Table 2. Intra-rater agreements were 0.68, 0.74 and 0.77. Inter-annotator agreements were 0.61, 0.63 and 0.65. The achieved Fleiss' kappa score was 0.45. Inter-annotator agreements as well as

---

[2] Except for the mentioned sample with five ratings.

the Fleiss' kappa score were evaluated based on samples with exactly one rating by each involved annotator.

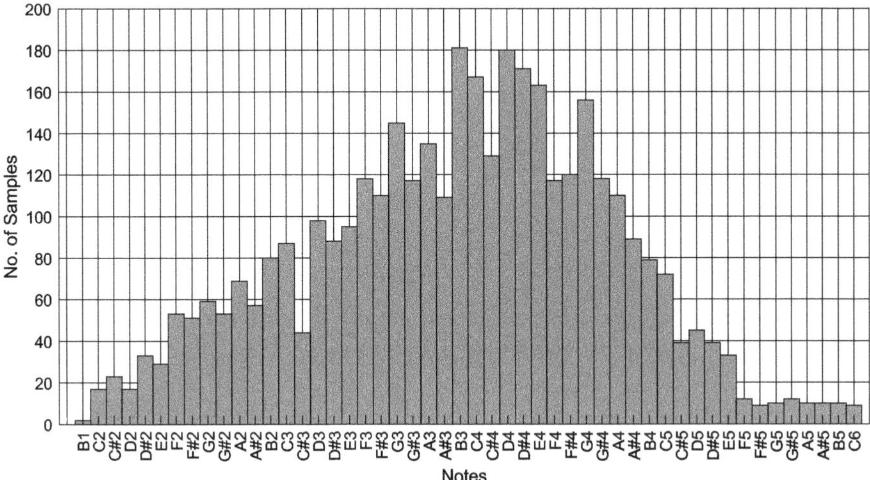

Fig. 1. Number of samples per note for the entire dataset. The highest note, a C6, was achieved in nominal Neutral. The lowest note, a B1, was achieved also in nominal Neutral.

Annotation Subsets. A substantial difference in accuracy was observed, when training the classifiers using the previously described annotation as ground truth, compared to using the nominal modes as a natural, alternative ground truth. To get additional insight into the impact of the annotation on the resulting classification accuracy, the following two subsets of the annotation were considered: A strong consensus subset, that only consisted of those samples rated identically by all annotators, i.e., all annotators assigned the same mode. Similarly, a weak consensus subset was constructed, where exactly one differing rating was allowed, that is, out of N ratings, at least N-1 were identical. The number of unique samples was 1843 in the strong consensus subset and 3355 in the weak consensus subset (multiplied by the number of recording devices). These subsets can be motivated by the previously mentioned idea of the center of a vocal mode as well as the observed disagreement of the annotators for certain samples. If, for example, a subject produced a mode, but partially deviated from its center, the distinction to another vocal mode can become less clear for this production. Especially in those cases, it is conceivable that the annotators gave contradicting ratings despite the acoustic features being quite similar. In such a case, classification algorithms may be confused by contradicting annotations; that is, annotations which give different class assignments despite identical features. If this was the case, a substantial improvement in classification accuracy is to be

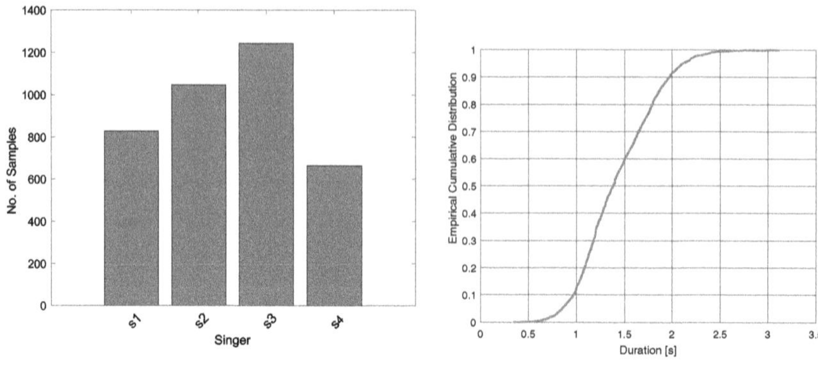

(a) Number of samples per subject.   (b) Distribution of the sample durations.

Fig. 2. Number of samples per subject and empirical cumulative distribution of the sample durations in seconds of all files of the dataset.

expected, at least for the strong consensus subset, seeing that one disagreeing rating can be seen as evidence for a deviation from a mode's center.

Dataset Statistics. Figure 1 depicts the number of samples per note in the dataset. Above E5 only a minor number of samples were recorded, largely due to limitations in vocal range of the subjects (partially also due to upper limits for the vocal modes). The subject s3 had a very low voice, achieving an F1 as his lowest note, the lowest among all participants. This subject's highest note was an E5. Similarly, the lowest and highest notes for subject s1 were a D#3 and C6, for subject s2 an E2 and a D#5, and for subject s4 an E3 and an E5. The number of samples per subject as well as the distribution of the sample duration in seconds is given in Fig. 2. The subject s3 had the largest vocal range, resulting in the largest number of samples. About 80 % of all recordings have a length between 1 and 2 s.

2.3   Baseline Classifiers

To provide baseline classification results, several classification algorithms were trained and evaluated using the recorded dataset, namely xgboost (XGB), a support vector machine (SVM), a k-nearest neighbor classifier (KNN), a random forest classifier (RF) and two convolutional neural networks (CNNs), namely ResNet18 and ResNet34 [8]. Both finetuning the ResNets initialized with pretrained ImageNet weights and training newly, randomly initialized ResNets were tested. Non-CNN classifiers were implemented using the python library scikit-learn and default parameters, except for the KNN, where the number of considered neighbors was varied, as well as the random forest, which used a maximum tree depth of 30. XGB was implemented using the code published by [5]. ResNets were implemented with pytorch 2.3.0 using default parameters if not stated otherwise.

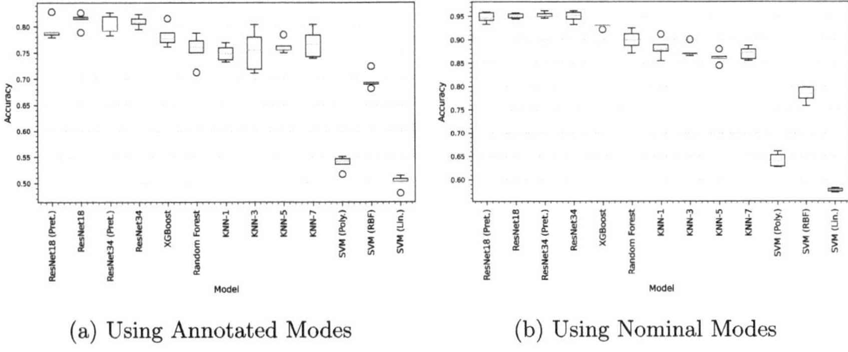

(a) Using Annotated Modes   (b) Using Nominal Modes

Fig. 3. Balanced accuracies on the test set across the 5-fold cross validation of all investigated classifiers using (a) the annotated vocal mode and (b) the nominal vocal mode as ground truth.

Data Preprocessing, Training and Evaluation. In all cases, waveforms of the samples were loudness normalized and subsequently energy normalized[3]. In general, Overdrive and Edge exhibit greater signal energy than Neutral and Curbing. However, in a practical situation, the distance to the microphone can vary, and consequently the energy of the recorded signal. This energy difference should therefore not be used, implicitly or explicitly, as a feature for classification. Additional energy normalization was motivated by observed slight differences in energy between the vocal modes even after loudness normalization. Initial learning rate was $5 \cdot 10^{-5}$ and the learning rate was reduced by a factor of 0.8 every three epochs if the training loss had not improved. Stochastic gradient descent served as optimization algorithm. The batch size was set to 128. Due to the recording method, each vowel production corresponds to several audio samples (recorded with different microphones. Therefore, splitting of the dataset into train and test subsets was performed such that no vowel production occurred in the training set that also occurred, recorded with a different microphone, in the test set. Splitting naively would distort the results, as the classifier would see augmented versions of training samples in the test set, i.e., the same production but different microphone. For all classifiers, a 90 %/10 % train/test split was used; that is, 90 % of the dataset was used to train the classifier, whereas the other 10 % was used solely for evaluating the trained model. The performance of the classifiers was measured using balanced accuracy across a 5-fold cross validation.

3   Results

Boxplots of the test balanced accuracies for all investigated classifiers across the 5-fold cross validation are depicted in Fig. 3. Figure 3a depicts results when

---

[3] Only for training the baseline classifiers. The samples in the dataset were stored as is.

Table 3. Balanced accuracies for all classifiers across the 5-fold cross validation on the test set and for the phone subset of the test set. Shown are the results for all four conditions considered: using the full annotation ("Annotated"), using the nominal modes as ground truth, using only the strong consensus subset and using the weak consensus subset.

Classifier	Annotated		Nominal		Strong Consensus		Weak Consensus	
	All [%]	Phones [%]	All [%]	Phones [%]	All [%]	Phones [%]	All [%]	Phones [%]
ResNet18 (Pret.)	79.5	78.9	94.7	93.7	90.6	89.2	81.9	81.2
ResNet18	**81.3**	**80.9**	95.2	94.2	89.4	86.5	**82.4**	**81.4**
ResNet34 (Pret.)	80.3	79.9	**95.3**	**94.4**	**90.9**	**89.8**	81.5	80.6
ResNet34	80.9	80.5	94.9	93.6	87.1	85.1	79.6	78.9
XGBoost	78.2	77.0	92.8	91.5	84.4	82.6	78.7	77.9
Random Forest	75.7	74.1	89.9	88.7	76.7	73.2	75.3	74.4
KNN-1	74.8	74.6	88.2	88.0	80.0	78.9	74.6	74.4
KNN-3	75.2	74.6	87.4	86.2	76.2	77.1	74.3	72.9
KNN-5	76.1	75.8	86.2	85.2	79.2	78.8	75.7	75.1
KNN-7	76.7	76.5	87.0	86.1	77.4	74.4	75.9	75.5
SVM (Poly.)	53.7	53.0	64.1	63.0	47.6	47.2	52.5	51.1
SVM (RBF)	69.6	67.9	78.2	75.9	80.9	77.4	70.5	68.2
SVM (Lin.)	50.2	48.8	57.7	56.4	47.0	45.9	48.2	47.3

Table 4. Aggregated test set confusion matrices across the 5-fold cross validation of the best performing models. Values are reported rounded to two decimal places.

(a) Full Annotation – ResNet18

	Neutral	Curbing	Overdrive	Edge
Neutral	2555	117	82	18
Curbing	159	908	77	88
Overdrive	92	55	803	131
Edge	24	68	146	1345

(b) Full Annotation – ResNet18

	Neutral	Curbing	Overdrive	Edge
Neutral	0.92	0.04	0.03	0.01
Curbing	0.13	0.74	0.06	0.07
Overdrive	0.09	0.05	0.74	0.12
Edge	0.02	0.04	0.09	0.85

(c) Nominal Modes – Pret. ResNet34

	Neutral	Curbing	Overdrive	Edge
Neutral	2514	35	32	1
Curbing	49	1191	20	39
Overdrive	12	13	1184	24
Edge	1	17	39	1470

(d) Nominal Modes – Pret. ResNet34

	Neutral	Curbing	Overdrive	Edge
Neutral	0.97	0.014	0.01	0.00
Curbing	0.04	0.92	0.02	0.03
Overdrive	0.01	0.01	0.96	0.02
Edge	0.00	0.03	0.01	0.96

using the annotation and Fig. 3b shows results when using the nominal modes as ground truth. Note the difference in y-axes' ranges. Corresponding test balanced accuracies, computed from the aggregated confusion matrices of the 5-fold cross validation, are given in Table 3. In the table, results for all four considered annotations/subsets - full annotation, nominal ground truth, strong consensus and weak consensus - are given. For all these cases, distinct accuracies are reported for the entire respective subset as well as the corresponding phones-only subsets. The ResNets outperform all other classifiers in every case with XGBoost being the best non-neural classifier. A significant difference in accuracy is observed between the classification using the annotation (Best accuracy: 81.3 %) and the classification based on the nominal vocal modes (Best accuracy: 95.3 %). Aggregated confusion matrices of the test set performance across the 5-fold cross validation are given for the ResNet18, using the full annotation, and the pretrained ResNet34, using the nominal modes as ground truth, in Table 4. To assess the performance across pitch, balanced accuracy of the ResNets (the respective best for the respective annotation) was evaluated for each half octave starting at C2-F2 and ending with approximately[4] F5-A#5. When using the annotation, balanced accuracy remained above 70 % between D2 and D5. It substantially reduced at the lowest end of the pitch range around C2. When using the nominal modes as ground truth, performance varied only slightly across pitch still achieving 90 % balanced accuracy at the lowest pitch range of C2-F2 and above 90 % for almost all half octaves. Lowest balanced accuracy of slightly above 70 % was observed for the half octave C5-F5.

4   Discussion

This work presented a newly collected dataset for automatic vocal mode classification together with several baseline classification results.

Dataset. While the collected dataset should be a reasonable dataset for the general task of automatic vocal mode classification due to the variety in notes and recording devices, additional singers (including amateurs), realistic musical content and singing techniques should be recorded to further address the problem of vocal mode classification. Additional singers help algorithms to generalize better to unseen ones. Realistic musical content, unlike major arpeggios as for this dataset, would be desirable. This means that the singers actually sing parts of real musical pieces. That way, a more nuanced coverage of the singers' voice and the way the vocal modes can be varied could be captured. However, actually performing such recording sessions is challenging, as singers and musical pieces have to match while maintaining balance between the vocal modes. Regarding singing techniques, the voice can be altered in its sound in many ways. Adding more or less air, making it brighter or darker sounding, and using more or less vibrato can drastically change the sound of a voice without changing the vocal mode. Such variations are not captured by the dataset but should be so in future work. Currently, a major obstacle in fully solving automatic vocal mode classification satisfactorily is obtaining a high quality ground truth. Some details of the annotation of this dataset are later discussed. As the Complete Vocal Institute itself showed [3], on average, even very experienced CVT students/teachers do

---

[4] The edges varied slightly depending on the cross validation split.

not exceed an accuracy of 80 % in acoustic vocal mode classification. However, [3] remains unclear about how their ground truth was obtained. By the description, it appears as if the nominal vocal mode served as ground truth. However, in general the nominal vocal mode can be different from the actual vocal mode, as, e.g., no proper hold was established for Curbing potentially resulting in Overdrive-like samples. A jury of highly experienced CVT-informed singers is required, out of which those with the highest test-retest ratings should be used to determine a definite ground truth similar to the procedure of [17]. However, considerable funding is required to achieve this.

Classifier Performance. While the achieved balanced accuracies of slightly above 80 % for the annotation, or up to about 95 % when using the nominal modes, is a good to very good performance, one cannot expect these accuracies to generalize to arbitrary, unseen samples due to the mentioned limitations of the dataset in terms of vocal traits. When using the full annotation, the observed decline in balanced accuracy for the lowest notes of the dataset, observed for all classifiers investigated, can be explained by the following reasons: The vocal modes according to CVT can be sung in a wide note range, with certain limits depending on vocal mode. Nonetheless, certain ranges are more common than others, resulting in the subjects being more accustomed to producing the vocal modes in certain ranges. Thus, not many, e.g., Curbing samples were recorded for the lowest note range. Furthermore is it uncommon to hear all vocal modes in very low vocal ranges. Because the sound properties typically change somewhat with changing pitch, it might be difficult to distinguish the vocal modes for very deep vocals without special training or at all. Further research is required to investigate this point.

Performance for the Different Subsets. The substantial decrease in balanced accuracy when using the annotation compared to using the nominal vocal modes might be evidence that the singers are better "judges" than the annotators/listeners. The singers had the advantage of being able to "carry" the vocal modes from the high range down to the lower range, as they were using major arpeggios in one go across their entire vocal range. This hypothesis is supported by the observed improvement of the classification accuracy, when moving from the full annotation to the strong consensus subset, where the classifiers achieved only slightly worse balanced accuracies, with the best classifier achieving a balanced accuracy of 90.9 % compared to 94.9 % for the nominal vocal mode problem and 80.8 % when using the full annotation. Also in line with this, is the considerably decreased classifier performance on the weak consensus subset. As the additional samples with slight disagreement between the annotators could introduce conflicting annotations - highly similar acoustic features get different ratings - it is no surprise to see a decrease in classifier performance. However, the comparison is somewhat impeded by the reduced number of samples for the strong consensus subset. The hypothesis, based on the observed classifier accuracies, then is, that the acoustic features distinguishing the vocal modes are consistently present in the recorded samples across most of the vocal ranges of the singers. This is suggested by the high balanced accuracy for the nominal vocal modes. However, these features are in general potentially either difficult to hear or the modes are rarely heard for certain ranges, leading to confusion by the annotators. If this is the case, then only additional annotators can ameliorate the situation. As little to no decrease in balanced accuracy was observed for the respective phone subsets, the classifiers should be somewhat robust to the used recording device and slight alterations of the room acoustics. This is useful for future smart-device applications. Prior tests, which used simpler convolutional neural networks and the nominal modes as ground truth, showed a considerable drop in balanced accuracy - more than 25 % for the MotorolaOne samples - for the phone subset, if the models were trained solely on the data recorded with the Behringer microphone.

Comparison to Related Work. The two published previous attempts at automatic vocal mode classification showed subpar classification performance. For [2], the classifier model - a two step decision tree based on vocal harmonics - was too simple. Additionally, the amount of data recorded was limited, with a single scale per vowel and four singers in total. For [18], which achieved 68.6 % balanced test accuracy for vocal mode classification using several features and the XGBoost algorithm, the amount of data, a crucial point in virtually all machine learning applications, was likely insufficient to achieve high accuracy. Each singer supplied only three different pitches. Additionally, recording was done by the subjects themselves, likely increasing classification complexity due to different recording conditions. The best comparison to [18] is our result when using the nominal modes as ground truth. For this, we achieved a balanced accuracy of 95.3 % using a pretrained ResNet34, aggregated across the 5-fold cross validation identically to [18]. This represents a gain of more than 25 % in balanced accuracy. Even using the exact same classifier that was used in [18], namely XGBoost, the gain in balanced accuracy is larger than 20 %, as we achieved a balanced accuracy of 92.8 % with XGBoost. However, due to the different datasets and slightly different ways the ground truth was obtained, a direct comparison is at least slightly unfair. It would be interesting to assess our models' performances on the data recorded for [18].

4.1   Quality of the Annotation

A Fleiss' kappa score of 0.45 indicates a substantial amount of disagreement between the annotators. Part of this disagreement stems from vocal modes being sung in lower registers. Here, both annotators and singers are possibly less experienced, as Overdrive and Edge, for instance, are most prominently heard in high notes, usually as part of a climax in a musical piece. This claim is substantiated by assessing the Fleiss' kappa score for a subset of the dataset, namely those resulting from ignoring the lowest K notes - and all corresponding samples - measured in semitones, of each singer (irrespective of the vocal mode), where K ranges from 0 semitones to 24 semitones. By increasingly ignoring lower notes, the Fleiss' kappa score steadily rises to around 0.54, about 0.09 higher than for the entire annotation. A tiny part of the Fleiss' kappa score could additionally be explained by human oversight in writing down the annotation. Excel sheets with four columns, one per vocal mode, were used, where an 'x' was used to mark the respective annotation. It is possible that in isolated instances, by accident, the wrong column was marked. Another reason for disagreements might be fatigue, as annotating large numbers of samples can be exhausting, such that later samples may have been rated with less scrutiny. Because the annotators created their annotation at home, the amount of samples rated in one session was not controlled. In the future, splitting annotation sessions into calibration - annotation - rest cycles could presumably improve annotation consistency. Calibration and rest cycles should allow to maximize intra-rater agreement and as such annotation consistency, i.e., giving the same rating for identical or virtually identical features. Unfortunately, neither [18] nor [3] reported a Fleiss' kappa score, making it difficult to draw comparisons to these publications. It can only be assumed to be similar based on their reported agreement between the nominal and the annotated modes, which are only slightly lower than the one reported in this work. Especially for lower pitches, additional annotators are of great interest. It is hoped that others, in particular the Complete Vocal Institute, are making attempts to contribute to our annotation, such that the quality of the annotation can be improved through a larger number of additional annotators. Crowd sourcing, both for recording samples and annotation, could be feasible, however, it was attempted to find additional singers through the CVT network to no avail.

4.2   Limitations

Due to a lack of interested professional singers, the main author also participated as a vocalist. As the recordings were supposed to be performed at the main authors' institute, only local professionals could realistically be invited.

While the main author did participate in the study by providing voice recordings in the same way the other subjects did, with s1 guiding this recording session, the main author did not influence the annotation itself. The main author did not provide an annotation. The individual annotations were created entirely during the private time of the three annotators without the presence of the main author. All annotators were explicitly encouraged to judge the anonymized samples according to their own personal judgment.

Therefore it is not assumed that the participation of the main author in the recording session impacted the annotation of the dataset.

5   Conclusion

This work presents a novel dataset for automatic vocal mode classification as well as baseline classification results. Four subjects (two male, two females) three of whom professional singers with extensive CVT-experience, were recorded resulting in 3,752 unique samples, each covering a single sustained note, covering the entire vocal range of the singers. These samples are made up of 1,605 nominal

Neutral, 642 nominal Curbing, 664 nominal Overdrive and 841 nominal Edge examples. In total, four microphones, including two smartphones, were used to record the subjects, resulting in more than 13,000 samples in total. Three annotators provided a ground truth annotation. The best balanced test set accuracy of 81.3 % was achieved by a ResNet18. The best balanced test set accuracy rose to 95.3 %, achieved by the ResNet34, if the nominal vocal modes were used as ground truth.

Acknowledgments. The authors would like to thank all invited subjects for contributing to our dataset. Furthermore we would like to thank Samantha Nicole Carbajal Pérez for the initial data segmentation of the Behringer recordings. The authors would also like to thank all anonymous reviewers for their helpful comments.

Disclosure of Interests. The authors declare no conflict of interest. This research did not receive any specific grant from funding agencies in the public, commercial, or not-for-profit sectors. Due to a lack of interested professional singers, the main author also participated as a vocalist.

References

1. Aaen, M., McGlashan, J., Christoph, N., Sadolin, C.: Deconstructing timbre into 5 physiological parameters: vocal mode, amount of metal, degree of density, size of larynx, and sound coloring. J. Voice (2021). https://doi.org/10.1016/j.jvoice.2021.11.013
2. Brixen, E., Sadolin, C., Kjelin, H.: Acoustical characteristics of vocal modes in singing. In: Proceedings of the 134th Audio Engineering Society Convention 2013 (2013)
3. Brixen, E., Sadolin, C., Kjelin, H.: The importance of onset features in listeners' perception of vocal modes in singing. In: Proceedings of the 137th Audio Engineering Society Convention 2014 (2014)
4. Brixen, E.B., Sadolin, C., Kjelin, H.: On acoustic detection of vocal modes. In: Proceedings of the 132nd Audio Engineering Society Convention 2012. Audio Engineering Society (2012)
5. Chen, T., Guestrin, C.: XGBoost: A scalable tree boosting system. In: Proceedings of the 22nd ACM SIGKDD International Conference on Knowledge Discovery and Data Mining, pp. 785–794. KDD '16, ACM, New York, NY, USA (2016). https://doi.org/10.1145/2939672.2939785
6. Complete Vocal Institute: Complete Vocal Technique. https://completevocalinstitute.com/complete-vocal-technique/. Accessed: 2026-01-23
7. Fantini, M., Fussi, F., Crosetti, E., Succo, G.: Estill voice training and voice quality control in contemporary commercial singing: an exploratory study. Logoped. Phoniatr. Vocol. **42**(4), 146–152 (2017)
8. He, K., Zhang, X., Ren, S., Sun, J.: Deep residual learning for image recognition. In: 2016 IEEE Conference on Computer Vision and Pattern Recognition (CVPR), pp. 770–778 (2016)
9. Henrich Bernardoni, N., et al.: Towards a common terminology to describe voice quality in western lyrical singing: contribution of a multidisciplinary research group. J. Interdisc. Music Stud. **2**(1&2), 71–93 (2008). https://hal.science/hal-00297248

10. Humphrey, E.J., et al.: An introduction to signal processing for singing-voice analysis: high notes in the effort to automate the understanding of vocals in music. IEEE Signal Process. Mag. **36**(1), 82–94 (2019). https://doi.org/10.1109/MSP.2018.2875133
11. Leppävuori, M., et al.: Characterizing vocal tract dimensions in the vocal modes using magnetic resonance imaging. J. Voice **35**(5), 804-e27 (2021)
12. McClellan, J.W.: A comparative analysis of speech level singing and traditional vocal training in the united states. Electr. Theses Dissertations (2011). https://digitalcommons.memphis.edu/etd/376
13. McGlashan, J., Aaen, M., White, A., Sadolin, C.: A mixed-method feasibility study of the use of the complete vocal technique (CVT), a pedagogic method to improve the voice and vocal function in singers and actors, in the treatment of patients with muscle tension dysphonia: a study protocol. Pilot Feasibility Stud. **9**(1), 88 (2023)
14. McGlashan, J., Thuesen, M.A., Sadolin, C.: Overdrive and edge as refiners of belting?: An empirical study qualifying and categorizing belting based on audio perception, laryngostroboscopic imaging, acoustics, LTAS, and egg. J. Voice **31**(3), 385.e11-385.e22 (2017). https://doi.org/10.1016/j.jvoice.2016.09.006
15. Miller, R.: The singing teacher in the age of voice science. in: on the art of singing. Oxford University Press (1996). https://doi.org/10.1093/acprof:osobl/9780195098259.003.0070
16. Sadolin, C.: Complete Vocal Technique. Bosworth Music (2013)
17. Saldías, M., Castro, C., Espinoza Catalán, V., Stoney, J., Quezada, C., Laukkanen, A.M.: Spectral features related to the auditory perception of twang-like voices. Logopedics, phoniatrics, vocology, pp. 1–18 (2024). https://doi.org/10.1080/14015439.2024.2345373
18. Sol, J., Aaen, M., Sadolin, C., Ten Bosch, L.: Towards automated vocal mode classification in healthy singing voice—an Xgboost decision tree-based machine learning classifier. J. Voice (2023)
19. Wilkins, J., Seetharaman, P., Wahl, A., Pardo, B.: Vocalset: A singing voice dataset. In: Proceedings of the International Society for Music Information Retrieval (ISMIR) Conference, pp. 468–474 (2018)
20. Yoo, J.C., Han, T.H.: Fast normalized cross-correlation. Circ. Syst. Sig. Process. **28**(6), 819–843 (2009). https://doi.org/10.1007/s00034-009-9130-7

A Novel Diffusion Model Based Approach for Sleep Music Generation

Timo Hromadka, Kevin Monteiro[(✉)], and Sam Nallaperuma-Herzberg

University of Cambridge, Cambridge, UK
`kidrm2@cam.ac.uk`

Abstract. Sleep disorders, particularly insomnia, and mental health conditions affect a significant fraction of adults worldwide, posing serious mental and physical health risk. Music therapy offers promising, low-cost, and non-invasive treatment, but current approaches rely heavily on expert-curated playlists, limiting scalability and personalization. We propose a low-cost generative system leveraging recent advances in diffusion models to synthesize music for therapy. We focus on insomnia and curate a dataset of waveform sleep music to generate audio tailored to sleep. To ensure real-world feasibility, we optimize our system for training and use on a single GPU, balancing quality and efficiency through extensive ablation studies. We show through subjective human evaluations that our generated music matches or outperforms existing baselines in both perceived quality and relevance to sleep therapy, while using only a fraction of the computational cost.

Keywords: Diffusion Models · Music Therapy · Insomnia · Mental Health · Sleep Music Generation

1   Introduction

It is estimated nearly a third of the world population suffers from insomnia [2], with one in ten suffering from chronic insomnia [11,40]. Mental health conditions such as depression and anxiety are also known to be comorbid with insomnia [3,35], and together they contribute to many lifestyle diseases with severe impact on health such as lessened metabolic, immunologic, and cardiovascular health. Additionally, sleep disruption increases susceptibility to numerous chronic conditions [4,13,31,38]. The progressive decline in sleep duration in the public is considered a public health epidemic [7] and most common treatments for insomnia are drug based, which have often been associated with adverse side effects.

Music therapy is emerging as an effective, non-invasive, and low-cost solution in the treatment of sleep and mental health disorders [17,21,26,32], as there is strong evidence that music can improve sleep quality [10,21,29,47] as well as the recovery approach in mental health care [26]. However, producing music for therapy on-demand is underexplored [39]. Existing systems are static, lacking the adaptability to cater to individual needs. Moreover, these approaches rely

P. Machado et al. (Eds.): EvoMUSART 2026, LNCS 16523, pp. 129–144, 2026.
https://doi.org/10.1007/978-3-032-24350-8_9

on expert therapists to curate music selections, limiting their accessibility and scalability for broader public use.

To address the up-and-coming field of music therapy generation, this paper focuses on low-cost sleep music generation that can be both trained and deployed efficiently, making it accessible for wider use, and operates on publicly available sleep music. To summarize, our main contributions are as follows: **(a)** we are the first to focus on generative sleep therapy using waveform audio, a more expressive and available audio data format, that allows for greater adaptability to personal preferences, surpassing the limitations of previous MIDI-based or algorithm-based approaches; **(b)** we are the first to apply the BigVGAN vocoder to music generation, demonstrating its superior performance despite being originally trained for speech synthesis; **(c)** we propose multiple model architectures that offer various trade-offs between generation quality and computational efficiency, with our fastest models leveraging a latent diffusion framework combined with the BigVGAN vocoder. We specifically focus on training models on a single consumer GPU, employing a minimal latent diffusion architecture and carefully chosen hyperparameters to maximize efficiency without compromising performance.; **(d)** we curate a specialized dataset of sleep music and perform detailed feature analyses on both the dataset and our model's generated samples, demonstrating how our system captures distinctive characteristics of sleep music; and **(e)** through comprehensive objective and subjective evaluations, including human listener studies, we show that our models outperform existing methods in terms of music relevance to sleep therapy. We achieve these improvements while maintaining smaller model size and reduced computational requirements compared to other state-of-the-art music models.

2   Background in Computer Music Generation

Previously, music intervention studies often relied on generic sleep music with no expert curation [29,39,47]. While some studies improve upon this by involving expert-selected music [10,21], it has been shown that music preference is individualistic and could have an impact on music therapy [5,51]. Moreover, there still remains a critical gap: across nearly all studies there is a severe lack of analysis on the specific sleep music characteristics [29], including melody, harmony, rhythm, etc. Implementing personalized music interventions on a large scale remains costly and logistically challenging.

Music generation methods can be broadly categorized into MIDI-based and waveform-based approaches. Transformers [46] have been widely employed to sequentially model MIDI notes, treating the task as a language modeling problem [20,22,37,54]. However, MIDI only provides instructions for synthesizing sound; it does not directly produce audio and requires additional processing through instruments or software synthesizers to generate sound, thus lacking the expressiveness and fidelity required for generating full, synchronized soundtracks. As such, MIDI datasets need to often be created manually, in contrast to being able to use readily available sleep music in the audio (waveform) format.

On the other hand, learning from raw audio (waveform) provides both flexibility and ease of access to training data. Instead of needing to build music from MIDI instructions, music can be directly created from waveforms, allowing for more naturally sounding and expressive results. Additionally, this approach benefits from a large amount of sleep music available as waveforms, whereas MIDI versions of these songs are rarely accessible.

Specifically for sleep music, only two music generation approaches have been put forth so far. The first uses MIDI-based music generation [53], while the other utilizes an algorithmic approach using randomization and Markov Chains to construct music elements [45]. Thus, no works up to date have focused on generating expressive sleep music from waveform audio. As a result, existing methods lack the ability to produce music that is both highly expressive, natural-sounding, and adaptable to users' preferences, limiting their effectiveness. Nevertheless, we cover the most popular approaches in literature for general music generation.

Recently, diffusion models have shown exceptional performance in generative AI tasks involving waveform audio. Existing approaches in the literature can be categorized as follows: **(a)** Spectrogram-based Diffusion Models, which process spectrogram data using encoders such as VAEs and employ a neural vocoder to convert spectrograms back into audio [6,18,19,28,52], or operate directly on waveforms [27,43,44]; **(b)** Diffusion Transformers (DiT), which generate audio by operating in the latent space, and use a combined diffusion model and transformer approach [1,33]; **(c)** Auto-regressive Language Models (most commonly transformers), which generate audio based on trained codecs [8,12,25,48, 49]; and **(d)** Alternative Generative Methods, such as Consistency Autoencoders (CAE) [36].

However, in the field of music generation, few released models are able to accommodate use on a single consumer GPU, often requiring large amounts of compute to train and run for inference. Our research addresses this limitation by developing lightweight models that can be trained and deployed for inference on a single consumer GPU.

3   Proposed Approach

Our framework consists of four main components: **(a)** the Waveform Processor, **(b)** Variational Autoencoder (VAE), **(c)** Diffusion Model, **(d)** and Vocoder. Figure 1 depicts the overall pipeline of the proposed approach.

3.1   Framework Components

The **Waveform Processor (a)** starts the pipeline by taking raw audio waveforms $x \in \mathbb{R}^{T_s}$ from the training dataset, where T_s is the length of the audio signal in samples (e.g. an audio file with a sampling rate of 44.1kHz has 44,100 samples per second). To optimize computational efficiency, the audio is downmixed to mono and resampled to 22 kHz, unless stated otherwise. The processed waveform is then transformed into a mel-spectrogram $\mathbb{X}_m \in \mathbb{R}^{T_{px} \times F_{mb}}$,

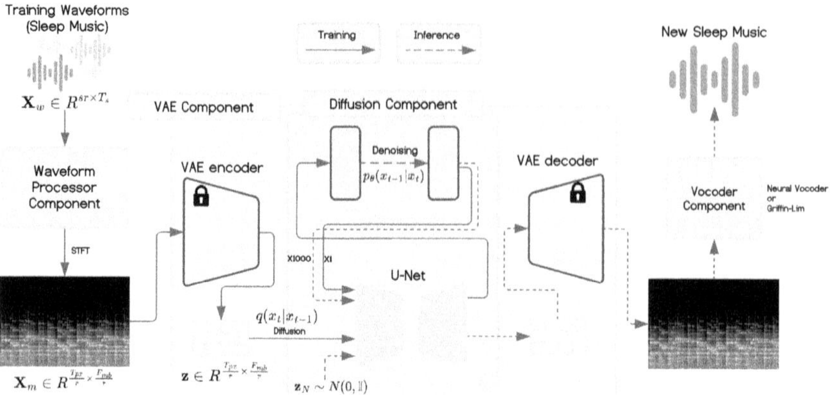

Fig. 1. Diagram of our proposed framework, which consists of four components: Waveform Processor, VAE, Diffusion, and Vocoder. Raw audio (waveform) of length T_s from our sleep dataset is downsampled to sr = 22kHz and converted to mono. It is then transformed into mel-spectrograms (X_m) with a time (x) resolution of T_{px} and a frequency (y) resolution of F_{mb}. A pre-trained VAE encodes the mel-spectrogram into a compressed 2-dimensional latent space (z) with a compression ratio r. During training (bold arrows), the diffusion model processes the batch once to predict the added noise in training samples, computing the loss for backpropagation. During inference (dashed arrows), the process starts with Gaussian noise and performs 1000 steps of ancestral sampling to generate a novel sample, which is passed through the VAE decoder to reconstruct the mel-spectrogram. Finally, the Vocoder converts the mel-spectrogram back into audio.

resolution and F_{mb} is the number of mel-frequency bins. The mel-spectrograms are sliced to the appropriate length for the next stage.

Next, the **Variational Autoencoder (VAE) (b)** encodes mel-spectrograms into a compact latent space, enabling efficient training and inference. Specifically, $\mathbb{X}_m \in \mathbb{R}^{T_{px} \times F_{mb}}$ is encoded into $\mathbf{z} \in \mathbb{R}^{C \times \frac{T_{px}}{r} \times \frac{F_{mb}}{r}}$, where C is the latent channel size and r is the downsampling factor. The decoder reverses this process, reconstructing a mel-spectrogram from the latent representation. Once the VAE is trained, its parameters are frozen during the training and inference stages of the diffusion model. The specific configuration for our VAE model is based on the Stable Diffusion repository [41] and the VAE architecture has an embedding dimension of 1, meaning the latent space is compressed to a single channel.

Then, the **Diffusion Model (c)** generates novel samples $\mathbf{z}' \in \mathbb{R}^{C \times \frac{T_{px}}{r} \times \frac{F_{mb}}{r}}$ from Gaussian noise, either in the pixel space (if no VAE is used), or in latent space (if the VAE is used). Using a U-Net backbone, the model progressively denoises the input noise through a reverse diffusion process over N iterative steps. During training, the model learns to predict the noise added at each step of the forward diffusion process. In inference, it generates novel samples by iteratively reversing this process, starting from pure Gaussian noise.

Finally, the **Vocoder Component (d)** converts the generated mel-spectrograms $\mathbf{X}'_{\mathrm{mel}} \in \mathbb{R}^{T_{px} \times F_{mb}}$ back into waveform audio $\mathbf{X}' \in \mathbb{R}^{T_s}$. Since phase information of mel-spectrograms is no longer present, this approach is not straightforward. We experiment with two approaches to make the conversion back to audio: the deterministic Griffin-Lim algorithm and the BigVGAN neural vocoder (available from the official NVIDIA BigVGAN GitHub repository). Specifically, we use the v2 model version (`bigvgan_v2_44khz_128band_256x`) which was pretrained on a diverse dataset, originally for speech synthesis, but we show it performs well for the music generation tasks.

3.2   Training Dataset

To train our model, we selected a suitable dataset of sleep music. We used samples from the publicly available Spotify Sleep Playlist Dataset [42], filtering it for songs explicitly labelled with "sleep" as one of their genres to exclude upbeat tracks. After filtering, the dataset has 19,000 30-second samples, amounting to ~158 h of music. For brevity, this paper often refers to this dataset as SSD (**S**potify **S**leep **D**ataset).

3.3   Hyperparameters and Training Details

This section describes important hyperparameters and technical details of the relevant components used in our approach. An overview of the parameters is shown in Table 1.

4   Evaluation Setup

We perform objective and subjective evaluation for the generated music and compare our model to recent baseline models.

4.1   Objective Assessments

In our experiments, generated samples are evaluated using the Fréchet Audio Distance (FAD) metric [23]. FAD measures the similarity between the distributions of two sets of audio tracks, one generated and one reference, by computing the Fréchet Distance [16], also known as the Wasserstein-2 distance, between their respective embeddings. We leverage the library and incorporate improved parameters for FAD calculation from [14], addressing their recent findings that traditional FAD metrics may not reliably align with human judgment. Specifically, the widely used VGGish [15] embedding model has been shown to inadequately reflect human perception. Instead, we leverage the CLAP music model backbone [50], which offers a stronger alignment with human evaluations, and comes in two variations: `clap-laion-audio` and `clap-laion-music`. This is in line with other new works such as [30,34], which also adopt the CLAP model backbone to compute embeddings for FAD calculation. To remain consistent

Table 1. Hyperparameters for each component. While most parameters are fixed throughout the paper following standard literature practices, others are the focus of the paper and vary during individual experiments and are marked as ' Variable '.

Component	Hyperparameter	Value
Waveform Processor	Sampling Rate	22 kHz or 44 kHz
	Mel-spec Pixel Width	512 px or 2048 px
	Mel Bands	Variable
	Hop Length	Variable
	FFT Size	Variable
VAE	KL Divergence Reg. Term	1×10^{-6}
	Discriminator	After 50,001 steps
	Base Channel Width	32
	Channel Multipliers	$[1, 2, 4, 4]$
	Input/Output Channels	1
	Residual Blocks per Level	2
	Learning Rate	4.5×10^{-6}
	Batch Size	32
	EMA	inv. gamma $= 1.0$, power $= \frac{3}{4}$, max decay $= 0.9999$
	Compression Rate	Variable
Diffusion Model	Down Blocks	[DownBlock, DownBlock, DownBlock, DownBlock, AttnDownBlock, DownBlock]
	Up Blocks	[UpBlock, AttnUpBlock, UpBlock, UpBlock, UpBlock, UpBlock]
	Out Block Channels	$[128, 128, 256, 256, 512, 512]$
	Learning Rate	10^{-4} w/ 500 warmup steps
	Adam Optimizer	$\beta_1 = 0.95, \beta_2 = 0.999, \epsilon = 10^{-8}$, weight decay $= 10^{-6}$
	EMA	inv. gamma $= 1.0$, power $= \frac{3}{4}$, max decay $= 0.9999$
	Training Steps	200k Steps
	Noising Parameters	$\beta_1 = 10^{-4}, \beta_T = 0.02$
	Noising Schedule	Cosine
	Inference Steps	1000 DDPM steps or 100 DDIM steps
	Batch Size	16 or 8
Vocoder	Neural Vocoder	bigvgan_v2_44khz_128band_256x
	Griffin-Lim Iterations	32

with recent literature, we nonetheless still report VGGish scores for comparison. For brevity, from now on we refer to FAD scores calculated using each of the three models as FAD_{cla}, FAD_{clm}, and FAD_{vgg}, respectively.

We use the FMA Pop dataset [9] as the reference set for FAD calculation to evaluate music quality, following recommendations for generative music tasks [14]. The FMA Pop dataset provides a more reliable benchmark for assessing music quality compared to the commonly used MusicCaps dataset [1]. Next, to specifically evaluate how closely our generated samples align with sleep music characteristics, we compute FAD scores using our curated Spotify Sleep Dataset as a reference set. Throughout this paper, we report FAD scores for both reference sets of FMA Pop and the Spotify Sleep Dataset. To prepare our reference sets for FAD calculation, we download the FMA Pop dataset [9] from the official repository.

4.2 Subjective Assessments

We conduct a subjective human study following the same procedure as in existing studies [8,24,27], to evaluate the generated music.

Human participants were asked to rate: **(a)** the overall perceived quality of the generated audio samples (**Qual**), and **(b)** the relevance of the generated audio samples to sleep music (**Rel**), both on a scale of 1 to 100. We leverage the Amazon Mechanical Turk platform to recruit participants and ensure they are paid at least the UK national minimum wage. Noisy annotations and outliers are dropped, such that responses from participants who did not listen to the full audio samples and/or annotators who rated the reference audio samples less than 85 are discarded. All audio samples were also normalized to -20.0 dBFS for fairness.

4.3 Baseline Models

Based on the findings from our literature review, we select AudioLDM [28] and MusicGen [8] as baseline models for comparison with our proposed model. We choose these models as they have publicly available APIs to generate samples and also offer configurations with small enough model sizes, which is in line with our goal of developing lightweight models for resource-constrained environments. We deploy these respective baseline models using the publicly available implementation on HuggingFace and generate samples by providing the prompt "relaxing sleep music perfect for sleep therapy" as well as other inputs based on the respective author's recommendations. Sleep music generated by our model is filtered (similar to how the best candidates are selected in the official implementation of AudioLDM [28]) to ensure the quality of the presented samples. We compute individual FAD scores [14] for these generated samples and select a subset with the top scores for further stages.

5 Evaluation Results

In this section, we first present a comparison between different model configurations followed by our objective and subjective evaluation results. We then include a brief analysis of the musical alignment between samples generated by our model and sleep music by comparing extracted musical features.

5.1 Model Configuration Comparison

Our ablation experiments showed that we can improve FAD scores by small architectural tweaks. To further enhance performance, we replace the Griffin-Lim algorithm for mel-spectrogram-to-audio conversion with a neural vocoder. We showcase all the main models and parameters side by side in Table 2.

Each long-sample model, corresponding to 2048×128-pixel mel-spectrograms, was trained for approximately 72 h, while each short-duration model, using 512×128-pixel spectrograms, was trained for around 36 h. All models were trained on a single A100 GPU. We selected these training durations to remain compute efficient, and to roughly have the same amount of training steps as other small models in the literature. Latent models performed worse than pure mel-spectrogram diffusion models, with noticeable improvements observed when employing the BigVGAN vocoder. The impact of the vocoder is more pronounced for shorter samples, suggesting that the generated mel-spectrograms are of higher quality. BigVGAN consistently outperforms the Griffin-Lim algorithm across all configurations, particularly when paired with a VAE during the diffusion process. This highlights BigVGAN's ability to mitigate reconstruction artifacts inherent in latent models. The strongest-performing setup combines a mel-spectrogram diffusion process with the BigVGAN vocoder. In terms of efficiency, VAE-based configurations are markedly faster, with generation times reduced from 49.75 to 3.44 s for longer samples and from 12.13 to 1.44 s for shorter samples. The BigV-GAN vocoder enhances sample quality while introducing only minimal computational overhead, further solidifying its advantage over the Griffin-Lim algorithm. Additionally, VAE models achieve faster-than-real-time generation. Across *all* experiments, the DDIM sampler is also capable of faster than real-time generation, however, this paper focuses on reporting results using the DDPM sampler. These findings demonstrate the feasibility of utilizing diffusion models for real-time music generation systems. We select the overall best performing model to carry forward into the human evaluation and feature analysis section, which is a mel-spectrogram based diffusion model using a BigVGAN Vocoder producing samples of $\sim$12 s (*Mel-spec + BigVGAN*).

5.2 Comparison with the Baselines

Both objective and subjective results are reported in Table 3. We first compare our proposed model with the selected baseline models using objective metrics. When using FMA Pop as the reference set, our model achieves FAD scores comparable to MusicGen-S while using only about 1/3 of the number of parameters. On the other hand, when using the Spotify Sleep Dataset as reference, the proposed model considerably out performs the baseline models demonstrating better objective alignment with sleep music characteristics. Next, we present the mean opinion score and 95% confidence intervals from the human evaluation study (subjective results). Again, the proposed model performs similar to MusicGen-S in both audio quality and perceived relevance to sleep music and outperforms AudioLDM-S across the same metrics.

Table 2. FAD Score Evaluation on the FMA_Pop Dataset for Models With Different Configurations and Sample Sizes. Smaller Sample Sizes (512×128) and Longer Sample Sizes (2048×128) are indicated in the Table. The Mel-spectrogram + 4× VAE + BigVGAN configuration is omitted as training was unable to converge within the allocated training time.

Configuration	Sample Length, secs+ kHz	Sampling Time DDPM/ DDIM (s)	Model Size (#Params/MB)	FMA Pop Reference Set		SSD Reference Set	
				FAD_{cla}	FAD_{clm}	FAD_{cla}	FAD_{clm}
2048×128 Samples (Longer)							
Mel-spec	23.8 (22kHz)	49.75/4.975	113M/455MB	0.824	0.917	0.446	0.481
Mel-spec (hl512 nfft1024)	47.5 (22kHz)	49.75/4.975	113M/455MB	0.949	0.904	0.697	0.736
Mel-spec + 4× VAE	23.8 (22kHz)	3.44/0.344	113+ 1.3 M/ 455+ 5 MB	1.154	1.190	0.696	0.719
Mel-spec + BigVGAN	11.9 (44kHz)	50.93/6.155	113+ 112 M/ 455+ 451 MB	**0.607**	**0.630**	**0.357**	**0.460**
Mel-spec + 4× VAE + BigVGAN	11.9 (44kHz)	4.62/1.524	113+ 1.3 + 112 M/ 455+ 5 + 451 MB	-		-	-
512×128 Samples (Shorter)							
Mel-spec	5.9 (22kHz)	12.13/1.213	113M/455MB	0.988	1.012	0.692	0.628
Mel-spec (hl512 nfft1024)	11.9 (22kHz)	12.13/1.213	113M/455MB	0.807	0.879	0.484	**0.506**
Mel-spec + 4× VAE	5.9 (22kHz)	1.44/0.144	113+ 1.3 M/ 455+ 5 MB	1.119	1.129	0.894	0.760
Mel-spec + BigVGAN	3.0 (44kHz)	11.585/1.343	113+ 112 M/ 455+ 451 MB	**0.806**	**0.807**	0.613	0.671
Mel-spec + 4× VAE + BigVGAN	3.0 (44kHz)	1.645/0.349	113+ 1.3 + 112 M/ 455+ 5 + 451 MB	0.832	0.821	**0.478**	0.541

We also observe high FAD values on the Spotify Sleep Dataset (100 samples) when using FMA Pop as the reference set. The FMA Pop set consists of studio recordings of pop songs while the Spotify Sleep Dataset consists of sleep music. Sleep music refers to audio that is typically instrumental and calming. It often features slow tempos, and may incorporate nature sounds such as rain, ocean waves, wind, etc. This difference in type of music between the Sleep Dataset and the FMA Pop set could explain the divergence seen, underscoring the importance of genre-specific reference sets and evaluation metrics when assessing generative models for specialized tasks such as sleep music generation.

Table 3. Subjective (Qual and Rel) and Objective (FAD) comparison between the baseline models (AudioLDM, MusicGen) and our proposed model. FAD scores are computed using FMA Pop and the Spotify Sleep Dataset as reference as indicated. Lower FAD scores indicate greater similarity to the reference set and are typically preferred. Higher Qual and Rel, from the subjective evaluation surveys, indicate better perceived audio quality and relevance to sleep music and are therefore better.

Model (100 samples)	#Params	Qual↑	Rel↑	FMA Pop Reference Set			SSD Reference Set		
				FAD_{cla}	FAD_{clm} ↓	FAD_{vgg}	FAD_{cla} ↓	FAD_{clm} ↓	FAD_{vgg} ↓
Sleep Dataset (human composed)	-	94.72 ± 0.81	92.31 ± 2.23	0.704*	0.827*	11.656*	0.027	0.022	0.336
AudioLDM-S	185M	66.14 ± 5.20	65.07 ± 5.59	**0.642**	0.834	**5.483**	0.864	0.825	9.069
MusicGen-S	300M	83.08 ± 3.34	82.41 ± 3.74	0.851	**0.825**	10.272	0.616	0.693	4.212
Proposed	**115M**	**83.70** ± 3.23	**85.74** ± 3.03	0.823	0.849	11.609	**0.251**	**0.416**	**2.782**

Our model has considerably less parameters and required far less training time when compared to the baseline models. We train our proposed model for 2 days on one A100 GPU with a batch size of 8, circa 200k steps, and similarly for our VAE (200k steps on a single A100 GPU). MusicGen-S trains for 1.5 million steps on 32 GPUs, and AudioLDM's VAE alone is trained on 1.5 million steps on a single GPU. Not only do we match performance but also surpass in terms of computational requirements.

On the whole, the proposed model outperforms AudioLDM-S and achieves performance that is comparable to, if not better than, MusicGen-S.

5.3   Musical Feature Analysis

While FAD scores provide an objective measure of how closely our generated music matches the distribution of real sleep music (with FAD_{cla} and FAD_{clm} values as low as 0.25 and 0.42, respectively), we further assess the musical similarity through feature analysis. This allows us to verify that our model not only produces high-quality music but also captures the distinctive characteristics of sleep music, setting it apart from other genres.

We extract 34 musical features for each sample across six categories: Articulation, Energy, Spectrum, Rhythm, Bass and Treble, and Noise- or Tone-like qualities (see Appendix for details). All audio samples, including generated and reference tracks, are truncated to 11.9 s to match the duration of samples from our chosen BigVGAN-based model.

We compare the musical features of our generated samples against those of real sleep music, as well as 10 other non-sleep genres from the GTZAN dataset, to highlight their distinctiveness. To visualize the closeness of audio samples in feature space, we plot the samples in 2D TSNE space in Fig. 2.

The generated sleep samples are heavily clustered with actual sleep music namely for all feature sets except for Articulation and Energy. Respectively,

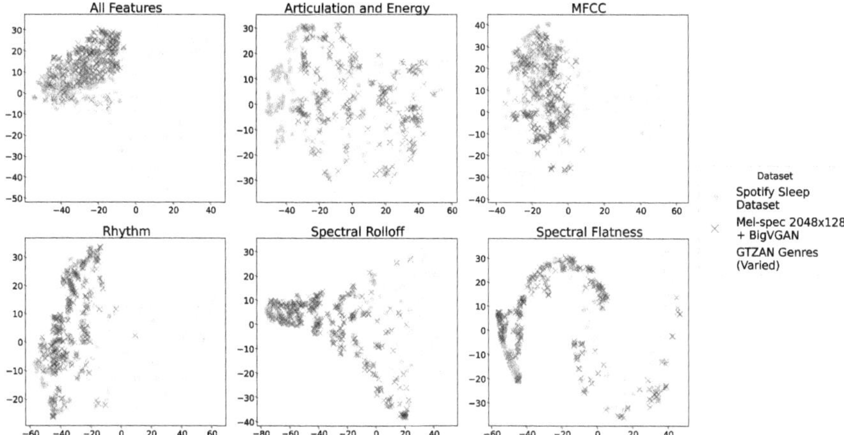

Fig. 2. We visualize in 2d TSNE space (perplexity set to 30) the samples generated by our model (2048×128 Mel-Spec model with BigVGAN), along with the training dataset (Spotify Sleep Dataset), as well as many other genres such as hip hop, jazz, classical, etc. from the GTZAN datasets. The generated sleep samples (black X) are heavily clustered with actual sleep music (gray circles) across all feature sets except Articulation and Energy. (Color figure online)

this means that our generated music matches real sleep music on a natural human auditory perception level (MFCC), has matching temporal placement of notes/structures (Rhythm), matching timbral characteristics (such as bass and treble tones) (Spectral Rolloff), and is quantified by a similar amount of tone-like vs. noise-like music (Spectral Flatness).

However, we observe less distinct clustering for the Articulation and Energy features, as these values are widely scattered across all genres. This suggests that Articulation and Energy are less effective at discriminating between genres compared to other features.

	All Features		Articulation and Energy		MFCC		Rhythm		Spectral Rolloff		Spectral Flatness	
	Spotify Sleep Dataset	bigvgan 2048 128	Spotify Sleep Dataset	bigvgan 2048 128	Spotify Sleep Dataset	bigvgan 2048 128	Spotify Sleep Dataset	bigvgan 2048 128	Spotify Sleep Dataset	bigvgan 2048 128	Spotify Sleep Dataset	bigvgan 2048 128
Spotify Sleep Dataset	0.00	0.44	0.00	0.23	0.00	0.24	0.00	0.04	0.00	0.11	0.00	0.02
bigvgan 2048 128	0.44	0.00	0.23	0.00	0.24	0.00	0.04	0.00	0.11	0.00	0.02	0.00
gtzan: blues	0.92	0.71	0.26	0.08	0.61	0.49	0.38	0.37	0.33	0.24	0.01	0.01
gtzan: classical	0.44	0.57	0.07	0.29	0.35	0.38	0.10	0.09	0.12	0.04	0.00	0.02
gtzan: country	0.84	0.62	0.19	0.10	0.48	0.32	0.30	0.29	0.39	0.31	0.06	0.04
gtzan: disco	1.29	1.04	0.27	0.07	0.67	0.50	0.53	0.52	0.55	0.45	0.14	0.12
gtzan: hiphop	1.46	1.18	0.51	0.31	0.68	0.51	0.65	0.63	0.58	0.49	0.17	0.15
gtzan: jazz	0.71	0.57	0.11	0.17	0.43	0.32	0.25	0.24	0.33	0.24	0.04	0.02
gtzan: metal	1.36	1.12	0.25	0.13	0.98	0.85	0.20	0.19	0.43	0.32	0.08	0.06
gtzan: pop	1.64	1.36	0.53	0.33	0.66	0.50	0.59	0.58	0.75	0.66	0.27	0.25
gtzan: reggae	1.27	1.08	0.26	0.12	0.56	0.41	0.73	0.71	0.57	0.49	0.12	0.10
gtzan: rock	1.05	0.82	0.20	0.12	0.64	0.49	0.32	0.31	0.45	0.36	0.09	0.07

Fig. 3. Euclidean distance of group mean sums across features. Following from the TSNE visualization, we can see that MFCC, Rhythm, and Spectral Rolloff features are either the closest to the Spotify Sleep Dataset, or second closest followed only by classical music.

To evaluate the similarity between the visualized 2D TSNE clusters and the discriminative power of various features, we calculated the mean of each feature cluster for each dataset and analyzed the pairwise differences between sample sets. The heatmaps in Fig. 3 reveal that the generated dataset aligns most closely with the sleep dataset for MFCC and Rhythm features, with a significant margin. For Spectral Rolloff, the generated set ranks second, following classical music. In the case of Spectral Flatness, this feature is noticeably less disciminating as the values are not as spread. Nonetheless the generated dataset ties for second place with country and jazz with classical music ranking first.

This strongly suggests that our generated samples closely resemble sleep music, not only in terms of FAD scores, but also in their audio feature profiles. This alignment highlights the potential of generative AI tools as effective aids in insomnia treatment.

6   Conclusion and Future Work

In this work, we developed lightweight generative models tailored specifically for sleep music. Objective and subjective evaluations show that our models produce high-quality audio with respect to generation artifacts, often outperforming other approaches in the literature. We also demonstrate the successful application of the BigVGAN vocoder for music generation, achieving high fidelity. Our experiments examined key design choices for model architectures, balancing efficiency (training and sampling speed) with output quality. The results indicate that our lightweight models generate sleep music with strong resemblance to real examples, supported by low Fréchet Audio Distance (FAD) scores and similarities across acoustic and audio features. Key findings include the following: **(a)** using a curated sleep music dataset enables our models to achieve superior quality and sleep-music relevance as rated by human listeners with significantly fewer parameters compared to existing methods; **(b)** a detailed feature analysis shows that our model's generated music aligns most closely with features of real sleep music, as compared to those of other genres; **(c)** pretrained BigVGAN vocoders, originally designed for speech, are capable of high-quality music generation; **(d)** alternative mel-spectrogram configurations (e.g., non-standard `hop_length`, `n_fft`, and `mel_bands`) outperform conventional literature settings; and **(e)** confirming that VAE compression exhibits diminishing returns, with excessive compression degrading audio quality, as indicated by higher FAD scores. Based on these findings, we select a middle ground of $4\times$ compression for lightweight diffusion model training, balancing efficiency with audio quality. Future research focuses on investigating which musical features best support specific sleep phases, integrating user data to enable adaptive real-time music generation that corresponds to a user's sleep state and deploying the proposed method in a real world study with insomnia patients. Continuous generation techniques such as successive conditioning or outpainting offer promising directions.

Acknowledgments. This work is funded by the Accelerate Programme for Scientific Discovery, University of Cambridge (Grant/Award Number: NRAG 646) and Impact Acceleration Accounts of EPSRC (Grant/Award Number: NRAG 852).

Disclosure of Interests. The authors have no competing interests to declare that are relevant to the content of this article.

References

1. Agostinelli, A., et al.: MusicLM: generating music from text (2023). https://doi.org/10.48550/arXiv.2301.11325. Accessed 2025/07/14
2. Bhaskar, S., Hemavathy, D., Prasad, S.: Prevalence of chronic insomnia in adult patients and its correlation with medical comorbidities. J. Family Med. Primary Care **5**(4), 780–784 (2016). https://doi.org/10.4103/2249-4863.201153
3. Blank, M., Zhang, J., Lamers, F., Taylor, A.D., Hickie, I.B., Merikangas, K.R.: Health correlates of insomnia symptoms and comorbid mental disorders in a nationally representative sample of us adolescents. Sleep **38**(2), 197–204 (2015)
4. van Cauter, E., et al.: Impact of sleep and sleep loss on neuroendocrine and metabolic function. Horm. Res. **67**(Suppl. 1), 2–9 (2007). https://doi.org/10.1159/000097543
5. Chang, E.T., Lai, H.L., Chen, P.W., Hsieh, Y.M., Lee, L.H.: The effects of music on the sleep quality of adults with chronic insomnia using evidence from polysomnographic and self-reported analysis: a randomized control trial. Int. J. Nurs. Stud. **49**(8), 921–930 (2012). https://doi.org/10.1016/j.ijnurstu.2012.02.019
6. Chen, K., Wu, Y., Liu, H., Nezhurina, M., Berg-Kirkpatrick, T., Dubnov, S.: MusicLDM: enhancing novelty in text-to-music generation using beat-synchronous mixup strategies (2023). https://doi.org/10.48550/arXiv.2308.01546. Accessed 2025/07/14
7. Watson, N.F., et al.: Consensus Conference Panel, In: Joint Consensus Statement of the American Academy of Sleep Medicine and Sleep Research Society on the Recommended Amount of Sleep for a Healthy Adult: Methodology and Discussion. Sleep **38**(8), 1161–1183 (2015). https://doi.org/10.5665/sleep.4886
8. Copet, J., et al.: Simple and Controllable music generation (2023). https://doi.org/10.48550/arXiv.2306.05284. Accessed 2025/07/14
9. Defferrard, M., Benzi, K., Vandergheynst, P., Bresson, X.: FMA: a dataset for music analysis (2017). https://doi.org/10.48550/arXiv.1612.01840. Accessed 2025/07/14
10. Dickson, G.T., Schubert, E.: Music on prescription to aid sleep quality: a literature review. Front. Psychol. **11** (2020). https://doi.org/10.3389/fpsyg.2020.01695. Publisher: Frontiers
11. Ellis, J., et al.: Chronic insomnia disorder across Europe: expert opinion on challenges and opportunities to improve care. Healthcare **11**(5), 716 (2023). https://doi.org/10.3390/healthcare11050716
12. Evans, Z., Parker, J.D., Carr, C.J., Zukowski, Z., Taylor, J., Pons, J.: Long-form music generation with latent diffusion (2024). https://doi.org/10.48550/arXiv.2404.10301. Accessed 2025/07/14
13. Gebara, M.A., et al.: Effect of insomnia treatments on depression: a systematic review and meta-analysis. Depress. Anxiety **35**(8), 717–731 (2018). https://doi.org/10.1002/da.22776

14. Gui, A., Gamper, H., Braun, S., Emmanouilidou, D.: Adapting frechet audio distance for generative music evaluation. In: ICASSP 2024 - 2024 IEEE International Conference on Acoustics, Speech and Signal Processing (ICASSP), pp. 1331–1335 (2024). https://doi.org/10.1109/ICASSP48485.2024.10446663. ISSN: 2379-190X
15. Hershey, S., et al.: CNN architectures for large-scale audio classification (2017). https://doi.org/10.48550/arXiv.1609.09430. Accessed 2025/07/14
16. Heusel, M., Ramsauer, H., Unterthiner, T., Nessler, B., Hochreiter, S.: GANs trained by a two time-scale update rule converge to a local nash equilibrium (2018). https://doi.org/10.48550/arXiv.1706.08500. Accessed 2025/07/14
17. Huang, J., Antonsdottir, I.M., Wang, R., Li, M., Li, J.: Insomnia and its non-pharmacological management in older adults. Curr. Geriatrics Rep. (2023). https://doi.org/10.1007/s13670-023-00397-1
18. Huang, Q., et al.: Noise2Music: text-conditioned music generation with diffusion models (2023). https://doi.org/10.48550/arXiv.2302.03917. Accessed 2025/07/14
19. Huang, R., et al.: Make-an-audio: text-to-audio generation with prompt-enhanced diffusion models (2023). https://doi.org/10.48550/arXiv.2301.12661. Accessed 2025/07/14
20. Huang, Y.S., Yang, Y.H.: Pop music transformer: beat-based modeling and generation of expressive pop piano compositions (2020). https://doi.org/10.48550/arXiv.2002.00212. Accessed 2025/07/14
21. Jespersen, K.V., Koenig, J., Jennum, P., Vuust, P.: Music for insomnia in adults. Cochrane Database System. Rev. **2015**(8), CD010459 (2015). https://doi.org/10.1002/14651858.CD010459.pub2
22. Jiang, J., Xia, G.G., Carlton, D.B., Anderson, C.N., Miyakawa, R.H.: Transformer VAE: a hierarchical model for structure-aware and interpretable music representation learning. In: ICASSP 2020 - 2020 IEEE International Conference on Acoustics, Speech and Signal Processing (ICASSP), pp. 516–520 (2020). https://doi.org/10.1109/ICASSP40776.2020.9054554. ISSN: 2379-190X
23. Kilgour, K., Zuluaga, M., Roblek, D., Sharifi, M.: Fréchet Audio Distance: a reference-free metric for evaluating music enhancement algorithms. In: Interspeech 2019, pp. 2350–2354. ISCA (2019). https://doi.org/10.21437/Interspeech.2019-2219
24. Kreuk, F., et al.: AudioGen: textually guided audio generation (2023). https://doi.org/10.48550/arXiv.2209.15352. Accessed 2025/07/14
25. Lan, G.L., et al.: High fidelity text-guided music editing via single-stage flow matching (2024). https://doi.org/10.48550/arXiv.2407.03648. Accessed 2025/07/14
26. Lee, J., Thyer, B.A.: Does music therapy improve mental health in adults? A review. J. Hum. Behav. Soc. Environ. **23**(5), 591–603 (2013)
27. Li, P., Chen, B., Yao, Y., Wang, Y., Wang, A., Wang, A.: JEN-1: text-guided universal music generation with omnidirectional diffusion models (2023). https://doi.org/10.48550/arXiv.2308.04729. Accessed 2025/07/14
28. Liu, H., et al.: AudioLDM: text-to-audio generation with latent diffusion models (2023). https://doi.org/10.48550/arXiv.2301.12503. Accessed 2025/07/14
29. Loewy, J.: Music therapy as a potential intervention for sleep improvement. Nat. Sci. Sleep **12**, 1–9 (2020). https://doi.org/10.2147/NSS.S194938
30. Manor, H., Michaeli, T.: Zero-shot unsupervised and text-based audio editing using ddpm inversion (2024). https://doi.org/10.48550/arXiv.2402.10009. Accessed 2025/07/14
31. Medic, G., , Micheline, W., , Hemels, M.E.: Short- and long-term health consequences of sleep disruption. Nat. Sci. Sleep **9**, 151–161 (2017). https://doi.org/10.2147/NSS.S134864

32. Mohamad Zamani, N.A., Omar, N., Huda Azmi, N.D.: Insomnia audio therapy mobile application with music recommender system. Math. Sci. Inform. J. **3**(1), 29–38 (2022). https://doi.org/10.24191/mij.v3i1.18264

33. Ning, Z., et al.: DiffRhythm: blazingly fast and embarrassingly simple end-to-end full-length song generation with latent diffusion (2025). https://doi.org/10.48550/arXiv.2503.01183. Accessed 2025/10/01

34. Novack, Z., McAuley, J., Berg-Kirkpatrick, T., Bryan, N.J.: DITTO: Diffusion inference-time T-optimization for music generation (2024). https://doi.org/10.48550/arXiv.2401.12179. Accessed 2025/07/14

35. Palagini, L., Hertenstein, E., Riemann, D., Nissen, C.: Sleep, insomnia and mental health. J. Sleep Res. **31**(4), e13628 (2022)

36. Pasini, M., Lattner, S., Fazekas, G.: Music2Latent: consistency autoencoders for latent audio compression (2024). https://doi.org/10.48550/arXiv.2408.06500. Accessed 2025/07/14

37. Qu, X., et al.: MuPT: a generative symbolic music pretrained transformer (2024). https://doi.org/10.48550/arXiv.2404.06393. Accessed 2025/07/14

38. Ramar, K., et al.: Sleep is essential to health: an American academy of sleep medicine position statement. J. Clin. Sleep Med. **17**(10), 2115–2119 (2021). https://doi.org/10.5664/jcsm.9476

39. Richter, M.: Towards the development of music as an intervention for Insomnia treatment: a research synthesis (2019). https://doi.org/10.24382/1211. https://pearl-prod.plymouth.ac.uk/handle/10026.1/14697. Publisher: University of Plymouth

40. Riemann, D., et al.: Insomnia disorder: state of the science and challenges for the future. J. Sleep Res. **31**(4), e13604 (2022). https://doi.org/10.1111/jsr.13604

41. Rombach, R., Blattmann, A., Lorenz, D., Esser, P., Ommer, B.: High-resolution image synthesis with latent diffusion models (2022). https://doi.org/10.48550/arXiv.2112.10752. Accessed 2025/07/14

42. Scarratt, R.J., Heggli, O.A., Vuust, P., Jespersen, K.V.: The audio features of sleep music: universal and subgroup characteristics. PLoS ONE **18**(1), e0278813 (2023). https://doi.org/10.1371/journal.pone.0278813. Publisher: Public Library of Science

43. Schneider, F.: ArchiSound: audio generation with diffusion (2023). https://doi.org/10.48550/arXiv.2301.13267. Accessed 2025/07/14

44. Schneider, F., Kamal, O., Jin, Z., Schölkopf, B.: Mousai: text-to-music generation with long-context latent diffusion (2023). https://doi.org/10.48550/arXiv.2301.11757. Accessed 2025/07/14

45. Tulilaulu, A., paalasmaa, j., waris, m., toivonen, h.: sleep musicalization: Automatic Music Composition from Sleep Measurements. In: Hollmén, J., Klawonn, F., Tucker, A. (eds.) Advances in Intelligent Data Analysis XI, p. 392–403. Springer, Berlin, Heidelberg (2012). https://doi.org/10.1007/978-3-642-34156-4_36

46. Vaswani, A., et al.: Attention is all you need (2023). https://doi.org/10.48550/arXiv.1706.03762. Accessed 2025/07/14

47. Wang, C.F., Sun, Y.L., Zang, H.X.: Music therapy improves sleep quality in acute and chronic sleep disorders: a meta-analysis of 10 randomized studies. Int. J. Nurs. Stud. **51**(1), 51–62 (2014). https://doi.org/10.1016/j.ijnurstu.2013.03.008

48. Wu, H., et al.: Towards audio language modeling – an overview (2024). http://arxiv.org/abs/2402.13236. Accessed 2025/07/14

49. Wu, H., et al.: Codec-SUPERB @ SLT 2024: a lightweight benchmark for neural audio codec models (2024). http://arxiv.org/abs/2409.14085. Accessed 2025/07/14

50. Wu, Y., et al.: Large-scale Contrastive language-audio pretraining with feature fusion and keyword-to-caption augmentation (2024). https://doi.org/10.48550/arXiv.2211.06687. Accessed 2025/07/14
51. Yamasato, A., et al.: How prescribed music and preferred music influence sleep quality in university students. Tokai J. Exp. Clin. Med. **45**(4), 207–213 (2020)
52. Yang, D., et al.: Diffsound: discrete diffusion model for text-to-sound generation (2023). https://doi.org/10.48550/arXiv.2207.09983. Accessed 2025/07/14
53. yang, j., min, c., mathur, a., kawsar, f.: sleepgan: towards personalized Sleep Therapy Music. In: ICASSP 2022 - 2022 IEEE International Conference on Acoustics, Speech and Signal Processing (ICASSP), pp. 966–970 (2022). https://doi.org/10.1109/ICASSP43922.2022.9747033. ISSN: 2379-190X
54. Zhang, N.: Learning adversarial transformer for symbolic music generation. IEEE Trans. Neural Netw. Learn. Syst. **34**(4), 1754–1763 (2023). https://doi.org/10.1109/TNNLS.2020.2990746. Conference Name: IEEE Transactions on Neural Networks and Learning Systems

Prompt and Circumstances: Evaluating the Efficacy of Human Prompt Inference in AI-Generated Art

Khoi Trinh[1]([✉]), Scott Seidenberger[1], Joseph Spracklen[2],
Raveen Wijewickrama[2], Bimal Viswanath[3], Murtuza Jadliwala[2],
and Anindya Maiti[1]

[1] University of Oklahoma, Norman, OK, USA
`{khoitrinh,seidenberger,am}@ou.edu`
[2] University of Texas at San Antonio, San Antonio, TX, USA
`joseph.spracklen@my.utsa.edu,`
`{raveen.wijewickrama,murtuza.jadliwala}@utsa.edu`
[3] Virginia Polytechnic Institute and State University, Blacksburg, VA, USA
`vbimal@vt.edu`

Abstract. The emerging field of AI-generated art has witnessed the rise of prompt marketplaces, where creators can purchase, sell, or share prompts to generate unique artworks. These marketplaces often assert ownership over prompts, claiming them as intellectual property. This paper investigates whether concealed prompts sold on prompt marketplaces can be considered as bona fide intellectual property, given that humans and AI tools may be able to sufficiently infer the prompts based on publicly advertised sample images accompanying each prompt on sale. Specifically, our study aims to assess (i) how accurately humans can infer the original prompt solely by examining an AI-generated image, with the goal of generating images similar to the original image, and (ii) the possibility of improving upon individual human and AI prompt inferences by crafting human-AI combined prompts with the help of a large language model (LLM). Although previous research has explored AI-driven prompt inference and protection strategies, our work is the first to incorporate a human subject study and examine collaborative human-AI prompt inference in depth. Our findings indicate that while prompts inferred by humans and prompts inferred through a combined human-AI effort can generate images with a moderate level of similarity, they are not as successful as using the original prompt. Moreover, combining human- and AI-inferred prompts using our suggested merging techniques did not improve performance over purely human-inferred prompts.

1 Introduction

Artificial Intelligence (AI) has transformed creative expression by enabling anyone to generate visually rich and conceptually coherent artworks from text

P. Machado et al. (Eds.): EvoMUSART 2026, LNCS 16523, pp. 145–160, 2026.
https://doi.org/10.1007/978-3-032-24350-8_10

prompts. Central to this process are text-to-image (`txt2img`) models such as Midjourney [15], DALL-E [23], Stable Diffusion [25], and GLIDE [17]. These systems combine a language encoder (e.g., CLIP [18]) with a diffusion-based image generator trained on large text-image datasets (e.g., LAION-5B [26]), translating textual descriptions into high-quality images.

Prompts are the primary means of steering such models and are often crafted through iterative trial and error [36]. Their effectiveness has given rise to professional "prompt engineering" and online marketplaces where creators buy and sell carefully designed prompts that yield distinctive artistic styles. Because well-crafted prompts are difficult to reproduce, many platforms treat them as proprietary information.

Recent copyright guidance clarifies that fully AI-generated works lack protection, but human – AI collaborations may qualify if they include meaningful human authorship [32–34]. Consequently, prompt marketplaces assert intellectual property rights over prompts [21,22], raising new questions about ownership and vulnerability to *prompt inference*—the act of reconstructing or approximating an original prompt from its generated image. Prior work has discussed automated prompt stealing attacks and countermeasures [27,29,39], yet little is known about how accurately humans or human – AI teams can perform such inference.

This work investigates three main research questions: *(i) how accurately can humans infer the text prompt behind an AI-generated image, (ii) to what extent do factors such as prompt length, complexity, or image generation model influence prompt inference results, and (iii) can large language models assist humans in producing prompts that yield visually similar results?* To address them, we conduct a human-subject study assessing participants' ability to infer prompts from AI-generated images and introduce a novel distribution-level evaluation method based on the Kolmogorov – Smirnov (KS) test to determine when two prompts produce statistically indistinguishable image distributions. This offers a rigorous alternative to single-image similarity comparisons and better accounts for the stochasticity of generative models.

Our results show that while humans can partially infer the subjects of prompts, accurately capturing stylistic modifiers remains difficult. AI-augmented human inferences do not outperform purely human attempts, suggesting that prompt-based intellectual property remains relatively resilient to reverse engineering under current techniques.

2   Background and Related Work

2.1   Prompt Marketplaces and Prompt Inference in AI Art

The popularity of `txt2img` models has spurred the rise of prompt marketplaces such as PromptBase, PromptHero, and CivitAI, where users buy, sell, and share text prompts for generating high-quality AI art (Fig. 1). These platforms operate on commission-based models and treat prompts as creative intellectual property [21,22], recognizing their role as both artistic tools and commercial assets.

Fig. 1. Example PromptBase listing selling a prompt associated with a specific artistic style.

Effective prompts substantially improve image quality [6,30], leading to a small but growing economy of "prompt engineering" professionals and collectors.

2.2 Prompt Inference, Subjects, and Modifiers

Prior work has examined both the vulnerability of prompts to inference attacks and methods to protect them. Several studies have shown that machine learning models can reverse-engineer or approximate hidden prompts [2,7,27,37], while others propose defensive techniques such as backdoor injection, data poisoning, and anomaly detection [29,39]. Yet these efforts focus mainly on automated systems rather than on human capabilities. If humans can accurately infer prompts through visual reasoning, the practical value of such defenses may be limited.

Recent work on human-AI co-creation [1,11] suggests that collaborative prompting may enhance creativity, but its potential for reconstructing prompts has not been studied. Understanding this human-AI interplay is critical for assessing whether collaborative inference threatens the originality or ownership of text prompts.

Prompts generally consist of a *subject*, defining the image's central theme (e.g., "cat," "forest," "robot"), and one or more *modifiers*, describing stylistic or contextual details such as lighting, texture, emotion, or medium. These linguistic components jointly steer the generative model toward specific visual outcomes. As shown in Fig. 2, even small modifier changes can substantially alter the resulting imagery, underscoring the complexity of prompt interpretation and inference.

Fig. 2. Image generations using SDXL with prompts containing the same subject (cat) and two modifiers (pixel art, dark colors).

3 Research Questions and Hypotheses

Inferring prompts behind AI-generated images is challenging for both humans and AI models [2,8,10,14,27] due to the complex interplay between visual content and linguistic nuance. Figure 2 illustrates how subtle modifier changes for the same subject can yield markedly different images. Humans rely on subjective interpretation when describing such content, often diverging in how they reconstruct subjects and stylistic cues. Understanding this human reasoning—alone or with AI support—remains largely unexplored and motivates our study.

A **key methodological contribution** of this work is a principled way to determine when two prompts generate statistically indistinguishable images. We treat each prompt as a *distribution* of outputs rather than a single instance, drawing multiple samples across seeds and comparing the resulting similarity-score distributions. Using the two-sample Kolmogorov – Smirnov (KS) test, we test whether these distributions differ significantly. This non-parametric, model-agnostic framing captures stochastic variability in txt2img models and extends prior single-image heuristics. It aligns with two-sample testing approaches such as kernel-MMD and classifier-based tests [4,9]. To our knowledge, **formalizing prompt inference as a distributional equivalence problem and evaluating it with the KS test is novel in this domain**. Details appear in Sect. 5.2 and Fig. 6.

Research Questions.

- **RQ1**: To what extent can humans accurately infer the original text prompts used to generate AI art, based solely on the visual content of the images?
- **RQ2**: How do specific factors such as subject complexity, modifier ambiguity, or image generation model influence prompt inference performance?
- **RQ3**: Can large language models (LLMs) improve human prompt inference through collaborative, combined prompt construction?

Hypotheses.

- **H1**: Humans can infer prompts with moderate success, especially recognizing subjects, but struggle with nuanced modifiers [10].
- **H2**: Inference success correlates with prompt characteristics (length, adjective and clause counts) and the underlying generation model, consistent with prior findings that richer descriptions aid interpretation [16].
- **H3**: Combining human-inferred prompts with AI-generated suggestions (e.g., CLIP Interrogator + GPT-4) will yield higher similarity scores than human inference alone [1, 12].

These questions and hypotheses inform our study design and evaluation (Sect. 4 and Sect. 5.2), enabling us to examine not only the mechanics of human and AI prompt inference but also its implications for prompt ownership and protection in commercial marketplaces.

4   Study Design and Participants

4.1   Prompt and Image Datasets

We generated sample images from two prompt types: *Controlled* and *Uncontrolled*. The two sets differ in the structure and constraint of the prompts, not in the availability of ground-truth text. In both cases, the original prompts are known to the authors and are used to generate the target images' ISM distributions for evaluation.

Controlled prompts provided a baseline for evaluating recognition of common subjects and modifiers across models and participant groups, while uncontrolled prompts introduced greater diversity and served as open-ended inference tasks.

Controlled Dataset. The controlled set was built from prompts scraped from Lexica[1], a public prompt-sharing platform. Using a *Selenium*-based script and *spaCy* parsing, we analyzed over 100,000 prompts to identify frequent subjects: *man, woman, astronaut, cat,* and *robot.* We then paired these with 121 stylistic modifiers drawn from Midjourney's keyword repository [38], spanning lighting, mood, medium, color palette, and perspective categories. Random combinations (1 – 5 modifiers per subject) yielded 100 controlled prompts.

We refer to this dataset as *controlled* because both the subject and the modifier components of each prompt are explicitly known and constrained by the experimental design. This structure enables direct comparison between the original prompt and participant-inferred prompts, and facilitates analysis of which prompt components (e.g., subjects versus modifiers) are more readily recovered.

Uncontrolled Dataset. The uncontrolled set consisted of 100 complete prompts randomly sampled from PromptHero, covering a wide stylistic range. These prompts were used to generate images for Part II of the survey, 25 per model across four `txt2img` systems.

---

[1] https://lexica.art, last accessed 2026/01/24.

4.2   Analyzed `txt2img` Models

We focused on four widely used models representing both base and fine-tuned architectures:

- **MidJourney v5.0** [15]: closed-source diffusion model producing highly stylized 1024×1024 images via Discord interface.
- **Stable Diffusion XL (SDXL)** [20,25]: open-source transformer-based model capable of 1024×1024 generation.
- **DreamShaper XL**: fine-tuned on SDXL for photo-realistic fantasy imagery, maintaining native 1024×1024 resolution.
- **Realistic Vision v5** [5]: fine-tuned on Stable Diffusion 1.5 for realistic photography; outputs were upscaled from 512×512 to 1024×1024 using *High-res.fix* and ScuNET-PSNR for fair comparison.

All target images and participant-submitted prompt generations were produced with the following settings shown in Table 1.

Table 1. Generation parameters used across models.

Parameter	Value
Resolution (Stable Diffusion XL and DreamShaper XL)	1024 × 1024
Resolution (Realistic Vision 5)	512 × 512 → 1024 × 1024 (upscaled)
CFG Scale	5
Sampling Steps	40
Sampling Method	Euler
Upscaling Settings for RV5	
Upscaler	ScuNET PSNR
Denoising Strength	0.4
Restore Faces	On
Hires. Fix	On
Midjourney	Parameters not user-controllable; generations used default model settings

At the study's start (late 2023), these models were the most active on major prompt marketplaces. DALL-E was excluded due to frequent version changes that affected replicability.

4.3   Pre-survey: Demographics and Familiarity

A total of 230 participants (on-campus and online via Amazon Mechanical Turk) completed the survey between December 2023 and March 2024. Most were aged 18 – 44, with a slight male majority (59%), and 16% reporting an arts background. Figure 3 visualizes this breakdown.

Participants' familiarity with generative AI tools was moderate overall: roughly one-third were *"Slightly Familiar"* with image generation tools, with similar familiarity levels across text, audio, and video applications (Fig. 4). This mix provided a balanced range of prior experience for evaluating prompt inference performance.

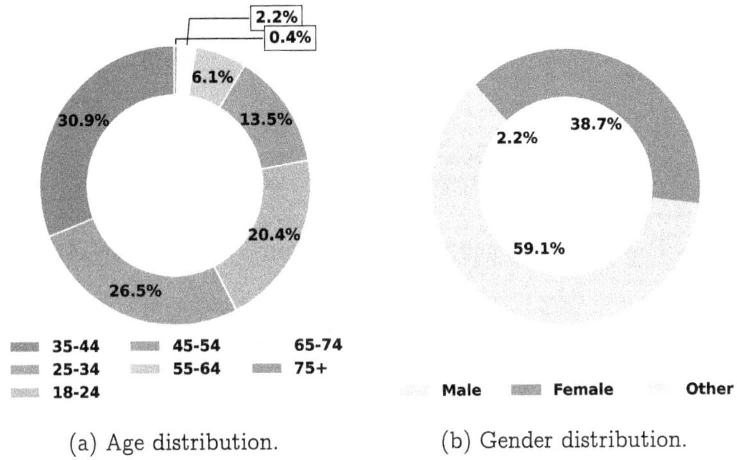

(a) Age distribution.   (b) Gender distribution.

Fig. 3. Demographic distribution of survey participants.

4.4   Survey Design and Flow

For our main user study to assess the ability of humans to infer prompts from AI-generated images, we design and implement a custom tool for our study interface. This tool dynamically loads and displays images together with the instructions, and can capture participant responses to these target images. Figure 5 shows an example of a question that participants saw.

Part I (Uncontrolled Dataset). Participants viewed five AI-generated images and wrote full-sentence prompts describing each, specifying both subject and modifiers. This task measured open-ended inference and linguistic creativity.

Part II (Controlled Dataset). Participants viewed another five images from the controlled set and generated descriptive prompts. Because the true prompts were known, these responses enabled direct comparison of inferred versus original text. An example interface is shown in Fig. 5.

Responses were filtered post-survey: entries were excluded if they lacked a subject or modifier, contained random or blank text, or were non-English. After filtering, we retained 1,141 valid responses for controlled and 1,145 for uncontrolled sets, forming the basis for subsequent evaluation.

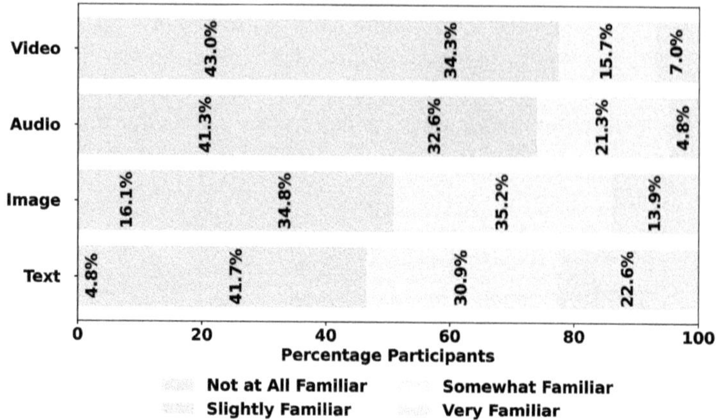

Fig. 4. Participants' familiarity levels with different generative AI tools.

Question: Enter the best fitting prompt

Fig. 5. An example of a survey question where the participant must type in a prompt for the given target image.

5   Experimental Setup

5.1   Image Similarity Metrics

We evaluate prompt inference accuracy by comparing the images generated from participants' inferred prompts to those from the original prompts using several image similarity metrics (ISMs). These metrics capture both perceptual and semantic relationships, providing a comprehensive measure of how closely inferred prompts reproduce the intended visuals.

Image Hash. We use the `imagehash` library[2] to compute perceptual hashes and compare them via Hamming distance. A smaller distance indicates higher similarity between the image generated from the inferred prompt and the original reference image. This metric yields a discrete similarity score ranging from 0 (most similar) to 64 (least similar).

Learned Perceptual Image Patch Similarity (LPIPS). LPIPS [40] quantifies perceptual similarity in deep feature space, reflecting human visual judgments more accurately than pixel-level metrics. Lower LPIPS values indicate greater resemblance between compared images in terms of structure, texture, and color composition. This metric yields a continuous similarity score from 0 (most similar) to 1 (least similar).

Image Embedding Similarity (CLIP Score) . The CLIP score [35] measures cosine similarity between text and image embeddings produced by OpenAI's ViT-L/14 and ViT-B/32 CLIP models[3]. Higher scores imply stronger text – image alignment. We use SentenceTransformers [24] wrappers for consistent embedding generation and joint computation of semantic and visual similarity. This metric yields a continuous similarity score from 0 (least similar) to 1 (most similar).

5.2   Quantifying Successful Inferences

To account for the stochasticity of `txt2img` models, we evaluate prompt inference at the distribution level rather than per-image. For each prompt, we generate 200 reference images (original prompt) and 50 inferred images (participant prompt) across random seeds, treating each set as a sample from an underlying distribution.

We then compare the similarity-score distributions of inferred versus reference generations using the **two-sample Kolmogorov – Smirnov (KS) test** [13,28]. The KS test, being non-parametric and sensitive to both distribution shape and location, robustly captures differences across generative variability. A prompt inference is considered a **hit** if no statistically detectable difference is observed between the two distributions at the 95% confidence level ($p > 0.05$) [3]. Importantly, we do not treat "hit" cases as evidence that two prompts are interchangeable; rather, we interpret these cases as the prompt differences not readily detectable under our evaluation procedure. The **hit rate** is then defined as the proportion of such successful inferences.

---

[2] https://pypi.org/project/ImageHash/, last accessed 2026/01/24.
[3] https://huggingface.co/openai/clip-vit-large-patch14, last accessed 2026/01/24.

Figure 6 illustrates LPIPS distributions for generations from varying prompt types. As prompt specificity decreases, similarity distributions diverge, indicating reduced correspondence between inferred and original prompts.

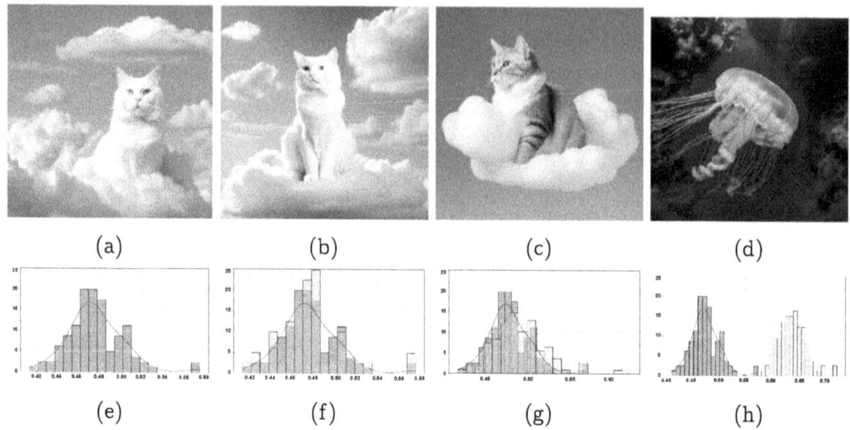

Fig. 6. LPIPS distributions for images generated from different prompts using SDXL. (a,b) identical prompts; (c) less specific; (d) different subject. Divergence in (f–h) reflects decreasing similarity between inferred and reference prompts.

This distribution-aware evaluation extends prior single-image heuristics, offering a reproducible, model-agnostic framework for quantifying prompt inference success. Midjourney outputs were excluded from these analyses due to seed-control limitations inherent to its closed-source interface.

5.3   Human AI Combined Inference

To assess collaboration effects, we combined human-inferred prompts with AI-generated prompts from the CLIP Interrogator [2], which produces descriptive text aligned with an input image's CLIP embeddings. Each human – AI prompt pair was merged using GPT-4 [19], instructed to synthesize a single prompt of up to 25 words without adding new concepts. For example:

> **Instruction:** Combine these two prompts into a 25-word description without extra details.
> 1. man dressed in steampunk in a steampunk factory
> 2. a man in a steampunk suit and top hat standing in front of a giant clock with gears, Bastien L. Deharme, fantasy art
>
> **GPT-4 Response:** a man in a steampunk suit and top hat stands in a factory surrounded by gears and clocks, embodying a fantasy art portrait.

Capping prompt length ensured conciseness and prevented duplication of keywords that could bias generations. Using an LLM for merging, rather than simple concatenation, preserved linguistic coherence and balance between human and AI contributions. These combined prompts were later compared with human-only and AI-only prompts to evaluate synergy effects (see Sect. 6.3).

6   Results and Analysis

We report findings following the setup in Sect. 5. We first measure human performance on prompt inference, then evaluate human-AI combined prompting, and finally analyze factors that influence success, including prompt type, model, and linguistic structure.

6.1   RQ1: Human Performance on Prompt Inference Tasks

For both the controlled and uncontrolled dataset, we quantify performance using the hit-rate criterion in Sect. 5.2 across ImageHash, LPIPS, and CLIP variants. Similarity score distributions from participant-inferred prompts are compared to those generated from the corresponding original prompt. These results are summarized in Table 2.

Table 2. Successful prompt inference (hits) by metric and target prompt set.

Similarity Metric	Controlled Hit Rate (%)	Uncontrolled Hit Rate (%)
ImageHash	53.29	53.72
LPIPS	22.89	9.77
CLIP B32	7.28	7.56
CLIP L14	7.05	7.21

Hit rates differ by metric. ImageHash is highest (about 53%) but captures shallow visual cues only and yields discrete distance values, which can inflate success and produce false positives relative to human judgment [31]. LPIPS is lower, especially for uncontrolled prompts, reflecting the difficulty of replicating fine perceptual details from open-ended images. CLIP metrics are lowest, indicating that matching semantic intent is hardest. Given these limitations, we exclude ImageHash from subsequent analyses and focus on LPIPS and CLIP.

H1 is partially supported. Participants often recover subjects but struggle with modifiers, which reduces LPIPS and CLIP hit rates. Differences between controlled and uncontrolled sets indicate that open-ended prompts increase difficulty.

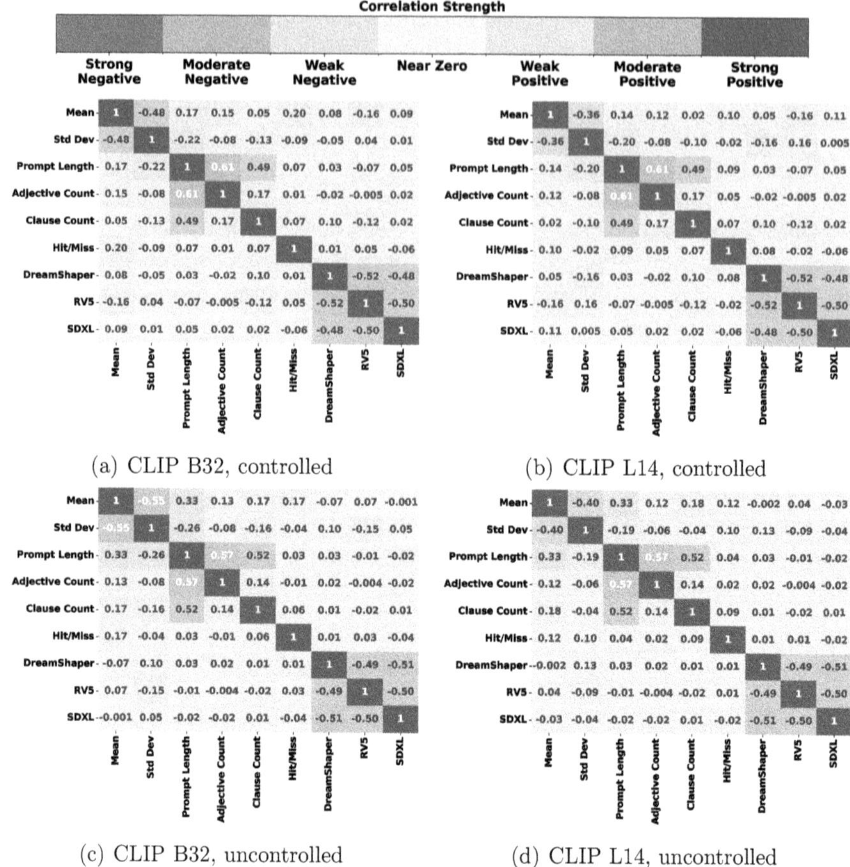

(a) CLIP B32, controlled (b) CLIP L14, controlled

(c) CLIP B32, uncontrolled (d) CLIP L14, uncontrolled

Fig. 7. CLIP correlations with prompt features, similarity statistics, hit outcomes, and model identity. Binary variables use point-biserial associations. One-hot indicators are mutually exclusive.

6.2 RQ2: Impact of Prompt Linguistic Features and Model Choice

We examine correlations between linguistic features, similarity statistics, hit outcomes, and model identity for CLIP and LPIPS.

Figure 7 shows consistent linguistic patterns across CLIP variants: longer prompts correlate with more adjectives and clauses; mean CLIP has weak positive associations with success, while higher mean alignment tends to coincide with lower variance. Model effects are modest but consistent. DreamShaperXL associates with slightly better success and lower variability, while RV5 shows greater variability and weaker success. SDXL is largely neutral.

For LPIPS (Fig. 8), mean LPIPS correlates negatively with success (controlled $r = -0.28$, uncontrolled $r = -0.05$), so lower LPIPS relates to more hits. Mean LPIPS also has small negative correlations with length, adjectives, and

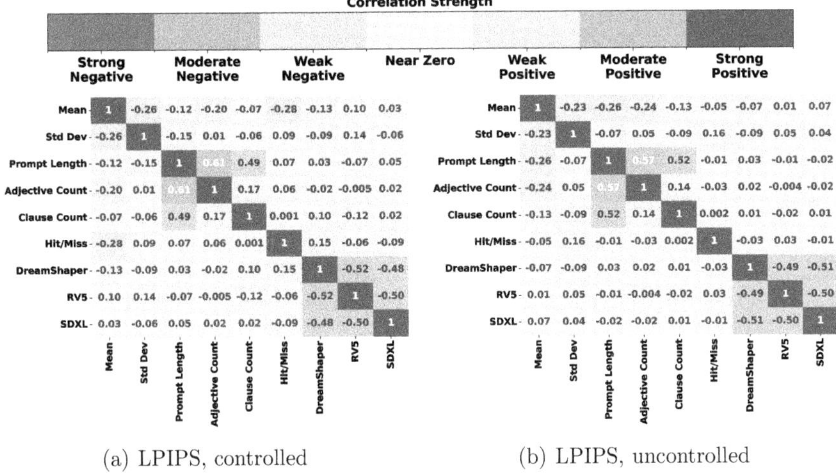

(a) LPIPS, controlled   (b) LPIPS, uncontrolled

Fig. 8. LPIPS correlations with prompt features, similarity statistics, hit outcomes, and model identity.

clauses. LPIPS variance has weak positive correlation with success (controlled $r = 0.09$, uncontrolled $r = 0.16$). DreamShaperXL shows the strongest positive relation with success (controlled $r = 0.15$) and the most negative relation with mean LPIPS (controlled $r = -0.13$, uncontrolled $r = -0.07$). RV5 tends toward higher variability (controlled $r = 0.14$, uncontrolled $r = 0.05$).

H2 is partially supported. Linguistic richness and model choice have modest effects. DreamShaperXL generations are more stable and perceptually aligned, which aids inference. RV5 is more variable. Trends are consistent across controlled and uncontrolled sets.

6.3   RQ3: Human-AI Combined Prompt Inference

We compare hit rates for human-only prompts with human-AI combined prompts formed via CLIP Interrogator plus GPT-4 merging (Sect. 5.3). Results are in Table 3. ImageHash is reported for completeness. Combined prompts improve ImageHash but not LPIPS or CLIP. This indicates better low-level resemblance without improved perceptual or semantic alignment.

H3 is not supported. The current merging strategy does not outperform human-only prompting on perceptual or semantic metrics. Future collaborative methods should target alignment with LPIPS and CLIP, not shallow similarity.

Table 3. Hit rates for human-only and human–AI combined prompts.

Similarity Metric	Human Only Hit Rate (%)	Human–AI Combined Hit Rate (%)
ImageHash	53.29	60.62
LPIPS	22.89	18.59
CLIP B32	7.28	5.89
CLIP L14	7.05	5.77

6.4   Discussion of Key Findings

Humans can often infer core subjects, especially in controlled settings, but struggle to recover modifiers and stylistic intent. Metric choice matters: ImageHash is optimistic and misaligned with human judgment [31], while LPIPS and CLIP are stricter and reveal lower success for open-ended prompts. Linguistic richness and model choice show modest but consistent effects; DreamShaperXL tends to produce more stable, aligned generations that are somewhat easier to approximate. Human-AI prompt merging increased shallow similarity only and reduced LPIPS and CLIP performance. Overall, results suggest that precise semantic alignment remains difficult in unconstrained contexts, which provides some resilience for prompt-based intellectual property.

Conclusion

This study presents a human-subject evaluation of prompt inference in AI-generated images, with implications for content authorship, web-based generative tools, and the emerging economy of prompt marketplaces. Our findings show that while human participants can infer broad subject matter from an image, they often struggle with accurately reconstructing modifiers and stylistic intent, particularly in unconstrained contexts. This limitation becomes more apparent when evaluated with perceptual and semantic similarity metrics such as LPIPS and CLIP, which better reflect alignment with the original prompt.

Notably, human-AI combined prompts did not outperform human inferences alone. Instead, naïve merging strategies often diluted semantic coherence, resulting in lower similarity scores under meaningful evaluation metrics. These findings suggest that while generative tools enable accessible creativity, effective collaboration between human users and AI systems remains a design challenge. Moreover, participant background, including AI familiarity, had limited impact on success rates, implying that prompt inference performance is more strongly influenced by task and system-level constraints than by user expertise.

Overall, these results provide evidence that prompts are not easily reverse-engineered from generated outputs. This supports the relative resilience of prompt-based intellectual property in current generative workflows. For web-based platforms hosting generative models and prompt-sharing ecosystems, this raises key considerations around content provenance, authorship claims, and secure prompt design.

References

1. Brade, S., Wang, B., Sousa, M., Oore, S., Grossman, T.: Promptify: text-to-image generation through interactive prompt exploration with large language models. In: ACM UIST (2023)
2. https://huggingface.co/spaces/pharmapsychotic/CLIP-Interrogator . ((web)). Accessed 2026/01/24
3. Fisher, R.A.: Statistical methods for research workers. In: Breakthroughs in statistics: Methodology and distribution, pp. 66–70. Springer (1970)
4. Gretton, A., Borgwardt, K.M., Rasch, M.J., Schölkopf, B., Smola, A.: A kernel two-sample test. J. Mach. Learn. Res. **13**(1), 723–773 (2012)
5. SG161222/Realistic-Vision-V5.1-noVAE. https://huggingface.co/SG161222/Realistic-Vision-V5.1-noVAE. ((web)). Accessed 2026/01/24
6. Lanz, J.A.: inside the lucrative new business of selling AI prompts. https://decrypt.co/137689/lucrative-new-business-selling-ai-prompts. ((web)). Accessed 2026/01/24
7. Li, J., Li, D., Xiong, C., Hoi, S.: Blip: Bootstrapping language-image pre-training for unified vision-language understanding and generation. In: ICML (2022)
8. Liu, V., Chilton, L.B.: Design guidelines for prompt engineering text-to-image generative models. In: Proceedings of the 2022 CHI Conference on Human Factors in Computing Systems, pp. 1–23 (2022)
9. Lopez-Paz, D., Oquab, M.: Revisiting classifier two-sample tests. arXiv preprint arXiv:1610.06545 (2016). Accessed 2026/01/24
10. Lu, Z., et al.: Seeing is not always believing: benchmarking human and model perception of AI-generated images. Adv. Neural. Inf. Process. Syst. **36**, 25435–25447 (2023)
11. Lyu, Y., Wang, X., Lin, R., Wu, J.: Communication in human–AI CO-creation: perceptual analysis of paintings generated by text-to-image system. Appl. Sci. **12**(22) (2022)
12. Mahdavi Goloujeh, A., Sullivan, A., Magerko, B.: Is it AI or is it me? Understanding users' prompt journey with text-to-image generative ai tools. In: Proceedings of the 2024 CHI Conference on Human Factors in Computing Systems, pp. 1–13 (2024)
13. Massey, F.J., Jr.: The kolmogorov-smirnov test for goodness of fit. J. Am. Stat. Assoc. **46**(253), 68–78 (1951)
14. MidJourney: methexis-inc / img2prompt. https://replicate.com/methexis-inc/img2prompt ((web)). Accessed 2026/01/24
15. MidJourney: Midjourney Model Versions. https://docs.midjourney.com/docs/model-versions ((web)). Accessed 2026/01/24
16. Moeßner, P., Adel, H.: Human vs. AI: a novel benchmark and a comparative study on the detection of generated images and the impact of prompts. In: Proceedings of the 1stWorkshop on GenAI Content Detection (GenAIDetect), pp. 47–58 (2025)
17. Nichol, A.Q., et al.: Glide: towards photorealistic image generation and editing with text-guided diffusion models. In: ICML (2022)
18. CLIP: Connecting text and images. https://openai.com/research/clip (2021). Accessed 2026/01/24
19. OpenAI, R.: GPT-4 Technical Report. arXiv preprint (2023). Accessed 2026/01/24
20. Podell, D., et al.: Sdxl: Improving latent diffusion models for high-resolution image synthesis. arXiv preprint arXiv:2307.01952 (2023). Accessed 2026/01/24
21. https://promptbase.com/tandcs . ((web)). Accessed 2026/01/24

22. Terms of Service | Promptrr.io. https://promptrr.io/terms-of-service/ ((web)). Accessed 2026/01/24
23. Ramesh, A., et al.: Zero-shot text-to-image generation. In: ICML (2021)
24. Reimers, N., Gurevych, I.: Sentence-BERT: sentence embeddings using Siamese BERT-networks. arXiv preprint arXiv:1908.10084 (2019). Accessed 2026/01/24
25. Rombach, R., Blattmann, A., Lorenz, D., Esser, P., Ommer, B.: High-resolution image synthesis with latent diffusion models. In: IEEE/CVF CVPR (2022)
26. Schuhmann, C., et al.: Laion-5B: an open large-scale dataset for training next generation image-text models. Adv. Neural. Inf. Process. Syst. **35**, 25278–25294 (2022)
27. Shen, X., Qu, Y., Backes, M., Zhang, Y.: Prompt stealing attacks against text-to-image generation models. In: 33rd USENIX Security Symposium (USENIX Security 24), pp. 5823–5840 (2024)
28. Smirnov, N.: Table for estimating the goodness of fit of empirical distributions. Ann. Math. Stat. **19**(2), 279–281 (1948)
29. Struppek, L., Hintersdorf, D., Kersting, K.: Rickrolling the artist: injecting invisible backdoors into text-guided image generation models. arXiv preprint arXiv:2211.02408 (2022). Accessed 2026/01/24
30. Timothy, M.: Are Premium AI Prompts Worth the Money? https://www.makeuseof.com/should-you-buy-ai-prompts/ ((web)). Accessed 2026/01/24
31. Trinh, K., Seidenberger, S., Wijewickrama, R., Jadliwala, M., Maiti, A.: A picture is worth a thousand prompts? Efficacy of iterative human-driven prompt refinement in image regeneration tasks. In: Kwok, J. (ed.) Proceedings of the Thirty-Fourth International Joint Conference on Artificial Intelligence, IJCAI-25. pp. 10189–10197. International Joint Conferences on Artificial Intelligence Organization (2025). https://doi.org/10.24963/ijcai.2025/1132. AI, Arts & Creativity
32. United States copyright office: copyright registration guidance: works containing material generated by artificial intelligence (2023). https://www.federalregister.gov/documents/2023/03/16/2023-05321/copyright-registration-guidance-works-containing-material-generated-by-artificial-intelligence. Accessed 2026/01/24
33. United States copyright office: copyright and artificial intelligence part 2: copyrightability (2025). https://www.copyright.gov/ai/Copyright-and-Artificial-Intelligence-Part-2-Copyrightability-Report.pdf. Accessed 2026/01/24
34. United States court of appeals: thaler v. perlmutter - copyright rejection of AI-generated art (2025). https://media.cadc.uscourts.gov/opinions/docs/2025/03/23-5233.pdf. Accessed 2026/01/24
35. Wang, J., Chan, K.C., Loy, C.C.: Exploring clip for assessing the look and feel of images. In: AAAI Conference on Artificial Intelligence. vol. 37 (2023)
36. Wang, J., et al.: Review of large vision models and visual prompt engineering. Meta-Radio. 100047 (2023)
37. Wu, Y., Yu, N., Li, Z., Backes, M., Zhang, Y.: Membership inference attacks against text-to-image generation models. arXiv preprint arXiv:2210.00968 (2022). Accessed 2026/01/24
38. Wulfken, W.: MidJourney Styles and Keywords Reference. https://github.com/willwulfken/MidJourney-Styles-and-Keywords-Reference/tree/main/Pages/MJ_V4/Style_Pages/Just_The_Style ((web)). Accessed 2026/01/24
39. Zhai, S., Dong, Y., Shen, Q., Pu, S., Fang, Y., Su, H.: Text-to-image diffusion models can be easily backdoored through multimodal data poisoning. In: Proceedings of the 31st ACM International Conference on Multimedia, pp. 1577–1587 (2023)
40. Zhang, R., Isola, P., Efros, A.A., Shechtman, E., Wang, O.: The unreasonable effectiveness of deep features as a perceptual metric. In: IEEE/CVF CVPR (2018)

Fluid Body: An Adaptive Embodied Sonification System for Cross-Cultural Performance

Yuting Xue[1]([✉]) [iD] and Yueshen Wu[2] [iD]

[1] City University of Hong Kong, Hong Kong, Hong Kong, SAR of China
`yuting.xue@my.cityu.edu.hk`
[2] Royal College of Art, London, UK
`yueshen.wu@network.rca.ac.uk`

Abstract. *Fluid Body* is an experimental, body-driven instrument with which participants can use a compact classical ballet arm-posture vocabulary to play a digital Guzheng. The system implements an interactive machine-learning mapping that combines a posture classifier to label each recognized posture, with a regression model for continuous control of granular sound synthesis parameters. Six ballet arm positions are associated with six edited Guzheng sound events, enabling participants to trigger distinct phrase regions through posture changes while continuously modulating the timbre in real time. We use intercultural co-adaptation in a dynamic, operational way. Rather than claiming literal translation between ballet and Guzheng traditions, the system creates a performable correspondence space where a codified movement vocabulary can be used to explore a traditional instrument's playing through an adaptive mapping. *Fluid Body* therefore contributes a concrete hybrid mapping design for embodied sonification in an intercultural context, demonstrating how machine learning can serve as a creative translator between different cultural artistic expressions.

Keywords: Cross-cultural Performance Art · Supervised ML-based System · Embodied Sonification

1 Introduction

Machine learning (ML) has increasingly emerged as a generative framework within computational creativity, extending beyond its technical role to act as an active collaborator in artistic creation. It is increasingly viewed as embodied media that engages with art-making's performative and sensory aspects [16].

Granular synthesis is a sound generation technique that breaks an audio signal into short fragments called "grains," which typically last around 5 to 100 milliseconds. By rearranging these grains through parameter control, effects like

---

Y. Xue and Y. Wu—Equal contribution.

© The Author(s), under exclusive license to Springer Nature Switzerland AG 2026
P. Machado et al. (Eds.): EvoMUSART 2026, LNCS 16523, pp. 161–175, 2026.
https://doi.org/10.1007/978-3-032-24350-8_11

time-stretching and pitch shifting can be achieved while retaining the source's material identity [15]. Parameter mapping sonification describes systematic mappings from data to sound synthesis parameters, and has increasingly been applied in creative contexts where mapping design itself becomes an artistic medium. [11]. Within this paradigm, combining embodied sensing, ML-based mapping, and sound synthesis has become a productive route for developing new musical interfaces and performance systems that distribute agency across performers and algorithms. [13].

Prior ML-based creative sonification systems have used movement data to classify musical genres [7]; generate sound structures from gesture [3]; and recognize emotional expressions in music performance [23]. In addition, ML has played an important role in cultural preservation, enabling traditional performance practices to be recorded, modeled, and reinterpreted as data, thereby contributing to the digital continuity of intangible cultural heritage [14]. These studies demonstrate how computer systems serve as innovative mediators between bodily expression, musical structure, and cultural representation.

Building upon these works, *Fluid Body* is an embodied sonification system that uses a codified vocabulary of ballet arm postures to engage with Guzheng-derived sound materials (Fig. 1). This project invites participants to "play" the digital Guzheng through ballet-inspired movements in a 180-degree curved projection setup. Based on Laban Movement Analysis (LMA), six ballet arm positions are linked with six related Guzheng sound events. In each state, additional motion features drive post-effects to extend timbral variation. This system uses a hybrid mapping learned through demonstration in an interactive ML platform: a classification model provides discrete posture-state selection to switch sound events, while a regression model provides continuous control over granular synthesis playback position.

Instead of treating culture as static heritage, we use intercultural co-adaptation in an operational sense. Through an adaptive ML mapping, this system transforms the movement vocabulary from one traditional practice to another sonic material. This work contributes (i) a creative strategy for selecting and pairing postures and sound events, (ii) a detailed hybrid classification-regression mapping design for granular synthesis playback control, and (iii) formative evidence focused on interaction legibility, controllability, and perceptual distinctiveness.

2   Related Work

2.1   Embodied Sonification

The notion of "Borrowed Gestures" [8] introduces a gestural vocabulary of non-musical bodily movements, including dance, sign language, and expressive motion, and links them with nuanced instrumental gestures. This approach deploys a movement sonification and highlights it as integral to musical ontology rather than simply external commands that control sound.

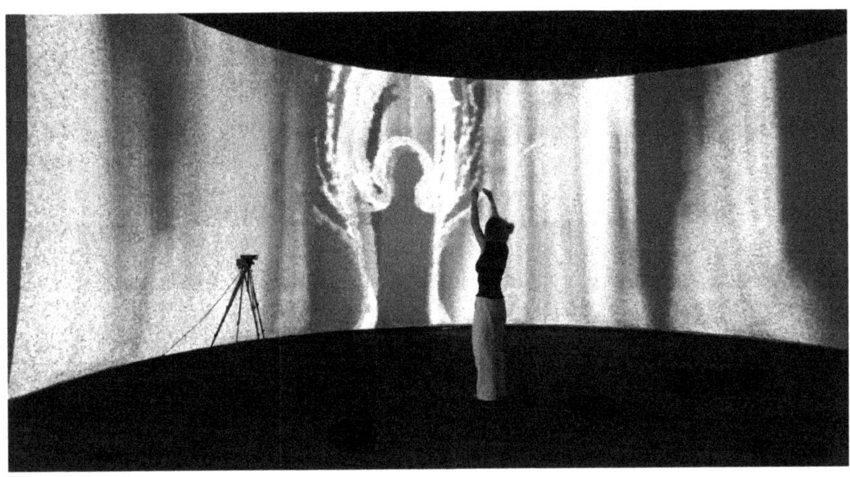

Fig. 1. The participant is performing a ballet posture in *Fluid Body* with the corresponding visual feedback presented.

Somacoustics [4] extends this principle through participatory performance: audiences physically engage with the blindfolded artist, whose gestures are sonified into a four-channel synthesized soundscape. Audiences play the artist's "instrument-body" and dance together based on social-bodily and auditory interactions. The *Embodied sounds Visual Interaction Electromyogram* [19] also uses gesture-to-sound mapping middleware to turn signals from the performer's muscles into synthesized sounds, exploring biosignals as digital musical instruments with contemporary ML and synthesis modules.

2.2   Dance and Cultural Preservation

A few studies have investigated the potential of computational tools to revitalize creativity and preserve culture. Afro-Brazilian performances such as the dance of *Caboclo de Pena* [5] employ motion-perception systems and real-time audio processing to explore the body as a resonant system of culture and sound. Similarly, choreographic game technologies have been used to record intangible cultural-heritage dances [22]. The movement trajectories of the Azeri folk dance are transformed into spatial dance data for interactive sonification. Combined with gamified interactions, this work enables the systematic preservation and reproduction of dance-driven music. *Chang-E* [21] develops a large-scale motion-capture dataset for the Dunhuang Dance, a classical dance in Chinese murals. The program employs ML models to recreate movement, and aligns the reconstructed dance segments with musical beats to re-establish temporal and aesthetic continuity between dance and music.

To sum up, existing systems reveal how machine learning enables adaptive mappings between physical and sonic parameters, positioning the body as an expressive medium and a source of sound. However, they seldom consider movement vocabulary with cultural encoding and symbolic meaning as the algorithmic interface for sound generation. Unlike prior works, *Fluid Body* does not treat the body merely as a signal source nor the movement as a monocultural static heritage. It combines the symbolic geometry of ballet with Guzheng-derived sound materials through granular synthesis, positioning posture as an adaptive element within an evolving aesthetic framework. In this way, machine learning supports an algorithmic correspondence between culturally situated forms, allowing aesthetic meaning to emerge through ongoing bodily–sonic interaction.

3   Fluid Body

Fluid Body presents an embodied, ML-based instrument that translates ballet movements into the tonal and timbral language of the traditional Chinese Guzheng. It is exhibited in a 180-degree curved projection space that surrounds the participant's frontal field of view (Fig. 2). To interpret the rhythmic and textural features of Guzheng and ballet, the system visualizes the participant's body silhouette with shadow and fog. The stylized mountains and clouds change in real time to respond to the sounds generated by movement. The palette is constrained to black and white, emphasizing the essence of landscape scenery. The system considers the human body as a computing interface where movement is a structural and expressive part of producing sound. Within this framework, the project explores how machine learning can establish adaptive mappings between culturally coded postures and sonic parameters to form an intercultural performance environment.

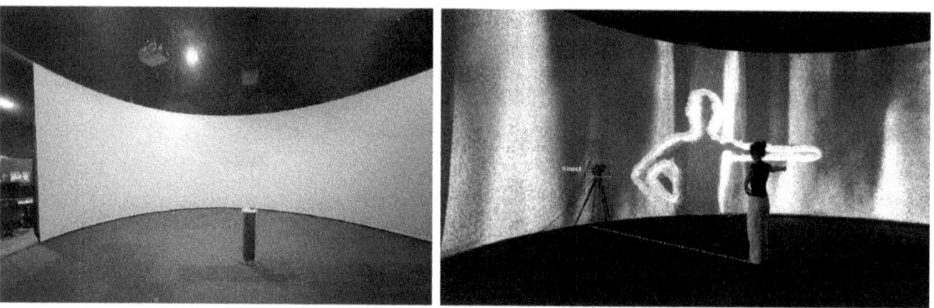

Fig. 2. The shoot from experience setup in the 180-degree gallery space.

3.1   Design Concept

Guzheng is a traditional Chinese plucked string instrument with 21 strings capable of generating seven notes across four octaves. The player plucks the strings with the right hand, coordinating with the left hand to modulate the pitch and add ornamentation. Finger techniques include "vibrato," where the string is rocked up and down to create a wavering pitch; "portamento," a smooth sliding transition connecting the pitches in a continuous motion; and "glissando," a rapid sweep across several strings that produces a flowing cascade of notes [17]. Ballet is a highly codified Western dance organized around four spatial directions (*en bas, en avant, à la seconde, en haut*) and five arm positions plus one basic "Bras bas" [20].

Although Guzheng and ballet come from different cultural traditions, both rely on embodied principles of rhythm, modulation, and fluidity. Both emphasize the continuity of gesture, controlled flow, and emotional nuance, prioritizing subtle modulation over raw power. Guzheng players often feature vibrato and dynamic rhythmic changes reflecting the natural flow of emotion. Similarly, ballet dancers utilize breath, suspension, and release to shape emotion. Neither strictly adheres to an external beat; instead, both practices cultivate an internalized sense of pulse shaped by breathing, emotion, and narrative rhythm. This makes them well-suited for algorithmic reinterpretation as co-evolving aesthetic systems.

3.2   Use of Laban Movement Analysis to "Decode" Ballet Postures

To connect ballet movement with Guzheng sound while preserving their cultural characteristics, the system adopts Laban Movement Analysis (LMA) as an interpretive framework [18]. Four analytical dimensions, "Space," "Shape," "Effort," and "Body," provide the model with structured parameters to decode six ballet arm postures that relate to Guzheng phrases.

The "Space" Dimension is used to describe four spatial directions of arms in classical ballet. In the system, these directions are treated as spatial vectors and mapped to different pitch registers of the Guzheng, ranging from low to high. Downward (*en bas*) movement is associated with lower and darker pitch regions, while upward (*en haut*) movement is linked to higher and brighter registers. Forward (*en avant*) and sideways (*à la seconde*) orientations occupy intermediate ranges. Through this mapping, movement direction becomes a means of navigating the Guzheng's pitch space.

The "Shape" Dimension corresponds to six classical arm positions, which establish momentary forms of stability in motion. Ballet uses a limited set of clearly defined arm positions that establish visual clarity and structure. These shapes are mapped to different Guzheng phrases with distinct timbral and melodic characteristics. Enclosing positions route the output toward sustained and connected material, while open and expansive shapes route it toward brighter and more segmented material. This makes the phrase structure respond to the body's geometry.

The "Effort" Qualities contain weight, time, space, and flow that describe the dynamic texture of motion. Differences in movement weight correlate with spectral density and harmonic thickness. Stronger motion produces denser sound, while lighter motion results in thinner textures. Flow connects to the ongoing changes parameter: bound movement producing more stable and constrained sound behavior, while free movement allows for wider pitch variation and smoother modulation.

The "Body" Category shows how the arms, shoulders, and torso work together. Similar to how the hands coordinate and control the string resonance in Guzheng playing, this aspect supports mappings between upper-body organization and the timbral shaping. The system analyzes how motion energy spreads across the body and links it to parameters that affect spatial depth and texture over time. This reinforces the intended coupling between embodied form and Guzheng-derived sound.

4   Embodied Sonification System

The system is comprised of three sections: "motion capture and data processing"; "supervised ML model training"; and "granular synthesis based parameter mapping sonification and visual generation" (Fig. 3). A Kinect V2 sensor first captures a participant's skeleton, streaming to TouchDesigner to process, and sends features to Wekinator, an interactive supervised ML environment through Open Sound Control (OSC). Wekinator is trained to associate specific output values with input feature sets using built-in classification and regression models [9]. Predicted outputs return to TouchDesigner via OSC for additional mapping, and are then forwarded to Ableton Live via "TD-Ableton" Package to control Granulator 3, a modulatable granular synthesizer [1,10] together with effects including "Resonator," "Grain Delay," and "Hybrid Reverb."

Leveraging Wekinator's learned mappings, we assign distinct Guzheng phrases and processing states to individual ballet postures; once the model is trained, the system renders a continuously morphing soundscape responsive to the participant's movement. Finally, using a virtual audio interface to route audio between applications, the sonic output is fed back into TouchDesigner for visual generation. This multi-application OSC pipeline produces a unified, low-latency audio-visual experience; the following sections detail each stage.

4.1   Motion Capture and Data Processing

We use Kinect V2 for raw motion data capture. Microsoft's body tracking stack fits an articulated model to the depth image and returns 25 joints per person, which has made Kinect V2 a durable choice for interactive performance systems [2,12].

For the six target hand and arm postures illustrated in Fig. 4, we extracted the two-dimensional (x, y) coordinates of seven skeletal joints from the motion

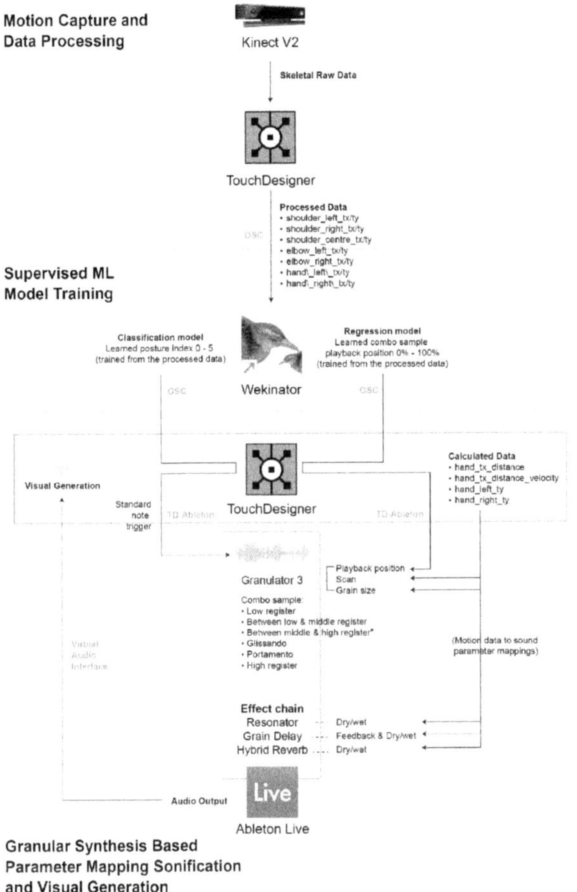

Fig. 3. System overview.

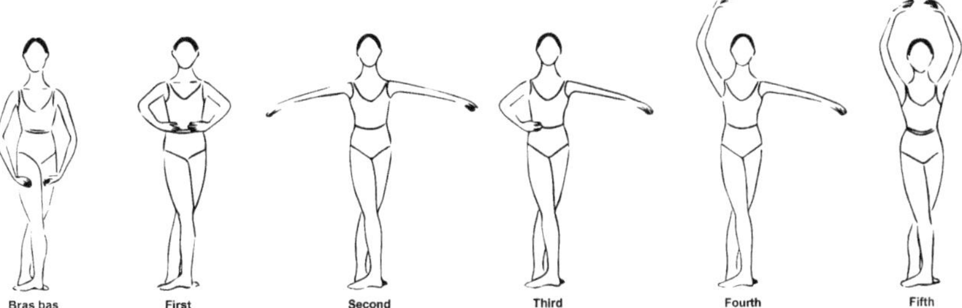

Fig. 4. The illustration of ballet arm positions.

capture stream, including the left, right, and center shoulders; left and right elbows; and left and right hands. Accordingly, a total of 14 feature channels were obtained: "shoulder_left_tx/ty," "shoulder_right_tx/ty," "shoulder_centre_tx/ty,"   "elbow_left_tx/ty,"   "elbow_right_tx/ty," "hand_left_tx/ty," and "hand_right_tx/ty." During capture, the Kinect V2 is mounted so that the optical center is 1200 mm above the floor and 2000 mm from the performer. Table. 1 summarizes the 14-channel skeletal data recorded in different ballet postures. In the current prototype, the performer presents each posture for approximately 3 to 6 s, yielding 180 to 360 samples per class at TouchDesigner's default 60 fps. This short recording period aims to support faster training iteration and reduce posture drift during holds. Silhouette masks are also extracted for the subsequent visual pipeline.

Table 1. Data collected from the ballet posture recording session.

Ballet posture		Bras bas	1st	2nd	3rd	4th	5th
Batch size of collected data		203	261	208	223	233	332
Movement duration		3s 383ms	4s 350ms	3s 466ms	3s 716ms	3s 883ms	5s 533ms
Data range	shoulder_left_tx	(-0.123, -0.111)	(-0.123, -0.093)	(-0.118, -0.108)	(-0.114, -0.057)	(-0.116, -0.072)	(-0.116, -0.101)
	shoulder_right_tx	(0.205, 0.217)	(0.189, 0.205)	(0.207, 0.228)	(0.225, 0.235)	(0.193, 0.225)	(0.194, 0.210)
	shoulder_centre_tx	(0.058, 0.071)	(0.058, 0.070)	(0.071, 0.076)	(0.066, 0.078)	(0.038, 0.067)	(0.037, 0.067)
	elbow_left_tx	(-0.156, -0.136)	(-0.239, -0.156)	(-0.370, -0.319)	(-0.327, -0.098)	(-0.194, -0.146)	(-0.266, -0.169)
	elbow_right_tx	(0.265, 0.268)	(0.223, 0.294)	(0.333, 0.461)	(0.448, 0.472)	(0.408, 0.447)	(0.273, 0.428)
	hand_left_tx	(-0.086, -0.046)	(-0.349, 0.029)	(-0.676, -0.488)	(-0.488, -0.011)	(-0.129, -0.083)	(-0.190, -0.065)
	hand_right_tx	(0.125, 0.167)	(0.083, 0.252)	(0.482, 0.733)	(0.727, 0.741)	(0.645, 0.730)	(0.156, 0.645)
	shoulder_left_ty	(-0.059, -0.055)	(-0.056, -0.018)	(-0.020, -0.018)	(-0.043, 0.009)	(-0.037, 0.054)	(0.004, 0.060)
	shoulder_right_ty	(-0.060, -0.059)	(-0.059, -0.019)	(-0.033, -0.025)	(-0.032, -0.025)	(-0.032, -0.015)	(-0.015, 0.047)
	shoulder_centre_ty	(-0.006, 0.004)	(0.004, 0.013)	(0.016, 0.022)	(0.014, 0.021)	(0.020, 0.033)	(0.024, 0.047)
	elbow_left_ty	(-0.277, -0.252)	(-0.252, -0.072)	(-0.081, -0.057)	(-0.094, -0.020)	(-0.028, 0.290)	(-0.123, 0.295)
	elbow_right_ty	(-0.267, -0.248)	(-0.256, -0.052)	(-0.107, -0.067)	(-0.129, -0.102)	(-0.123, 0.012)	(-0.091, 0.264)
	hand_left_ty	(-0.501, -0.439)	(-0.439, -0.035)	(-0.042, 0.009)	(-0.037, 0.125)	(0.150, 0.604)	(-0.075, 0.604)
	hand_right_ty	(-0.513, -0.437)	(-0.437, -0.103)	(-0.106, -0.036)	(-0.132, -0.100)	(-0.157, 0.209)	(-0.002, 0.569)

4.2   Supervised ML Model Training

After feature extraction, we learn the mapping from ballet postures to Guzheng playback control using interactive supervised learning in Wekinator. Wekinator enables artists to construct mappings by demonstration within a performer-in-the-loop workflow: users stream feature data over OSC, label desired outputs, trigger short training runs, and redeploy results immediately [9]. In this work, we use two of Wekinator's model families: a classifier for discrete posture state, and a regressor for continuous control. While Wekinator provides multiple algorithms, the current prototype uses the default k-Nearest Neighbor classifier and the default neural-network regressor for rapid iteration with minimal configuration.

Our system requires: (i) posture recognition to support discrete state changes that are perceptually legible, and (ii) continuous control to enable smooth timbral evolution within and between each posture. In gestural Digital Musical Instrument (DMI) research, the use of classification for discrete-event selection or mode switching and regression for continuous-parameter control has been prevalent [6]. We therefore implement a hybrid mapping trained on the same 14-channel feature stream. The classifier outputs one of six posture labels, used as a state index that determines when the system should retrigger playback. In parallel, the regressor outputs a continuous value mapped to the granular engine's "scan position" within a concatenated Guzheng "combo buffer," enabling subtle variation during posture holds, such as breathing or deviations from the canonical pose, to produce smooth timbral change. This model combination supports both legibility and interpolation within a single sensing and control system.

Table 2. Mapping information of ballet postures and sound events.

Ballet posture	Related sound events	Index	Original source	Selection & Duration	Percentage of the combo sample
Bras bas	Low register	0	*Autumn Moon Over Han Palace*	00:15:66 to 00:24:91 from the source; 9s25 in total	0%
First	Between low and middle register	1	*Lin Chong Flees by Night*	00:29:92 to 00:42:92 from the source; 13s in total	12%
Seond	Between middle and high register	2	*High Mountains and Flowing Water*	01:14:03 to 01:28:71 from the source; 14s67 in total	29%
Third	Glissando	3	*Lin Chong Flees by Night*	00:02:03 to 00:14:66 from the source; 12s63 in total	48%
Fourth	Portamento	4	*Autumn Moon Over Han Palace*	00:00:00 to 00:14:94 from the source; 14s94 in total	64.5%
Fifth	High register	5	*High Mountains and Flowing Water*	02:24:59 to 02:37:09 from the source; 12s5 in total	84%

Source audio consists of three classic Guzheng pieces: *Autumn Moon Over the Han Palace* (汉宫秋月) demonstrates the elegant, melancholic style of traditional court music; *Lin Chong Flees by Night* (林冲夜奔) portrays dramatic tension drawn from classical literature; and *High Mountains and Flowing Water* (高山流水) conveys the philosophical ideal of harmony between humanity and nature. All materials are tuned to D major. From each piece, we extract two 9–15 second snippets, known as "sound events," aligned to the six target postures. These events are placed end-to-end to form a single combo buffer (Table. 2). During playback, the regressor controls the scan position within this buffer. When the predicted posture label changes, the system sends a new MIDI note-on. This action restarts the granular engine at the corresponding segment, allowing for smooth transitions between posture-linked states.

Our current prototype prioritizes expressive interaction over exact posture matching, so we did not frame posture recognition as a benchmark classification task using an external dataset or report formal accuracy metrics. Instead, we used formative validation within a performer-in-the-loop workflow, which aligns with interactive ML practice. After each short training run, we redeployed the models immediately and then rehearsed scripted sequences of posture holds and changes. We checked whether the posture state remained stable during holds, how often the system mistakenly triggered, whether the regression output changed smoothly as it controlled the scan position, and whether the full pipeline responded smoothly from sensing to ML to audio and visuals. When failures such as sudden jumps in scan position happened, demonstrations near unclear transitions were added or adjusted. Smoothing in the control mapping was also adjusted until the behavior was stable enough for interaction. We treat this development-time validation as suitable for a design probe, since the main goal is reliable and interpretable behavior in artistic use.

Table 3. Mapping of hand movement data and sound effect parameters in Ableton.

Movement and spatial dimensions	Sample range from the movement data	Mapping sent to Ableton
Speed of arm extension movement	hand_tx_distance_velocity $\in$ (5, 0)	Scan $\in$ (0, 2)
	hand_tx_distance_velocity $\in$ (5, 0)	Grain size $\in$ (800, 2000)
Amplitude of arm extension	hand_tx_distance $\in$ (0, 1.5)	Resonator dry/wet $\in$ (0, 0.95)
		Hybrid reverb dry/wet $\in$ (0, 0.95)
Vertical displacement of the left arm	hand_left_ty $\in$ (-0.7, 0.6)	Grain delay feedback $\in$ (0, 0.95)
Vertical displacement of the right arm	hand_right_ty $\in$ (-0.7, 0.6)	Grain delay dry/wet $\in$ (0, 0.95)

4.3   Granular Synthesis Based Parameter Mapping Sonification and Visual Generation

Sound generation is grounded in granular synthesis, where many short "grains" are drawn from recorded material and recombined through parameter control to produce textures ranging from pitched sounds to dense sound masses, while retaining the source's material identity [15]. We follow parameter mapping sonification, whereby motion features are transformed to control synthesis parameters; this requires normalizing features to meaningful ranges, interpolating updates to avoid discontinuities, and bounding parameters to preserve stability [11]. Concretely, the regressor output controls playback position within the combo buffer, while additional motion data computed in TouchDesigner are mapped directly to a small set of post-effects in Ableton, including Resonator, Grain Delay, and Hybrid Reverb, to enrich timbre and reflect ballet's spatial qualities (Table. 3). In combination with the trained posture mapping, this produces an ever-drifting Guzheng soundscape shaped by the performer's upper-body movement.

Visual generation is implemented in TouchDesigner. A silhouette mask is extracted and composited with simple real-time image operators to create an ink-painting-like persistence in which motion traces are briefly retained and displaced by noise, providing a legible indication of how movement shapes the sound. Audio output from the sound engine is routed back into TouchDesigner to drive additional audio-reactive visuals, such as wave-like line graphics, closing the audio-visual feedback loop. In this way, the visual layer reinforces performer agency, as each posture and transition leaves both a visible trace and a corresponding sonic articulation in real time.

5   Discussion and Future Directions

This project presents a performer-in-the-loop framework that combines ballet postural features with granular-synthesis controls via supervised ML mappings. Pairing posture-level classification with continuous regression over an aggregated audio buffer produces a sonically legible, morphable control surface suitable for parameter mapping. Accordingly, this discussion emphasizes qualitative behavior, design decisions, and pathways towards a systematic evaluation.

Although the current prototype employs multiple software environments and virtual audio routing, the architecture is modular and can be reproduced using substitute components. In general, the system decomposes into: (i) body tracking, (ii) feature extraction in Table. 1, (iii) a supervised machine-learning mapping combining classification and regression algorithms, (iv) granular synthesis, and (v) real-time visual rendering. In principle, Kinect V2 can be replaced by any body tracker that provides comparable joint estimates. Likewise, the classification–regression mapping can be reimplemented in a scripted ML environment, Granulator 3 can be substituted with alternative granular engines (e.g.,in Max/MSP), and the visuals can be created in other real-time rendering environments such as Unreal Engine.

5.1   Formative Observations

Pilot engagements with a small participant group suggest that even without formal ballet training, participants can produce flowing audio-visual textures in the Guzheng performance through embodied movement. The audience performed six ballet arm gestures in sequence, and the corresponding data were recorded in TouchDesigner and plotted over time (27.6 s in total). The purpose of this analysis is to provide descriptive, system-level evidence of mapping stability

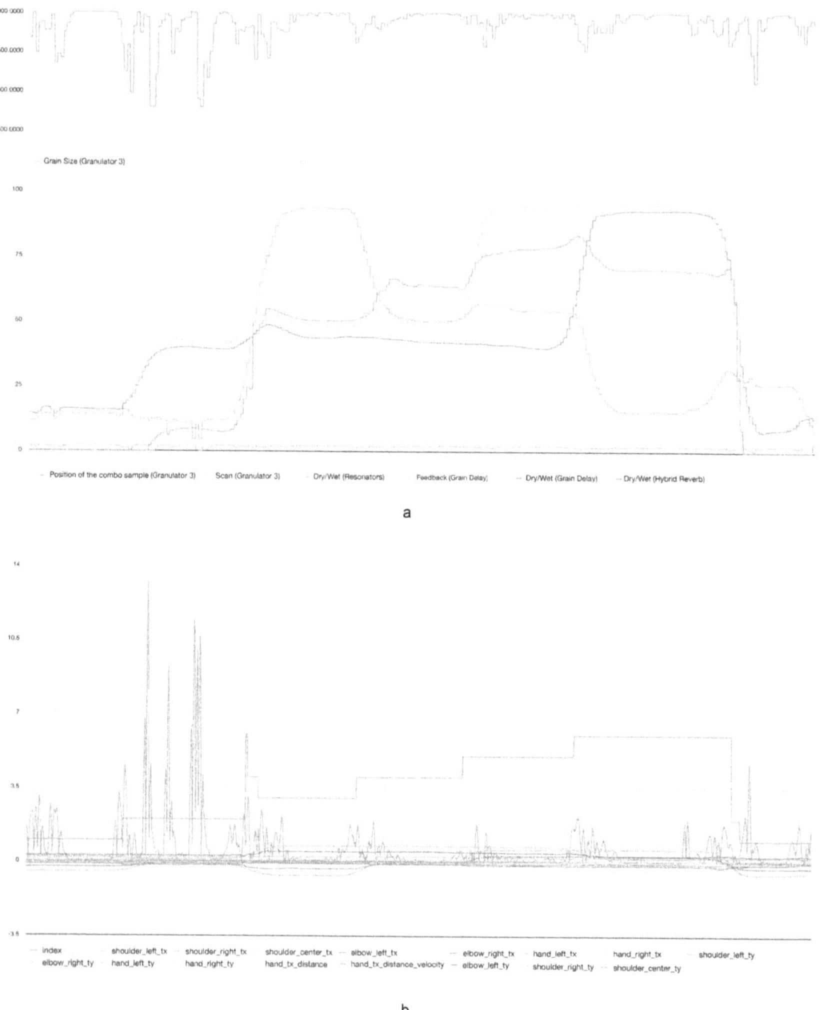

Fig. 5. Parameter feature behavior across time frame. a): Effects chain parameters in Ableton. b): Kinect-derived motion data.

and responsiveness under a realistic interaction. Figure 5 presents audio-engine parameters in Ableton, including combo sample playback position, "Grain Size" and "Scan" in Granulator 3, and "Dry/Wet (Resonators)," "Feedback (Grain Delay)," "Dry/Wet (Grain Delay)" and "Dry/Wet (Hybrid Reverb)." Alongside the motion data originated from the Kinect, reflecting the relationships between motion traces and audio responses. As seen in the above figure, small peaks in hand-to-hand distance and hand speed usually occur before changes in the posture index. These steps correspond to ramps in the sample playback position and in the effects chain, which produce smooth transitions in the audio texture.

During posture holds, the index shows stable dwell periods, while "Scan" and "Grain Size" fluctuate slightly due to the limited frame rate; by contrast, the audio effects settle into bounded plateaus. This separation of time scales, including fast, granular micro-variation against slower, capped envelopes in the effects, preserves the Guzheng timbral identity within a posture and supports the reported sense of legibility. For instance, when the "Dry/Wet (Resonators)" control reaches its upper bound, even minor fluctuations in "Scan" or "Grain Size" can produce large spectral changes. The figure also indicates that the effects chain depends strongly on different postures, which may reduce longer-term sonic variation. These observations justify the present use of clamping, ramping, and smoothing and provide concrete insights for further development.

5.2 Future Development

While the Guzheng phrases and ballet postures anchor the experience in a culturally and corporeally specific space, the current mapping may still risk overfitting when executed outside the canon. In the future, we have several directions to follow:

(1) Temporal modeling and phrase expansion. We plan to incorporate dynamic time warping (DTW) within the supervised ML model and enlarge the Guzheng library with additional culturally salient phrases. Combining continuous-movement recognition with our classification and regression scheme should render "in-between" gestures more musically sensible.

(2) Platform consolidation. We plan to integrate key functions into one environment, or into a scripted and containerized bundle. This should make the system easier to use, more stable, and easier to reproduce for other practitioners.

(3) Evaluation framework. We aim to develop a two-track evaluation that adds to the current formative observations with controlled measures. The first track will consist of specific tasks to measure: (i) posture recognition and stability during position holds and transitions; (ii) timbral discriminability across postural states; (iii) agency and control through time-to-target tasks. The second track will involve open-ended studies. In this part, we will conduct think-aloud sessions and semi-structured interviews during free exploration to gain insights into perceived agency, expressivity, effort, and intercultural interpretation. These two sections offer both reliable evidence for the ML-mediated mapping and credible accounts of creative use.

The current outcome is a design probe that demonstrates the idea is feasible and offers a clear interaction language for AI-assisted performance. Next, we will refine the probe step by step and turn it into a research instrument. This instrument should support perceptual measurements and careful descriptions of artistic practice. This will help us explain not only that the system works, but also how it works and for whom it works best.

6   Conclusion

Fluid Body creates a supervised ML-mediated performance interface that translates codified ballet language to Guzheng sound events. This enables participants to "play" a Guzheng-inspired soundscape with ballet movement in real time. The project establishes a dynamic correlation between physical movement, sound texture, and visual feedback based on LMA and hybrid ML models. It demonstrates how machine learning can serve as a creative catalyst rather than a technical tool. Participants are able to shape the texture of sound with their bodies using parameter mapping sonification, independent of ballet training. This process redefines musical performance as a dialogue between motion and sound, tradition and reinterpretation. Contextualizing Chinese Guzheng in the setting of Western ballet turns it into a living "intercultural instrument," where cultural identity is performed through embodied negotiation instead of static representation.

In practice, participants used clear, recognizable positions to trigger sound events. Meanwhile, small, continuous movements were added to shape texture, density, and spatial depth. In this paper, we report formative evidence from pilot engagements and plotted traces in the system, showing stable posture-state switching, continuous timbral control, and responsive audio-visual coupling under realistic interaction. These observations support a two-level interaction design: one level sets the fundamental tone and pitch of each sound event, and the other level supports dynamic timbre shaping in real time. This introduces diversity and a degree of randomness that reflect each performer's personal style and narrative.

This work also shows how machine learning and bodily movement analysis can collaborate as artistic creation tools within the scope of artificial intelligence mediated systems. It opens a pathway toward embodied instruments that position machine learning as a medium for aesthetic and cultural co-creation. Beyond its immediate artistic output, the system reflects broader tendencies toward human-in-the-loop creativity.

References

1. Ableton AG: Ableton Live 12 Manual. Ableton AG, Berlin, Germany (October 2025). https://cdn-resources.ableton.com/resources/pdfs/live-manual/12/2025-10-16/live12-manual-en.pdf, Accessed 03 Nov 2025
2. Albert, J.A., Owolabi, V., Gebel, A., Brahms, C.M., Granacher, U., Arnrich, B.: Evaluation of the pose tracking performance of the azure kinect and kinect v2 for gait analysis in comparison with a gold standard: A pilot study. Sensors **20**(18), 5104 (2020)
3. Bakogiannis, K., Andreopoulou, A., Georgaki, A.: The development of a dance-musification model with the use of machine learning techniques under covid-19 restrictions. In: Proceedings of the 16th International Audio Mostly Conference, pp. 81–88 (2021)

4. Bomba, M.S., Dahlstedt, P.: Somacoustics: interactive body-as-instrument. In: Proceedings of the International Conference on New Interfaces for Musical Expression, pp. 95–100 (2019)

5. Camargo, A.V.A.: Dance and music as integrated practices in Afro-Brazilian performances: a technocultural study. Revista Brasileira de Estudos da Presença **13**, e129840 (2023)

6. Caramiaux, B., Tanaka, A.: Machine learning of musical gestures: Principles and review. In: Proceedings of the International Conference on New Interfaces for Musical Expression (NIME), pp. 513–518. Graduate School of Culture Technology, KAIST (2013)

7. Carlson, E., Saari, P., Burger, B., Toiviainen, P.: Dance to your own drum: Identification of musical genre and individual dancer from motion capture using machine learning. J. New Music Res. **49**(2), 162–177 (2020)

8. Cavdir, D., Wang, G.: Borrowed gestures: the body as an extension of the musical instrument. Comput. Music. J. **45**(3), 58–80 (2021)

9. Fiebrink, R., Trueman, D., Cook, P.R., et al.: A meta-instrument for interactive, on-the-fly machine learning (2009)

10. Henke, R.: Granulator iii (2023). https://roberthenke.com/technology/granulator3.html. Accessed 03 Nov 2025

11. Hermann, T., Hunt, A., Neuhoff, J.G., et al.: The sonification handbook, vol. 1. Logos Verlag Berlin (2011)

12. Kurillo, G., Hemingway, E., Cheng, M.L., Cheng, L.: Evaluating the accuracy of the azure kinect and kinect v2. Sensors **22**(7), 2469 (2022)

13. Nogueira, M.R., Menezes, P., Maçãs de Carvalho, J.: Exploring the impact of machine learning on dance performance: a systematic review. Int. J. Performance Arts Digital Media **20**(1), 60–109 (2024)

14. Reshma, M., Kannan, B., Raj, V.J., Shailesh, S.: Cultural heritage preservation through dance digitization: a review. Digital Appl. Archaeol. Cult. Heritage **28**, e00257 (2023)

15. Roads, C.: Microsound. The MIT Press (2002)

16. Scurto, H., Caramiaux, B., Bevilacqua, F.: Prototyping machine learning through diffractive art practice. In: Proceedings of the 2021 ACM Designing Interactive Systems Conference, pp. 2013–2025 (2021)

17. Shi, Y.: An Exploration of Seven Tunes Heard in China by Bright Sheng for Solo Cello. Ph.D. thesis (2024)

18. de Souza, A.L.: Laban movement analysis–scaffolding human movement to multiply possibilities and choices. In: Dance Notations and Robot Motion, pp. 283–297. Springer (2015)

19. Tanaka, A., Visi, F., Donato, B.D., Klang, M., Zbyszyński, M.: An end-to-end musical instrument system that translates electromyogram biosignals to synthesized sound. Comput. Music. J. **47**(1), 64–84 (2023)

20. Vaganova, A.: Basic principles of classical ballet: Russian ballet technique. Courier Corporation (2012)

21. Wang, Z., et al.: Chang-e: A high-quality motion capture dataset of Chinese classical dunhuang dance. ACM J. Comput. Cult. Heritage (2024)

22. Yadaei, A., Azadehfar, M.R., Alizadehashrafi, B.: Creation of melodic and rhythmic patterns based on cultural heritage: Sonification of Azerbaijani dance movements. J. Sound Music Games **3**(2–3), 1–27 (2022)

23. Zhang, Y., Li, M., Pan, S.: Deep learning-based emotion recognition algorithms in music performance. Scalable Comput.: Pract. Exper.e **25**(6), 4712–4719 (2024)

EvoLiveDJ: A LLM-Based Agentic Framework for Interactive Evolutionary Live Music Performance

Kamer Ali Yuksel[(✉)] and Hassan Sawaf

aiXplain Inc., San Jose, CA, USA
{kamer,hassan}@aixplain.com

Abstract. We present *EvoLiveDJ*, a framework that unifies large language model (LLM)–driven music generation with interactive evolutionary feedback in a live coding context. The system uses an LLM to generate and iteratively refine Strudel code, while audience listening behaviour serves as a real-time fitness signal that guides selection. Each generation produces multiple musical variants that are played in parallel and selectively bred through LLM-guided semantic crossover and multi-scale mutation informed by audience preference and the model's own critique. Analysis of captured generations reveals coherent evolutionary dynamics in symbolic musical code, including the emergence of stable and diverging lineages, multi-level mutation processes, and role-aware inheritance across rhythmic, melodic, and timbral structures. This hybrid approach combines the knowledge-driven musical competence of LLMs with the exploratory power of evolutionary search, enabling a continuous creative dialogue between AI and audience. The results point toward a new agentic paradigm of AI in music, which actively collaborates, self-evaluates, and evolves musical ideas in real time with human co-creators.

1 Introduction

Can an AI compose music like a human artist—not by producing a single output and stopping, but by improvising, evaluating, and refining ideas in a loop, responding to feedback from listeners? In this paper, we explore a system that achieves this by integrating a Large Language Model (LLM) with an interactive evolutionary music process. The LLM generates music as live-coding instructions (Strudel code), and the composition evolves over successive iterations, guided by audience feedback and the LLM's own hypotheses for improving the composition. The goal is to harness an LLM's knowledge of musical structure and style, while preserving the open-ended creativity and adaptation offered by evolutionary algorithms and human input. Recent advances in AI have produced powerful music generation models, yet most operate in a one-shot fashion—a user provides a prompt and the model returns a fixed audio piece. This yields quick results but limits creative control. By contrast, live coding and evolutionary approaches emphasize iterative development: music can be built layer-by-layer, tweaked, and reworked on the fly. Our approach bridges these paradigms. The LLM serves as an intelligent composer that can propose complex musical ideas in code, while an interactive genetic algorithm process allows those ideas to be continuously refined. In each generation, the LLM's compositions are treated as a population of variants; the audience's preferences determine which variants survive and combine to form the next musical iteration.

In this work, we present EvoLiveDJ, a fully integrated agentic framework that unites LLM-driven musical reasoning with interactive evolutionary search in a live coding context. The system composes, evaluates, and iteratively refines Strudel patterns by combining LLM-guided semantic variation with real-time, audience-driven fitness signals. We detail the overall architecture, describe how audience behaviour and LLM critique jointly guide generational transformation, and analyze the emergent evolutionary dynamics that arise in symbolic musical code. Through controlled simulations and live-audience demonstrations, we show that this hybrid loop enables rapid, musically coherent adaptation and produces identifiable lineages of evolving musical ideas. Our key contributions are fourfold:

- **LLM-Based Music Coding as a Structured Genome:** We demonstrate that an LLM can generate, analyze, and iteratively modify Strudel code that functions as a symbolic musical "genome", which encodes rhythm, harmony, timbre, and effects as modular and recombinable units. The LLM acts not only as a code generator but as a musically informed collaborator capable of applying role-aware transformations such as motif development, rhythmic alteration, parameter modulation, and structural variation.
- **LLM-Guided Semantic Crossover and Mutation:** Moving beyond random genetic operators, the LLM performs *semantic* variation. It recombines parent patterns by aligning musical roles (kick, hats, bass, FX, melody) and introducing musically coherent mutations informed by its own critique. This supports inheritance of rhythmic backbones, timbral blocks, and melodic motifs while introducing controlled mutations at multiple scales– micro-parametric drift, rhythmic microstructure changes, motif expansion, harmonic shifts, and structural evolution.
- **Interactive, Audience-Driven Live Performance:** We develop an end-to-end live performance system in which audience behaviour provides real-time fitness signals that determine which musical variants survive to the next generation. The framework connects Strudel's browser-based live coding engine to an LLM via the Model Context Protocol (MCP), supports multi-stream audio playback for parallel variants, and tracks listener engagement to guide evolutionary selection. It ensures smooth generational transitions, synchronized code–audio execution, structured prompting for critique and variation, and robust fallback strategies suitable for live performance.
- **Emergent Evolutionary Dynamics in Symbolic Musical Code:** We provide the first analysis of evolutionary behaviour emerging in an LLM-driven live coding environment. Across dozens of captured generations, we observe stable and diverging *lineages* (percussive, FX-driven, melodic– harmonic, and hybrid), coherent phylogenetic structures in code space, and a unified mutation taxonomy reminiscent of biological evolution. The LLM

preserves lineage identity while enabling exploratory drift and crossover, producing patterns of inheritance, divergence, and convergence.

2 Background and Related Work

2.1 LLMs for Music Generation

Large Language Models (LLMs) have primarily been developed for text, but recent work has begun to apply them to music by representing musical data in textual form. *ChatMusician* by Yuan et al. (2024) is a prime example: it trains a LLaMA2-based model on symbolic music notation (ABC notation) so that the LLM can "understand and generate music" as if it were a language [2]. By treating music as a second language, ChatMusician is capable of composing full-length structured music (e.g., with specified chords, melodies, or forms), even outperforming a GPT-4 baseline in quality [2]. This demonstrates that, with sufficient training, LLMs can acquire strong musical competence.

Strudel is a live-coding environment that generates music in real time using a text-based pattern language derived from TidalCycles [1]. Recent developments have connected LLMs to Strudel via the Model Context Protocol (MCP) server [3], enabling models such as GPT-4 or Claude to write, execute, and analyze Strudel code programmatically. The MCP server provides domain-specific tools for generating drum patterns and basslines, applying music-theory operations (e.g., transposition, chord progressions), and analyzing audio output [4,5]. This effectively gives the LLM a "musical API," allowing it to compose interactively in a conversational loop–a process termed "vibe coding." For example, when prompted with "Add a funky off-beat bassline" or "Make the hi-hats more complex," the LLM refines the code accordingly. Through this iterative exchange, the system builds music layer by layer with real-time feedback, transforming the LLM from a simple code generator into an active creative collaborator.

Our work builds on these advances by using an LLM as the central creative engine for music generation in code. Rather than training a new model from scratch, we utilize an LLM's pre-existing knowledge and augment it with a structured environment (Strudel's DSL and tools). This allows us to prompt the LLM to create musically valid code and even analyze the resulting music (e.g., checking spectrum or rhythmic complexity) via the provided tools. Importantly, while prior examples of LLM live coding have been human-driven (a human gives iterative prompts), we aim to partially automate the iteration: the LLM itself can identify areas to improve and generate new variants, as we describe later. This leans on the concept of *agentic creativity*, where an AI agent can carry out a sequence of creative tasks, perceive the outcome, and adjust its strategy [6]– exactly the behavior we encourage in our evolutionary loop.

2.2 Evolutionary Music Composition

One of the most notable public experiments in evolutionary music is *Darwin-Tunes* [7], which tested whether pleasant music could emerge from a Darwinian process driven by listener feedback. The system generated short looping audio sequences–"genes" initialized as random noise–and posted them online for users to rate. Highly rated loops reproduced by mixing and mutating, while poorly rated ones were discarded. Over hundreds of generations, the dissonant noise evolved into musically appealing loops guided by collective audience preferences [7]. Listener ratings improved markedly over the first ∼500 generations before stabilizing, suggesting convergence to a consistent, likable musical style. Audio analysis confirmed increasing harmonic and rhythmic structure in later generations [7], demonstrating that musical organization can emerge from random initial conditions under selective pressure.

Interactive Evolutionary Computation (IEC) extends this idea by using human evaluation as the fitness function in creative domains [8]. Traditionally, a user listens to several variants and manually rates or selects them–a process that quickly becomes fatiguing. To scale participation, researchers have leveraged crowds or live audiences. For example, the *DarwinTunes* study itself used online crowdsourced ratings [7], while T. Carson's *A More Perfect Union* (2018) demonstrated IEC in live performance [9,10]. In Carson's system, each audience member's smartphone played a looping melody (a population member), and the longer a listener kept a loop playing, the higher its fitness–assuming that sustained listening indicates preference [9,11]. Implemented via a web app and Tone.js synthesis engine, the system crossed over and mutated top-rated melodies to generate new ones in real time [10,11]. As Carson observed, "the audience becomes the performance" by actively shaping the work [10]. This echoes Karl Sims' earlier gallery experiments, where visitor presence time determined the fitness of evolving visual forms [12]. Carson's adaptation demonstrated that a distributed audience can collectively steer open-ended evolution toward pleasing results without any predefined goal [9].

While *DarwinTunes* and similar projects show that music can evolve to be more enjoyable, they also highlight key challenges: evolution is slow and demands sustained listener participation. Our approach accelerates this process by starting from musically informed material (LLM-generated code rather than noise) and by involving the AI in the variation process. Instead of random mutations, the LLM performs musically coherent recombination and transformation, combining the novelty of evolutionary exploration with the musicality of knowledge-driven generation. Our system builds on these interactive evolutionary frameworks but applies them to LLM-composed, code-based music. Each musical variant is represented as symbolic Strudel code specifying rhythm, melody, harmony, and timbre, rather than fixed audio loops. This symbolic representation allows flexible transformations and meaningful crossover, enabling the LLM to merge parent compositions coherently instead of splicing raw data. Moreover, by initializing with non-random seeds (e.g., stylistic prompts), musically coherent results can emerge within just a few generations–addressing the "cold start" problem seen in earlier systems [7]. Compared with direct deep-learning generation, our approach emphasizes creativity and novelty. As argued by Majidi and Toroghi [13], conventional generative models tend to replicate learned styles, while human-guided evolution can push creation beyond the model's training distribution. Audience-driven selection aligns with this principle: iterative feedback steers the AI toward emergent, surprising outcomes that satisfy aesthetic preference while avoiding overfitting to convention [13]. The interplay between stochastic variation and audience evaluation thus forms a powerful creative engine, promoting both coherence and novelty. In summary, our work synthesizes three key traditions–LLM-based musical reasoning, evolutionary art, and live participatory performance–into a unified framework that enables music to evolve dynamically via continuous collaboration between AI and the audience.

2.3 Comparison with Prior Approaches

Hereby, we note the contrast between our LLM-based approach and other AI music generators. Systems like Google's *MusicLM* or Meta's *AudioCraft* generate audio from text descriptions using trained neural networks, but they do not provide an interface for stepwise editing or user feedback mid-generation (they are "one-shot" black boxes). In comparison, by generating editable code, our system offers transparency and fine control. Users or the AI itself can tweak parameters (change a sample, increase a tempo, alter a note sequence) and immediately hear the effect, rather than having to regenerate from scratch [14]. This positions our system in a unique space combining granular control with AI-driven composition. Compared with end-to-end deep learning music systems such as *MusicLM* or *Jukebox*, our approach emphasizes adaptability and interactivity. A static generative model can render a high-quality track, but cannot adjust it dynamically if the audience's preferences shift. In contrast, *EvoLiveDJ* responds to audience input in real time, functioning more like a live improviser or DJ than a composer. This paradigm exemplifies collaborative creativity, where AI complements rather than replaces human agency [15]. Unlike earlier interactive GA systems such as Carson's *A More Perfect Union* [9,10], where crossover and mutation operated blindly on low-level data, our LLM-guided process performs *semantically informed variation*. The LLM understands musical constructs–rhythm, melody, harmony–and can recombine them coherently. This integration effectively hybridizes search-based and knowledge-based methods, resembling *memetic algorithms* where global search (evolution) is augmented by local optimization (LLM refinement). Consequently, convergence to high-quality, musically coherent states occurs within a few generations rather than dozens. Compared to human-driven interactive genetic algorithms such as *GenJam* [16], our system drastically reduces cognitive load and manual selection. Audience feedback is largely implicit–derived from listening behavior rather than explicit scoring–making participation effortless and organic. Furthermore, the inclusion of an LLM allows for semantically informed mutations. For example, during one trial, the LLM proposed in its critique: "Perhaps adding a jazz chord progression would enrich the harmonic content." In the following generation, it introduced a II–V–I progression over an existing beat, which participants described as "richer" and "more interesting." Such domain-informed interventions are beyond the capability of a random GA mutation, underscoring the value of hybridizing symbolic AI knowledge with evolutionary search.

3   Methodology: LLM-Evolved Musical Live Coding

Our methodology implements a human-in-the-loop evolutionary strategy where the LLM governs representation and variation, and the audience provides the fitness evaluation. The LLM maintains each generation within a musically coherent space, mitigating degenerative mutations, while the audience ensures aesthetic grounding in human preference. The full algorithm can be summarized as follows:

1. **Initialization:** Prompt LLM to generate an initial population of N musical code variants (optionally based on a style or genre).
2. **Audience Feedback:** Play all N variants in parallel for audience evaluation (via position, listening time, or explicit votes).
3. **Selection & Critique:** Determine parent variants with the highest fitness. Prompt LLM to analyze outcomes and hypothesize improvements.
4. **Crossover & Mutation:** Prompt LLM to produce N new variants incorporating feedback and guided variation.
5. **Repeat:** Continue from Step 2 until performance ends.

At the heart of the system is an LLM agent that can both generate new music code and analyze existing music code. All musical pieces in our system are represented as *Strudel* code–textual patterns that generate music when executed in the Strudel live coding environment. A Strudel code snippet can specify multiple concurrent voices (e.g., drum patterns, basslines, melodies) with transformations such as effects or scale changes. By using Strudel code as the *genome*, any generated individual can be immediately realized *phenotypically* as audible music by running it in the browser. The LLM, which operates in two primary modes below, is prompted in a structured way to produce Strudel code using the same syntax and functions a human live-coder would employ.

Generation Mode. Given a prompt or context (which may include descriptions of desired style or previous best individuals), the LLM outputs one or more Strudel code snippets representing new musical pieces. For the initial generation (Generation 0), the prompt may be provided by a human (e.g., "Create four variations of a funky drum and bass groove with slightly different basslines"). In later generations, the LLM's context is automatically generated by the system, summarizing which pieces were selected as parents and what audience feedback indicated (e.g., "Variant A was most popular for its melody; Variant B had good rhythm but was less liked"). The LLM then proposes new code for the next generation–effectively performing mutation and crossover.

Critique Mode. After each generation's music is played and audience feedback recorded, the LLM is asked to reflect on the results. It receives an analysis prompt such as: "Critique the last set of music variants. Which elements might have appealed to the audience? What could be improved or changed for the next generation?" The LLM articulates hypotheses (e.g., "The audience seemed to prefer the variant with a more energetic hi-hat pattern; adding more rhythmic variation may help the others"). This textual self-evaluation then informs the prompt for the next generation's creation. Engaging the LLM in a dialogue about the music allows it to simulate the role of a music producer or theorist, guiding the evolution with expert insight.

```
You are a world-class techno producer analyzing a Strudel pattern.
    Provide a concise but deep critique describing the loop's style, vibe
    , groove, low-end, clarity, texture, stereo field, energy curve, and
    overall commercial potential. Explain what works well and what limits
     the loop musically, then describe how it should evolve naturally in
    the next iteration without breaking its identity. Identify which
    layer or layers (kick, hats, snare, bass, synth, pad, fx) should be
    modified or replaced and explain the musical reason. Briefly comment
    on composition, timbre, groove, complexity, and creativity, and
    propose 4-6 practical improvement ideas focusing on subtle evolution,
     tension, microvariation, rhythmic refinement, spatial movement, or
    modulation. Finally, return a JSON dictionary containing a filtered
    subset of appropriate samples selected from the provided context.
```

In each generation, we maintain a population of N musical variants (typically $N = 4$ or 5 for live settings) that the LLM generates in batch. The resulting variants are then deployed for parallel playback. Each audience member can use a smartphone or a headphone system to toggle between live streams. A mobile web app streams all variants and allows users to solo one at a time–the variant they listen to the longest implicitly receives their vote. The app can also collect explicit feedback via ratings (e.g., thumbs up/down). After the audience has listened for a defined interval (e.g., one or two minutes per generation), the system performs selection. The fitness of each variant is computed based on engagement metrics (listening time or explicit votes) to select the parent individuals. For instance, if 70% of listeners stayed with variant A, 20% with B, and 10% with C, then A receives the highest fitness. After crossover and mutation, a new generation of N variant codes is produced. The process repeats: playback, audience feedback, selection, critique, and variation.

Listing 1.1. Mutation/Crossover System Prompt

```
You are an expert techno producer and Strudel live-coding specialist.
    Your task is always to generate musically coherent, loopable, DJ-
    usable Strudel patterns that evolve existing material without
    breaking syntax. Every output must contain exactly two blocks: first
    a samples({...}, 'github:tidalcycles/dirt-samples'); declaration, and
     second a single $: stack(...) expression containing the full pattern
    . No commentary or text outside these two blocks is allowed. Use only
     sample keys declared in samples(...) and never filepaths inside
    stack(...). All parentheses and quotes must be balanced, and exactly
    one stack(...) call must appear in the output. Avoid TR909., FX.,
    JavaScript constructs, play(), assignments, colons, or malformed <>
    or []. Use only safe Strudel transformations such as rev, slow, fast,
     density, gain, pan, linger, off, stut, echo, swing, every, sometimes
    , rarely, degradeBy, cutoff, resonance, shape, crush, hpf, lpf,
    vibrato, slide, n, add, scale, octave, voicings, superimpose, combine
    , palindrome, jux, zoom, append, hurry, ply, and speed. Maintain a
    techno-appropriate aesthetic emphasizing tight groove, consistent
    pulse, coherent energy, clarity, and incremental evolution.
```

Crossover. The LLM is prompted with the two parent codes (A and B) and instructed to "Create a new music pattern that combines elements of A and B." It may take the drum pattern from A and melody from B, merging them coherently. It can also blend stylistic traits (e.g., tempo or key) while ensuring musical consistency. The LLM effectively acts as a "recombination DJ," remixing parents meaningfully–preserving harmony, rhythm, and tempo.

Listing 1.2. Weighted Crossover User Prompt

```
Perform a weighted crossover of the given parent patterns, combining
    their musical characteristics in proportion to their audience
    preference weights. High-weight parents should determine the core
    rhythmic and structural identity including kick placement, snare grid
    , primary hats, and bassline motion. Medium-weight parents should
    influence articulation, accents, fills, and modulation, while low-
    weight parents contribute subtle microvariations and light textural
    details. Blend rhythmic phrasing, density, swing, sample usage,
    harmonic gestures, dynamics, spatial behavior, and modulation curves
    according to the weights, resolving conflicts by prioritizing high-
    weight structure while allowing medium and low weights to add nuance.
     The resulting pattern must be musically coherent, loopable, DJ-
    usable, and should introduce a mild and coherent improvement.
```

Mutation. The LLM is prompted to "Take the combined pattern and introduce a slight random or creative variation." Mutations may alter notes, add fills, or introduce modulation. The mutation rate can be tuned via prompt intensity (e.g., "make one small change" versus "several changes"). The LLM applies *informed* mutations–guided by its prior critique e.g. if it is noted that "Variant A could use more syncopation," it may deliberately introduce syncopated rhythms.

Listing 1.3. Mutation/Iterative Evolution User Prompt

```
Evolve the given pattern incrementally based on the reflection feedback,
    preserving its core musical identity while introducing exactly one
    subtle but meaningful improvement to groove, clarity, punch, energy,
    or DJ usability. Apply the reflection guidance selectively rather
    than globally, and modify or replace at least one layer to achieve
    the improvement. The resulting pattern must remain structurally
    coherent, minimal, stable, loopable, and safe for performance,
    without introducing syntax errors or runtime issues.
```

3.1 Implementation Details

The *EvoLiveDJ* system is implemented using a combination of web-based technologies and AI services. Below we outline the key components and design choices that enable real-time interactive evolution during live performances. We use the *Strudel Live Coding MCP Server* (Zujkowski, 2025) to bridge our LLM with the Strudel live coding environment [17]. This server is a Node.js application that automates a headless browser running `strudel.cc` and exposes an API via the Model Context Protocol (MCP) for command exchange. The LLM can issue commands such as `play` (to execute code in Strudel), `stop` (to halt playback), and invoke generation tools as described earlier [3,4,17]. The MCP server also provides analysis utilities via the `analyze` tool, enabling retrieval of frequency-domain features, rhythmic density, and other live audio parameters [5]. This allows us to bypass manual browser automation–the LLM effectively "types" in the Strudel editor and receives immediate feedback via the MCP interface.

The framework is model-agnostic and can interface with multiple large language models. For our prototype, we used `GPT-4` via the OpenAI API, given its strong code-generation capability and large context window, sufficient to handle multiple code variants and corresponding analyses. Prompts were carefully engineered to balance creativity and code validity. The LLM's responses are parsed for actionable suggestions (e.g., "add syncopation to bassline") and integrated into the subsequent generation prompts, closing the feedback loop. All components of the project–including the LLM prompts, audience interaction server, and example Strudel scripts–are maintained in a public GitHub repository (to be released). The repository also contains recorded examples of evolved musical pieces, detailed system documentation, and replication instructions.

For real-world performance, we developed a lightweight web application accessible via smartphones. This interface connects to a central server that streams audio for each musical variant–either through WebRTC or multiple parallel audio streams. The user interface provides: (1) toggle or selection buttons for each variant, (2) continuous tracking of user engagement, such as active listening time or switching frequency. All user interactions are logged and transmitted to the central server in real-time, providing the data necessary for computing fitness. At the end of each generation (or periodically during playback), the server aggregates client data. Depending on the experimental setup:

- **Listening Time as Fitness:** Fitness F_i for variant i is proportional to the cumulative listener-seconds across all users.
- **Voting-Based Fitness:** Fitness is computed from discrete user votes or the proportion of listeners tuned into each variant at a given cutoff time.

4   Experiments

To evaluate the *EvoLiveDJ* framework, we conducted both offline simulations–where researchers emulated an "audience" by programmatically assigning fitness according to predefined preferences–and a live demonstration with volunteer listeners. These trials aimed to assess three core aspects of the system: (1) the LLM's ability to generate valid and diverse Strudel code variants, (2) the responsiveness of the evolutionary loop to audience-driven selection pressure, and (3) the musical coherence and quality of the evolved outputs. Across test cases, the LLM (GPT-4 in our prototype) reliably produced valid Strudel code, with correctness reinforced by the Strudel MCP's domain-specific analysis tools. When errors did occur–typically minor syntactic issues such as unbalanced parentheses or undefined sample keys–they were automatically detected by MCP's structured error reporting and subsequently corrected by the LLM [17]. Iterative observations revealed meaningful adaptation rather than random variation. In one sequence, the LLM began with a straightforward rock drum beat; after repeated audience preference for syncopated rhythms, later generations incorporated ghost notes and off-beat snares. This suggests that the LLM integrated evaluative cues into its generative strategy. Real-time operation introduced practical latency constraints. During the live demonstration, GPT-4 responses averaged 3–5 s per prompt, which remained acceptable for transitions between generations. To ensure robustness, fallback mechanisms replayed the best prior variant or switched to cached code in case of network or model failure. For large-scale shows, maintaining backups and enabling human override (e.g., manual triggering or editing of code) prevents silence or stalled playback.

A live demonstration with eight participants used three variants per generation and approximately one minute of listening time per round. Adaptive behaviour emerged almost immediately. Rather than drifting arbitrarily, the system began to differentiate lineages based on listener engagement. Variants featuring *stable percussive structure, syncopated hi-hats,* and *energetic spatial modulation* consistently attracted more attention than more experimental or sparse alternatives. This early preference created selective pressure that shaped subsequent generations. By the second generation, the LLM had already begun performing *semantic crossover*, recombining the rhythmic backbone of a high-fitness percussive pattern with timbral or textural material from other candidates. In parallel, mutation introduced *incremental refinements*—for example by adding off-beat accents, subtle stutter effects, or restrained `perlin`-based panning. These changes preserved the identity of the favoured groove while exploring nearby variants in code space. Over successive generations, the population retained diversity but gradually converged toward a *high-tempo, beat-driven style* characterized by a coherent kick–snare grid, light rhythmic ornamentation, and spatial movement. Listener engagement across the top variants stabilized after only a few rounds, echoing the "fitness plateau" reported in *DarwinTunes* [7], but achieved here far more rapidly due to the LLM's structured initialization and musically meaningful mutation strategies. Participants reported feeling directly involved in shaping the music's evolution, noting that their listening choices influenced the direction of the system. This resonates with Carson's observation that audience-guided evolutionary systems heighten emotional engagement by creating a sense of co-authorship [10]. A minority of listeners initially found channel switching to be a non-obvious representation of preference; clearer onboarding or optional explicit rating controls may improve usability in larger installations. Real-time visualization of fitness and generational progression could further enhance the participatory experience.

Subjective evaluations indicated that while the resulting patterns were not polished compositions, they consistently exhibited musical coherence and stylistic identity. Because the initial population already contained *structured rhythmic and harmonic material*—rather than random noise—the system avoided the extended cold-start phase typical of earlier evolutionary music systems such as *DarwinTunes*, which required hundreds of generations before producing listenable loops [7]. EvoLiveDJ instead produced coherent musical states within only a few evolutionary cycles. At the same time, the LLM's training data naturally biases it toward familiar stylistic templates and mainstream harmonic idioms. Without countermeasures, this can lead to a gradual *homogenization of musical outcomes*, especially when audience preferences reinforce conventional structures. To preserve exploratory capacity, prompt-level diversification strategies proved useful—for example periodically encouraging the LLM to introduce rhythmic irregularities, explore non-Western scales, or combine unconventional effect modules. Injecting occasional "wild mutations" helped maintain population diversity and prevented premature convergence into narrow stylistic attractors. Balancing *stability* and *novelty* proved essential. When the LLM was permitted unrestricted freedom, the system occasionally exhibited abrupt stylistic jumps between generations, disrupting the sense of continuous evolution. Introducing soft constraints instructing the LLM to "preserve the core musical identity while varying secondary elements" led to smoother, more comprehensible evolutionary paths. This behaviour reflects the logic of hybrid evolutionary–memetic strategies [18,19], where global exploration is tempered by local, context-sensitive refinement. Ultimately, EvoLiveDJ demonstrates that integrating audience preference with LLM-driven semantic mutation yields adaptive and musically coherent patterns, while retaining the creative unpredictability of evolutionary art.

5 Evolutionary Dynamics

A key outcome of *EvoLiveDJ* is the emergence of recognisably *evolving* musical structures. Within an interactive evolutionary loop, the LLM generates Strudel patterns that exhibit inheritance, recombination, mutation, and stylistic convergence–phenomena analogous to biological evolution expressed through symbolic musical code. Rather than drifting randomly, generated variants cluster into coherent *lineages* shaped jointly by audience preference and LLM critique. Analysis of over seventy captured patterns reveals three recurrent families (Fig. 1):

1. **Percussive lineages**, anchored by a stable kick–snare–hi-hat scaffold, with variation driven by articulation, density, and spatialisation.
2. **FX-driven lineages**, dominated by timbral evolution–echo depth, stutter intensity, filtering, and spatial modulation.
3. **Melodic–harmonic lineages**, built from Strudel's melodic DSL and evolving through motif expansion, repetition, and ornamentation.

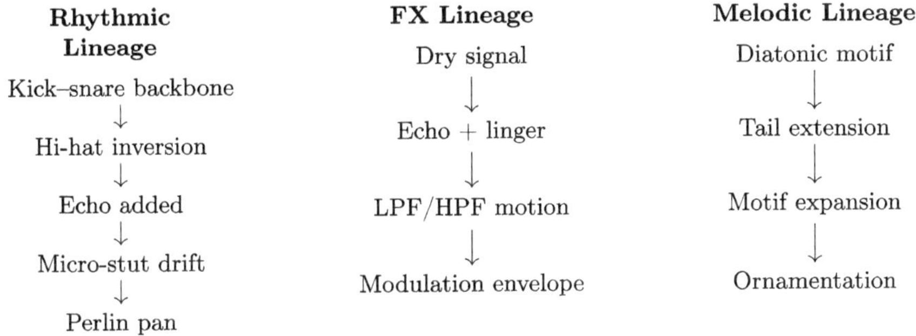

Fig. 1. Three evolutionary lineages shown side-by-side: rhythmic, FX, and melodic.

Later generations frequently exhibit *hybrid* forms that blend these families, such as a percussive backbone supporting a developing melodic contour. Across all families, we observe three consistent mechanisms of musical evolution: (1) inheritance through stable layer structures, (2) semantically meaningful crossover across musical roles, and (3) multi-scale mutation ranging from micro-parametric drift to structural reorganisation. Unlike naive textual crossover–which risks breaking syntax–the LLM performs *role-aware* recombination. Drum layers, hat layers, bass layers, and FX chains are treated as modular semantic units. When multiple parent patterns share structural similarity (e.g., identical kick grids but divergent hats), the LLM merges them coherently. A descendant may inherit the kick–snare backbone from the highest-fitness parent, hi-hats recombined from two parents using rhythmic alignment, and FX modules inherited intact as transferrable timbral "regulators." For example, combining s("bd   bd") with s(" sd   ") preserves metrical identity, while borrowing the hi-hat gesture s("hh*4").every(2,rev) refreshes surface detail. Echo blocks (e.g., .echo(.21,.11,.26)) and spatial modulations frequently propagate intact across generations. Crossover between rhythmic and harmonic material is also common: a stable drum genome layered with a melodic contour such as n("0 2 4 6 7 6 4 2").scale("<c3:major>") yields hybrids integrating rhythm and pitch.

Across captured lineages, musical evolution proceeds along two dominant axes: **Rhythmic intensification** (hats become increasingly syncopated, dense, and modulated) and **Timbral–spatial enrichment** (echo depth grows, stutter patterns drift, and perlin-based panning widens the stereo field). Despite these accumulating changes, lineages maintain a stable rhythmic identity, highlighting the LLM's ability to preserve structural coherence while performing exploratory

Table 1. Unified taxonomy of mutation processes observed in EvoLiveDJ lineages.

Mutation Category	Description and Musical Effect
Micro-mutation	Subtle expressive tweaks (e.g., gain, hat velocity, timing offsets, mild pan(perlin.range()) drift) that shape groove and timbre without changing global structure.
Parametric drift	Gradual shifts in echo, stut, linger, or hpf/lpf (e.g., stut(2,1/8,0.18) → stut(3,1/16,0.12)). Modulates spaciousness, brightness, and rhythmic microstructure.
Rhythmic-microstructure	Local rhythmic edits such as increased hat density, higher step resolution, reversed fragments, or deeper stutters. Increases perceived drive, syncopation, and energy.
Layer insertion	Addition of new layers (e.g., glitches, percussive accents, gabber hits) that increase rhythmic and/or timbral complexity.
Layer deletion	Removal of redundant hats, FX blocks, or overcrowded layers to improve clarity and reduce mix clutter.
Motif expansion	Melodic sequences grow from short diatonic kernels into extended or ornamented contours (e.g., "0 2 4 6 7*2 6 4 6"). Enhances melodic expressiveness.
Harmonic/modal shifts	Changes to scale, pitch range, octave, or mode alter tonal colour while preserving rhythmic identity.
Structural evolution	Mutations to long-range operators such as every, sometimes, or linger, affecting phrasing, tension curves, and periodicity.

Table 2. Representative micro-evolution diffs across generations.

Parent variant	Offspring variant
```@@ hats-layer @@``` ```- s("hh*4")```	```// added reversible hi-hats``` ```+ s("hh*4").every(2,rev)```

```@@ fx-layer @@```
``` s("bd sd").gain(1)```
	```// spatial widening mutation```
	```- s("bd sd")```
	```+ s("bd sd").echo(.21,.11,.26)```

```@@ hats-layer @@```
```- s("hh*4").every(2,rev)```
	```// rhythmic microstructure```
	``` mutation```
	```+ s("hh*4").every(2,rev).stut```
	``` (2,1/8,0.18)```

```@@ hats-layer @@```
```- s("hh*4").every(2,rev).stut```
```   (2,1/8,0.18)```
	```// added spatial modulation```
	```+ s("hh*4").every(2,rev).stut```
	``` (3,1/16,0.12)```
	```.pan(perlin.range(-0.35,0.35))```

variation. Mutation in *EvoLiveDJ* operates across multiple scales and affects different musical structures. Table 1 summarizes the mutation processes observed across these lineages, while Table 2 shows representative parent–offspring code diffs illustrating micro-mutations and parametric drift in practice. These mechanisms jointly produce a phylogeny where percussive and melodic lineages diverge early, the FX lineage evolves orthogonally, and hybrids arise through crossover. Rhythmic backbones show strong inheritance; FX modules propagate intact with drift; and melodic material evolves through expansion and ornamentation. The system thus organises musical transformations into coherent evolutionary trajectories guided simultaneously by audience preference and LLM-directed variation.

6   Conclusion

We have presented *EvoLiveDJ*, a framework that unites the generative capabilities of Large Language Models (LLMs) with the adaptive search mechanisms of evolutionary computation for live, audience-interactive music creation. The system leverages an LLM to generate, critique, and refine Strudel live-coding patterns, while real-time audience engagement provides the selective pressure that shapes each generation. Our experiments demonstrate that an LLM can act as an autonomous creative agent, performing semantic crossover, role-aware inheritance, and multi-scale mutation rather than relying on naive or syntactically fragile genetic operators. This hybrid process leads to rapid musical adaptation guided by human preference. Analysis of captured generations reveals coherent evolutionary dynamics in symbolic musical code, including the emergence of distinct rhythmic, FX-driven, and melodic–harmonic lineages; hybridization through semantic recombination; and a structured taxonomy of mutations ranging from micro-parametric drift to structural evolution. These patterns indicate that the system does not merely iterate on code, but organizes musical material into stable and diverging trajectories analogous to biological evolution. Audience feedback functions as an aesthetic fitness signal, while the LLM maintains coherence by preserving core rhythmic or harmonic identity across generations.

Our findings validate several key outcomes:

- **Autonomous Creative Agency:** The LLM produced coherent, evolving musical variations and integrated evaluative cues, effectively functioning as a self-reflective creative agent within an evolutionary loop.
- **Organic Audience Integration:** Implicit listening behaviour served as a natural fitness metric, enabling real-time participatory evolution without demanding explicit scoring or annotation from listeners.
- **Emergent Evolutionary Structure:** Evolution proceeded through identifiable lineages, semantic crossover, and multi-level mutation processes, demonstrating that LLM-evolved genomes support structured musical phylogenies.
- **Rapid Adaptation:** Because musical genomes begin from structured LLM-generated seeds rather than random noise, convergence toward preferred musical states emerges within just a few generations.

References

1. McLean, A.: Making programming languages to dance to: Live coding with Tidal. In: Proceedings of the International Conference on New Interfaces for Musical Expression (NIME 2014), pp. 449–454 (2014)
2. Yuan, W., Yang, T., Yang, L., et al.: ChatMusician: Understanding and Generating Music with Large Language Models. arXiv preprint arXiv:2401.11504 (2024)
3. Park, D., Bhat, S., Pine, A., Zujkowski, J.: The Model Context Protocol (MCP): enabling ai co-creation through structured interaction. In: Proceedings of the International Conference on Computational Creativity (ICCC 2020), pp. 140–147 (2020)
4. Hawley, A., Zujkowski, J.: Strudel: a browser-native live coding environment for rhythm and sample-based music. In: Proceedings of the International Conference on Live Coding (ICLC 2022), pp. 32–39 (2022)
5. Fiebrink, R., Trueman, D., Cook, P.R.: Real-time interactive supervised learning. In: Proceedings of the International Conference on New Interfaces for Musical Expression (NIME 2011), pp. 465–468 (2011)
6. Karimi, H., et al.: Agentic creativity: Large language models as creative agents. arXiv preprint arXiv:2310.03631 (2023)

7. MacCallum, R.M., Mauch, M., Burt, A., Leroi, A.M.: Evolution of music by public choice. Proc. Natl. Acad. Sci. **109**(30), 12081–12086 (2012). https://doi.org/10.1073/pnas.1203182109

8. Takagi, H.: Interactive evolutionary computation: fusion of the capabilities of EC optimization and human evaluation. Proc. IEEE **89**(9), 1275–1296 (2001). https://doi.org/10.1109/5.949485

9. Carson, T.: A More Perfect Union: Composition with audience-controlled smartphone speaker array and evolutionary computer music. Master's thesis, Wesleyan University (2018)

10. Carson, T.: A More Perfect Union. In: Proceedings of the 5th Web Audio Conference (WAC 2019), Trondheim, Norway, pp. 73–78 (2019)

11. Sims, K.: Galápagos: creating organic artwork through interactive evolution. IEEE Comput. Graphics Appl. **17**(4), 10–19 (1997). https://doi.org/10.1109/38.595266

12. Sims, K.: Interactive evolution of equations for procedural models. In: The Visual Computer **9**, 466–476 (1993). https://doi.org/10.1007/BF01901069

13. Majidi, T., Toroghi, R.M.: Conserving human creativity with evolutionary generative algorithms: a case study in music generation. arXiv preprint arXiv:2406.05873 (2024)

14. Agostinelli, A., Copet, J., Kreuk, F., et al.: MusicLM: Generating Music From Text. arXiv preprint arXiv:2301.11325 (2023)

15. Davis, N., Völkel, T., Loughridge, J., et al.: Collaborative creativity with generative AI: Models, metaphors, and opportunities. In: Creativity and Cognition (C&C '23), pp. 1–14. ACM, New York (2023). https://doi.org/10.1145/3591196.3592984

16. Biles, J.A.: GenJam: a genetic algorithm for generating jazz solos. In: Proceedings of the International Computer Music Conference, pp. 131–137. ICMA, San Francisco (1994)

17. Zujkowski, J.: Strudel live coding MCP server. In: Proceedings of the International Conference on Live Coding (ICLC 2025), to appear (2025)

18. Storn, R., Price, K.: Differential evolution – a simple and efficient heuristic for global optimization over continuous spaces. J. Global Optim. **11**(4), 341–359 (1997). https://doi.org/10.1023/A:1008202821328

19. Moscato, P., Cotta, C.: A gentle introduction to memetic algorithms. In: Corne, D., Dorigo, M., Glover, F. (eds.) New Ideas in Optimization, pp. 105–144. McGraw-Hill, London (1999)

AI Co-Artist: A LLM-Powered Framework for Interactive GLSL Shader Animation Evolution

Kamer Ali Yuksel[(✉)] and Hassan Sawaf

aiXplain Inc., San Jose, CA, USA
`{kamer,hassan}@aixplain.com`

Abstract. Creative coding and real-time shader programming are at the forefront of interactive digital art, enabling artists, designers, and enthusiasts to produce mesmerizing, complex visual effects that respond to real-time stimuli such as sound or user interaction. Despite the rich potential of tools such as GLSL, the steep learning curve and the requirement for programming fluency pose substantial barriers for newcomers and even experienced artists without a technical background. In this paper, we present *AI Co-Artist*, a novel interactive system that harnesses the capabilities of large language models (LLMs), specifically GPT-4, to support the iterative evolution and refinement of GLSL shaders through a user-friendly, visually-driven interface. Drawing inspiration from the user-guided evolutionary principles pioneered by the Picbreeder platform, our system empowers users to evolve shader art using intuitive interactions, without needing to write or understand code. AI Co-Artist serves as both a creative companion and a technical assistant, allowing users to explore a vast generative design space of real-time visual art. Through comprehensive evaluations, including structured user studies and qualitative feedback, we demonstrate that AI Co-Artist significantly reduces the technical threshold for shader creation, enhances creative outcomes, and supports a wide range of users in producing professional-quality visual effects. Furthermore, we argue that this paradigm is broadly generalizable. By leveraging the dual strengths of LLMs—semantic understanding and program synthesis—our method can be applied to diverse creative domains, including website layout generation, architectural visualizations, product prototyping, and infographics. We also explore whether human curators in the interactive process could be replaced or augmented with multimodal vision-language models acting as autonomous aesthetic judges to allow closed-loop evolution.

1   Introduction

The rise of procedural graphics and shader programming has enabled a revolution in digital creativity, particularly in interactive and generative art. Shaders, small programs that run on the GPU, offer fine-grained control over visual outputs at

P. Machado et al. (Eds.): EvoMUSART 2026, LNCS 16523, pp. 192–206, 2026.
https://doi.org/10.1007/978-3-032-24350-8_13

Fig. 1. Screenshot of the *AI Co-Artist* interface showing initial audio-reactive shaders, user selection UI, and audio upload controls. Each canvas renders a distinct LLM-generated shader responsive to the uploaded audio file.

Fig. 2. Multiple variants generated from a single parent shader with an interactive evolution. Each square represents a distinct audio-reactive fragment shader generated by the LLM based on user-selected parents. Users view these live-rendered shaders and act as curators by selecting visually compelling ones. The interface encourages the exploration of aesthetic diversity, enabling users to steer the creative direction.

the pixel or vertex level. Yet, this power is often offset by the complexity of the required programming knowledge, which deters many creatives from engaging directly with the medium. Even those with moderate experience in computer graphics may find the trial-and-error nature of shader design challenging, especially when experimenting with reactive or time-varying effects. Parallel to this development, the field of AI-assisted creativity has undergone a transformation driven by LLMs. Trained on massive corpora of natural language and source code, LLMs like GPT-4 have demonstrated remarkable capabilities in tasks such as code synthesis, documentation, transformation, and even creative tasks such as poem writing or art description. Importantly, LLMs are no longer just code completion tools; they are capable of autonomously generating novel, complex programs from vague or incomplete prompts. This unlocks an opportunity for collaborating with AI in creative workflows.

In *AI Co-Artist*, we leverage the generative and interpretive abilities of LLMs to overcome the barriers inherent in shader design. The system positions the user as a creative curator: they select visual outputs they find appealing, and the LLM evolves new shader code based on these preferences. Unlike traditional design tools, the user does not need to specify exact changes or understand the underlying codebase. Instead, they guide the system through aesthetic preference, making the process more accessible, exploratory, and engaging. This interactive evolution process mirrors natural creative cycles—brainstorm, prototype, refine—augmented by an AI assistant capable of producing diverse, structurally valid code variations. The result is a co-design paradigm where both the human and AI contribute meaningfully: the human sets the aesthetic trajectory, while the AI fills in technical details and explores adjacent possibilities. Moreover, because LLMs understand a wide range of domains and programming languages, this framework can be easily extended beyond shaders. It provides a glimpse into a broader future of AI-human collaboration in creative tasks.

This paper presents a fully working implementation of this paradigm in the domain of GLSL fragment shaders. By coupling real-time visual feedback, audio-reactivity, and evolutionary AI-driven code generation, *AI Co-Artist* transforms the shader authoring experience into a playful and intuitive exploration of aesthetic space. Figures 1 and 2 provide an overview of the system in action, showing the web interface and an example population of offspring shaders generated from a selected parent during interactive evolution. The system not only generates functional code—it evolves a creative conversation between the user's visual preferences and the AI's expressive potential, realized through structured code transformations like crossover and mutation. A working prototype of the system is available in the supplementary material for experimentation. The HTML can be run locally by anyone interested in exploring the system. To activate the GPT-4-powered shader evolution, users must insert their own OpenAI `API_KEY` in the appropriate variable in the HTML file. Our contributions are four-fold:

- **An Interactive Evolutionary Framework for Shader Design:** We introduce a user-guided, LLM-powered system that evolves GLSL fragment shaders through visual selection and automatic code transformation, removing the need for manual programming.
- **LLM-Orchestrated Shader Crossovers and Mutations:** We design prompt-based workflows that allow LLMs to perform crossover and mutation operations on GLSL shaders, yielding valid and visually distinct variants without human intervention.
- **Seamless and Intuitive User Experience:** A web-based interface enables users to browse, select, evolve, and preview shaders effortlessly. Fullscreen previews, progress feedback, and shader download options support creative exploration.
- **A Generalizable Co-Creation Paradigm:** Though focused on shaders, our approach exemplifies a broader class of AI-human creative collaboration tools that can be adapted to other design domains such as music synthesis, generative visuals, or architectural forms.

2 Background and Related Work

Our work lies at the intersection of interactive evolutionary computation (IEC), AI-assisted code generation, and human-computer creative collaboration. Interactive Evolutionary Computation (IEC) leverages human evaluation to guide evolutionary algorithms, making it particularly effective in domains where quantitative fitness functions are challenging to define, such as art and design. [1] provides an extensive survey on IEC applications, highlighting its success in addressing complex, real-world problems. A foundational early work in expression-based evolutionary graphics is Karl Sims' SIGGRAPH paper on evolving images, textures, and animations via symbolic program representations and interactive selection [2]. An influential example of collaborative IEC is **Picbreeder**, an online platform that enables users to collaboratively evolve images through selective breeding [11]. Users choose images they find visually compelling, which are then mutated and combined to produce new offspring. Earlier web-based interactive evolutionary art systems also existed, such as *International Interactive Genetic Art II* (1994–1996), which explored online, user-driven evolutionary image breeding [3]. In addition, Tatsuo Unemi's **SBArt** line of systems represents an early and sustained thread of evolutionary art software, with **SBArt4** supporting automated and interactive evolution of abstract images and animations, including daily evolved WebGL animations published online since 2011 [4,5]. Other efforts in evolutionary art have leveraged genetic algorithms for creative generation. For instance, **GenJam** is an interactive genetic algorithm that evolves jazz solos in collaboration with a human performer [12]. Similarly, the **NeuroEvolution of Augmenting Topologies (NEAT)** algorithm has been applied to evolve network topologies for various creative tasks [10]. [6] discusses the application of genetic programming in evolving artistic designs, emphasizing the role of human aesthetic evaluation in guiding the evolution process.

The advent of Large Language Models (LLMs) has revolutionized code generation and transformation. Models such as OpenAI's **Codex** have demonstrated remarkable capabilities in generating code snippets from natural language prompts [13]. **GitHub Copilot**, powered by Codex, assists developers by autocompleting code and suggesting alternative implementations, effectively serving as an AI pair programmer. Tools like **Code Llama** offer AI-driven code completion across multiple programming languages, enhancing developer productivity [14]. Importantly, LLM-assisted evolutionary exploration of creative artifacts (including shader-like programs) has also been demonstrated in Joel Simon's **Lluminate**, which frames LLMs as variation operators within an evolutionary search process [15]. Our approach differs in its emphasis on an end-to-end, real-time *interactive* shader evolution experience: a browser-based multi-canvas WebGL population with live audio reactivity, user-in-the-loop curation, and an empirical usability study comparing against a standard shader authoring baseline. A recent development in AI-assisted programming is the concept of **vibe coding**, introduced by Andrej Karpathy in February 2025. Vibe coding refers to an AI-dependent programming technique where a person describes a problem in natural language prompts to an LLM tuned for coding, which then generates the corresponding software. This approach shifts the programmer's role from manual coding to guiding, testing, and refining the AI-generated code, making coding more accessible to those without extensive expertise.

Building upon these advancements, **AI Co-Artist** introduces a new design pattern that combines the generative capacity of LLMs with human aesthetic judgment in a loop of iterative refinement. In this system, the user acts as a creative curator, selecting visual outputs they find visually appealing, while the LLM evolves new shader code based on these preferences. Unlike traditional design tools, the user does not need to specify exact changes or understand the underlying codebase. Instead, they guide the system through aesthetic preference, making the process more accessible, exploratory, and engaging. This interactive evolution process mirrors natural creative cycles—brainstorm, prototype, refine—augmented by an AI assistant capable of producing diverse, structurally valid code variations. The result is a co-design paradigm where both the human and AI contribute meaningfully: the human sets the aesthetic trajectory, while the AI fills in technical detail and explores adjacent possibilities. Moreover, because LLMs understand a wide range of domains and programming languages, this framework can be easily extended beyond shaders, providing a glimpse into a broader future of AI-human collaboration in creative tasks.

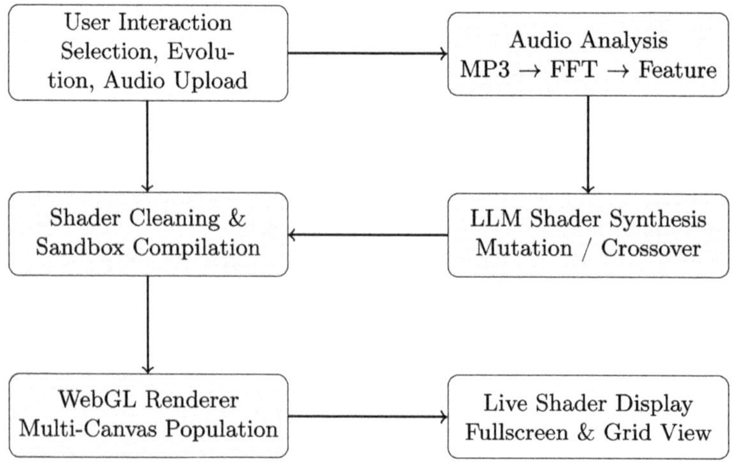

Fig. 3. System architecture of AI Co-Artist. Users interact through shader selection and audio upload. Audio is processed into a normalized scalar feature and passed to the WebGL renderer, while selected shaders are provided for LLM crossover or mutation. Generated shaders are compiled, validated and inserted back into the system.

3   System Architecture

The system architecture of AI Co-Artist is composed of three major components: a WebGL-based real-time shader rendering engine, a client-side audio processing module using Tone.js, and a server-mediated GPT-4 interface for shader generation. Upon loading, the application presents a grid of shader canvases, each initialized with a randomly selected or pre-defined shader from a curated seed bank. Each canvas is independently animated, with time and audio analysis passed as uniforms. Audio files are uploaded by users in MP3 format. Using Tone.js, the system performs a real-time FFT decomposition of the signal, extracting frequency domain features. These features are normalized and provided to each shader via the `u_audio` uniform, enabling dynamic, beat-reactive visual effects. User interaction occurs through selection. Users click on shaders they find aesthetically pleasing. Selected shaders are used as parents in either mutation (single parent) or crossover (multi-parent) generation operations. When triggered, the selected shader(s) are packaged into a structured natural language prompt and sent to the GPT-4 API. The LLM response is parsed to extract GLSL code, which is then compiled using WebGL. If the shader compiles successfully and renders without error, it replaces one of the non-selected shaders in the population. If compilation fails, up to five retry attempts are made with regenerated shaders. A notable design challenge is maintaining diversity while avoiding code bloat or performance degradation. To address this, shaders are constrained to a maximum code length, and rendering time per frame is capped. The system also supports full-screen previews and hides UI components during immersive exploration. The application runs in the browser without backend dependencies (except for LLM calls), ensuring privacy, low latency, and accessibility.

3.1 WebGL-Based Real-Time Shader Rendering Engine

AI Co-Artist is implemented as a browser-native, fully client-side system that integrates WebGL-based rendering, real-time audio analysis, and an LLM-driven shader evolution backend. The architecture is intentionally lightweight and portable, requiring no server-side computation beyond calls to the OpenAI API. The system transforms a user's aesthetic preferences into an iterative evolutionary loop in which GLSL fragment shaders are dynamically generated, mutated, validated, rendered, and curated. Figure 3 illustrates the complete data and control flow among the major components. At the core of AI Co-Artist is a multi-canvas WebGL rendering engine that maintains a population of concurrently animated shaders. Each shader variant is rendered inside its own `<canvas>` element and is associated with an independent WebGL context, program, and rendering loop. The vertex shader is a minimal pass-through program that draws a full-screen quad, while the fragment shaders are supplied by the LLM or the curated base set. To ensure consistency across variants, the system automatically wraps each generated fragment shader with a standard interface that injects the uniforms `iTime`, `iResolution`, and `u_audio`, followed by a unified entry point that forwards execution to `mainImage`. Each shader is compiled on-the-fly using WebGL, and invalid shaders are detected before they reach the user interface through an off-screen sandbox compilation stage.

3.2   Client-Side Audio Processing and Feature Extraction

Every shader variant runs in its own animation loop using `requestAnimationFrame`. During each frame, the system updates the uniform values according to the current canvas resolution, elapsed time since initialization, and audio-reactive feature described below. This design maintains high frame rates and smooth animations, even when managing a population of more than a dozen simultaneously evolving shaders. A long-press gesture enables fullscreen rendering for immersive exploration, while short clicks toggle selection, enabling user-driven curation. Audio reactivity is implemented using the Web Audio API and Tone.js. When the user uploads an MP3 file, the system constructs an analysis pipeline in which a `Tone.Player` streams audio through a 32-band FFT analyser. For each rendering frame, the analyser produces a frequency-domain vector, from which the system computes an aggregate audio energy value. This value is normalized into a continuous scalar in the range $[0, 1]$, which is injected into every shader as the uniform `u_audio`. Because this parameter modulates visual parameters without affecting temporal logic, shaders remain numerically stable while achieving expressive, beat-synchronous visual dynamics. The audio-reactive design also ensures that generated shaders do not require knowledge of the audio subsystem, keeping them portable and self-contained. Each shader receives the audio-reactive scalar $a_t = f(\text{FFT}_t)$ derived from FFT analysis, which modulates color, motion, or structural parameters. This term affects the rendered phenotype but not the underlying genotype (shader code). Thus, the mapping phenotype(S, a_t) is dynamic and time-varying, influencing user evaluation and hence the evolutionary trajectory.

3.3   GPT-4–Mediated Shader Generation and Evolution Module

AI Co-Artist maintains a population of 14 shader variants, each represented by a canvas, shader program, and metadata describing its selection state and fullscreen mode. Selection is performed solely through short-click interactions, thereby making aesthetic curation the dominant mode of user engagement. The `Evolve` button becomes active once a non-empty subset of variants is selected. When the user invokes evolution, the system determines whether to apply mutation (single parent) or crossover followed by mutation (multiple parents). Unselected variants are replaced with newly generated offspring, while selected shaders persist as elite individuals, enabling directional exploration of the visual design space. To ensure robustness, newly created shader code is cleaned, validated syntactically, and compiled in a sandbox before becoming part of the population. The system retries generation up to a configurable number of attempts if compilation fails. This prevents unusable shaders from entering the population and sustains a smooth interactive exploration experience. The evolutionary process is powered by GPT-4 through structured prompts. For each evolution event, the system constructs either a mutation or crossover prompt, embedding one or more parent shaders inside a natural language description that requests structural preservation, artistic coherence, and syntactically valid GLSL code.

The LLM is queried with a moderately high sampling temperature to encourage aesthetic diversity, and the response is parsed to extract raw shader code without explanations or formatting. The resulting fragment shader is checked for compilation correctness before being admitted into the population. This LLM-mediated pipeline functions as a semantic mutation operator, enabling complex transformations that exceed traditional stochastic mutations while maintaining validity and expressivity.

4   Methodology

AI Co-Artist is designed as a fully interactive, web-based system that combines WebGL shader rendering, real-time audio reactivity, and LLM-powered shader synthesis into an evolutionary art creation experience. The methodology consists of five major components: (1) initialization and population of shader variants, (2) real-time audio feature extraction, (3) user interaction and selection, (4) shader evolution via LLM-driven crossover and mutation, and (5) shader replacement and iteration. Upon loading, the system initializes a shader population with a predefined number of GLSL fragment shaders, either from a curated base set or from generated mutations. The base shaders are designed to be visually diverse and audio-reactive, leveraging the uniform variable `u_audio` derived from the audio input. If the total number of initial shaders (defined as `POPULATION_SIZE`) exceeds the number of base shaders, additional shaders are generated via mutation using an OpenAI LLM (e.g., GPT-4). Each shader is rendered in an independent WebGL canvas element using a common vertex shader and a dynamically compiled fragment shader. All canvases are embedded in a responsive grid layout for comparative visual exploration. For each new shader variant, the system dynamically compiles the GLSL code using WebGL and updates the rendering canvas. All shaders are sandboxed per canvas, ensuring isolation and parallel rendering. The canvas can be toggled to full-screen mode through a long press (or hold gesture), enabling immersive exploration. Failed shaders are excluded from rendering to maintain visual consistency. Successfully compiled shaders update the respective slot, effectively evolving the population. All generated shaders undergo syntax validation and WebGL compilation checks. If a generated shader fails, the system retries up to a maximum number of attempts (`MAX_ATTEMPTS`) before falling back to a base shader.

4.1   Interactive Evolution Loop

AI Co-Artist implements an evolutionary process in which users guide the selection process while large language models (LLMs) perform semantic mutation and crossover to generate new shaders. In this section, we present a formal model of this interaction, capturing the human—AI feedback loop, population dynamics, and LLM-based transformation operators. This formalization provides a foundation for analyzing stability, diversity, and convergence properties of the system. Let $\mathcal{P}_t = \{S_t^{(1)}, S_t^{(2)}, \ldots, S_t^{(N)}\}$ denote the shader population at iteration t, where each $S_t^{(i)}$ is a GLSL fragment shader program represented as a string. In AI Co-Artist, $N = 14$. Each program is compiled and rendered independently within the WebGL engine and evaluated visually by the user. Because shaders share a fixed interface (uniforms, main entry point), they occupy a constrained but expressive program space $S \in \mathcal{S} \subseteq \text{GLSL}$ where $\mathcal{S}$ is the subspace of fragment shaders compatible with the system's rendering pipeline. At each iteration, the user selects an arbitrary non-empty subset $\mathcal{E}_t \subseteq \mathcal{P}_t$ representing elite individuals whose visual characteristics the user wishes to preserve. Selection is therefore non-numeric, qualitative, and subjective, reflecting aesthetic judgment rather than fitness evaluated by a predefined objective. Unselected shaders constitute the replaceable set $\mathcal{R}_t = \mathcal{P}_t \setminus \mathcal{E}_t$. The complete interaction cycle is therefore:

$$\underbrace{\mathcal{P}_t \xrightarrow{\text{render}} \text{visuals}}_{\text{WebGL}} \xrightarrow{\text{user selection}} \mathcal{E}_t \xrightarrow{\text{LLM operators}} \mathcal{P}_{t+1}$$

This defines a mixed-initiative evolutionary system in which humans provide qualitative selection pressure and the LLM executes semantic variation. Unlike classical evolutionary algorithms, where mutation and crossover are stochastic and local, the LLM enables globally meaningful transformations informed by learned priors from code and artistic corpora. The result is an evolutionary process that is expressive, user-driven, and semantically rich, enabling high-level creative exploration while preserving shader validity across generations. The evolution cycle is fully user-driven. The user begins by selecting one or more shaders that they find visually appealing. This selection triggers the evolutionary logic: (1) if one shader is selected, mutation is applied, (2) if multiple shaders are selected, crossover is applied first to synthesize a parent, followed by mutation to generate variants, and (3) non-selected shaders are then replaced by these new variants. This loop allows continuous aesthetic refinement through intuitive curation, enabling an active participation in the process. Each iteration allows the user to progressively guide the population toward a desired aesthetic space.

4.2   LLM as Evolution Operators

The core evolution mechanism is driven by GPT-4 (or equivalent) through structured prompt engineering. Depending on user selection, the system applies one of two evolutionary strategies:

- **Mutation**: When a single shader is selected, it is passed to the LLM along with a prompt to generate a creatively modified but structurally similar variant. This promotes stylistic exploration while maintaining coherence.

  You are a world-recognized artist and shader programmer. Create an innovative and improved variant of the provided GLSL fragment shader code. Preserve its structure while creatively enhancing visual dynamics. Provide only the GLSL shader code without explanations.
- **Crossover**: When multiple shaders are selected, their source codes are included in a prompt asking the LLM to synthesize a coherent hybrid that

blends stylistic and structural attributes. The result serves as a new base for further mutation.

> *You are a renowned shader artist. Create a cohesive hybrid shader by blending the visual attributes of the following GLSL shaders. Return only the resulting GLSL shader code without any explanations.*

Given the selected set $\mathcal{E}_t$, the system determines whether to apply mutation or crossover. For a single selected shader S, mutation produces a semantic variant $M_\theta(S) = S'$, where M_θ is the LLM transformation parameterized by model weights θ. Mutation introduces stylistic and structural changes while preserving functional integrity. For a set of parents $\{S_1, S_2, \ldots, S_k\}$, crossover produces a blended hybrid $C_\theta(S_1, \ldots, S_k) = S^\star$, where $S^\star$ inherits conceptual patterns from multiple parents. The hybrid is typically further refined by mutation $S' = M_\theta(S^\star)$. The LLM thus serves as a high-level, knowledge-aware recombination engine. New shaders replace only unselected individuals. Formally, at each iteration:

$$\mathcal{P}_{t+1} = \mathcal{E}_t \cup \{M_\theta(S_j^\star) \mid j = 1, \ldots, |\mathcal{R}_t|\},$$

where $S_j^\star$ is either a mutated elite shader (single-parent case) or a crossover-derived hybrid (multi-parent case). This ensures that user-selected programs persist across generations while the remainder of the population explores new aesthetic directions. The success of the system hinges on carefully crafted prompts that steer the LLM toward generating visually diverse, syntactically correct, and semantically meaningful shaders. Prompts are designed to be minimal yet expressive, invoking the LLM's creativity without overwhelming its token window:

The **mutation prompt** takes a single shader as input and asks the model to improve upon it. It emphasizes creativity, aesthetic refinement, and structural preservation. The intent is to nudge the model to explore adjacent variants of a known good shader, adding new elements or modifying dynamics without straying too far. The **crossover prompt** is more exploratory. It provides multiple shaders and asks the model to create a coherent hybrid. This invites the LLM to blend stylistic and structural elements, leading to more novel and diverse outputs. These prompts are tested across dozens of iterations and continuously refined. We found that explicitly invoking the model's artistic role ("world-recognized artist") improved the coherence and boldness of outputs. Encouraging structure preservation reduced the rate of compilation errors.

5   Discussion

AI Co-Artist employs a novel form of semantic evolution in which mutation and crossover are delegated to a large language model rather than implemented through classical stochastic operators. This design choice yields transformations that are syntactically valid, aesthetically meaningful, and context-aware in ways that conventional symbolic mutation often fails to achieve. In this section, we examine the functional behavior, structural characteristics, and emergent properties of the two LLM-driven operators, drawing parallels to evolutionary computation while highlighting their distinct advantages.

5.1   Semantic Mutation via LLM Transformation

Traditional mutation operators in evolutionary art systems perturb numerical constants, insert noise patterns, or randomly modify syntactic structures. However, such operations often produce invalid GLSL shaders or degenerate outputs. By contrast, AI Co-Artist formalizes mutation as a *semantic rewriting* operation performed by the LLM. Given a parent shader S, the mutation operator M produces a variant S' by generating a meaning-preserving but visually distinct reformulation:

$$S' = M_\theta(S),$$

where M_θ denotes the LLM transformation under model parameters θ.

Empirically, the LLM exhibits several characteristic mutation modes:

- **Parameter Mutations.** The model frequently alters numeric constants, oscillation frequencies, color weights, or smoothing factors. These adjustments preserve the underlying structure while creating perceptible aesthetic variation.
- **Spatial Structure Mutations.** The LLM modifies coordinate warping expressions, distance estimators, or rotation matrices, thereby producing new spatial patterns such as ripples, spirals, or domain-warped textures.
- **Color Mapping Mutations.** A common mode involves introducing new gradient mappings, nonlinear color transforms, or palette functions. These changes yield substantial variation while retaining compatibility with the core rendering logic.
- **Temporal and Audio Reactivity Enhancements.** The model occasionally introduces modulation terms involving `iTime` or `u_audio`, producing richer animations. Because the prompts explicitly encourage "enhancing visual dynamics," such behavior is frequent.

Unlike random perturbations, these mutations are coherent and intentional, guided by the LLM's internal representation of artistic structure. This makes each mutation "high-level" in the sense that it preserves semantic intent even when rewriting code substantially.

5.2   Crossover as Semantic Blending

In classical evolutionary computation, crossover combines substrings or parse trees from two parent genomes. However, fragment shaders do not lend themselves naturally to syntactic crossover: swapping arbitrary code blocks often breaks variable scope, function dependencies, or numerical stability. The LLM overcomes this by performing crossover as a *conceptual blending* operation:

$$S_{\text{hybrid}} = C_\theta(S_1, S_2, \ldots, S_n),$$

where C_θ instructs the model to "blend" or "combine" visual characteristics, enabling it to merge conceptual elements across parents.

Analysis of generated hybrids reveals several consistent crossover behaviors:

- **Palette Transfer.** The LLM often inherits the color palette from one shader while using the spatial structure of another. For example, a fractal pattern from S_1 may be recolored using the neon palette from S_2.
- **Structural Fusion.** The model combines different structural motifs by integrating their coordinate transformations, noise functions, or domain-warping techniques. This results in visually complex hybrids that often exceed the expressive capacity of a single parent.
- **Behavioral Blending.** The temporal or audio-reactive behaviors of multiple parents may be fused, e.g., adopting the beat-driven pulse from one parent while retaining rotation dynamics from another.
- **Function Merging.** The LLM occasionally merges helper functions (e.g., noise, fbm, or pattern functions) into a unified shader, rewriting variable names as necessary to preserve correctness.

This semantic crossover is particularly powerful because it reflects the LLM's understanding of the functional and visual intent of each codebase, rather than blindly recombining syntax. The resulting hybrids are typically well-formed and visually rich, avoiding the brittleness associated with structural crossover operators in DSL-based genetic programming.

5.3   Emergent Properties of LLM-Driven Evolution

LLM-based mutation and crossover introduce several emergent phenomena that do not arise in classical evolutionary computation:

- **High-Level Exploration.** The space of possible mutations spans high-level artistic concepts, such as symmetry, contrast, or motion, rather than simple numeric perturbations.
- **Implicit Domain Knowledge.** The model frequently employs advanced shader patterns—domain warping, ray-marching, layered noise—without being explicitly programmed to do so.
- **Robustness to Structural Variations.** Because the LLM preserves interfaces and rewrites code holistically, shaders remain compilable across generations, ensuring evolutionary stability.
- **Non-Local Variation.** Mutations may involve coordinated changes across multiple lines of code, producing consistent behavior changes that would be improbable through random mutation.

These behaviors position the LLM not merely as a mutation engine, but as a domain-savvy creative collaborator capable of producing meaningful aesthetic jumps while respecting execution constraints.

5.4   Comparison to Classical Evolutionary Operators

Table 1 summarizes the contrast between LLM-driven and classical evolutionary operators. The LLM approach offers semantic coherence, high-level creativity, and structural reliability, at the cost of increased computational overhead and reliance on external model inference. In summary, the mutation and crossover operators in AI Co-Artist leverage LLMs to perform meaning-aware transformations that emulate human-like shader editing. This enables a richer and more reliable evolutionary process than traditional genetic operators, and plays a central role in the system's ability to support open-ended creative exploration.

Table 1. Comparison of classical and LLM-driven evolutionary operators.

Aspect	Classical Mutation/Crossover	LLM-Driven Mutation/Crossover
Granularity	Local numeric/syntactic perturbation	Global semantic rewriting
Validity	Often fragile, frequent syntax errors	High validity due to holistic rewriting
Creativity	Low to moderate	High, guided by learned artistic priors
Domain Knowledge	None	Implicit from training data
Computational Cost	Low	Higher (LLM inference)
User Control	Indirect	Direct via prompt and selection

6   User Experience Study

AI Co-Artist is designed around the principle of human-in-the-loop creativity, allowing users to guide shader evolution solely through visual preferences. The system minimizes technical friction while maximizing aesthetic agency. Upon launching the interface, the user is greeted with a grid of audio-reactive shaders rendered in real-time. Each shader is displayed in its canvas, and all are animated simultaneously. Users can upload their own MP3 files to drive the animations. Once an audio file is uploaded, all shaders are synchronized to the audio's beat and frequency structure. Each canvas is interactive:

- **Short click or tap:** Selects or deselects the shader. A white border highlights selected shaders.
- **Long press (1.5 s):** Triggers fullscreen preview of the shader, hiding UI elements for an immersive experience. Releasing the press exits fullscreen.

These interactions support rapid browsing, comparative evaluation, and focused inspection of visual effects. Once the user selects one or more shaders, the `Evolve` button is enabled. Upon clicking it:

1. The selected shaders are used to generate a new parent (via crossover and mutation).
2. The system then applies mutations to this parent shader.

3. All non-selected shaders in the population are replaced by new mutated variants.

This allows the user to retain visually compelling outputs while evolving the rest of the population. Selected shaders can be downloaded via the `Download` button. This saves all selected GLSL codes into a plain text file, allowing users to reuse, remix, or publish their creations outside the platform. This interaction model enables both novices and professionals to steer complex shader evolution using nothing but aesthetic judgment, eliminating the need for code manipulation while maintaining rich creative control.

To evaluate the effectiveness and usability of AI Co-Artist, we conducted a structured user study involving 50 participants recruited from digital art communities, shader coding forums, and university courses. Participants were divided into two balanced cohorts: shader novices (no coding experience, n = 30) and shader experts (moderate to advanced GLSL proficiency, n = 20). Each participant was asked to complete two tasks in randomized order: (1) create as many visually interesting audio-reactive shaders as possible in 25 min using AI Co-Artist, and (2) perform the same task using Shadertoy.com without AI assistance. We measured quantitative metrics, including: (1) Average number of completed shaders, (2) Time to first visually pleasing result, (3) Compilation error rate, and User satisfaction (Likert scale 1–5). Novice users created an average of 4.2 shaders with AI Co-Artist vs 0.6 with Shadertoy. Experts created 6.8 vs 2.9. Time to first viable output was reduced by over 60% for both groups. Satisfaction averaged 4.7/5 (AI Co-Artist) vs 2.8/5 (Shadertoy). Compilation errors occurred in less than 3% of generations after retries. Qualitative feedback indicated that users appreciated the "collaborative feeling" with the AI. Novices described the system as "magical" and "fun," while experts saw it as an ideation partner. Many participants requested extensions into other creative domains.

7   Conclusion

AI Co-Artist demonstrates how LLMs can be effectively integrated into creative workflows through interactive evolution. By combining real-time shader rendering, audio reactivity, and LLM-generated code, the system provides a novel platform for visual exploration. Users serve as curators, while the LLM serves as an ideation engine. Our user study confirms that AI Co-Artist improves productivity, reduces learning curves, and enhances satisfaction across user types. It stands as a compelling case for human-AI collaboration in digital art, and offers a template for similar systems in adjacent domains. AI Co-Artist lowers the barrier to shader art, democratizing access to a domain that was previously limited to technically skilled individuals. By providing a co-creative AI agent, it enables a broader audience—including musicians, students, and designers—to explore procedural visual art without programming expertise.

References

1. Takagi, H.: Interactive evolutionary computation: fusion of the capabilities of EC optimization and human evaluation. Proc. IEEE **89**(9), 1275–1296 (2001). https://doi.org/10.1109/5.949485
2. Sims, K.: Artificial evolution for computer graphics. In: Proceedings of the 18th Annual Conference on Computer Graphics and Interactive Techniques (SIGGRAPH '91), pp. 319–328. ACM, New York (1991). https://doi.org/10.1145/122718.122752
3. Mount, J., Reilly, S.N., Witbrock, M.: International interactive genetic art II (web-based interactive evolutionary art exhibit, Carnegie Mellon University, 1994–1996. https://johnmount.github.io/mzlabs/GeneticArt/. Accessed 11 Nov 2025
4. Unemi, T.: SBArt4 for an automatic evolutionary art. In: Proceedings of the IEEE Congress on Evolutionary Computation (CEC), Brisbane, Australia (2012). https://unemi.vivian.jp/sbart/4/. Accessed 11 Nov 2025
5. Unemi, T.: SBArt4 Daily Evolved Animation on WebGL (online resource, active since 2011). https://unemi.vivian.jp/sbart/4/DailyWebGL/. Accessed 11 Nov 2025
6. Lewis, M.: Evolutionary visual art and design. In: The Art of Artificial Evolution, pp. 3–37. Springer, Heidelberg (2008)
7. Bentley, P.J.: Evolutionary Design by Computers. Morgan Kaufmann, San Francisco (1999)
8. Romero, J., Machado, P. (eds.): The Art of Artificial Evolution: A Handbook on Evolutionary Art and Music. Springer, Heidelberg (2007)
9. Machado, P., Romero, J., Greenfield, G.: Artificial Intelligence and the Arts. Springer, Cham (2021)
10. Stanley, K.O., Miikkulainen, R.: Evolving neural networks through augmenting topologies. Evol. Comput. **10**(2), 99–127 (2002). https://ieeexplore.ieee.org/document/6790655. Accessed 11 Nov 2025
11. Secretan, J., Beato, N., D'Ambrosio, D.B., Rodriguez, A., Campbell, J., Stanley, K.O.: Picbreeder: a case study in collaborative evolutionary exploration of design space. Evol. Comput. **19**(3), 373–403 (2011)
12. Biles, J.A.: GenJam: a genetic algorithm for generating jazz solos. In: Proceedings of the International Computer Music Conference, pp. 131–137 (1994)
13. Chen, M., et al.: Evaluating large language models trained on code. arXiv preprint arXiv:2107.03374 (2021)
14. Wang, L., Zhang, Z., Chowdhery, A., Li, X., Lepikhin, D., et al.: Code llama: open foundation models for code. arXiv preprint arXiv:2308.12950 (2023)
15. Simon, J.: Lluminate (LLM-assisted creative exploration project). https://www.joelsimon.net/lluminate. Accessed 11 Nov 2025

Short Talks

Music in the Age of Artificial Intelligence: Meaning and Creativity From a Complex-Systems Perspective

Güncel Gürsel Artıktay[(✉)][iD]

Harran University, Şanlıurfa, Turkey
artiktay@harran.edu.tr

Abstract. This theoretical article approaches the impact of generative artificial intelligence in music not merely as "a new tool for composition/production", but as a multiplier that reconfigures—both in speed and in scale—a sociotechnical regime of selection operating across the tool–interface relationship, platform infrastructures, listener practices, and legal/regulatory frameworks. The text distinguishes end-to-end automation (where, with a single command, decisions regarding form, harmony, arrangement, performance, and post-production are generated largely by the model) from partial/auxiliary task automation and argues that prompting-centred workflows redistribute expertise, time, and responsibility. From the perspective of complexity science, it is proposed that musical meaning and creative value are not "fixed" qualities embedded within the work, but rather emerge within processes of emergence (the derivation of higher-level patterns from lower-level interactions), feedback, and path dependence (the accumulation of small early differences into long-term inequalities), through interactions between production practices and platform circulation. The study further discusses how listening contexts shaped by provenance and labelling information may reframe listeners' judgements via the effort heuristic (the inference that 'more effort = higher value'), and how, in content economies with low reproduction costs, trust may be re-established through costly commitments (costly commitments: hard-to-imitate trust signals involving labour/risk). In conclusion, the article suggests that keeping the unit of analysis in musicology solely at the level of the work is insufficient; instead, there is a need for mixed-method research programmes that relate micro (perception), meso (production), and macro (platform selection/circulation) levels.

Keywords: Complex systems · Complexity science · Musical meaning · Emergence · Musical creativity · AI and music

1 Introduction

Artificial intelligence is becoming increasingly embedded in the workflows of the creative industries, restructuring how art is conceived, produced, distributed,

P. Machado et al. (Eds.): EvoMUSART 2026, LNCS 16523, pp. 209–223, 2026.
https://doi.org/10.1007/978-3-032-24350-8_14

and experienced. Advances in text-to-music, style transfer, vocal/performance synthesis (i.e., the generation of human-like singing or instrumental performance), and harmony automation (i.e., semi-automated suggestions for accompaniment/chord decisions) have triggered a rapid shift in music production from performance-centred practices towards design- and prompting-oriented workflows. The product/interface layer is decisive in translating these capabilities into end-user experience. For example, with Stable Audio 2.0, Stability AI has enabled the generation of coherent tracks of up to three minutes at 44.1 kHz stereo resolution from a single natural-language prompt. Stable Audio 2.5, alongside improving inference speed and quality, has also introduced additional control affordances such as context-aware song completion (inpainting: regenerating a selected time interval of an uploaded recording—or its continuation—in a manner consistent with the surrounding musical context) (Stability AI, 2024, 2025). Suno (Suno) and Udio (Udio), meanwhile, have popularised prompting-centred production through promises of creation 'within seconds' and low-threshold web interfaces that can be used with minimal technical knowledge.

It is important not to reduce this diffusion solely to the end-to-end automation of finished songs (end-to-end automation), that is, the system's making of all intermediate decisions via a single command, without human intervention. Rather, machine learning more often assumes partial and auxiliary tasks across the production and analysis pipeline—for instance stems separation, transcription, mixing/mastering assistance, and accompaniment/chord suggestions. Such supportive augmentations redistribute where expertise, time, and responsibility concentrate within the production chain without eliminating human agency: while part of the performance contribution is delegated to automation, the relative weight of decision layers such as goal specification, prompt/parameter selection, and quality control increases.

Industry adoption indicators likewise point to a system in transition. For instance, MIDiA Research's consumer survey for the first quarter of 2024 (Q1 2024) reported that only 5% of listeners indicated having ever used generative AI to create a song or lyrics (MIDiA Research, 2024). The International Music Summit (Business Report 2025) states that, throughout 2024, 10% of consumers used generative AI to make music or lyrics, and that 60 million people engaged with music-creation applications (International Music Summit [IMS], 2025).

This transition cannot be considered independently of the clusters of risks platforms face. On the one hand, rights uncertainty increases: the possibility of output grounded in unlicensed data, content that imitates an artist's identity/voice, and questions of how copyright income should be distributed to which stakeholders and by what criteria—i.e., problems of allocating royalty shares—come to the fore simultaneously. On the other hand, catalogue integrity comes under pressure; spam uploads, manipulation of listening and discovery metrics, and the consequent 'recommendation pollution' of recommender systems with 'unnecessary content' can degrade a platform's content quality. Finally, in terms of user trust, distancing from or negative perceptions of AI content in certain contexts may produce 'perceptual penalties', such as reduced engagement, thereby directly affecting platform performance. In its statement dated 11 September 2025, Deezer reported that a substantial portion of daily deliveries to the platform consisted of fully AI-generated tracks; however, these tracks accounted for only a small share of total streams. Deezer further reported that a notable proportion of streams of fully AI-generated tracks were detected as manipulative and were therefore excluded from both royalties and recommender systems (Deezer, 2025). Accordingly, under this policy set, the asymmetry between accelerated supply and comparatively slower growth in consumption/visibility is an expected outcome. At precisely this point, listeners' responses to 'knowledge of artificiality' (knowledge that a track has been generated by AI) and the way context is constructed become decisive. Audience research and experimental studies indicate that the emotional impact of AI-generated music is not a fixed attribute; rather, it varies depending on prior knowledge of a track's provenance, how the content is framed and labelled, and the scenario in which listening occurs. For example, in experimental designs where knowledge about composer identity is manipulated, listeners' liking ratings can systematically decrease when they believe a piece has been composed by AI (Shank et al., 2023). By contrast, audiovisual experiments drawing on biometric measures have reported that AI music can increase arousal (e.g., pupil dilation), while emotional valence (its position on the pleasant–unpleasant axis) may not change significantly in some cases (Fišer et al., 2025). These findings suggest that, rather than compressing the discussion into binary judgements such as 'Is AI music good or bad?', listener perception should be explained through ecological models that jointly account for the full set of conditions shaping the listening experience. A central gap in the existing literature is the excessive emphasis on singular genius narratives and on intrinsic work properties, alongside the under-modelled roles of recommendation environments, provenance/traceability, and institutional rules of selection. This article argues that AI accelerates and augments creative production; however, from a complexity-science perspective, it emphasizes that creative value and musical meaning still undergo emergence within multilayered feedback processes that remain embodied, contextual, and dependent on social acceptance. Moreover, it is not new for tools to shape production: recording practices (edit, comp, overdub, time/pitch correction), sampling, and programmable tools (DAWs, plug-ins, presets) have long displaced production from being 'the work of a single person', already rendering the idea of 'singular agency' contestable. What changes markedly with generative AI is the capacity to produce large numbers of variants at speed, for these to enter circulation through platform algorithms, and to operate in tandem with accompanying infrastructures for provenance and production tracking.

2   Complexity-Science Perspective

Complex system is an open system in which patterns emerge from the local interactions of many components, where these patterns cannot be reduced to the sum of the parts, and which engages in continuous energy–matter–knowledge exchange with its environment. In such systems, dynamics are often non-linear; organisation occurs without central control, and feedback loops continually reshape system behaviour (Mitchell, 2009; Holland, 2014). Accordingly, explanation should focus not on isolating individual variables to establish causality, but on the configuration of connections among components and the interaction architecture produced by this configuration; for this reason, single-cause explanations are structurally inadequate in many cases.

However, complexity should not be conflated with complicated systems, because the distinction directly determines which explanatory and predictive regime is reasonable. Complicated systems may be multi-component, yet the relations among components are relatively well specified and loosely coupled. For this reason, one can decompose the system into subparts, design and test each part independently, and then recombine them to achieve predictably functioning performance; control and debugging also proceed in a more orderly fashion thanks to this modularity. By contrast, in complex systems, components are tightly coupled; as the system adapts to its environment, feedback continually recalibrates its operation. The system therefore carries its "past" into the present: small initial differences can amplify over time, narrowing subsequent trajectories (path dependence) and rendering forecasts that rely on producing a single precise value for a given moment structurally fragile (Ladyman & Wiesner, 2020; Sigahi & Sznelwar, 2024). For this reason, in complex domains the analytical aim is not to generate a single deterministic prediction, but to construct scenarios—descriptions of possible futures within the constraints of a path-dependent possibility space—and to make uncertainty explicit as part of decision and action processes (Byrne, 2005; 2024).

Nevertheless, complexity does not mean complete chaos. Chaotic systems are often deterministic, in the sense that their dynamics are specified by a rule or equation. Yet sensitivity to initial conditions means that very small initial differences grow over time, which in practice constrains reliable long-term prediction (Lorenz, 1963). Complex systems may sometimes exhibit chaotic behaviour; however, complexity does not necessarily entail chaos. Many complex systems operate in a structure that is not wholly entangled but can be decomposed into subsystems. This modular structure partially prevents a change in one part from 'bringing down' the entire system at once. Moreover, systems are often layered: while local rules operate at lower levels, more general patterns emerge at higher levels. Thus, even if it is difficult to predict every detail of a system with precision, it remains possible to generate short-term/local forecasts for certain parts and reasonable bounds for the system's overall behaviour (Simon, 1962; Strogatz, 1994). In this sense, Weaver's concept of "organised complexity" remains useful for describing systems in which many factors are simultaneously operative, yet the web of relations remains traceable (Weaver, 1948).

Complexity science is a transdisciplinary research field that investigates how order arises in complex systems composed of many interconnected components. The field's backbone consists of themes such as interdependence, self-organisation, non-linear interactions, and emergence. The methodological core of the complexity approach encompasses network thinking and non-linear dynamics, as well as theories of knowledge and computation, and approaches to fitness and multiagent modelling (Mitchell, 2009; Strogatz, 1994). The approach does not, however, rest on proclaiming a single universal "law of complexity". On the contrary, it treats the combined use of different explanatory levels and methods as legitimate, depending on the context of the phenomenon under study: at times foregrounding network patterns, at times dynamic processes, and at times qualitative analyses. For this reason, rather than locking itself into a single "correct" theory, the field proceeds with a plural toolkit that adjusts—according to context—our assumptions about what we count as part of the system and how we justify knowledge (Byrne, 2005; Capra & Luisi, 2014; Cilliers, 2016; Ladyman & Wiesner, 2020; Sigahi & Sznelwar, 2024).

3   Emergent Nature of Music

Music is a multi-layered, context-sensitive phenomenon. It is the product of interactions operating simultaneously across acoustic (the organization of sound), cognitive–embodied (perception, expectation), sociocultural (meaning, norms), and technological (instruments, software, platforms) levels. Accordingly, it is more appropriate to understand music not as a set of "things", but as a set of relationships and interactions. A complexity-science perspective likewise aims not to attribute behaviour to a single cause, but to track interactions, contextual constraints, and feedback concurrently. In music research, this perspective enables the examination and understanding of the complexity of musical experience through interaction networks spanning multiple levels. Such a pluralistic view is well suited to analysing how actors at different scales—from human composers/performers and listeners to data-driven recommender systems and platform design—jointly produce shared patterns; it is also necessary for grasping music's contemporary sociotechnical ecology. In this light, it is illuminating to draw an ontological distinction between the musical work and processes of musical experience/production. Musical experience is an emergent process constituted over time within an interaction network in which the organisation of sound, perception and expectation, cultural meanings, and affect continually nourish one another. The work, by contrast, is akin to the moment at which this flow "freezes and becomes visible" within a particular context: a result crystallised within a specific spatiotemporal arrangement. For this reason, rather than treating works as immutable, inviolable entities, it is more accurate to conceive of them as relatively stable nodes produced by processes that are reconstructed anew through each performance, recording, listening event, and circulation. In an era of AI-supported production and consumption, this ontological distinction becomes even more pronounced. Data-trained generative models have moved beyond symbolic schemes that follow pre-written rules, becoming multi-layered structures that learn regularities in timbre, rhythm, harmony, and form from data; in practice, they therefore function like systems capable of recognising and reconstructing "style". Models operating in the raw-audio domain (e.g., Jukebox) can sustain timbre and an expressive contour more continuously over minutes, while text-to-music models (e.g., MusicLM), when constructing a piece's overall character from a textual description, may in some cases maintain what might be called "long-horizon flow"—not merely sounding good for a few seconds, but sustaining, over 1–3 min, a coherent sense of sectional form, a tension–release trajectory, repetition–variation relationships, and an energy curve—while in other cases weakening this through drift, abrupt turns, or loop-like repetition (Agostinelli et al., 2023; Dhariwal et al., 2020). This shift converts human–machine interaction into a productive loop: the human sets the goal, specifies constraints, and evaluates the output; the machine, by expanding the space of options, generates many possible realisations of the same musical idea. As a result, creative agency ceases to be a property that "emerges from within" a single subject; instead, it becomes a distributed network of responsibility shared across decisions, tools, data regimes, and platform conditions. In the age of AI, the ontology of the musical work increasingly becomes intertwined with the transparency and traceability of the production chain; that is, the question "What is the work, and where does it begin and end?" now rests not only on the final file that emerges, but also on the processes that produced that file. Content provenance standards such as Content Credentials (C2PA) associate— via digital signatures—knowledge about a content item's production life cycle ("who did what, when, and with which tool") with the content; thus, if the content or this record is altered subsequently without being documented, such intervention can become detectable during verification. The aim is not to prohibit intervention, but to render its trace visible; through such verifiable traces, it becomes possible to describe with clearer boundaries which data, tools, and human decisions the work rests upon (C2PA, 2025). It should be noted, however, that such systems do not automatically declare that "the content is correct"; rather, they perform a more limited but important function: they indicate that provenance knowledge is well-formed, that the record has not been corrupted, and that the signing party can be authenticated within a relationship of trust. Within this framework, the boundaries of the work are redrawn not through a singular "artefact", but through technical documentation that renders visible the relationships among layers of data, model, interface, and human intervention. The musical work thus appears not only as an "output", but as a process node defined through verifiable operational traces.

4 From Genius to Ecology: Redefining Musical "Creativity"

In an era of AI-supported production and consumption, a complexity-science lens makes it untenable to conceive of 'creativity' in music as a singular, essential quality that suddenly appears within the mind of a genius. Creativity is, rather, a pattern that emerges over time from interdependencies among individuals and the tools they use; institutions and the rules governing circulation, datasets, models, and interfaces; and the responses of listener communities. For this reason, instead of reducing the 'creative agent' to the composer alone, it is more explanatory to conceptualize it as a multi-component ecology extending from instruments and software to data and models; from platform design and recommendation mechanisms to listener communities and copyright/legal regimes.

Musical creativity often emerges at the ecosystem's 'threshold conditions': in the tensions between what a listener expects and what surprises them; between the familiarity of repetition and the freshness of variation; and between established patterns and deliberate departures from them. Today's production regime operates, in broad terms, along a data →model →interface →platform →listener →law/regulation chain. Accordingly, 'new combinations' do not arise solely from model design; the cultural pools from which data are assembled and how they are represented, the ways interfaces steer users, and platforms' circulation rules (recommendation logics, visibility thresholds, policy and metadata regimes) also determine the conditions of novelty.

Within this framework, novelty often functions less as invention ex nihilo than as a combinatorial process: existing patterns are copied, recombined, and synthesized into new patterns (Eagleman & Brandt, 2019). 'Creative' products therefore tend to emerge not through radical ruptures, but through small yet cumulative deviations sustained at the limits of tolerance for uncertainty. Outputs that are overly similar do not attract attention and are not selected; those that are overly different, by contrast, struggle to take hold at the scale of a community. The success of musical creation depends on the capacity to sustain this balance—namely, to carry familiarity and novelty simultaneously by establishing sufficient alignment with the network's key nodes, such as listeners, performers, and institutions.

Against this background, AI tools function less as agents that 'invent' creativity on their own than as speed-and-scale multipliers that expand the interaction space: more trials, faster variation, broader search. The result is that creative responsibility does not concentrate in a single individual, but is organized in distributed fashion across decisions, tools, platform conditions, and rules. In other words, in many cases what is 'creative' is not a singular subject, but the configuration itself—an assemblage composed of human–tool–platform–rule combinations.

Creativity is not merely a matter of 'producing something new'; it is also a matter of legitimation—of that novelty being recognized as creative, accepted, and put into circulation within the field. Put differently, whether a production is deemed 'creative' is constituted less by an assumed essence carried within the work than by who is allowed to speak, which criteria are treated as valid, and which institutions confer approval. From a Foucauldian perspective, the label 'creative' is not a neutral description affixed to production, but a relational attribution operating within knowledge/power relations and bounded by a field's internal norms: how critics, curators, academia, industry actors, award mechanisms, media, and platforms construct the category of 'creative' and to whom they allocate it becomes decisive (Foucault, 2013).

On the consumption side, circulation regimes—such as platforms' recommender systems, interface flows that steer users, and royalty/content policies—markedly affect how visible creative knowledge becomes and which contents are foregrounded. This means that creativity is evaluated not only by 'what is produced', but also by 'how it is circulated and how it is received'. A track's fate is not reducible to the tastes of individual listeners; it is shaped together with population-level selection, pathways of diffusion within networks, and the dynamics of the attention economy (who allocates how much time to what). As production and consumption draw closer together, it becomes more explanatory to treat creativity not as a property emanating from singular genius, but as an ecological process emerging from the interaction of many actors; correspondingly, the means of measuring the 'creative' shift from fixed, universal criteria towards evaluative frameworks that vary by context. Considering these factors, AI—through both symbolic operations and statistical learning—can generate new combinations that are not present verbatim in the training data but are nonetheless derived from it. In some contexts, such combinations may strike listeners as unexpectedly 'apt' or 'convincing'. Yet describing an output as 'creative' seldom follows automatically from a model's generative capacity; it operates more like an attribution completed when human judgement and the field's processes of legitimation come into play—an attribution that emerges afterwards at the level of knowledge. In other words, AI opens a potential space for meaning making; 'creative meaning' is granted through the recognition of this potential within criticism, institutions, community norms, and practices of circulation. From a complexity-science standpoint, musical creativity in the age of AI must be considered on two planes simultaneously. First, creation is a multi-agent process that emerges over time within interdependencies among humans, tools, platforms, institutions, and norms; thus, 'creativity' is not a property that can be enclosed within a single mind, but a pattern produced by an ecology. Second, the judgement 'creative' is an epistemic decision issued at the intersection of embodied perception and experience, social judgement, and platform regimes that organize circulation—in other words, a knowledge claim established through criteria and justifications ('we count this as creative'). For this reason, creativity is less an 'essential quality' than the selection, circulation, and stabilization—under specific conditions—of trials and variations sustained within networks. What is counted as 'creative' is the capacity to produce patterns that not only come into being but also can hold at the scale of a community and generate meaning through repeated encounters.

5   Reconstructing "Meaning"

In an era of AI-supported production and consumption, 'meaning' in music cannot be reduced to the relationship between the composer and the work. What produces and transforms meaning is an ecology of sociotechnical elements, including data-intensive production tools, platforms' recommendation and curation mechanisms, interface designs, copyright/content policies, and the field's institutional practices. At the ontological level, musical meaning does not stand as something 'given' at a single level; it is an emergence arising from continuously operating feedback among embodied performance, sonic material, social context, and technological interfaces. For this reason, it is more explanatory to think about musical experience not in terms of 'things' but in terms of 'relations': timbral, rhythmic, and harmonic patterns are not shaped solely by local rules (e.g., the internal norms of a particular style), but also by an interaction order that extends across the network as a whole (who produces/listens to/circulates what, where, and under which conditions). When this framework is carried to the biocultural level, musical meaning can be conceived as potential knowledge: the more effectively the regularities and differences in a work or performance align with the listener's reservoirs of knowledge (genre knowledge, cultural memory, listening habits, embodied modes of accompaniment), the more clearly the direction and intensity of meaning become articulated. Ontology thus attaches not only to 'musical structure', but also to a multi-layered knowledge ecology jointly constituted by the perceiving subject and the environment that sustains the structure. At the epistemological level, complexity points to a knowledge regime that cannot be exhausted by a single method or metric. Knowledge of musical meaning requires the joint mobilization of qualitative interpretation and quantitative modelling, sensitivity to local cases and network-level mapping, and the integration of experimental findings with theoretical abstraction. Hence, rather than seeking a single universal law, it becomes critical to identify patterns—such as path dependence, feedback, and centralization—that recur across levels. AI-supported ecosystems of production and consumption sharpen this picture, particularly on the consumption side. Recommender systems and curation algorithms redistribute listener exposure, thereby altering—at the level of the 'selection environment'—the conditions under which meaning emerges which contents are encounterable, which are repeatedly rendered visible, and which persist in circulation are closely tied to these mechanisms. Experimental research on cultural markets has shown that small initial differences can amplify over time into large inequalities of visibility, and that 'success' depends not only on intrinsic qualities but also on early exposure and position within a network (Salganik et al., 2006; Yucesoy & Barabási, 2016). For this reason, epistemic practice in musicology should not remain limited to analyzing works; it must also model the patterns produced by algorithmic distribution, user interactions, and platform economies. At this point, the issue of costly commitments occupies a pivotal place on both ontological and epistemological maps. According to signaling theory, reliable signals are typically costly: they are difficult to imitate and provide verifiable cues regarding the signaler's capacity and intent (Spence, 1973; Zahavi, 1975). Findings suggesting that communities that impose a 'cost threshold' at the level of collective action can be more robust and more integrated support this intuition at the level of cultural organization (Iannaccone, 1992; Sosis & Bressler, 2003). In interpersonal trust contexts, credible commitments—such as taking risks and putting one's own resources at stake—can render statements of intent more persuasive (Ohtsubo & Watanabe, 2009). Applied to the musical field, behaviors such as assuming the risks of live performance, increasing transparency in the production process (e.g., traceable provenance records), labor-intensive craft practices, or open sharing of data/code can function as costly signals of 'human-led' contribution. Under conditions in which production and reproduction have become artificially cheap, such signals acquire a more visible function for differentiation and the formation of trust within a community. Meaning, accordingly, is nourished not only by sonic patterns but also by the architecture of commitments undertaken by the actors who carry those patterns. This view suggests that recipients may use perceived levels of labour and risk in the production process as an indirect indicator of aesthetic value (the effort heuristic) (Kruger et al., 2004). Indeed, provenance knowledge that a work was produced with AI can, for some listeners, weaken liking responses by lowering perceived effort and creativity, and by triggering a sense of threat to beliefs about the uniqueness of human creativity (Dunham et al., 2025; Heimstad et al., 2025; Millet et al., 2023; Shank et al., 2023). This may therefore be one reason why songs known to have been made with AI receive lower liking ratings. Taken together, these findings make it possible to conceptualize musical meaning in the age of AI across three levels. Ontologically, meaning is not something that 'emanates' from a single source; it is a distributed property of a musical–ecological network, an emergence produced jointly through the interaction of sonic material, embodied performance, social norms, and algorithmic interfaces. Epistemologically, for this reason, 'knowing meaning' cannot be resolved through a single method: it calls for a mixed-method and reflexive approach that combines deep reading at the level of individual works/events with modelling of circulation–selection patterns at network and community scales. Normatively, to sustain trust in content economies with low reproduction costs, costly commitments become critical, because they generate hard-to-imitate signals that rebuild mechanisms of trust. This triadic framework neither reduces musical meaning to a 'fixed essence inside the work' nor abandons it to 'purely arbitrary interpretation'; instead, it positions meaning as an emergence shaped by interactions and commitments, helping to explain why communities may still find certain productions 'worth investing in' even when platforms redistribute visibility.

6   Conclusions

The primary aim of this article is to conceptualize the impact of generative artificial intelligence (GenAI) in the musical domain not merely as "a new tool for composition/production", but as a speed-and-scale multiplier that reconfigures a sociotechnical regime of selection operating across the tool–interface–platform–listener–law/regulation chain. A complexity-science perspective enables this transformation to be read not by reducing it to a single cause (e.g., "model quality has improved" or "creativity is over"), but as an ecology jointly produced by non-linear interactions, feedback loops, and multi-level dependencies. Within this framework, the central conclusion is as follows: musical meaning and creative value are not fixed properties embedded in the work as an "essence", but emergent outcomes arising from interactions among production practices, platform visibility, provenance/traceability knowledge, labelling/framing, and listener acceptance. These emergent outcomes require holding together two planes that the article has insisted on distinguishing from the outset. First, on the production side, the practical impact of GenAI is in many cases not reducible to "end-to-end automation via a single command"; more decisive is the redistribution of expertise, time, and responsibility across the production chain. Prompting-centered workflows shift part of performative gestural labour into interface and parameter decisions, while increasing the weight of decision layers such as goal setting, constraint specification, and quality control. Accordingly, the claim that "human influence has disappeared" is, in most contexts, an overgeneralization both empirically and conceptually; a more realistic picture is the reorganization of decisions within distributed creative agency. Second, on the consumption side, "how what is produced is put into circulation" has become as decisive as what is produced. Platforms' recommender systems, interface flows, and content policies redistribute listener "exposure", transforming—at the level of the selection environment—the conditions under which creative value emerges. The critical point here is path dependence, which is common in complex systems: exposure differences that initially appear small can, through feedback, amplify into long-term inequalities of visibility. For this reason, the category "creative" is constituted in practice not only through aesthetic judgements, but also through institutional–technical variables such as a platform's visibility thresholds, metadata regime, and capacity to counter manipulation. In other words, in the age of GenAI the object of musicology is not only the "work", but the circulation regime that selects, repeats, and renders the work visible. A second major conclusion advanced by the article is that the layers of provenance/traceability and labelling/framing are no longer merely "auxiliary metadata", but an interpretive key that actively constitutes the listening experience. If the same sonic content, when presented with the knowledge "produced with AI", can in some contexts elicit lower liking and value attribution, then what requires discussion is less the music's "essential" quality than the sense-making regime triggered by provenance knowledge in the listener. This regime often operates through the effort heuristic: recipients may use the labour/effort presumed to stand behind a production as an indirect indicator of value. As GenAI lowers the cost of reproduction, this heuristic becomes more fragile, because large numbers of "convincing" productions can be replicated at low cost. Consequently, the re-establishment of trust and differentiation depends not only on detection technologies, but on redesigning the signals that indicate what is trustworthy. At this juncture, the concept of "costly commitments" offers a framework that strengthens the article's normative discussion (without sliding into legal counsel). From the standpoint of signalling theory, reliable signals are often costly and are credible precisely because they are hard to imitate. In a musical ecosystem, this may take the form of assuming the risks of live performance, increasing transparency in the production process, maintaining verifiable provenance records, labour-intensive craft practices, and process documentation. The argument here is not that "AI is bad"; rather, when reproduction costs fall, the design of trust signals becomes more critical for communities' ability to distinguish productions they regard as "worth investing in". If such signals (the labelling regime, provenance infrastructure, and platform policy set) cannot be established, the risks faced by the system (rights uncertainty, the erosion of catalogue integrity, recommendation pollution, and user trust) may become mutually reinforcing feedback loops, pulling down the ecosystem's overall quality. A third, and more stringent, conclusion follows: in the age of GenAI, if musicology keeps its unit of analysis solely at the level of text/score/sonic artefact, it will systematically miss a decisive part of the phenomenon. This is because the primary stage on which meaning and value emerge is increasingly platform-based encounters. Methodologically, this implies that no single paradigm (purely qualitative hermeneutics or purely quantitative metricism) will suffice; a multi-level, mixed-methods research programme is required. Here, "mixed methods" does not mean simply placing two methods side by side, but designing an approach capable of relating, within a single theoretical model, micro-level listener perception, meso-level production practices, and macro-level circulation/selection patterns. Finally, because this study deliberately offers a theoretical framework without collecting primary data, it is necessary to indicate clearly the forward-looking lines along which the framework can be empirically tested and refined. At least three research trajectories are particularly fruitful:

Framing/labelling experiments. Experimental designs comparing, across genres and cultures, how different presentations of provenance knowledge (AI/hybrid/human) affect liking, attributions of creativity, perceived labour, and trust. (The key question is not "Is AI worse?", but "In which context does which label trigger which judgement?")

Platform-level circulation analysis. Network analyses and/or multi-agent simulations examining how early exposure differences amplify, how manipulation injects "noise" into recommender systems, and how policy interventions alter selection dynamics. Cultural-market experiments have already shown that small initial differences can increase inequality and unpredictability; platform ecologies may further strengthen this mechanism.

Provenance infrastructure and trust-signal studies. Interdisciplinary (HCI/STS/musicology) research on how traceability systems (as much as the technical standard itself) are displayed in interfaces, how they are read by listeners, and under which conditions they can fulfil a "costly commitment" function. In sum, in the GenAI music ecosystem, resolving questions of "creativity" and "meaning" requires neither a return to romantic genius narratives nor a wholesale surrender of production to an automation logic. A more robust approach is to treat music as a sociotechnical ecology in which patterns emerge within distributed networks of agency, are shaped by platforms' selection conditions, and require new signalling mechanisms to rebuild trust. The contribution of this article is to move the discussion beyond aesthetic binaries (good/bad; human/AI) and to propose a higher-resolution conceptual map for explaining the conditions under which musical value emerges.

Disclosure of Interests. The author declares no conflicts of interest.

References

Agostinelli, A., et al.: MusicLM: Generating music from text (2023). (arXiv:2301.11325) [Preprint]. arXiv. https://arxiv.org/abs/2301.11325. Accessed 28 Jan 2026

Ardeliya, V., Taylor, J., Wolfson, J.: Exploration of artificial intelligence in creative fields: generative art, music, and design. Int. J. Cyber IT Serv. Manage. **4**(1), 40–46 (2024). https://doi.org/10.34306/ijcitsm.v4i1.149

Byrne, D.S.: Complexity, configurations and cases. Theory, Culture Society **22**(5), 95–111 (2005). https://doi.org/10.1177/0263276405057194

Byrne, D.S.: Scenarios—using the complexity frame of reference to inform the construction of available futures in the possibility space. Front. Compl. Syst. **2**, 1306328 (2024). https://doi.org/10.3389/fcpxs.2024.1306328

C2PA. (2025, May 1). Content credentials: C2PA technical specification (Version 2.2). https://spec.c2pa.org. Accessed 28 Jan 2026

Capra, F., Luisi, P.L.: The systems view of life: a unifying vision. Cambridge University Press (2014)

Cilliers, P.: Critical complexity. Routledge (2016)

Deezer Newsroom, 11 September 2025. 28% of all delivered music is now fully AI-generated [Community post]. https://community.deezer.com/. Accessed 28 Jan 2026

Dell'Anna, F.: On the enactive approaches to musical emotions. Front. Psychol. **12**, Article 716499 (2021). https://doi.org/10.3389/fpsyg.2021.716499

Dhariwal, P., Jun, H., Payne, C., Kim, J. W., Radford, A., Sutskever, I.: Jukebox: A generative model for music (2020). (arXiv:2005.00341) [Preprint]. arXiv. https://arxiv.org/abs/2005.00341, last accessed 2026/01/28

Dunham, R.L., van Kleef, G. A., Stamkou, E.: The threat of synthetic harmony: The effects of AI vs. human origin beliefs on listeners' cognitive, emotional, and physiological responses to music. Computers in Human Behavior: Artificial Humans. Advance online publication (2025). https://doi.org/10.1016/j.chbah.2025.100205

Eagleman, D., Brandt, A.: . Yaratıcı tür: İnsan yeniliğinin anatomisi [The runaway species: How human creativity remakes the world]. Domingo Yayınları. (Original work published 2017) (2019)

Favela, L.H.: Cognitive science as complexity science. WIREs Cogn. Sci. **11**(5), e1525 (2020). https://doi.org/10.1002/wcs.1525

Fernández, J.J., Vico, F.: AI methods in algorithmic composition: a comprehensive survey. J. Artif. Intel. Res. **48**, 513–582 (2013). https://doi.org/10.1613/jair.3908

Fišer, N., Martín-Pascual, M., Andreu-Sánchez, C.: Emotional impact of AI-generated vs. human-composed music in audiovisual media: A biometric and self-report study. PLOS ONE **20**(6), e0326498 (2025). https://doi.org/10.1371/journal.pone.0326498

Foucault, M.: Kelimeler ve şeyler: İnsan bilimlerinin bir arkeolojisi [The order of things: An archaeology of the human sciences]. İmge Kitabevi Yayınları. (Original work published 1966) (2013)

Gałuszka, P.: The influence of generative AI on popular music: fan productions and the reimagination of iconic voices. Media, Culture & Society, **47**(3), 603–612 (2024). https://doi.org/10.1177/01634437241282382

Heimstad, S.B., Wien, A. H., Gaustad, T.: Machine heuristic in algorithm aversion: Perceived creativity and effort of output created by or with artificial intelligence. Computers in Human Behavior: Artificial Humans. Advance online publication (2025). https://doi.org/10.1016/j.chbah.2025.100190

Hiller, L.A., Isaacson, L.M.: Experimental music: Composition with an electronic computer. McGraw-Hill (1959)

Holland, J.H.: Complexity: A very short introduction. Oxford University Press (2014)

Iannaccone, L.R.: Sacrifice and stigma: Reducing free-riding in cults, communes, and other collectives. J. Polit. Econ. **100**(2), 271–291 (1992)

International Music Summit. (2025). IMS business report 2025 [Industry report]. https://www.internationalmusicsummit.com/news/ims-business-report-2025. accessed 2026/01/28

Kruger, J., Wirtz, D., Van Boven, L., Altermatt, T.W.: The effort heuristic. J. Exp. Soc. Psychol. **40**(1), 91–98 (2004). https://doi.org/10.1016/S0022-1031(03)00065-9

Ladyman, J., Wiesner, K.: What is a complex system? Yale University Press (2020). https://doi.org/10.12987/9780300256130-006

Lopez, J. (2024, April 3). Introducing Stable Audio 2.0. Stability AI. https://stability.ai/news/stable-audio-2-0. Accessed 28 Jan 2026

Lorenz, E.N.: Deterministic nonperiodic flow. J. Atmos. Sci. **20**(2), 130–141 (1963). https://doi.org/10.1175/1520-0469(1963)020%3C0130:DNF%3E2.0.CO;2

Martin, S. (2025, September 10). Stability AI introduces Stable Audio 2.5, the first audio model built for enterprise sound production at scale. Stability AI. https://stability.ai/news/stability-ai-introduces-stable-audio-25-the-first-audio-model-built-for-enterprise-sound-production-at-scale, last accessed 2026/01/28

MIDiA Research, 6 June 2024. GenAI music platforms are niche among consumers—is being too easy the problem? https://www.midiaresearch.com/blog/gen-ai-music-platforms-are-niche-among-consumers-is-being-too-easy-the-problem. Accessed 28 Jan 2026

Millet, K., Buehler, F., Du, G., Kokkoris, M.D.: Defending humankind: Anthropocentric bias in the appreciation of AI art. Comput. Hum. Behav. **143**, 107707 (2023). https://doi.org/10.1016/j.chb.2023.107707

Mitchell, M. (2009). Complexity: A guided tour. Oxford University Press

Mozer, M.C.: Neural network music composition by prediction: Exploring the benefits of psychophysical constraints and multiscale processing. Connect. Sci. **6**(2–3), 247–280 (1994)

Ohtsubo, Y., Watanabe, E.: Do sincere apologies need to be costly? Test of a costly signaling model of apology. Evol. Hum. Behav. **30**(2), 114–123 (2009)

Pachet, F., Roy, P., Carré, B. (2020). Assisted music creation with Flow Machines: Towards new categories of new (arXiv:2006.09232) [Preprint]. arXiv. https://doi.org/10.48550/arxiv.2006.09232. Accessed 28 Jan 2026

Patel, A.D.: Music, language, and the brain. Oxford University Press (2008)

Patel, A.D.: Evolutionary music cognition: cross-species studies. In: Rentfrow, P.J., Levitin, D.J. (eds.) Foundations in Music Psychology: Theory and Research, pp. 459–501. MIT Press, Cambridge, MA (2019)

Salganik, M.J., Dodds, P.S., Watts, D.J.: Experimental study of inequality and unpredictability in an artificial cultural market. Science **311**(5762), 854–856 (2006)

Shank, D.B., Stefanik, C., Stuhlsatz, C., Kacirek, K., Belfi, A.M.: AI composer bias: Listeners like music less when they think it was composed by an AI. J. Exp. Psychol. Appl. **29**(3), 676–692 (2023). https://doi.org/10.1037/xap0000447

Sigahi, T.F.A.C., Sznelwar, L.I.: Between complexity and complicatedness: Typology for organizational systems. Kybernetes (2024). https://doi.org/10.1108/K-02-2023-0233

Simon, H.A.: The architecture of complexity. Proc. Am. Philos. Soc. **106**(6), 467–482 (1962)

Sosis, R., Bressler, E.R.: Cooperation and commune longevity: A test of the costly signaling theory of religion. Cross-Cult. Res. **37**(2), 211–239 (2003)

Spence, M.: Job market signaling. Quart. J. Econ. **87**(3), 355–374 (1973)

Strogatz, S.H.: Nonlinear dynamics and chaos. Addison-Wesley (1994)

van der Schyff, D., Schiavio, A.: Evolutionary musicology meets embodied cognition: Biocultural coevolution and the enactive origins of human musicality. Front. Neurosci. **11**, Article 519 (2017). https://doi.org/10.3389/fnins.2017.00519

Weaver, W.: Science and complexity. Am. Sci. **36**(4), 536–544 (1948)

Wilkins, R., Hodges, D., Laurienti, P., Steen, M., Burdette, J.: Network science and the effects of music preference on functional brain connectivity: From Beethoven to Eminem. Sci. Rep. **4**, 6130 (2014). https://doi.org/10.1038/srep06130

Yucesoy, B., Barabási, A.L.: Untangling performance from success. EPJ Data Sci. **5**(1), 17 (2016)

Zahavi, A.: Mate selection–a selection for a handicap. J. Theor. Biol. **53**(1), 205–214 (1975). https://doi.org/10.1016/0022-5193(75)90111-3

Generative Artificial Intelligence, Musical Heritage and the Construction of Peace Narratives: A Case Study in Mali

Nouhoum Coulibaly[1], Ousmane Ly[2], Michael Leventhal[1(✉)],
and Ousmane Goro[3]

[1] RobotsMali AI4D Lab, Bamako, Mali
{research,mleventhal}@robotsmali.org
[2] Faculté de Médecine et d'Odonto-Stomatologie (FMOS), Université des Sciences, Techniques et Technologies de Bamako (USTTB), Bamako, Mali
[3] Conservatoire des Arts et Métiers Multimédia Balla Fasseké, Bamako, Mali

Abstract. This study explores the capacity of generative artificial intelligence (Gen AI) to contribute to the construction of peace narratives and the revitalization of musical heritage in Mali. The study has been made in a political and social context where inter-community tensions and social fractures motivate a search for new symbolic frameworks for reconciliation. The study empirically explores three questions: (1) how Gen AI can be used as a tool for musical creation rooted in national languages and traditions; (2) to what extent Gen AI systems enable a balanced hybridization between technological innovation and cultural authenticity; and (3) how AI-assisted musical co-creation can strengthen social cohesion and cultural sovereignty. The experimental results suggest that Gen AI, embedded in a culturally conscious participatory framework, can act as a catalyst for symbolic diplomacy, amplifying local voices instead of standardizing them. However, challenges persist regarding the availability of linguistic corpora, algorithmic censorship, and the ethics of generating compositions derived from copyrighted sources.

Keywords: Generative artificial intelligence · Malian musical heritage · peace narratives · Suno AI · musical hybridization · national languages · social cohesion · cultural sovereignty

1 Introduction: Context and Problem Statement

Mali faces persistent challenges in terms of social cohesion. A country once renowned for its inter-ethnic respect and tolerance, now confronts an increasingly fragmented cultural identity threatening the tradition of peaceful co-existence. There is a national imperative to discover narrative innovations capable of recomposing the societal fabric. Music, an important, historic medium for transmitting identity and cultural mediation, sometimes appears insufficient to reach modern

P. Machado et al. (Eds.): EvoMUSART 2026, LNCS 16523, pp. 224–235, 2026.
https://doi.org/10.1007/978-3-032-24350-8_15

audiences or engage with the aspirations of younger generations when it remains confined to traditional forms.

The emergence of generative artificial intelligence (Gen AI) opens new possibilities for exploring the nexus of cultural heritage and innovation to produce a modern language of reconciliation imbued with local meaning. Suno AI, a platform for generating musical compositions from text prompts [4], was selected as a relevant tool for exploring this intersection of technology and tradition. The complementary use of ChatGPT and GEMINI for the generation and translation [4] of texts into national languages completed the technological framework for the experiment.

This study tested the hypothesis that Gen AI, used responsibly with cultural sensitivity [11], can revitalize musical heritage by fostering the emergence of indigenous, socially inclusive, and technologically contemporary peace narratives.

2   Methodology

2.1   Study Design and Research Questions

The study employed an empirical qualitative approach centered on a co-creation workshop, complemented by thematic analyses of the resulting musical productions and semi-structured interviews with participants. This approach allows us to understand creative processes, cultural perceptions, and the potential of the approach to simulate the creation of culturally-rooted peace narratives. The study was guided by three primary hypotheses: (H1) Gen AI tools can be effectively appropriated by participants with varying levels of technological literacy when provided with structured training; (H2) AI-generated musical compositions incorporating traditional Malian instruments and national languages will be perceived as culturally authentic by participants; (H3) The collaborative process of AI-assisted music creation can foster intergenerational and interinstitutional dialogue around themes of reconciliation.

2.2   Study Population

The "Voices of Reconciliation" workshop brought together 30 participants recruited using a stratified sample ensuring diversity: 10 ministerial representatives, 9 youth actors from the National Youth Council, 6 women's organizations, and 5 traditional music students from the Balla Fasséké Kouyaté Conservatory. This composition produced a plurality of institutional, generational, artistic and gender perspectives, essential for a credible co-construction of narratives. Participants were required to: (a) be at least 18 years of age; (b) have basic literacy in at least one national language; (c) demonstrate interest in either traditional Malian music or digital technologies; and (d) be available for the full three-day workshop duration. Exclusion criteria included inability to commit to the complete workshop schedule.

2.3   Workshop Structure

The workshop was structured in three phases:

Phase 1 — Initial training: familiarization with the principles of Gen AI applied to music, operation of ChatGPT, GEMINI and Suno AI, introduction to prompt engineering.

Phase 2 — Integration of tradition and modernity: development of prompts integrating characteristic traditional Malian sound motifs (kora, balafon, djembe) [7], into contemporary arrangements, experimentation with hybrid styles.

Phase 3 — Linguistic authenticity: creation of lyrics in national languages (Bambara, Fulfulde, Tamasheq, Songhai, Dogon), phonetic transcription and adaptation of prompts to respect tonal structures and local cultural references.

2.4   Creative Process and Prompt Engineering

The creative process followed an iterative methodology combining human artistic direction with AI generation capabilities. Each composition underwent multiple refinement cycles before achieving its final form.

Prompt Development Strategy: Participants were trained to construct prompts using a structured framework: [Genre/Style] + [Instrumentation] + [Tempo/Mood] + [Cultural References] + [Thematic Content]. Initial prompts were often generic (e.g., *African music with drums*), which yielded stereotypical outputs lacking Malian specificity.

Successful Prompt Examples: Through iterative refinement, participants developed more sophisticated prompts. For example, a successful prompt for the Afrobeat-Mandingo pattern was: *Afrobeat fusion with traditional Mande instruments, featuring kora melody over balafon ostinato, 125 BPM, brass section with Mandingue-style horn arrangements, djembe percussion, uplifting and celebratory mood.* This specificity produced outputs that participants recognized as culturally grounded.

Unsuccessful Approaches: Several prompt strategies proved ineffective: (a) requests for specific artist imitations were blocked by platform content policies; (b) prompts in national languages without phonetic guidance resulted in mispronunciations; (c) overly complex prompts combining more than three musical traditions produced incoherent outputs; (d) requests for the imzad (Tuareg instrument) often defaulted to generic violin sounds due to limited training data.

Iteration Statistics: On average, each of the 12 final compositions required 8–12 generation attempts before achieving participant satisfaction. The primary reasons for rejection included: inappropriate instrumentation (34%), tempo mismatches (22%), vocal quality issues (28%), and stylistic incongruence (16%).

2.5   Data Collection and Analysis

The data includes twelve original compositions produced in the workshop, analyzed thematically around the concepts of peace, forgiveness, memory, diversity, and cultural mediation. The semi-structured interviews explored perceptions of legitimacy, impact, limitations, and sustainability of the approach. The entire output of the workshop is available in this Google Drive folder.

Data Collection Instruments: Three complementary instruments were employed: (1) a post-workshop survey with 15 Likert-scale items and 5 open-ended questions administered to all willing participants (n=13); (2) semi-structured interviews lasting 20–30 minutes with a purposive sample of 8 participants representing each stakeholder group; and (3) systematic documentation of the creative process including all prompts attempted, generation outputs, and participant deliberations during selection.

Analysis Framework: Qualitative data were analyzed using thematic analysis following Braun and Clarke's six-phase framework [3]. Two researchers independently coded interview transcripts, with inter-rater reliability of 0.82 (Cohen's kappa [5]). Musical compositions were analyzed using an adapted version of Agawu's African music analysis framework [1], examining: rhythmic structures, melodic organization, instrumental choices, linguistic content, and symbolic references.

3   Results

3.1   Hybrid Musical Architectures

The analysis reveals three recurring hybridization patterns:

Afrobeat-Mandingo Pattern: 120–130 BPM rhythmic structure in 4/4 time signature, incorporating the kora and balafon as the main melodic carriers [12], while electric guitar, bass and brass and wind instruments (tenor saxophone, trumpet) provide the Afrobeat1 groove. Traditional passages (solo kora/balafon) alternate with energetic sections carried by electric instruments, creating a sophisticated progressive dynamic. This pattern is exemplified by this composition. This song speaks to the solidarity between the peoples of the Sahel, making reference to the AES, the Alliance of Sahel States, a political union created in 2023 by Mali, Niger and Burkina Faso to strengthen cooperation. The French lyrics are below:

> Un seul peuple, un seul cœur, AES
> Avançons sans peur, AES
> Dans l'unité, la paix va briller
> Cohésion sociale on va gagner !

> Jeunes et vieux marchons côte à côte
> Nos rêves ensemble, notre force éclate
> Chaque main qui s'élève construit demain
> Dans l'unité, on tient notre chemin

> One people, one heart, AES
> Let us move forward without fear, AES
> In unity, peace will shine

Social cohesion we will win!

Young and old, let us walk side by side
Our dreams together, our strength bursts forth
Each hand that rises builds tomorrow
In unity, we keep moving forward

Reggae-sabar Model: bipartite architecture featuring an initial reggae section (soft rhythm guitar, deep bass, light percussion) creating a meditative atmosphere, followed by a modulation towards Senegalese sabar rhythms characterized by energetic percussion (djembe, balafon, groove drums). This musical narrative structure symbolizes a journey from individual reflection to collective celebration. Through linguistic diversity, this multi-language passage calls on people to join hands in collective celebration.

Communal Harmony: communal harmony across languages as exemplified by this composition weaving together 5 languages.:

K'an bɛ se, k'an bɛ ka hakili don (Bambara)
Jam tun, jam tun, no ɓe ngoodi (Fulfulde)
Boro foori nda siini, bara koyra béeri (Songhay)
Assalam, assalam, akal negh ad tew (Tamasheq)
Main dans la main, bâtissons l'avenir (French)

Let us be united, let us share wisdom
Peace everywhere, may peace guide us
Reconciliation and peace build the country
Peace, peace, may our land live
Hand in hand, let us build the future

Meditative Tuareg Model: slow tempo (80–110 BPM), Tuareg pentatonic scales, dominance of the imzad (monocord bowed string instrument) and the tende (single skin hand drum central to Tuareg culture). This model evokes a spiritual and contemplative mood [13], with progressive mystical rises embodying the nomadic traditions of the desert. The Tamasheq and French lyrics present an invocation to peace for all the people of Mali. Here is the composition and Tamasheq lyrics.

Alâfiyet n Sahel
Alâfiyet, alâfiyet
Sahel wa n alâfiyet
Kel Tamasheq d wassaghmar
Amassa n tafrawt
Taggalt n Mali
Alâfiyet! Alâfiyet!
Sahel wa n alâfiyet
Taggalt d wassagher
Mali d Sahel
Alâfiyet ar abadan

Peace of the Sahel
Peace, peace
Sahel of peace
Tuaregs and their brothers
Children of freedom
Unity of Mali
Peace! Peace!
Sahel of peace
Unity and brotherhood
Mali and Sahel
Peace for eternity

3.2   Linguistic and Semantic Strategies

Four major discursive strategies structure the compositions.

Stratified Multilingualism: the systematic use of several national languages within a single work, forging a sense of collective identity. For example, *"Yɛrɛ ka sɛbɛ"* (Our Hearts are One) intersperses French with the primary language, Bambara, while *"Farafina Kelen Yan"* (One Africa) alternates between Bambara, Fulfulde, Songhay, and Tamasheq.

Reconciliation Lexicon: recurring lexical field structuring the compositions around key terms: *bɛn* (Bambara: union), *faraŋɔkan* (Bambara: together), *alâfiyet* (Tamasheq: peace), *kelenya* (Bambara: unity), *wassagher* (Tamasheq: brotherhood), *taggalt* (Tamasheq: union), *duddal* (Fulfulde: reconciliation), *jam* (Fulfulde: peace). These terms carry a symbolic charge beyond their simple literal translation.

Unifying Geographical Imagery: encompassing geographical references symbolically reconstituting the territorial integrity and national identity of Mali ("From Kayes to Gao, from Ségou to Timbuktu", "Kidal to Ménaka", "From the Sahel to Bamako"), countering narratives of fragmentation. Here is a composition, with French lyrics, illustrating this discursive strategy:

Je suis le Mali, Tu es le Mali, Nous sommes le Mali
Ensemble, nous ferons le Mali
Je suis le Mali, Tu es le Mali, Nous sommes le Mali
Debout pour le Mali

Notre combat, c'est le Mali
Ensemble, pour toujours, le Mali!
De Tombouctou à Sikasso
Une voix qui s'élève

Pour la paix, pour la terre
Pour l'avenir qu'on rêve
Nos mains unies, nos cœurs sans peur
Pour le Mali, pour l'honneur ...

I am Mali, You are Mali, We are Mali
Together, we will build
I am Mali, You are Mali, We are Mali
Standing proud for Mali

Our struggle is for Mali
Together, forever – Mali!
From Timbuktu to Sikasso
One voice rises up

For peace, for our land
For the future of our dreams
Our hands united, our hearts without fear
For Mali, for our honor ...

Sahelian Organic Metaphors: images taken from the local environment such as the (Niger) river, *ténéré n alâfiyet* (Tamasheq: desert of peace), *azaluy n alâfiyct* (Tamasheq: caravan of peace), and the palaver tree, anchoring the discourse of peace in daily experience and the Malian collective imagination.

3.3   Narrative Structures

The compositions present a common narrative architecture in four phases:
Invocation Phase: instrumental introductions and first verses establishing a historical or spiritual context, invoking ancestors, traditions or sacred geography.
Diagnostic Phase: central verses identifying challenges without explicitly naming them, evoking "what has been broken" or calling for overcoming fears.
Collective Resolution Phase: refrains sung in chorus proposing unity, reconciliation, fraternity, with a repetitive character aimed at anchoring memory.
Spiritual Elevation Phase: musical bridges and final sections adopting a spiritual or transcendent register, with invocations such as *Ya Allah! Alâfiyet!* (Tamasheq: O Allah! Peace!) or projections towards the future such as *Alâfiyet ar abadan* (Tamasheq: Peace Forever) as can be heard in this song:

3.4   Instrumentation and Cultural Significance

The organological analysis reveals a symbolic hierarchy:
Identity Instruments: kora, balafon and djembe appear in 90% of the compositions, functioning as markers of Mande authenticity. with the dominance of the balafon and the djembe [2,13]. Here is an illustration.
Mediation Instruments: brass and wind instruments and electric guitar create sounds familiar to young urbanites [8] while respecting traditional rhythmic structures. Here is an example, *"Aw bismillah"*.
Specific Regional Instruments: imzad and tahardent (Tuareg lute) in Tamasheq compositions, tam-tam and sabar percussion in Senegalese-style variants. Here is an example.

3.5   Vocal Performance

The compositions systematically present a responsorial and antiphonal structure inherited from griot traditions [1], fulfilling several functions: participatory (choral format facilitating collective appropriation), pedagogical (repetition reinforcing memorization), and symbolic (polyphony materializing harmony in diversity) [6].

The analysis of vocal timbres reveals the use of distinct registers for specific effects: deep male voices for solemn invocations, female voices for passages of hope and spiritual elevation, and childlike voices in certain refrains symbolizing the future.

3.6   Technological Appropriation

Examination of the prompts used as the workshop progressed revealed a growing sophistication, allowing the specification of traditional musical scales, control of instrumental balance, integration of prosodic indications [9] respecting linguistic tones, and programming of fluid stylistic transitions. This progression testifies to a process of appropriation [8] where the AI tool becomes an instrument at the service of a controlled cultural vision.

3.7   Social and Perceptual Effects

Participants reported a strengthening of cultural belonging a renewed confidence in the possibilities of dialogue, and a heightened awareness of languages and traditions as positive resources. The workshop fostered institutional recognition: the works were promoted during the Semaine Nationale de la Réconciliation (National Week of Reconciliation - SENARE 2025), confirming their institutional and social resonance.

3.8   Participant Survey Analysis

To complement the qualitative observations, a post-workshop survey was conducted among 13 participants to quantify their satisfaction and perceptions of the creative process. Given the small sample size (n=13), these quantitative findings should be interpreted as exploratory indicators rather than statistically generalizable results. Participant survey results are illustrated in Fig. 1 and Fig. 2.

The results revealed a high satisfaction rate (average score: 4.15/5), with 77% having previously used AI tools and 100% reporting a positive change in perception of generative AI applied to music. Approximately 85% felt that the compositions authentically reflected Malian musical heritage, and 80% stated they felt capable of creating new pieces autonomously or with minimal assistance after the training.

Survey Limitations: Several methodological limitations should be acknowledged: (1) the response rate of 43% (13/30) may introduce self-selection bias; (2) participants who remained for the survey may have been those most engaged

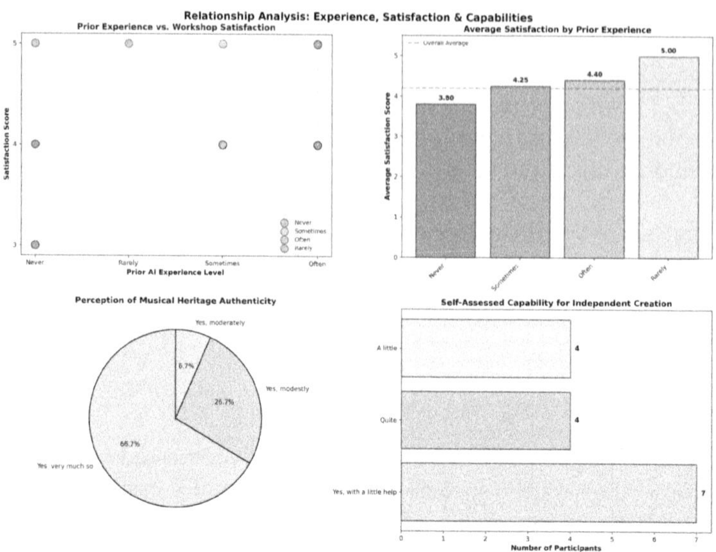

Fig. 1. Relationship analysis showing prior AI experience vs. workshop satisfaction, average satisfaction by experience level, perception of musical heritage authenticity, and self-assessed capability for independent creation (n=13).

with the process; (3) the survey was administered immediately post-workshop, potentially capturing enthusiasm rather than sustained impact; (4) social desirability bias may have influenced responses given the institutional nature of participants.

Qualitative comments emphasized the fusion of tradition and modernity as the most enriching part of the experience, and many noted that the use of national languages reinforced their cultural pride. However, participants also identified limitations, such as short workshop duration, imperfect pronunciation in local languages, and the need for more realistic instrument reproduction.

4   Discussion

4.1   Transformative Potential of Gen AI

The results provided preliminary evidence that Gen AI, used in a culturally respectful participatory framework, may serve to catalyze the construction of peaceful narratives by linking ancestral memory and shared futures.

This approach is in line with contemporary reflections on cultural sovereignty and the role of technologies in preserving national creative identities. "AI must be a lever of cultural sovereignty, putting technology at the service of creators and not the other way around". [10] Applied to the Malian context, this perspective positions Gen AI as a tool for amplifying local heritage and voices, contributing to a symbolic sovereignty where technology values endogenous cultures rather than standardizing them.

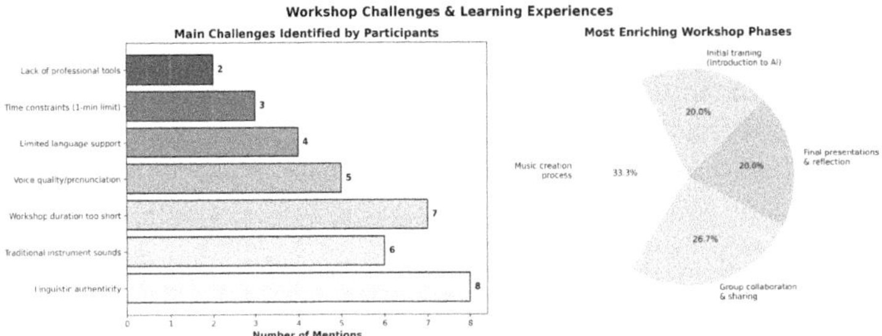

Fig. 2. Workshop challenges identified by participants (left) and most enriching workshop phases (right). Linguistic authenticity and traditional instrument reproduction emerged as primary technical limitations.

Alternative Explanations: The positive outcomes observed may be attributable to factors beyond the AI tools themselves. The workshop format brought together diverse stakeholders in a structured collaborative environment, which may independently foster dialogue and social cohesion regardless of the technological medium. The novelty effect of interacting with AI systems may also temporarily elevate enthusiasm. Additionally, the institutional visibility of the project (culminating in SENARE 2025 promotion) may have reinforced positive perceptions. Future research should employ control conditions to isolate the specific contribution of AI-assisted creation versus collaborative music-making more broadly.

4.2 Limits Identified

In the course of the workshop, participants encountered certain classes of limitations.

Insufficient Linguistic Corpora: some national languages lack sufficient corpora in technological tools, posing problems of transcription and tonal recognition.

Algorithmic Censorship: certain words (e.g., *Niger, soko*) are automatically flagged or suppressed by moderation algorithms due to negative associations in other linguistic contexts.

Authenticity-Innovation Tension: the challenge of preserving traditional sound while innovating structurally, without falling into superficial hybridization.

Ethical and Legal Issues: participants expressed concerns about the origin of training data, intellectual property, and the remuneration of holders of traditional musical heritage.

4.3 Implications for Public Policy

Visibility of the workshop results on a national stage engendered a broader discussion with national authorities on the implications of the experiment. Potential translational implications for public policy included:

Institutional Integration: incorporation of Gen AI into national cultural programs with a clear ethical and legal framework.
Educational Curricula: integration into music schools and cultural institutions to develop technological and creative skills.
Linguistic Infrastructure: support for the construction of linguistic and musical corpora specific to national languages.

5   Conclusion

This exploratory study suggests the potential of generative artificial intelligence, properly framed and culturally contextualized, to contribute to heritage revitalization and the construction of peace narratives. Analysis of the twelve compositions produced reveals notable sophistication in musical hybridization, layered multilingualism, and progressive narrative construction.

Beyond musical production, even on the scale of a limited experiment, we have observed multidimensional transformative effects: identity strengthening, intergenerational and interinstitutional dialogue, and technological empowerment ensuring the replicability of the process. The institutional recognition at SENARE 2025 testifies to the social and political relevance of this approach.

This research is part of the reflections on digital cultural sovereignty, positioning Gen AI as a medium for promoting endogenous cultures. The initiative opens up future research perspectives: longitudinal studies assessing sustainable impact, comparative research with other post-conflict contexts, exploration of integration with other forms of cultural expression.

References

1. Agawu, K.: The African Imagination in Music. Oxford University Press, New York (2016)
2. van Beek, W.: The Dancing Dead: Ritual and Religion among the Kapsiki/Higi of North Cameroon and Northeastern Nigeria. Oxford University Press, Oxford (2012)
3. Braun, V., Clarke, V.: Using thematic analysis in psychology. Qual. Res. Psychol. **3**(2), 77–101 (2006). https://doi.org/10.1191/1478088706qp063oa
4. Casini, L., Vila, L.C., Dalmazzo, D., Kaila, A.K., Sturm, B.L.T.: Data-driven analysis of text-conditioned AI-generated music: A case study with suno and udio. arXiv preprint arXiv:2509.11824 (2025)
5. Cohen, J.: A coefficient of agreement for nominal scales. Educ. Psychol. Measur. **20**(1), 37–46 (1960). https://doi.org/10.1177/001316446002000104
6. Cousin, M.: Tambor de crioula et tambor de mina, expressions musicales rituelles afro-descendantes de são luis do maranhão (brésil). créolité, identité, "culture de résistance". Cahiers d'ethnomusicologie **27**, 341 (2014). https://journals.openedition.org/ethnomusicologie/2224, online bibliographic notice (OpenEdition)
7. Goro, O.: Musique dans la culture Guru en pays dogon: Voix et Voies de l'oju. Ph.D. thesis, Conservatoire des Arts et Métiers Multimédia Balla Fasseké, Bamako (2025)

8. Hannecart, C.: Les transmissions hybrides de la pratique musicale. Diversité **173**, 81–86 (2013)
9. Migliore, O.: Analyser la prosodie musicale du punk, du rap et du ragga français (1977–1992). Ph.D. thesis, Université Paul-Valéry Montpellier 3 (2016)
10. Ministère de la Culture (France): IA et création musicale, vers un nouveau modèle? https://www.culture.gouv.fr/actualites/ia-et-creation-musicale-vers-un-nouveau-modele (2025). Accessed 20 Jan 2026
11. Pram, L.J., Morreale, F.: Opening musical creativity? embedded ideologies in generative-AI music systems. arXiv preprint arXiv:2508.08805 (2025)
12. Rouget, G.: Un film expérimental: Batteries dogon. L'Homme **5**(2), 126–132 (1965)
13. Schaeffner, A.: Introduction à musique et danses funéraires chez les dogons de sanga. L'Homme **177–178**, 207–250 (2006)

Crystallizing Semantics: Mapping the Journey of Word Meaning in Language Models

Himanshu Dwivedi[(✉)]

Directorate of Education, Delhi, India
`himanshu.dwivedi@delhi.gov.in`

Abstract. Understanding how language models (LMs) acquire and refine semantic meaning during training is an open challenge in natural language processing (NLP). Prior work has largely focused on static evaluations of pretrained models, providing limited insight into the temporal dynamics of meaning representation. In this study, we conduct a fine-grained longitudinal analysis to date of semantic similarity tracking in autoregressive language models. Using models from the Pythia family (70M and 410M parameters), we systematically evaluate six complementary metrics—AUPRC (Area Under Precision-Recall Curve), AUROC (Area Under Receiver Operating Characteristic Curve), Kendall's Tau, KT-Cosine, KTKendall, and Spearman correlation—across three established human-judgment datasets (SimLex-999, WordSim-353, MTurk-771). Our results reveal consistent patterns: (i) early training stages exhibit sharp improvements in alignment with human similarity judgments, (ii) larger models (410M) achieve consistently higher correlations than smaller models (70M), and (iii) certain metrics (e.g., KTKT) peak early before gradually declining, indicating non-monotonic dynamics of semantic structure formation. We release a largescale dataset of over 160,000 evaluation traces (JSON, CSV, and plots) covering 80,000 training steps, enabling reproducibility and further exploration of semantic dynamics. To our knowledge, this is the first dataset capturing step-by-step semantic evolution across model scales. These findings not only extend prior work on meaning dynamics in LLMs but also provide a valuable benchmark resource for the community.

Keywords: Language models · Semantic representation · Training dynamics · Word similarity · Pythia · Embedding evaluation · Rank-based metrics

1 Introduction

The study of how word meaning emerges in neural language models has long been central to understanding both natural language processing (NLP) and cognitive modeling. Word embeddings—dense vector representations of words—have provided a powerful lens into how models capture semantic relationships [1, 2]. Early distributional methods such as Word2Vec and GloVe [demonstrated that geometric relations in vector space can align with human judgments of similarity and analogy, sparking a vast body of research on embedding quality and evaluation benchmarks [3–5].

© The Author(s), under exclusive license to Springer Nature Switzerland AG 2026
P. Machado et al. (Eds.): EvoMUSART 2026, LNCS 16523, pp. 236–251, 2026.
https://doi.org/10.1007/978-3-032-24350-8_16

Recent work has shifted focus from static embeddings to training dynamics: when and how word meaning crystallizes during pretraining. Notably, Liu et al. [6] analyzed checkpoints of the Pythia language models (70M–410M parameters) [7], tracking the evolution of word embeddings relative to human similarity judgments (WordSim353) and static baselines such as GloVe. Their analysis used rank-based metrics, including a novel "KT-on-KT" (KTKT) measure, to argue that semantic structure emerges surprisingly early in training, stabilizing after only a small fraction of total steps. They further showed that larger models converge more cleanly, and that traditional cosine-based similarity may obscure important structural mismatches with human judgments.

While influential, this prior work faced several limitations: the evaluation relied almost exclusively on a very small subset of WordSim353 ($\approx$96 word pairs surviving tokenizer overlap), making statistical robustness a concern; only a single baseline embedding set (GloVe) was considered; and comparisons focused narrowly on a few rank-based measures without systematically examining discriminative performance metrics such as AUROC and AUPRC. Furthermore, the analysis was confined to a single family of Pythia models [7] and primarily to the embedding layer, leaving open questions about generality, metric sensitivity, and broader dataset alignment.

This study extends prior analyses of semantic formation in language models in four key ways. First, we broaden the evaluation beyond WordSim-353 by incorporating SimLex-999 and MTurk-771, two benchmarks designed to capture more fine-grained and diverse human similarity judgments. Second, we introduce a set of complementary evaluation metrics that combine rank-based approaches (Spearman [8], Kendall [9], KTcos [6], KTKT) with discriminative measures (Area Under the Receiver Operating Characteristic Curve (AUROC) and Area Under the Precision–Recall Curve (AUPRC)), enabling a more holistic assessment of semantic structure. Third, we conduct scaling comparisons between two representative Pythia models (70M and 410M parameters), allowing us to examine how model size influences semantic stabilization and alignment. Finally, we analyze full training trajectories using multiple checkpoints, systematically preserving intermediate results to characterize convergence patterns and early stabilization effects.

Our results confirm and extend earlier claims: semantic structure indeed emerges early and stabilizes rapidly, with larger models converging faster and more consistently. However, we also observe that discriminative measures (AUROC, AUPRC) continue to improve well beyond the point of stabilization in rank-based metrics, suggesting a two-phase process where core semantic relations emerge early, but fine-grained discriminability continues to sharpen later in training. Moreover, our experiments show that KTKT diverges while KTcos converges, reinforcing the view that different similarity metrics capture complementary aspects of representation quality.

Taken together, this work provides a richer and more reliable account of how word meaning develops during pre-training. By combining broader datasets, multiple evaluation metrics, and cross-model scaling, we replicate and extend prior findings under broader evaluation settings. These insights contribute to the interpretability of large language models, with implications for efficient training, model selection, and early stopping strategies.

2 Related Work

2.1 Semantic Representation in Neural Language Models

Early distributional approaches such as Word2Vec, GloVe, and fastText [10] established that lexical semantics can be modeled by vector similarity, showing alignment with human similarity judgments on benchmarks like WordSim-353. These methods provided static embeddings but offered little insight into the dynamics of how word meaning emerges during training.

2.2 Probing Neural Representations

The rise of large pretrained transformers [11–14] sparked a wave of research probing model internals. Studies on BERT and GPT families explored semantic encoding via correlation with benchmarks [15, 16] and downstream probes [17]. However, most analyses are static, considering only the final trained model.

2.3 Dynamics of Meaning During Training

A smaller but growing body of work has explored semantic development. Saphra et al. [18] studied how RNNs acquire structure over time. The Pythia project released training checkpoints specifically to encourage research on interpretability across scales.

The most directly relevant prior study is the work by [6]. They tracked the embedding layer of five Pythia models (14M–410M params) across training checkpoints, comparing similarity scores with human judgments from WordSim-353. They [6] analyzed the embedding layers of Pythia models across training checkpoints and reported that semantic structure emerges early and stabilizes quickly. They demonstrated that rank-based metrics, particularly KTKT, reveal discrepancies between human similarity judgments and learned embeddings that cosine similarity obscures. Their results further showed that larger models converge more smoothly and form more stable geometries. However, their evaluation relied on a very small subset of WordSim-353 (96 usable pairs), limiting statistical robustness and generalizability.

2.4 Our Contribution

Our contributions are as follows:

1. **Expanded benchmarks.** We evaluate semantic similarity using three datasets—WordSim-353, SimLex-999, and MTurk-771—reducing selection bias and capturing broader semantic phenomena.
2. **Metric diversity.** In addition to KTcos and KTKT, we incorporate AUROC, AUPRC, Spearman, Kendall, and cosine-based diagnostics to provide a more nuanced evaluation.
3. **Scaling dynamics.** We analyze two representative Pythia models (70M and 410M parameters) across 80,000 training steps, balancing analytical depth with computational feasibility.

By broadening datasets and metrics while carefully controlling computational budget, our work addresses the key limitations of the work by [6] and provides a more robust account of semantic development in transformers. Table 1 situates our study relative to prior work on semantic representation learning.

Table 1. Model Configurations.

Study	Models	Checkpoints	Datasets	Metrics	Key Findings	Limitations
Word2Vec / GloVe	static	—	WordSim353, MEN, etc	Cosine	capture similarity	no dynamics
Saphra et al. [18]	RNN	many	language modeling	intrinsic probes	syntax first, semantics later	small scale
Liu et al. [6]	Pythia 14M–410M	154 ckpts	WordSim 353 (96 pairs)	KTcos, KTKT	Early stabilization, rank mismatch	tiny eval set
This work	Pythia 70M, 410M	80k steps	WS353, SimLex, Mturk	6 metrics (AUPRC, AUROC, Spearman, Kendall, KTcos, KTKT)	early semantics + richer evaluation	Budget constraints, subset of models
Word2Vec / GloVe	static	—	WordSim 353, MEN, etc	Cosine	capture similarity	no dynamics

3 Methodology

3.1 Models

We base our analysis on the Pythia family of transformer language models, which were designed explicitly to facilitate interpretability studies. In this work, we focus on two representative scales:

- EleutherAI/pythia-70m (12 layers, 70M parameters)
- EleutherAI/pythia-410m (24 layers, 410M parameters)

Both models were trained on the Pile dataset [19], a large and diverse English text corpus, and we track their development over 80,000 training steps. For each checkpoint (every ~10,000 steps), we extract embedding representations for evaluation.

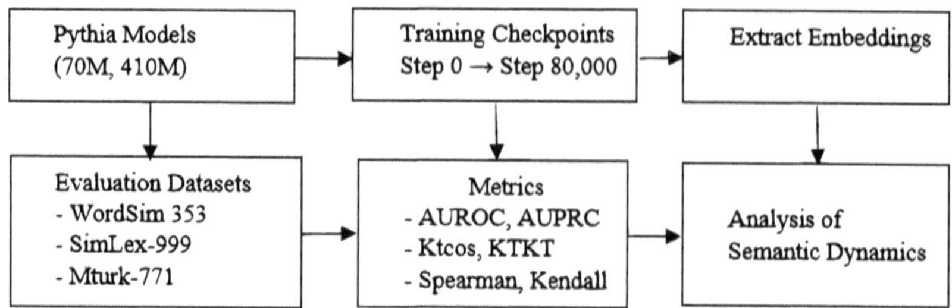

Fig. 1. Experimental Pipeline for Tracing Word Meaning Dynamics

3.2 Benchmark Datasets

To evaluate semantic knowledge, we use three widely adopted human-annotated benchmarks:

- WordSim-353: 353 word pairs annotated for semantic similarity.
- SimLex-999: 999 word pairs designed to test genuine similarity rather than relatedness.
- MTurk-771: 771 pairs rated via crowdsourcing.

These datasets provide complementary views of lexical semantics.

Table 2. Summary of benchmarks—#pairs, focus, annotation method

Dataset	# Pairs	Focus	Annotation Method
WordSim-353	353	General relatedness	Human similarity/relatedness ratings by students and researchers Finkelstein et al. [3]
SimLex-999	999	True similarity (not just relatedness)	Expert-designed and crowdsourced ratings focusing on substitutability (Hill et al. [4]
MTurk-771	771	Word relatedness	Crowdsourced judgments via Amazon Mechanical Turk Halawi et al. [5]
WordSim-353	353	General relatedness	Human similarity/relatedness ratings by students and researchers Finkelstein et al. [3]
SimLex-999	999	True similarity (not just relatedness)	Expert-designed and crowdsourced ratings focusing on substitutability (Hill et al. [4]

3.3 Evaluation Metrics

A key contribution of our work is expanding the set of evaluation metrics beyond simple correlation. We use six complementary measures:

1. AUPRC (Area Under Precision-Recall Curve) – captures alignment quality in imbalanced similarity distributions.
2. AUROC (Area Under Receiver Operating Characteristic Curve) – evaluates ranking consistency across thresholds.
3. Spearman's Rank Correlation – measures monotonicity of similarity judgments.
4. Kendall's Tau – alternative rank correlation, more robust to ties.
5. KT-Cosine – cosine similarity applied to ranked lists, capturing global order preservation.
6. KT-Kendall – Kendall-on-Kendall (ranking similarities within embeddings, then comparing ranks).

This richer metric suite allows us to disentangle nuances of representational development, e.g., whether a model improves at global ranking vs. local pairwise consistency.

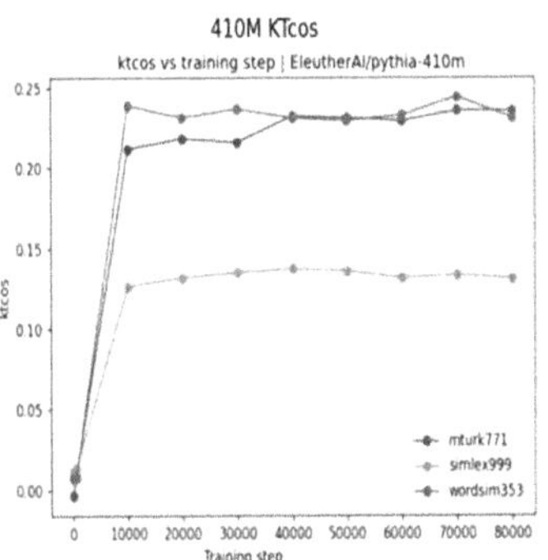
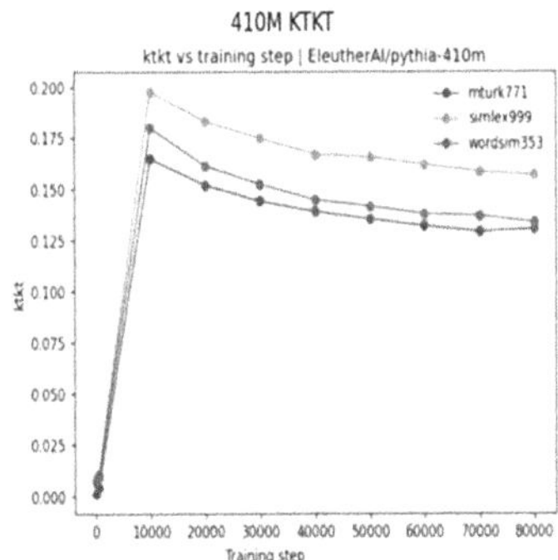

Fig. 2. Comparison of KTcos and KTKT on Pythia-410M. KTcos suggests convergence toward stability, while KTKT reveals divergence from human similarity judgments, highlighting the importance of metric choice.

The selected metrics serve complementary purposes rather than redundant evaluation. Rank-based measures (Spearman, Kendall, KTcos, KTKT) assess global ordering and structural consistency of similarity judgments, while AUROC and AUPRC quantify discriminative capacity between high- and low-similarity word pairs. Using both classes allows us to distinguish early stabilization of semantic structure from later improvements in fine-grained discrimination.

3.4 Experimental Pipeline

Figure 1 provides an overview of the experimental pipeline, from checkpoint selection and embedding extraction to metric computation and visualization. Our workflow proceeds as follows:

1. **Checkpoint Sampling:** Extract embeddings at steps {0, 10k, 20k, ..., 80k}.
2. **Word Pair Projection:** For each benchmark, compute cosine similarities between word embeddings.

3. **Metric Computation:** Compare predicted similarities to human ratings using the six metrics.
4. **Result Aggregation:** Store outputs as JSON logs, CSV summaries, and plots for reproducibility.

All experiments were implemented in Python 3.10, using:

- PyTorch [20] for model handling
- Hugging Face Transformers [21] for pretrained checkpoints
- NumPy/Pandas/Matplotlib for data analysis and visualization

Our Jupyter notebook provides a complete reproducible environment.

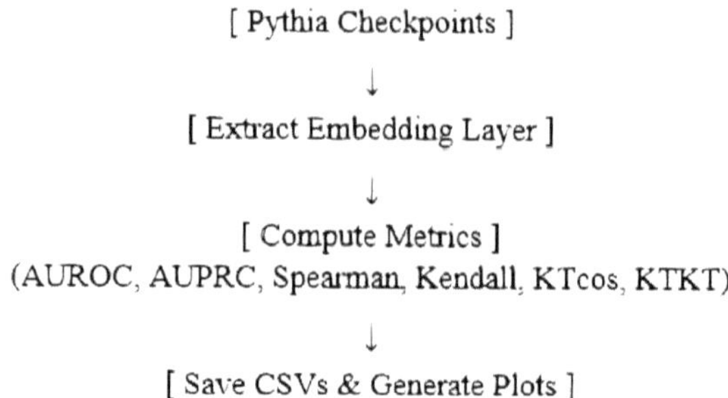

Fig. 3. Flowchart of experimental pipeline

4 Methodology

4.1 Models

To assess the alignment of model-internal representations with human semantic judgments, we used three standard benchmarks:

- **WordSim-353 (WS353):** consists of 353 word pairs annotated by human judges for semantic similarity and relatedness and has been widely used in early embedding research (e.g., Word2Vec, GloVe). Following prior work, we applied tokenizer-based filtering to ensure vocabulary coverage. After filtering, approximately 300 word pairs were retained, with the exact number varying slightly by model and checkpoint.
- **SimLex-999:** 999 word pairs annotated specifically for similarity (not relatedness). Provides a more rigorous test, penalizing loose associations and emphasizing true semantic similarity.
- **MTurk-771**: 771 word pairs with crowd-sourced ratings. Offers broader coverage and complements the biases of WordSim-353 and SimLex-999.

Each dataset provides human similarity judgments on a numeric scale. To ensure comparability across sources, we min–max normalized gold scores to the range [0, 1]. For reproducibility, we stored all normalized versions as CSV files in our artifact directory.

4.2 Models and Checkpoints

We focused on two transformer models from the EleutherAI Pythia suite:

- **Pythia-70M:** a small-scale model (12 transformer layers, ~70M parameters).
- **Pythia410M:** a mid-scale model (24 layers, ~410M parameters).

Both models were trained autoregressively on The Pile dataset and released with checkpoints at every 1,000 steps, up to 143,000 steps. For computational efficiency, we restricted analysis to every 1,000 steps up to 80,000.

At each checkpoint, we extracted the input embedding matrix (model.get_input_embeddings().weight) for the tokenizer vocabulary. Table 2 summarizes the semantic similarity benchmarks used in our evaluation.

4.3 Models and Checkpoints

To quantify semantic alignment between model embeddings and human gold judgments, we employed a suite of metrics:

1. **Spearman's ρ**: rank correlation between model-predicted similarities and human scores.
2. **Kendall's τ**: rank correlation focusing on pairwise ordering consistency.
3. **KTcos (Kendall-on-Cosine)**: cosine similarity of word pairs $\rightarrow$ ranked $\rightarrow$ compared to gold rankings with Kendall's τ.
4. **KTKT (Kendall-on-Kendall)**: ranks within embeddings compared across words, then ranked again, capturing structural mismatches.
5. **AUROC (Area Under Receiver Operating Characteristic Curve)**: measures discriminative ability across thresholds.
6. **AUPRC (Area Under Precision–Recall Curve)**: particularly sensitive to imbalanced similarity distributions.

This expanded suite allowed us to test both rank-based (Spearman, Kendall, KTcos, KTKT) and discriminative (AUROC, AUPRC) perspectives, addressing the limitation of the work by [6] using only rank-based metrics.

4.4 Experimental Protocol

- **Word Vector Extraction**: For each dataset pair (word1, word2), embeddings were retrieved. Multi-token words were averaged across constituent subwords.
- **Similarity Computation**: Cosine similarity between word embeddings.
- **Evaluation**: Computed correlation and discriminative metrics vs. normalized gold scores.
- **Logging**: All metrics were saved per-model, per-dataset, per-checkpoint in CSV files. Progress plots were generated (e.g., AUROC/Spearman over steps).
- **Reproducibility**: Intermediate results (metrics.json, PNG plots) were stored under structured directories: plots/{model}/{step}/{dataset}/.

5 Results and Analysis

5.1 Overall Trends

Across both Pythia-70M and Pythia-410M, we observe that semantic alignment with human judgments emerges very early in training. For most metrics (Spearman, Kendall, AUROC, AUPRC), performance improves sharply within the first 10–20k steps and then stabilizes. This confirms and extends prior findings [6] that word meaning "crystallizes" early.

However, unlike the base study (which reported plateauing after ~5k steps), our results show continued finegrained improvements up to ~40k steps, suggesting that larger evaluation sets (SimLex-999, MTurk-771) capture later refinements in semantic structure that WordSim-353 alone could not.

5.2 Model Scale Effects

Model size strongly impacts stability and alignment quality.

- **Pythia-70M**: Exhibits noisy metric curves, with frequent oscillations even after 40k steps. While it reaches reasonable alignment (Spearman $\rho \sim$ 0.35–0.40), fluctuations suggest shallow embedding geometry and high sensitivity to training dynamics.
- **Pythia-410M**: Shows smoother convergence, consistently outperforming the smaller model across all metrics. Final scores (Spearman $\rho \sim$ 0.50, AUROC > 0.80, AUPRC > 0.75) indicate robust, humanaligned semantic structure.

This scaling effect mirrors earlier observations in static embeddings [1, 2] and extends them to temporal dynamics: larger models not only perform better but also stabilize faster. As shown in Fig. 3, larger models exhibit smoother trajectories and faster stabilization across training, highlighting the role of scale in semantic alignment.

5.3 Metric-Specific Insights

Rank-based correlations such as Spearman's ρ and Kendall's τ increase rapidly during early training and then plateau, indicating early stabilization of relative semantic ordering. In contrast, KTKT reveals a gradual divergence from human judgments, while KTcos suggests convergence, reinforcing the sensitivity of evaluation outcomes to metric choice. Discriminative measures (AUROC and AUPRC) show continued improvement beyond the plateau of rank-based metrics, indicating ongoing refinement in separating highly similar from weakly related word pairs.

As illustrated in Fig. 2, KTcos suggests smooth convergence toward stability, whereas KTKT exposes increasing divergence from human similarity judgments as training progresses.

5.4 Dataset Effects

Performance patterns differ subtly across datasets:

- **WordSim-353**: Highest variance across steps; confirms sensitivity to small size.
- **SimLex-999**: More stable; reflects gradual alignment with true similarity rather than loose associations.
- **MTurk-771**: Provides a middle ground, confirming generalizability across annotator sources.

Using multiple datasets addresses a core limitation of the work by [6] (only 96 usable pairs after tokenization).

5.5 Visualization of Semantic Trajectories

Beyond scalar metrics, we generated plots of AUROC/AUPRC/Spearman progression. Both models show trajectories that rise rapidly, then flatten. For Pythia-410M, curves are smoother and consistently higher, demonstrating the stabilizing effect of scale. Figure 4 shows the progression of Spearman and Kendall correlations across training steps for both models. Figure 5 presents AUROC and AUPRC trajectories, highlighting continued gains in discriminative performance beyond early stabilization.

5.6 Limitations of Results

This study has several limitations. First, analysis was restricted to the input embedding layer, leaving open how semantic representations evolve in deeper layers. Second, although checkpoints were analyzed up to 80k steps, later phases of training may introduce additional refinements. Third, while broader than prior work, the evaluation datasets remain limited in size and scope compared to modern contextual benchmarks. These constraints may affect the generality of conclusions and motivate future extensions.

6 Discussion

6.1 Early Semantic Crystallization

Our findings reinforce the conclusion of the work by [6] that word meaning emerges early in training. Both 70M and 410M models show sharp improvements within the first 10–20k steps, followed by stabilization. This suggests that the embedding geometry required for semantic similarity is acquired quickly, and subsequent training focuses on higher-level structure (syntax, discourse, reasoning). Figure 6 schematically summarizes this two-stage behavior, illustrating early stabilization of semantic structure followed by later refinement of discriminative capacity.

However, unlike the base paper, we observe continued gradual improvements up to 40k steps, particularly in discriminative metrics (AUROC, AUPRC). This nuance was invisible in earlier work, likely because their reliance on WordSim-353 limited statistical power. Our results thus refine, rather than overturn, the claim of early crystallization.

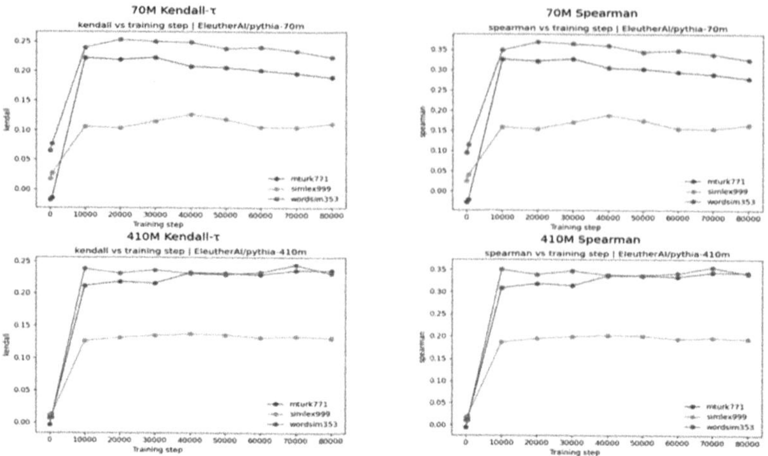

Fig. 4. Spearman/Kendall progress curves for both models

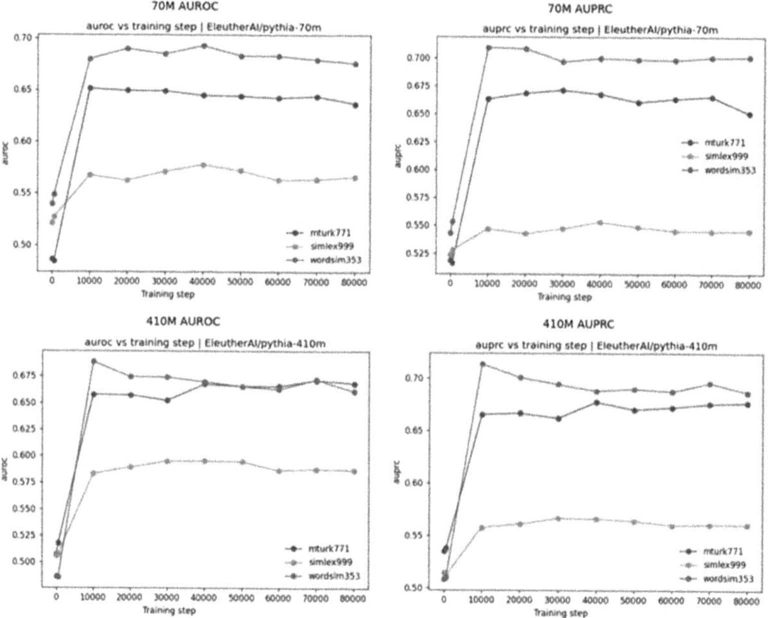

Fig. 5. AUROC/AUPRC curves

6.2 Scaling and Stability

The contrast between Pythia-70M and Pythia-410M highlights a scaling law:

- Small models form noisy, unstable meaning representations.
- Larger models converge faster and produce more stable semantic geometries.

This aligns with scaling literature in language models [23, 24] and suggests that semantic alignment with human judgments is another capability governed by scale.

6.3 Metric Comparisons

Our inclusion of AUROC and AUPRC provides new evidence that rank correlation alone underestimates progress. Rank metrics plateau, but discriminative metrics show that models keep improving in their ability to separate semantically related from unrelated pairs.

Furthermore, while KTKT corroborates the insight of the work by [6] that cosine similarity hides structural mismatches, our broader dataset coverage indicates the divergence is less severe than originally suggested. In other words, the instability of KTKT vs. human judgments is partly an artifact of small, noisy evaluation sets. A direct comparison with Liu et al. [6] is provided in Table 3.

6.4 Implications for Representation Learning

These findings have two practical implications for representation learning. First, since meaningful semantic structure emerges within the first 20,000 training steps, early-stage checkpoints may already be sufficient for applications that rely primarily on lexical similarity, such as thesaurus construction or knowledge graph bootstrapping. Second, the results highlight the importance of evaluation design: reliance on small benchmarks such as WordSim-353 alone risks over-interpreting artifacts of limited test sets, underscoring the need for larger and more diverse similarity datasets.

6.5 Limitations

This study has several limitations. First, the analysis is restricted to the input embedding layer, whereas semantic refinement may occur hierarchically across deeper transformer layers. Second, checkpoint coverage extends only up to 80,000 training steps, and later stages of training may introduce additional dynamics such as specialization or improved handling of rare words. Third, although the evaluation expands beyond prior work, newer representation similarity measures such as Centered Kernel Alignment (CKA) [25] and SVCCA/PWCCA [26, 27] were not explored and may provide complementary insights. Finally, all evaluated benchmarks consist of isolated word pairs; contextual similarity tasks (e.g., SCWS, WiC, CoSimLex) would better capture polysemy and usage-dependent meaning.

6.6 Future Directions

Several directions emerge naturally from this work. One extension would involve analyzing semantic dynamics across transformer layers to better understand where semantic versus syntactic information crystallizes during training. Another direction is cross-model generalization, replicating the analysis on other open model families (e.g., LLaMA, Mistral) to test whether early semantic stabilization is a universal phenomenon.

Finally, future work could formalize embedding trajectory properties—such as length, curvature, and radiality—as predictors of downstream performance, moving beyond scalar similarity metrics.

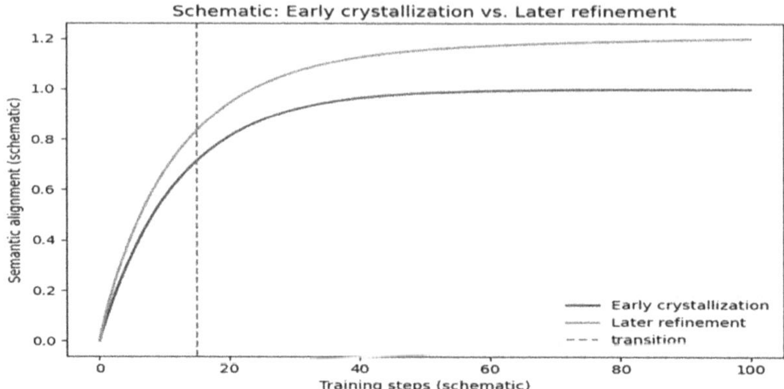

Fig. 6. Conceptual schematic illustrating early semantic crystallization followed by later discriminative refinement.

Table 3. Comparing what our work adds vs. the work by [6] (datasets, metrics, model sizes, findings).

Aspect	Liu et al. [6]	Our Work (This Paper)
Datasets	WordSim-353 (96 pairs after filtering)	WordSim-353, SimLex-999, MTurk-771 (much larger + diverse)
Models	Pythia 14M → 410M (5 sizes, all checkpoints)	Pythia 70M & 410M (full training, efficient subset)
Metrics	Cosine, KTcos, KTKT	Cosine, KTcos, KTKT, AUROC, AUPRC, Spearman, Kendall
Findings	Meaning forms early; KTKT diverges from humans	Early crystallization confirmed; richer metrics show more nuanced trends, esp. Discrimination (AUROC/AUPRC)
Limitations	Tiny eval set; single dataset; only rank metrics	Broader evaluation; modern metrics; more generalizable, though still embedding-layer focus

7 Conclusion

In this study, we traced the development of semantic meaning in language models by analyzing embedding dynamics across training checkpoints of the Pythia family. Building on the pioneering work by [6], we expanded the evaluation along three dimensions:

datasets, metrics, and model scale. By incorporating SimLex-999 and MTurk-771 alongside WordSim-353, we reduced reliance on small-scale benchmarks. By introducing discriminative measures such as AUROC and AUPRC in conjunction with rank-based metrics, we obtained a more nuanced view of semantic alignment and refinement. Finally, direct comparison between Pythia-70M and Pythia-410M highlighted the stabilizing effect of model scale on semantic structure. Together, these results reinforce the view that semantic structure crystallizes early during training while continuing to undergo finer-grained refinement. Our findings confirm the early crystallization of meaning, with most semantic structure forming within the first 20k training steps, but also reveal continued incremental improvement in discriminative capacity, previously overlooked due to limited evaluation. Larger models demonstrate faster convergence and greater stability, underscoring the role of scale in semantic learning.

These results have both theoretical and practical implications: they refine our understanding of when meaning emerges during training, provide methodological lessons for evaluating semantic similarity, and suggest that monitoring early checkpoints can serve as an efficient diagnostic in large-scale pre-training.

At the same time, our work highlights ongoing challenges, including the need for contextual evaluations, deeper layer-wise analyses, and broader cross-model comparisons. We view this study as a step toward a more systematic understanding of how human-aligned semantics develop in neural language models, and invite the community to build upon our methodology with richer datasets, metrics, and interpretability tools.

Ethics Statement.   This work is intended as a research contribution and does not involve human subjects, personal data, or sensitive information beyond the use of publicly available datasets (WordSim-353, SimLex-999, MTurk-771). All datasets are widely used in the NLP community under academic-use terms, and we adhered to their licensing conditions.

The models analyzed in this study (Pythia-70M and Pythia-410M) are openly released checkpoints from EleutherAI, trained on The Pile, a publicly available corpus. We did not perform additional training or data collection.

We emphasize that our findings are intended for scientific analysis of representation dynamics, not for deployment or downstream applications. No claims are made regarding the suitability of these models for end-user systems, and care should be taken to avoid over-interpreting correlation with human judgments as an indication of safety or reliability in applied settings.

References

1. Mikolov, T., Chen, K., Corrado, G., Dean, J.: Efficient estimation of word representations in vector space. arXiv:1301.3781 (2013). Accessed 20 Aug 2025
2. Pennington, J., Socher, R., Manning, C.D.: GloVe: global vectors for word representation. In: Proceedings of EMNLP, pp. 1532–1543 (2014)
3. Finkelstein, L., et al.: Placing search in context: the concept revisited. ACM Trans. Inf. Syst. **20**(1), 116–131 (2002)
4. Hill, F., Reichart, R., Korhonen, A.: SimLex-999: evaluating semantic models with genuine similarity estimation. Comput. Linguist. **41**(4), 665–695 (2015)
5. Halawi, G., Dror, G., Gabrilovich, E., Koren, Y.: Large-scale learning of word relatedness with constraints. In: Proceedings of KDD (2012)

6. Liu, S., Ye, Y.: Tracing the development of word meaning during training. Stanford CS224N custom project report, Stanford University (2025). https://web.stanford.edu/class/cs224n/fin alreports/256989451.pdf. Accessed 20 Aug 2025

7. Biderman, S., Schoelkopf, H., Anthony, Q., Gao, L., et al.: Pythia: a suite for analyzing large language models across training and scaling. arXiv:2304.01373 (2023)

8. Spearman, C.: The proof and measurement of association between two things. Am. J. Psychol. **15**(1), 72–101 (1904)

9. Kendall, M.G.: A new measure of rank correlation. Biometrika **30**(1–2), 81–93 (1938)

10. Bojanowski, P., Grave, E., Joulin, A., Mikolov, T.: Enriching word vectors with subword information. Trans. Assoc. Comput. Linguis. **5**, 135–146 (2017)

11. Vaswani, A., et al.: Attention is all you need. In: Guyon, I., et al. (eds.) Advances in Neural Information Processing Systems, vol. 30, pp. 5998–6008 (2017)

12. Devlin, J., Chang, M.W., Lee, K., Toutanova, K.: BERT: pre-training of deep bidirectional transformers for language understanding. In: Proceedings of NAACL-HLT 2019, pp. 4171–4186 (2019)

13. Radford, A., Wu, J., Child, R., Luan, D., Amodei, D., Sutskever, I.: Language models are unsupervised multitask learners. OpenAI Technical Report (2019). https://cdn.openai.com/better-languagemodels/language_models_are_unsupervised_multitask_learners.pdf. Accessed 20 Aug 2025

14. Brown, T.B., et al.: Language models are few-shot learners. In: Advances in Neural Information Processing Systems (NeurIPS 2020), vol. 33, pp. 1877–1901 (2020)

15. Reif, E., et al.: Visualizing and measuring the geometry of BERT. In: Advances in Neural Information Processing Systems (NeurIPS 2019), vol. 32, pp. 8594–8603 (2019)

16. Ethayarajh, K.: How contextual are contextualized word representations? Comparing the geometry of BERT, ELMo, and GPT-2 embeddings. In: Proceedings of EMNLP 2019, pp. 55–65 (2019)

17. Wang, A., Singh, A., Michael, J., Hill, F., Levy, O., Bowman, S.R.: GLUE: a multi-task benchmark and analysis platform for natural language understanding. In: Proceedings of EMNLP 2018, pp. 353–355 (2018)

18. Saphra, N., Lopez, A.: Understanding learning dynamics of language models with SVCCA. In: Proceedings of NAACL-HLT 2019, pp. 3257–3267 (2019)

19. Gao, L., et al.: The Pile: An 800GB dataset of diverse text for language modeling. arXiv: 2101.00027 (2021)

20. Paszke, A., et al.: PyTorch: an imperative style, high-performance deep learning library. In: Advances in Neural Information Processing Systems (NeurIPS 2019), vol. 32, pp. 8024–8035 (2019). https://arxiv.org/abs/2101.00027. Accessed 20 Aug 2025

21. Wolf, T., et al.: Transformers: state-of-the-art natural language processing. In: Proceedings of the 2020 Conference on Empirical Methods in Natural Language Processing: System Demonstrations, pp. 38–45 (2020)

22. Kaplan, J., McCandlish, S., Henighan, T., Brown, T.B., et al.: Scaling laws for neural language models. arXiv:2001.08361 (2020). Accessed 20 Aug 2025

23. Hoffmann, J., Borgeaud, S., Mensch, A., Buchatskaya, E., et al.: Training compute-optimal large language models. arXiv:2203.15556 (2022). Accessed 20 Aug 2025

24. Kornblith, S., Norouzi, M., Lee, H., Hinton, G.: Similarity of neural network representations revisited. In: Proceedings of ICML (2019)

25. Raghu, M., Gilmer, J., Yosinski, J., Sohl-Dickstein, M.: SVCCA: singular vector canonical correlation analysis for deep learning dynamics and interpretability. In: Advances in Neural Information Processing Systems (2017)

26. Morcos, A.S., Raghu, M., Bengio, S.: Insights on representational similarity in neural networks with canonical correlation. In: Advances in Neural Information Processing Systems (2018)
27. Radinsky, K., Agichtein, E., Gabrilovich, E., Markovitch, S.: A word at a time: computing word relatedness using temporal semantic analysis. In: Proceedings of the 20th International Conference on World Wide Web (WWW 2011), pp. 337–346 (2011)

Probing for Advanced Music Theory Concepts in Generative Music Models

Derek Kwan[(✉)] and Patrick J. Donnelly

Oregon State University, Corvallis, OR 97330, USA
{kwand,donnellp}@oregonstate.edu

Abstract. Western music theory concepts underlie the compositional processes of much of the music consumed today and consequently serve as the conceptual underpinnings of much of the training data for generative music models. Applying the interpretability technique of probing, Wei et al. [58] studied how well basic musical concepts such as scales, time signatures, and simple triadic harmonic progressions are encoded in the latent spaces of the Jukebox and MusicGen generative models. Through constructing datasets isolating these concepts, they trained simple classifiers to classify elements of these concepts based on their latent representations. We extend their work by creating datasets focused on five higher-level and more advanced music theory concepts: polyrhythms, dynamics, seventh chords, mode mixture, and secondary dominants. Performing experiments with the same Jukebox and MusicGen models, we train classifiers on the datasets' latent representations. We find that these advanced concepts are encoded in their latent spaces and that secondary dominant classification is the most difficult of our probing tasks.

Keywords: Generative Music · Music Theory · Probing

1 Introduction

Generative AI has quickly become a prominent aspect across the field of music. In 2025, streaming music platforms reported a large influx of music characterized as AI-generated or spam. Deezer categorized around 30,000 daily uploads to their platform as spam [47] while Spotify labeled around 75 million songs likewise over a year [50]. Furthermore, a growing number of commercial generative AI music tools have become available to consumers, including Suno [52], Udio [22], Stability AI's Stable Audio [20], and Adobe's Generate Soundtrack [57].

Although they may enable the mass generation of what Spotify labels "spammy tracks" [50], generative music models, particularly those which allow text- or metadata-conditioned generation, offer novel ways of interacting with the music composition process. With these models, aspiring composers can dictate their musical ideas through natural language instead of playing an instrument or directly manipulating symbolic music representations. This added layer of abstraction, combined with generators trained on audio rather than symbolic notation, raises questions about the use of Western musical theory concepts that typically guide traditional means of composition.

© The Author(s), under exclusive license to Springer Nature Switzerland AG 2026
P. Machado et al. (Eds.): EvoMUSART 2026, LNCS 16523, pp. 252–270, 2026.
https://doi.org/10.1007/978-3-032-24350-8_17

Wei et al. [58] addressed this line of inquiry by creating the SynTheory dataset featuring the Western music theory concepts of notes, major and minor scales, intervals, triad chords, major and minor chord progressions, tempi, and time signatures. Investigating OpenAI's `Jukebox` [18] and Meta's `MusicGen` [15], they extracted the generative music models' latent representations of the dataset and trained probing classifiers to discriminate types of the music theory concepts. As probing classifiers measure the mutual information between a concept's latent representation and its class [6], high degrees of mutual information between a dataset's latent representations and its concept classes suggest that the concept is encoded within a model's latent space. A model that does not encode these concepts may still be able to generate aesthetically pleasing music. However, as concepts such as harmony can be described in terms of syntax [46], this music may sound like gibberish within the context of Western music.

In this work, we extend the approach of [58] to explore other high-level and advanced music theory concepts through creating datasets[1] and associated classification tasks centered around these concepts. Our expansion explores the music theory concepts of polyrhythms, dynamic expression, seventh chords, mode mixture, and secondary dominants. With polyrhythms, we expand on [58]'s tempo dataset and explore the ability of a model to encode not just the absolute rate of one but the relative rates of two simultaneous rates of constant pulses. We extend the previous work's focus on pitch and rhythmic concepts towards loudness with the dynamics dataset. With seventh chords, we expand on the concept of the previous triad chord dataset to investigate the encoding of four-note chords. While the previous work investigated the ability to classify chord progressions, we investigate the higher-level concept of classifying types rather than identities of chord progressions with the mode mixture and secondary dominants datasets.

We follow [58] by extracting layer outputs for these datasets from `MusicGen`'s audio encoder `EnCodec`, `MusicGen-small`, `MusicGen-medium`, `MusicGen-large`, and `Jukebox`. We train single-hidden-layer MLP probing classifiers on each classification task. Expanding on [58]'s probing techniques, we also train a neural network with no hidden layers to examine linear relationships within each task.

2 Related Works

The technique of training classifier and regression probes on neural network (NN) latent spaces has been applied across a wide variety of network architectures and applications. Examples of this include studying the internal changes in image latent representations within Convolutional Neural Networks by measuring their classification accuracies [3,4] and assessing the ability of various graph neural network and transformer architectures to encode molecular structures [2]. Large Language Models have been a particularly active target of probing research. This includes probing for lower-level linguistic concepts in LSTMs through part-of-speech and semantic tagging tasks [7] as well as high-level linguistic concepts

---

[1] Available at https://huggingface.co/datasets/derekxkwan/syntheory_plus, last accessed 22 Jan 2026.

including hyperbole [48] and metaphor [1] in the intermediate word presentations of BERT [17], RoBERTa [36], and ELECTRA [14] language models.

The past five years have seen a growing body of research related to probing NN latent spaces with general music information retrieval (MIR) and music theory tasks. In the general MIR domain, Castellon et al. [12] used the MagnaTagATune [33], GTZAN [53,54], Giantsteps-MTG[2], and Giantsteps [30] datasets to probe Jukebox embeddings with the tasks of music tagging, music genre classification, key detection, and emotion recognition. For example, the authors probed Jukebox with the Giantsteps dataset for key detection and achieved an accuracy score of 66.7% over 24 classes (12 key centers with major or minor modes) [12]. Their focus was to demonstrate the viability of Jukebox embeddings as representations useful for other, non-generative MIR tasks such as those generated by the MERT [35] music understanding model. However, they also demonstrated the encoding of general music concepts in these embeddings, inspiring others, such as Wei et al. [58], to probe for musical concepts in other generative models. Koo et al. [31] continue this line of research of music concept probing by focusing on the same tasks as [12] with the addition of instrument recognition on MusicGen embeddings from MUSDB [45] and MoisesDB [43] datasets. These authors [32] later expanded on this work through using intervention techniques to encourage the presence of certain instruments in generated music, preceding the work on finer-grained music generation control by Facchiano et al. [21] and Panda et al. [41]. In work that seeks to do the opposite of probing by discovering concepts in models' latent spaces, Singh et al. [49] used the MusicSet [39] dataset and sparse autoencoders to discover latent concepts in MusicGen-large and -small embeddings. They also devised a pipeline to automatically assign labels to latent concepts using Google's Gemini Flash 1.5, pretrained audio classifiers from Essentia [8], and the CLAP text-audio embedding model [59].

With respect to probing for music theory concepts, Foscarin et al. [23] created music theory concept-focused MIDI datasets to probe NN layer outputs, like Wei et al. [58], but focused their experiments on the work of classical music composers. Training a ResNet-50 model to classify an excerpt's piano-roll representation by composer, they extracted its penultimate layer embeddings from their datasets focused on the higher-level music theory concepts of alberti bass, "difficult-to-play music", and contrapuntal texture. Given binary classification tasks, they adapted the Testing with Concept Activation Vectors (TCAV) technique by Kim et al. [28] and calculated vectors corresponding to each concept. They used the CAVs to perturb the embeddings of music used to train the ResNet-50 model. By calculating the directional derivatives of these perturbations with respect to the model's output, they obtained a measure of how significant a concept is to a composer's music.

Focusing on generative models, Ma and Xia [38] created a dataset of triads to probe MusicGen attention heads. Elaborating on [58], they supplemented their training with interventions to supplement the task of classifying chord quality

---

[2] https://github.com/GiantSteps/giantsteps-mtg-key-dataset, last accessed 22 Jan 2026.

with chord root information. Ma et al. [37] created datasets to probe the intermediate layer outputs of `MusicGen` models as well as `MERT` [35] on the tasks of chord quality and chord root. Additionally, they used the NSynth dataset [19] to probe these representations on the tasks of pitch and timbre classification. Elaborating on the basic probing classification technique, they investigated how modifying `MusicGen-small` layer outputs affected pitch and chord root classification.

Facchiano et al. [21], Panda et al. [41], and Zhao et al. [60] use probing classifiers to derive fine-grained control of music generation. [21] use `GPT-o1` to produce sets of prompts for `MusicGen-melody` to generate music in fast and slow tempi as well as bright and dark timbres. Calculating the embedding means at a given layer for each of these sets, they use the resulting tempo and timbre steering vectors to control their respective aspects in generation. Using the SynTheory [58] and GTZAN [53] datasets, [41] calculate steering vectors using the `DiffMean` technique and linear probes to intervene on generated music genre and instrumentation. [60] used SynTheory to both probe `MusicGen-large` using kernel ridge regression and control generation using recursive feature machines.

3   Datasets

We extend Wei et al.'s SynTheory [58], which focused on tempi, time signatures, notes, scales, intervals, triad chords, and triad chord progressions, by constructing datasets featuring five advanced musical concepts (see Table 1). Two datasets focus on the pitch-independent concepts of **polyrhythms** and **dynamics** expression patterns. The other three datasets focus on the harmony-related concepts of **seventh chords**, **mode mixture**, and **secondary dominants**.

To generate musical examples for our datasets, we emulate [58] and procedurally render 4-second MIDI files using the Python `mido`[3] library. These MIDI files are realized as WAV files using `RustySynth`[4] with the `TimGM6mb`[5] soundfont. The MIDI instruments are chosen from five non-pitched percussive instruments for the non-pitch-centric datasets and from the 92 pitched MIDI instruments used by [58] for the pitch-centric datasets. The five percussion instruments slightly diverge from [58]'s choices in the interest of timbral variety and include "Agogo", "Woodblock", "Taiko Drum", "Melodic Tom", and "Snare Drum 1". Like [58], we construct MIDI files for each non-pitch-centric concept example with reverb levels of 0, 63, and 127 and randomized start-time offsets.

Polyrhythms are two streams of constant pulses occurring at separate coprime rates and are defined by their integer ratio [40]. Given coprime integers $r_i, r_j = 2, \ldots, 11$, the polyrhythms dataset (Fig. 1, top row) consists of 30,600 examples of polyrhythms with ratio $r_i : r_j$ $(r_i < r_j)$ realized over one 4/4 bar. We rendered these with instrument pairs i_i, i_j and considered assignments (i_i to r_i, i_j to r_j) and (i_j to r_i, i_i to r_j) as distinct but referring to the same polyrhythm

---

Table 1. The five datasets consist of 153,876 examples in total.

Dataset	Examples	Classes	Pitch-Centric	Balanced
Polyrhythms	30,600	34	No	Yes
Dynamics	41,580	7	No	No
Seventh Chords	35,328	8	Yes	Yes
Mode Mixture	19,872	2	Yes	Yes
Secondary Dominants	26,496	3	Yes	Yes

$r_i : r_j$. To ensure invariance of the probed concept to overall speed, we rendered each polyrhythm as one bar at 60 BPM, two at 120, and three at 180. Like [58], we introduce randomized offsets in start times of polyrhythm figures within the MIDIs. However, we limit two offset start times to values less than a 16th note at 60 BPM (250 ms) so that each example has at least one complete polyrhythm iteration. We also include the offset of 0 ms as well as two offsets with values up to 1,600 ms to ensure examples with the lowest r_i value of 2 have two sounding pulses. For consistency and to ensure an even distribution throughout the possible range, we set these values from a NumPy random number generator.

The **dynamics** dataset (Fig. 1, bottom row) focuses on classifying the seven dynamic patterns of a hairpin (a crescendo followed by a decrescendo [29]), "reverse hairpin" (a decrescendo followed by a crescendo), subito transition from soft to loud dynamic, subito transition from loud to soft, crescendo, decrescendo, and no dynamic change ("flat"). All examples are realized over one 4/4 bar at 60 BPM and consist of a constant stream of notes of quarter-note subdivisions two through eight. All patterns aside from "flat" are enumerated over all pairs of soft dynamics *pp*, *p*, and *mp* and loud dynamics *mf*, *f*, and *ff* in which the MIDI velocity values for each dynamic are obtained by dividing the maximum velocity value 127 by six, rounding when necessary. We enumerate over all six dynamics for the 1,890 total "flat" examples. For the first four pattern types, we denote the changing point of dynamic direction (the crest of loudness for hairpins and the change in dynamics for subito patterns) inflection points and realize them over the second, third, and fourth beats. These pattern types have 8,505 examples each. Crescendi and decrescendi occur over the full bar and have 2,835 examples each. We use curated offsets of 0 ms and two offsets limited to the maximal duration of an eighth note at 60 BPM (500 ms).

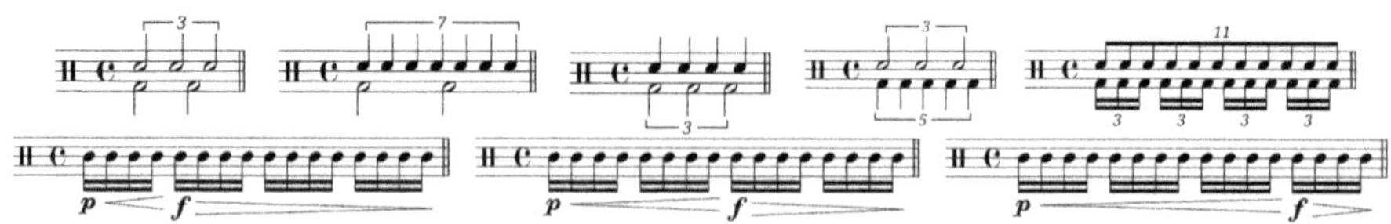

Fig. 1. The **top** row features the **polyrhythm** ratios 2:3, 2:7, 3:4, 3:5, and 11:12. The **bottom** row features the **dynamic** expression of a "hairpin" realized in sixteenth notes with inflection points at beats two, three, and four.

The **seventh chords** dataset (Fig. 2, top row) enables the classification task of discriminating between eight seventh chord qualities: major 7th, minor 7th, major-minor 7th, minor-major 7th, half-diminished 7th, fully diminished 7th, augmented major 7th (major 3rd, major 7th; known as the Lydian augmented chord [34]), and augmented minor 7th (major 3rd, minor 7th; associated with the whole-tone scale [34]). As with the triad dataset in [58], these chords are realized over all twelve pitch classes and all inversions (root, 1st, 2nd, 3rd) in closed voicing and in a register closest to middle C. We note that this process results in identical sounding samples for fully-diminished seventh chords, but leave these in for class balance. The chords are realized as four quarter notes at 60 BPM but, unlike [58], we use a sustain of 90% of a beat to clarify attacks.

The **mode mixture** and **secondary dominant** datasets focus on distinguishing characteristics of chord progressions. The **mode mixture** dataset (Table 2; Fig. 2, middle row) is focused on distinguishing between a major chord progression from SynTheory [58] and a version that substitutes iv for IV and ♭VI (a major chord built on the lowered sixth scale degree) for vi. The **secondary dominant** dataset (Table 3; Fig. 2, bottom row) consists of seventh chord progressions with circle-of-fifths root motion. Our task is to distinguish diatonic versions of progressions from versions with a secondary dominant or a tritone substitution (a substitution of a dominant seventh chord with another with its root a tritone away [34]). We focus on the major and minor progressions in Table 3 where ♭II^7 is a major-minor seventh rooted one semitone above the tonicized scale degree. Following [58], both datasets realize each chord in a progression in root position and as a quarter-note at 60 BPM. Each progression is realized over all twelve pitch classes and we transpose chords so that their root is within a tritone of C4. With each dataset, we include metadata such as the identity of a progression at various levels of detail to enable other classification tasks.

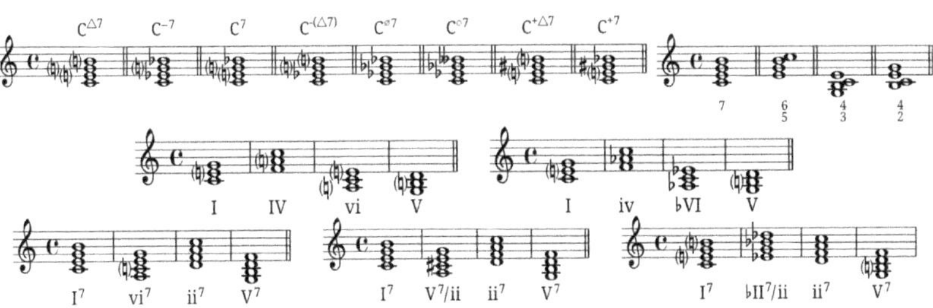

Fig. 2. The **top** row demonstrates the chord qualities in the **seventh chord dataset** as well as a major 7th chord in root position and 1st, 2nd, and 3rd inversions. The **middle** row shows I-IV-vi-V in original and **mode mixture** variants. The **bottom** row shows the I^7-vi^7-ii^7-V^7 major progression from the **secondary dominant** dataset with original, secondary dominant, and tritone substitution variants.

Table 2. We selected progressions from [58] and added **mode mixture** variants.

Original Progression	With Mode Mixture
I $-$ IV $-$ V $-$ I	I $-$ iv $-$ V $-$ I
I $-$ IV $-$ vi $-$ V	I $-$ iv $-$ $\flat$VI $-$ V
I $-$ V $-$ vi $-$ IV	I $-$ V $-$ $\flat$VI $-$ iv
I $-$ vi $-$ IV $-$ V	I $-$ $\flat$VI $-$ iv $-$ V
ii $-$ V $-$ I $-$ vi	ii $-$ V $-$ I $-$ $\flat$VI
V $-$ IV $-$ I $-$ V	V $-$ iv $-$ I $-$ V
V $-$ vi $-$ IV $-$ I	V $-$ $\flat$VI $-$ iv $-$ I
vi $-$ IV $-$ I $-$ V	$\flat$VI $-$ iv $-$ I $-$ V
IV $-$ I $-$ V $-$ vi	iv $-$ I $-$ V $-$ $\flat$VI

Table 3. Major (top) and minor (bottom) chord progressions follow circle-of-fifths root movement in two of the three classes in the **secondary dominant** dataset.

Original Progression	With Secondary Dominant	With Tritone Substitution
$I^{\Delta 7}$-vi^7-ii^7-V^7	$I^{\Delta 7}$-V^7/II-ii^7-V^7	$I^{\Delta 7}$-$\flat II^7$/II-ii^7-V^7
$I^{\Delta 7}$-$I^{\Delta 7}$-$IV^{\Delta 7}$-V^7	$I^{\Delta 7}$-V^7/IV-$IV^{\Delta 7}$-V^7	$I^{\Delta 7}$-$\flat II^7$/IV-$IV^{\Delta 7}$-V^7
$I^{\Delta 7}$-ii^7-V^7-$I^{\Delta 7}$	$I^{\Delta 7}$-V^7/V-V^7-$I^{\Delta 7}$	$I^{\Delta 7}$-$\flat II^7$/V-V^7-$I^{\Delta 7}$
$I^{\Delta 7}$-iii^7-vi^7-V^7	$I^{\Delta 7}$-V^7/VI-vi^7-V^7	$I^{\Delta 7}$-$\flat II^7$/VI-vi^7-V^7
i^7-$VI^{\Delta 7}$-$ii^{\emptyset 7}$-V^7	i^7-V^7/II-$ii^{\emptyset 7}$-V^7	i^7-$\flat II^7$/II-$ii^{\emptyset 7}$-V^7
i^7-i^7-iv^7-V^7	i^7-V^7/IV-iv^7-V^7	i^7-$\flat II^7$/IV-iv^7-V^7
i^7-$ii^{\emptyset 7}$-V^7-i^7	i^7-V^7/V-V^7-i^7	i^7-$\flat II^7$/V-V^7-i^7
i^7-$III^{\Delta 7}$-$VI^{\Delta 7}$-V^7	i^7-V^7/VI-$VI^{\Delta 7}$-V^7	i^7-$\flat II^7$/VI-$VI^{\Delta 7}$-V^7

4 Extracting Latent Representations in Models

We use our five datasets to probe for their corresponding music theory concepts in the latent spaces of OpenAI's `Jukebox` [18], the `small`, `medium`, and `large` variants of Meta's `MusicGen` model [15], and its audio encoder `EnCodec` [16]. The number of layers of each model's encoder (including experiment baselines) and the dimension of their extracted representations are shown in Table 4. We illustrate our entire experiment pipeline in Fig. 3. Following [58], we average over the sequence length dimension of each latent representation before probing.

`Jukebox` is a generative model which conditions on artist and genre with optional lyrics or preexisting audio to output novel audio. Its encoder encodes raw audio at three hierarchical levels of quantization, achieved through training separate VQ-VAE (vector quantized variational autoencoders) using various hop lengths [18]. Like [58], we utilize `jukemirlib`[6] to extract `Jukebox` layer outputs. `jukemirlib` takes as input dataset raw audio sampled at 44.1 kHz, encodes it

---

[6] https://github.com/rodrigo-castellon/jukemirlib, last accessed 22 Jan 2026.

Table 4. Model layer dimension (Dim.) and number of layers (N.L.) MG(A/S/M/L) = MusicGen-(Audio Codec/Small/Medium/Large), Concat. = concatenated mel, chroma, and MFCC features. (+1) refers the outputs of the decoder embedding layer [56].

	Mel	MFCC	Chroma	Concat.	MGA	MGS	MGM	MGL	Jukebox
N.L.	1	1	1	1	1	24 (+1)	48 (+1)	48 (+1)	72
Dim.	768	120	72	960	128	1024	1536	2048	4800

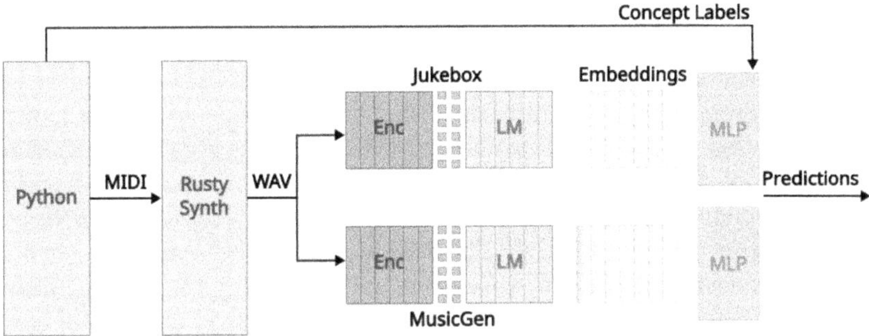

Fig. 3. This diagram illustrates our experimental process from MIDI creation to probing. Parallel pathways show the usage of different models for layer output extraction.

through its VQ-VAE, and extracts the layer outputs from the sparse transformer [13] corresponding to the topmost hierarchical quantization level.

MusicGen processes raw audio through the EnCodec audio encoder [16], which generates K codebook vectors per audio time step. These vectors are the result of residual vector quantization (RVQ) in which an input is quantized to entries in the first codebook and each successive codebook is learned from the previous quantization error. Given conditioning on T5-encoded text [44], these K codebook vectors are predicted using a "delay" interleaving approach [15]. Following [58], we use raw audio resampled to 32 kHz and processed through EnCodec alongside an empty input string for MusicgenForConditionalGeneration's[7] forward pass. We extract the resulting decoder hidden states (outputs of each decoder transformer layer including residual connection) as well as initial decoder embeddings. We apply this process to the MusicGen-small, -medium, and -large variants. Like [58], we also input our audio into MusicGen-large's encoder and extract the output before RVQ.

5   Experimental Design

We follow Wei et al.'s [58] splitting methodology and allocate 70%, 15%, and 15% of each dataset for training, validation, and testing, respectively. To aid reproducibility and consistency across different computational resources, we divide

---

[7] https://huggingface.co/docs/transformers/v4.57.0/en/model_doc/musicgen, last accessed 22 Jan 2026.

each dataset into 20 folds where the first 14 are used for training while the next three are used for validation and the last three for testing. The folds were determined using scikit-learn's [11,42] StratifiedKFold and we stratify based on task-defined classes for polyrhythms, dynamics, and seventh chords. For the mode mixture and secondary dominant datasets, we stratify based on specific chord progressions, which ensures both an even distribution among folds for chord progressions but also the mode mixture type present and the type of secondary dominant used.

Following [58], our classifier probe is a single-hidden-layer MLP in PyTorch with a hidden dimension size of 512 and ReLU activation. The MLP has an initial dropout operating on its inputs and dropout after the first layer's ReLU. We optimize the probe's weight parameters using cross-entropy loss and use normalized inverse-class-proportional weighting for the dynamics dataset experiments as its class proportions are unbalanced. In experiments involving data normalization, we reimplement scikit-learn's StandardScaler [11] in PyTorch, as our experiments run utilizing CUDA. The parameters of the scaler, which divides its input by the running element-wise mean and divides by the running element-wise standard deviation, are found through calculations over each training batch over the 100 training epochs. These parameters are frozen during validation and evaluation. Over the 100 epochs, we track accuracy over the validation set and stop training if the best score is not improved upon over the last 10 epochs. For each experiment, we save the model with the highest validation accuracy. For our unbalanced dynamics dataset, we instead track balanced accuracy.

We adopt [58]'s method of hyperparameter grid search for probes, making adjustments when necessary. When models have only one layer (all baselines and MusicGen's audio encoder), we search over learning rates $10^{-3}, 10^{-4}, 10^{-5}$; dropouts $0.25, 0.5, 0.75$; using or not using data normalization; batch sizes 64 and 256; and weight decay values of $0, 10^{-3}, 10^{-4}$ for the ADAM optimizer. For multi-layered model experiments (Jukebox and MusicGen-small, -medium, and -large), we fix all hyperparameters (a 10^{-3} learning rate, using data normalization, a batch size of 64, and no weight decay), aside from dropout, in order to reduce the number of trials needed. In these experiments, we search for the transformer layer (including decoder embedding layer [56]) whose output yields the best accuracy score. We search over dropout values, unlike [58] who fixed it to 0.5. We found fixing all hyperparameters led to MusicGen audio codec producing better validation accuracy scores over some datasets (particularly for polyrhythms) than multi-layer MusicGen models (particularly -small).

We perform experiments using the same baselines as [58] and duplicate their baseline extraction pipeline[8]. They extracted the MFCC, Constant-Q Chroma, and Mel Spectrogram features of the input using librosa and concatenated the mean and standard deviations of the zeroth-, first-, and second-order differences across frames. A fourth baseline is derived from concatenating all three baselines.

To explore the strength of linear relationships between our concepts in generative model latent space and their classes, we go beyond the experiment method-

---

[8] https://github.com/brown-palm/syntheory, last accessed 22 Jan 2026.

ologies of [58] and repeat our experiments using a neural network with no hidden layer. As these probes lack a hidden layer, we remove the ReLU activation function after the first and only layer of neurons but retain the initial dropout.

6   Results and Discussion

In this section, we provide the results from the single-hidden-layer MLP probe and linear probe experiments conducted over all models, baselines, and datasets. For comparison, we provide the results from Wei et al. [58] as Table 5.

6.1   MLP Probe Results

In our MLP results (Table 6), there is a marked difference in performance on the baseline features between our pitch-centric and non-pitch-centric datasets. The baseline experiments on the pitch-centric datasets achieved relatively high accuracies comparable to the generative model experiments. Chroma experiments showed the strongest performance, which can be attributed to the feature directly representing twelve semitones of equal temperament [10]. Although not directly representing musical semitones, mel spectrograms measure spectral power with bins also aligned to human pitch perception [51] and achieve the third strongest baseline performance. The weaker performance of MFCCs can perhaps be attributed to smaller latent dimension, as we follow [58] and collect 20 coefficients per frame compared to the mel's 128. Furthermore, MFCCs are focused on representing a mel spectrogram's shape [26,55] and thus may be less suited for chord-related tasks (particularly when generalizing over inversions). As the concatenated feature contains both mels and chroma, its experiments exhibit the strongest performance across all baselines over the pitch-centric datasets. For our non-pitch-centric datasets, MFCC experiments instead exhibit the strongest performance out of the non-concatenated baselines. As pitch is irrelevant in distinguishing polyrhythms or dynamic patterns, MFCCs rather than mels and chroma are better suited for their classification tasks as the features represent a frame's spectral energy with a smaller-dimension vector.

In our MLP experiments over four datasets and over all models (excluding `MusicGen`'s audio codec), we achieve accuracies over 94%. This is notable as although our evaluated datasets focus on more advanced and higher-level concepts than [58], we are still able to train MLP probes with comparable performance to probes trained on more fundamental concepts. Aside from the `MusicGen-medium` mode mixture results, our accuracies monotonically decrease from deeper models with larger-dimension latent spaces to shallower models with smaller-dimension spaces. This follows intuition as larger-dimension latent spaces and deeper models have more expressive capability and are better able to model a diverse set of concepts. Although [58] found that `MusicGen-small` representations yielded the highest `MusicGen` model accuracies on their fundamental concepts, our results align with Copet et al. [15] who found that the larger `MusicGen` models generally performed better on metrics such as Frechet Audio Distance [27] and the KL-Divergence between ground-truth and generated music tags.

Although our secondary dominant experiments produced relatively weaker results, we achieved scores that are nearly double that of chance and higher (≥ 0.667) over `Jukebox` and `MusicGen` models. `Jukebox`, the largest and deepest model, was the sole model to outperform baseline representations. We attribute these results to the difficulty of the dataset's task. Whereas the mode mixture experiments focus on detecting if major chord progressions borrow chords from their relative minor, the secondary dominant experiments involve classification of chord borrowing from outside a progression's major or minor mode. This consequently requires generalization over both major and minor chord progressions. Another potential source of difficulty could also be a prevalence of irregularly-resolving secondary dominants [25] in the generative models' training data.

Consistent with [58], we found that the `MusicGen` audio codec returned the lowest non-baseline accuracy scores. We suspect that the representation's small dimension (128) compared to the other models (≥ 1024) may contribute to these results. Furthermore, `EnCodec` is a general purpose audio codec trained on non-music datasets (Common Voice [5], Audio Set [16,24]) in addition to a music dataset (MTG-Jamendo [9]). Thus, its representative capability is not specifically catered towards music concepts like pitch. It is also likely that averaging over the sequence length of its output (for model comparison) is not beneficial for classification tasks. Although `EnCodec` is the audio encoder for the `MusicGen`, `MusicGen` models benefit from training solely on music [15] as well as from calculating their latent representations from unaveraged, discretized `EnCodec` output.

Given the high triad chord probing accuracies achieved in [58], the high accuracies we achieved in seventh chords experiments are not surprising. The accuracies achieved in the mode mixture experiments are higher than [58]'s reported chord progression results over all models and nearly all baselines. Although our classification task operates at a higher level of abstraction, it is also a binary classification class compared to the 19 classes in [58]'s chord progression task. Although the tempo regression task is not directly comparable to our polyrhythm classification task, tempo is the most closely related concept in [58] to polyrhythms and can provide rough expectations of polyrhythm results. Given that polyrhythms characterize relative rather than absolute rates, it is not surprising that we achieved weaker performance in our metric (accuracy) and steeper performance decline in smaller models than [58]'s tempi R^2 scores.

The representations with the highest accuracies largely correspond to the middle or last half of a model's decoder layers but never the final layers (Table 7). This echoes [58]'s results probing for more fundamental music concepts. This also somewhat aligns with the findings in Facchiano et al. [21] for which `MusicGen-melody` layers 10 through 18 (0-indexed, 48 layers total) provided the best representations for tempo and brightness steering vectors. Although this range precedes any best `MusicGen-medium` and `-large` layers, our concepts involve higher-level musical aspects than tempo or timbre brightness. We also observe that the best `Jukebox` and `MusicGen-large` layers occur at equal or greater depths in the pitch-centric datasets than the non-pitch-centric

Table 5. These accuracy results are from [58]'s publication.

Model	Notes	Intervals	Scales	Chords	Chord Prog.	Tempi (R^2)	Time Sig.
Jukebox	0.951	0.995	0.978	0.997	0.971	0.993	1.000
MGL	0.866	0.972	0.905	0.989	0.901	0.965	0.905
MGM	0.851	0.983	0.863	0.989	0.870	0.956	0.883
MGS	0.897	0.995	0.949	0.990	0.942	0.969	0.911
MGA	0.729	0.965	0.383	0.879	0.330	0.947	0.677
Mel	0.712	0.995	0.897	0.988	0.723	0.785	0.827
MFCC	0.467	0.822	0.370	0.863	0.872	0.923	0.688
Chroma	0.954	0.820	0.989	0.994	0.869	0.847	0.672
Concat.	0.941	0.997	0.972	0.992	0.868	0.947	0.833

Table 6. These MLP probe micro-averaged accuracy scores result from evaluating on the test set given the best performing hyperparameters found in grid search.

Model	Polyrhythms	Dynamics	Seventh Chords	Mode Mixture	Secondary Dominants
Jukebox	1.0000	0.9990	0.9958	0.9980	0.9814
MGL	0.9998	0.9684	0.9764	0.9919	0.7560
MGM	0.9976	0.9527	0.9751	0.9802	0.6574
MGS	0.9935	0.9432	0.9724	0.9936	0.7137
MGA	0.7155	0.5376	0.6738	0.8916	0.5279
Mel.	0.4871	0.6572	0.8399	0.9671	0.6203
MFCC	0.7240	0.7720	0.5040	0.8100	0.5783
Chroma	0.3695	0.3933	0.9560	0.9973	0.8708
Concat.	0.7608	0.8600	0.9322	0.9879	0.7825

datasets but cannot make any justifying conclusions from these experiments alone.

6.2   Linear Probe Results

Aside from the dynamics dataset, we observe accuracy decreases over all models and baselines when substituting the MLP with the linear probe (Table 8). This decrease is more prominent with pitch-centric concepts than non-pitch-centric concepts, suggesting that the relationships between chord qualities and progression types and their latent representations are less linear than the relationships between specific rhythms and dynamics and their latent representations. The accuracy decrease from larger to smaller models (including the `MusicGen` audio codec) is more pronounced in the linear probe experiments than in the MLP experiments. As before, this decrease is more prominent in the pitch-centric concept experiments. This suggests that increased number of layers and larger-dimension latent spaces enable generative models to learn relationships between

Table 7. These layer indices (1-indexed) for single-hidden-layer MLP probe experiments correspond to the evaluation accuracy scores.

Model (n layers)	Polyrhythms	Dynamics	Seventh Chords	Mode Mixture	Secondary Dominants
Jukebox (72)	54	56	58	56	58
MGL (48)	32	33	42	43	40
MGM (48)	33	32	45	32	32
MGS (24)	9	16	13	10	12

latent representations and classes that are increasingly linear. It also suggests that pitch-centric concepts benefit most from this added dimension and depth.

Although we intuitively expect that tasks with a higher-level of abstraction with respect to their raw musical material may have relationships with their classes that are more complicated than linear, the largest overall drop is exhibited in the seventh chord dataset accuracies. It is unclear why this dataset's experiments demonstrate the largest declines. We observe that the secondary dominant dataset's results, which also involve seventh chords, exhibit declines in accuracy, but more experiments are needed to explore this potential connection.

The lack of decrease (and in some models, improvement) of accuracy scores from probing the dynamics dataset suggest a strongly linear relationship between the latent representations of the dataset examples and their respective classes. The poorer performance over several of the generative models in the MLP probe experiments compared to the linear probe experiments may suggest that the MLP probe is overparameterized for the classification task and overfit to the training data, despite efforts towards regularization and overfitting prevention through dropout and early stopping. Surprisingly, the score from probing the `MusicGen` audio codec decreased compared to the MLP probe. Again, this may be due to the process of averaging over the sequence length. It is noteworthy that out of the baseline experiments, the mel representation as well as the concatenated baseline (perhaps due to its inclusion of the mel representation) proved resilient to the probe substitution while the MFCC representation did not. This

Table 8. These micro-averaged accuracy scores for linear probes result from evaluating on the test set given the highest performing hyperparameters found in grid search.

Model	Polyrhythms	Dynamics	Seventh Chords	Mode Mixture	Secondary Dominants
Jukebox	1.0000	0.9992	0.9564	0.9970	0.9227
MGL	0.9847	0.9936	0.6863	0.8916	0.6684
MGM	0.9503	0.9825	0.5649	0.8701	0.6735
MGS	0.8481	0.9530	0.5915	0.7909	0.6153
MGA	0.2617	0.3850	0.2114	0.5532	0.3784
Mel.	0.2928	0.6484	0.2914	0.6600	0.4031
MFCC	0.1776	0.5964	0.2510	0.5955	0.4230
Chroma	0.1251	0.2259	0.1665	0.6945	0.5846
Concat.	0.4257	0.7879	0.3656	0.7966	0.6365

suggests that patterns in the mel frequency bins (as in the mel representation) have stronger linear relationships with dynamics classes than patterns in the coefficients describing their overall shape (as in MFCCs).

7   Limitations and Future Work

Our experiments were limited to one partition of the 20 folds for splits and one setting of RNG seeds due to limited computational resources. Although we performed class-based stratified sampling during fold creation and expect similar latent space data distributions in each fold, experiments with different partitions and seeds can provide insight into the statistical significance of our results. Ablation study experiments can provide further insights into generalization from our synthetic data to non-synthetic, real-world music examples. Although we have attempted to account for performance variability through generalizing over polyrhythm tempi and dynamic inflection points, we seek to further these efforts through systematically introducing unevenness in realizing polyrhythm as well as crescendi and decrescendi. For mode mixture and secondary dominants, we seek to investigate voicings in other inversions or systematic approaches to smooth voice leading between chord voices. Additionally, we anticipate that shuffling the chord order in these datasets and not averaging over embedding sequences could provide insights into how identities and types of chord progressions are encoded.

Although we have found that our dataset concepts are encoded in the latent spaces of these generative models, it is beyond the scope of this work to investigate if these models will generate distinct classes in these concepts. It is also beyond the scope of this work to investigate which text prompts or sequence of text prompts can be used to generate distinct concept classes, as we do not vary our text input nor artist or genre information when generating latent representations. From their encoding in latent space, we can only conclude that is possible to generate distinct concept classes and leave their generation for future work.

8   Conclusion

We have shown through the use of single-hidden-layer MLP probing classifiers that the concepts of polyrhythms, dynamics, seventh chords, mode mixture, and secondary dominants are encoded in the latent spaces of the generative music models `Jukebox` and `MusicGen-small`, `-medium`, and `-large`. Through linear probing classifiers, we have shown that there exist strong linear relationships between the dataset generative model encodings and their classes, particularly in the larger models. As these concepts are encoded in the latent spaces of these generative models, we have demonstrated that these models have the capacity to potentially generate distinct classes of these concepts. We have demonstrated through our polyrhythms and dynamics probing experiments that non-pitch-centric concepts beyond tempi and time signatures are encoded by these generative models. Through probing for mode mixture and secondary dominant concepts, we expanded on the work of [58] to demonstrate that higher-level music theoretical concepts are encoded by these models as well.

References

1. Aghazadeh, E., Fayyaz, M., Yaghoobzadeh, Y.: Metaphors in pre-trained language models: probing and generalization across datasets and languages. In: Muresan, S., Nakov, P., Villavicencio, A. (eds.) Proceedings of the 60th Annual Meeting of the Association for Computational Linguistics. vol. 1, pp. 2037–2050. Association for Computational Linguistics (2022). https://doi.org/10.18653/v1/2022.acl-long.144

2. Akhondzadeh, M.S., Lingam, V., Bojchevski, A.: Probing graph representations. In: Ruiz, F., Dy, J., van de Meent, J.W. (eds.) Proceedings of The 26th International Conference on Artificial Intelligence and Statistics. Proceedings of Machine Learning Research, vol. 206, pp. 11630–11649. PMLR (2023). https://proceedings.mlr.press/v206/akhondzadeh23a.html, Accessed 26 Oct 2025

3. Alain, G., Bengio, Y.: Understanding intermediate layers using linear classifier probes. https://doi.org/10.48550/arXiv.1610.01644, Accessed 29 Oct 2025

4. Alain, G., Bengio, Y.: Understanding intermediate layers using linear classifier probes. In: International Conference on Learning Representations (2017). https://openreview.net/forum?id=HJ4-rAVtl, Accessed 28 Oct 2025

5. Ardila, R., et al.: Common voice: a massively-multilingual speech corpus. In: Calzolari, N., et al., (eds.) Proceedings of the Twelfth Language Resources and Evaluation Conference, pp. 4218–4222. European Language Resources Association (2020). https://aclanthology.org/2020.lrec-1.520/, Accessed 30 Oct 2025

6. Belinkov, Y.: Probing classifiers: promises, shortcomings, and advances. Comput. Linguist. **48**(1), 207–219 (2022). https://doi.org/10.1162/coli_a_00422

7. Belinkov, Y., Màrquez, L., Sajjad, H., Durrani, N., Dalvi, F., Glass, J.: Evaluating layers of representation in neural machine translation on part-of-speech and semantic tagging tasks. In: Kondrak, G., Watanabe, T. (eds.) Proceedings of the Eighth International Joint Conference on Natural Language Processing (Volume 1: Long Papers), pp. 1–10. Asian Federation of Natural Language Processing (2017). https://aclanthology.org/I17-1001/, Accessed 26 Oct 2025

8. Bogdanov, D., et al.: ESSENTIA: an open-source library for sound and music analysis. In: Proceedings of the 21st ACM International Conference on Multimedia, New York, NY, USA, pp. 855–858. MM '13, Association for Computing Machinery (2013). https://doi.org/10.1145/2502081.2502229

9. Bogdanov, D., Won, M., Tovstogan, P., Porter, A., Serra, X.: The MTG-Jamendo dataset for automatic music tagging. In: Machine Learning for Music Discovery Workshop, International Conference on Machine Learning (ICML 2019). Long Beach, CA, United States (2019). http://hdl.handle.net/10230/42015, Accessed 30 Oct 2025

10. Brown, J.C.: Fundamental frequency tracking and applications to musical signal analysis. In: Beauchamp, J.W. (ed.) Analysis, Synthesis, and Perception of Musical Sounds: The Sound of Music, pp. 90–121. Springer (2007). https://doi.org/10.1007/978-0-387-32576-7_2

11. Buitinck, L., et al.: API design for machine learning software: experiences from the scikit-learn project. In: ECML PKDD Workshop: Languages for Data Mining and Machine Learning, pp. 108–122 (2013). https://hdl.handle.net/2268/154357, Accessed 28 Jan 2026

12. Castellon, R., Donahue, C., Liang, P.: Codified audio language modeling learns useful representations for music information retrieval. In: Proceedings of the 22nd International Society for Music Information Retrieval Conference, pp. 88–96. ISMIR (2021). https://doi.org/10.5281/zenodo.5624605, Accessed 31 Oct 2025

13. Child, R., Gray, S., Radford, A., Sutskever, I.: Generating long sequences with sparse transformers (2019). https://doi.org/10.48550/arXiv.1904.10509, Accessed 02 Nov 2025

14. Clark, K., Luong, M.T., Le, Q.V., Manning, C.D.: ELECTRA: pre-training text encoders as discriminators rather than generators. In: International Conference on Learning Representations (2019), https://openreview.net/forum?id=r1xMH1BtvB, Accessed 01 Nov 2025

15. Copet, J., et al.: Simple and controllable music generation. In: Oh, A., Naumann, T., Globerson, A., Saenko, K., Hardt, M., Levine, S. (eds.) Advances in Neural Information Processing Systems. vol. 36, pp. 47704–47720. Curran Associates, Inc. (2023). https://proceedings.neurips.cc/paper_files/paper/2023/hash/94b472a1842cd7c56dcb125fb2765fbd-Abstract-Conference.html, Accessed 21 Oct 2025

16. Défossez, A., Copet, J., Synnaeve, G., Adi, Y.: High fidelity neural audio compression. Trans. Mach. Learn. Res. (2023). https://openreview.net/forum?id=ivCd8z8zR2, Accessed 21 Oct 2025

17. Devlin, J., Chang, M.W., Lee, K., Toutanova, K.: BERT: pre-training of deep bidirectional transformers for language understanding. In: Burstein, J., Doran, C., Solorio, T. (eds.) Proceedings of the 2019 Conference of the North American Chapter of the Association for Computational Linguistics: Human Language Technologies (Long and Short Papers). vol. 1, pp. 4171–4186. Association for Computational Linguistics (2019). https://doi.org/10.18653/v1/N19-1423

18. Dhariwal, P., Jun, H., Payne, C., Kim, J.W., Radford, A., Sutskever, I.: Jukebox: a generative model for music (2020). https://doi.org/10.48550/arXiv.2005.00341, Accessed 21 Oct 2025

19. Engel, J., et al.: Neural audio synthesis of musical notes with WaveNet autoencoders. In: Precup, D., Teh, Y.W. (eds.) Proceedings of the 34th International Conference on Machine Learning. Proceedings of Machine Learning Research, vol. 70, pp. 1068–1077. PMLR (2017). https://proceedings.mlr.press/v70/engel17a.html, Accessed 02 Nov 2025

20. Evans, Z., Carr, C., Taylor, J., Hawley, S.H., Pons, J.: Fast timing-conditioned latent audio diffusion. In: Salakhutdinov, R., et al., (eds.) Proceedings of the 41st International Conference on Machine Learning. Proceedings of Machine Learning Research, vol. 235, pp. 12652–12665. PMLR (2024). https://proceedings.mlr.press/v235/evans24a.html, Accessed 27 Oct 2025

21. Facchiano, S., et al.: Activation patching for interpretable steering in music generation (2025). https://doi.org/10.48550/arXiv.2504.04479, Accessed 05 Oct 2025

22. Fink, C.: Udio AI music raises $10 million, $6.5 million for Spines AI, more Sora, cinematic AI. Forbes (2024). https://www.forbes.com/sites/charliefink/2024/04/11/udio-ai-music-raises-10-million-65-million-for-spines-ai-more-sora-cinematic-ai/, Accessed 28 Oct 2025

23. Foscarin, F., Hoedt, K., Praher, V., Flexer, A., Widmer, G.: Concept-based techniques for "musicologist- friendly" explanations in deep music classifiers. In: Proceedings of the 23rd International Society for Music Information Retrieval Conference, pp. 876–883. ISMIR (2022). https://doi.org/10.5281/zenodo.7316804, Accessed 29 May 2025

24. Gemmeke, J.F., et al.: Audio set: an ontology and human-labeled dataset for audio events. In: 2017 IEEE International Conference on Acoustics, Speech and Signal Processing (ICASSP), pp. 776–780 (2017). https://doi.org/10.1109/ICASSP.2017.7952261, ISSN: 2379-190X

25. Hutchinson, R.: Music Theory for the 21st-Century Classroom. University of Puget Sound (2025). https://musictheory.pugetsound.edu/mt21c/MusicTheory.html, Accessed 30 Oct 2025
26. Jensen, J.H., Christensen, M.G., Murthi, M.N., Jensen, S.H.: Evaluation of MFCC estimation techniques for music similarity. In: 2006 14th European Signal Processing Conference, pp. 1–5 (2006). https://ieeexplore.ieee.org/abstract/document/7071659, Accessed 02 Nov 2025
27. Kilgour, K., Zuluaga, M., Roblek, D., Sharifi, M.: Fréchet audio distance: a reference-free metric for evaluating music enhancement algorithms. In: Interspeech 2019, pp. 2350–2354 (2019). https://doi.org/10.21437/Interspeech.2019-2219
28. Kim, B., et al.: Interpretability beyond feature attribution: quantitative testing with concept activation vectors (TCAV). In: Proceedings of the 35th International Conference on Machine Learning. vol. 80, pp. 2668–2677. PMLR (2018). https://proceedings.mlr.press/v80/kim18d.html, Accessed 15 Oct 2025
29. Kim, D.H.S.: The Brahmsian hairpin. 19th-Century Music **36**(1), 46–57 (2012). https://doi.org/10.1525/ncm.2012.36.1.046
30. Knees, P., et al.: Two data sets for tempo estimation and key detection in electronic dance music annotated from user corrections. In: Proceedings of the 16th International Society for Music Information Retrieval Conference, pp. 364–370. ISMIR (2015). https://doi.org/10.5281/zenodo.1414996, Accessed 28 Oct 2025
31. Koo, J., Wichern, G., Germain, F.G., Khurana, S., Le Roux, J.: Understanding and controlling generative music transformers by probing individual attention heads. In: IEEE ICASSP Satellite Workshop on Explainable Machine Learning for Speech and Audio (XAI-SA) (2024). https://www.merl.com/publications/TR2024-032, Accessed 25 Oct 2025
32. Koo, J., Wichern, G., Germain, F.G., Khurana, S., Le Roux, J.: SMITIN: self-monitored inference-time intervention for generative music transformers. IEEE Open J. Sign. Process. **6**, 266–275 (2025). https://doi.org/10.1109/OJSP.2025.3534686
33. Law, E., West, K., Mandel, M.I., Bay, M., Downie, J.S.: Evaluation of algorithms using games: the case of music tagging. In: Proceedings of the 10th International Society for Music Information Retrieval Conference, pp. 387–392. ISMIR (Oct 2009). https://doi.org/10.5281/zenodo.1417647, Accessed 28 Oct 2025
34. Levine, M.: The Jazz Theory Book. Sher Music (1995)
35. Li, Y., et al.: MERT: acoustic music understanding model with large-scale self-supervised training. In: Kim, B., Yue, Y., Chaudhuri, S., Fragkiadaki, K., Khan, M., Sun, Y. (eds.) International Conference on Representation Learning. vol. 2024, pp. 12181–12204 (2024). https://proceedings.iclr.cc/paper_files/paper/2024/hash/33dffa2e3d2ab74a783d1a8c292f66d9-Abstract-Conference.html, Accessed 28 Oct 2025
36. Liu, Y., et al.: RoBERTa: a robustly optimized BERT pretraining approach (2019). https://doi.org/10.48550/arXiv.1907.11692, Accessed 01 Nov 2025
37. Ma, W., Li, X., Xia, G.: Do music LLMs learn symbolic concepts? a pilot study using probing and intervention. In: Audio Imagination: NeurIPS 2024 Workshop AI-Driven Speech, Music, and Sound Generation (2024). https://openreview.net/forum?id=uvzw0gS0Nn, Accessed 29 May 2025
38. Ma, W., Xia, G.: Exploring the internal mechanisms of music LLMs: a study of root and quality via probing and intervention techniques. In: International Conference on Machine Learning (ICML) Workshop on Mechanistic Interpretability (2024). https://openreview.net/forum?id=Kr6nkNa4TQ, Accessed 29 May 2025

39. McCallum, M.C., Korzeniowski, F., Oramas, S., Gouyon, F., Ehmann, A.F.: Supervised and unsupervised learning of audio representations for music understanding. In: Proceedings of the 23rd International Society for Music Information Retrieval Conference, pp. 256–263. ISMIR (2022). https://doi.org/10.5281/zenodo.7316644, Accessed 28 Oct 2025
40. Møller, C., Stupacher, J., Celma-Miralles, A., Vuust, P.: Beat perception in polyrhythms: time is structured in binary units. PLoS ONE **16**(8), e0252174 (2021). https://doi.org/10.1371/journal.pone.0252174
41. Panda, D., et al.: Fine-grained control over music generation with activation steering (2025). https://doi.org/10.48550/arXiv.2506.10225, Accessed 05 Oct 2025
42. Pedregosa, F., et al.: Scikit-learn: machine learning in Python. J. Mach. Learn. Res. **12**(85), 2825–2830 (2011). http://jmlr.org/papers/v12/pedregosa11a.html, Accessed 22 Jan 2026
43. Pereira, I., Araújo, F., Korzeniowski, F., Vogl, R.: MoisesDB: a dataset for source separation beyond 4-stems. In: Proceedings of the 24th International Society for Music Information Retrieval Conference, pp. 619–626. ISMIR (2023). https://doi.org/10.5281/zenodo.10265363, Accessed 22 Jan 2026
44. Raffel, C., et al.: Exploring the limits of transfer learning with a unified text-to-text transformer. J. Mach. Learn. Res. **21**(140), 1–67 (2020). http://jmlr.org/papers/v21/20-074.html, Accessed 31 Oct 2025
45. Rafii, Z., Liutkus, A., Stöter, F.R., Mimilakis, S.I., Bittner, R.: MUSDB18 - a corpus for music separation (2017). https://doi.org/10.5281/zenodo.1117372, Accessed 22 Jan 2026
46. Rohrmeier, M., Pearce, M.: Musical syntax I: theoretical perspectives. In: Bader, R. (ed.) Springer Handbook of Systematic Musicology, pp. 473–486. Springer (2018). https://doi.org/10.1007/978-3-662-55004-5_25
47. Sarmiento, I.G.: AI-generated music is here to stay. will streaming services like Spotify label it? National Public Radio (2025). https://www.npr.org/2025/08/08/nx-s1-5492314/ai-music-streaming-services-spotify, Accessed 30 Oct 2025
48. Schneidermann, N., Hershcovich, D., Pedersen, B.: Probing for hyperbole in pre-trained language models. In: Padmakumar, V., Vallejo, G., Fu, Y. (eds.) Proceedings of the 61st Annual Meeting of the Association for Computational Linguistics. vol. 4, pp. 200–211. Association for Computational Linguistics (2023). https://doi.org/10.18653/v1/2023.acl-srw.30
49. Singh, N., Cherep, M., Maes, P.: Discovering and steering interpretable concepts in large generative music models. In: NeurIPS 2025 Workshop AI4Music (2025). https://openreview.net/forum?id=jVSlJk5qNA, Accessed 22 Jan 2026
50. Spotify: Spotify strengthens AI protections for artists, songwriters, and producers (2025). https://newsroom.spotify.com/2025-09-25/spotify-strengthens-ai-protections/, Accessed 30 Oct 2025
51. Stevens, S.S., Volkmann, J.: The relation of pitch to frequency: a revised scale. Am. J. Psychol. **53**(3), 329–353 (1940). https://doi.org/10.2307/1417526
52. Suno Inc.: Suno model timeline & information (2025). https://help.suno.com/en/articles/5782721, Accessed 28 Oct 2025
53. Tzanetakis, G.: Manipulation, analysis and retrieval systems for audio signals. Ph.D thesis, Princeton University (2002)
54. Tzanetakis, G., Cook, P.: Musical genre classification of audio signals. IEEE Trans. Speech Audio Process. **10**(5), 293–302 (2002). https://doi.org/10.1109/TSA.2002.800560

55. Urbano, J., Bogdanov, D., Herrera, P., Gómez, E., Serra, X.: What is the effect of audio quality on the robustness of MFCCs and chroma features? In: Wang, H.M., Yang, Y.H., Lee, J.H. (eds.) Proceedings of the 15th International Society for Music Information Retrieval Conference (ISMIR 2014), Taipei, Taiwan, pp. 573–578. ISMIR (2014). https://doi.org/10.5281/zenodo.1416276, Accessed 19 Oct 2025
56. Vaswani, A., et al.: Attention is all you need. In: Advances in Neural Information Processing Systems. vol. 30. Curran Associates, Inc. (2017). https://proceedings.neurips.cc/paper/2017/hash/3f5ee243547dee91fbd053c1c4a845aa-Abstract.html, Accessed 31 Oct 2025
57. Weatherbed, J.: Adobe's new AI audio tools can add soundtracks and voice-overs to videos. The Verge (2025). https://www.theverge.com/news/807809/adobe-firefly-ai-audio-generate-soundtrack-speech, Accessed 30 Oct 2025
58. Wei, M., Freeman, M., Donahue, C., Sun, C.: Do music generation models encode music theory? In: Proceedings of the 25th International Society for Music Information Retrieval Conference, pp. 680–687. ISMIR (2024). https://doi.org/10.5281/zenodo.14877427, Accessed 19 Aug 2025
59. Wu, Y., Chen, K., Zhang, T., Hui, Y., Berg-Kirkpatrick, T., Dubnov, S.: Large-scale contrastive language-audio pretraining with feature fusion and keyword-to-caption augmentation. In: International Conference on Acoustics, Speech and Signal Processing (ICASSP), pp. 1–5. IEEE (2023). https://doi.org/10.1109/ICASSP49357.2023.10095969
60. Zhao, D., Beaglehole, D., Berg-Kirkpatrick, T., McAuley, J., Novack, Z.: Steering autoregressive music generation with recursive feature machines. https://doi.org/10.48550/arXiv.2510.19127, Accessed 23 Oct 2025

End-To-End Song Structure Segmentation via EncoderDecoder Network Architecture and Hand-Crafted Features

Phan Le Son[1][✉] [ID], Nghi Nguyen[2], and Lam Pham[3,4,5] [ID]

[1] Modeling Evolutionary Algorithms Simulation and Artificial Intelligence, Faculty of Electrical and Electronics Engineering, Ton Duc Thang University, Ho Chi Minh City, Vietnam
`phanleson@tdtu.edu.vn`
[2] Amanotes, Ho Chi Minh City, Vietnam
`nghi.nguyen@amanotes.com`
[3] Laboratory for Artificial Intelligence, Institute for Computational Science and Artificial Intelligence, Van Lang University, Ho Chi Minh City, Vietnam
`lam.phamdang@vlu.edu.vn`
[4] Vietnam Faculty of Information Technology, Van Lang School of Technology, Van Lang University, Ho Chi Minh City, Vietnam
[5] Vietnam Center for Digital Safety & Security, Austrian Institute of Technology (AIT), Vienna, Austria
`lam.pham@ait.ac.at`

Abstract. A song narrates a story through its lyrics, rhythm, and structure. Understanding the song structure is beneficial to provide valuable insights into the song's anatomy. Therefore, song structural segmentation, one of the main tasks of music information retrieval (MIR), has drawn attention and has been popular in the music industry with relevant and diverse applications of song cutting, song recommendation, song classification, etc.. However, automatically detecting the structure of a song is challenging because different music genres present different principles for song structure without a general rule. In this paper, we propose a deep learning model based on an encoderdecoder architecture, combined with hand-crafted acoustic features, for end-to-end song structure segmentation. We train and evaluate the proposed model using an 800-song self-collected dataset and a benchmark Beatles dataset, both covering a wide range of music genres. We also construct two baseline systems which use the conventional Laplacian segmentation (LS) method and a combination between Laplacian segmentation and multilayer perceptron network (LS-MLP) to compare with the proposed encoder-decoder network. Experimental results on the datasets demonstrate that our proposed encoder-decoder network outperforms two baselines, especially for detecting the boundaries of audio segments. Furthermore, our model proves effective to detect the label for each segment. The high performance evaluated on diverse music genres shows potential to apply the proposed model for music structure analysis applications.

Keywords: Song structural segmentation · Music information retrieval · Encoder-decoder architecture · Hand-crafted feature

P. Machado et al. (Eds.): EvoMUSART 2026, LNCS 16523, pp. 271–286, 2026.
https://doi.org/10.1007/978-3-032-24350-8_18

1   Introduction

In the field of music composition, music structuring is one of the techniques used by composers to narrate a story. Music structure analysis (MSA) involves organizing musical elements such as melody, harmony, rhythm, and timbre in a manner that reveals the underlying structure and form of a musical piece. Music composers leverage these techniques to produce a coherent narrative that helps listeners engage with the musical experience [1].

For example, the repetition of harmonic progressions (e.g., sequences of chords), particularly in the context of Western tonal music, enables artists to guide listeners through a journey and create dramatic narratives [2]. Interestingly, audiences are often able to perceive musical structures [3] and grasp the narrative of a piece even without formal musical training or cultural background. This suggests that structural cues are deeply embedded in the musical signal and are crucial for shaping the listening experience. From a compositional perspective, the hierarchical structure of a song–such as verses, choruses, and bridges–plays a key role in evoking and regulating listeners' emotions. By manipulating the length, order, and placement of musical segments, composers craft the dynamic arc of a piece. Therefore, an automatic tool for music structural segmentation can serve as a valuable aid for composers and music producers by providing insights into the structural patterns of existing music, assisting in music analysis, remixing, and generation tasks. Moreover, such tools can also support educational, musicological, and retrieval applications where structural understanding is essential.

In general, the song's structure can be classified into two levels: phrase-structure segmentation and function segmentation. Phrase-structure segmentation aims to detect short-duration segments based on melodic information such as a short motif. Meanwhile, functional segmentation is more widely studied and focuses on longer musical durations, referred to as music sections, which may last longer than one minute. In this work, we focus on the task of function segmentation. Given a song, the task of function segmentation helps separate the song into several sections. Each section is characterized by a label that presents the function of section such as Intro or Verse, etc.

In many musical pieces, each structural section (e.g., Verse, Chorus, Bridge) comprises a different number of beats, but the beats within a given section typically serve a similar musical function and, therefore, share the same structural label. Performing segmentation at the beat level allows finer-grained analysis and better alignment with musical timing, making it particularly useful for tasks such as music summarization, generation, and interactive applications. However, accurate segmentation at the beat level, especially near section transitions, is challenging. Since beats have variable temporal durations that depend on tempo and musical style, which introduces inconsistencies in feature representation and makes learning segment boundaries more difficult. Therefore, designing effective segmentation algorithms that operate at the beat level requires careful consideration of beat timing and context across diverse genres.

Regarding the techniques proposed for the MSA, the early and traditional approaches [4–8] used Fourier, Constant-Q Transform (CQT), chroma transform, etc. to transform an entire song into a two-dimensional spectrogram. Two consecutive column vectors represent two consecutive timestamps, which normally are beats in a music song. Meanwhile, the sequence of column vectors across the row presents the sequence of beats in chronological order. Given the sequence of beats in the spectrogram, the self-similarity matrix (SSM) is obtained by computing the similarity of pairs of column vectors. The significant difference between sub-matrices in SSM helps to indicate the boundary between two sections in the song. Recently, MSA has achieved potential performance by leveraging the power of Neural Networks (NN). In particular, the music song is first transformed into a spectrogram. Then, the spectrogram is fed into a NN to extract the sequence of audio embeddings. Finally, the sequence of audio embeddings is used to compute the SSM, indicating the boundary of audio segments. In other words, the NN serves to transform the spectrogram into a sequence of audio embeddings, where the features become notably distinct at the transition boundaries between adjacent sections. For example, the authors in [9] used a CNN-based network architecture to extract audio embeddings from the CQT spectrogram. Then, SSM is computed from consecutive audio embedding vectors to detect the boundary between audio sections. The boundary is identified by both harmonic/homogeneity and novelty points. Recently, authors in [10] leveraged a DNN architecture to extract harmonic embeddings and a transformer architecture to extract spectral-temporal audio embeddings. The paper proved that computing SSM from the audio embeddings extracted from the deep neural network is effective and helps to achieve a better performance compared with the traditional methods.

Also leveraging deep neural networks and deep learning techniques, in this paper, we propose an encoder-decoder network architecture for the MSA. Different from the systems mentioned above, the propose encoder-decoder architecture presents an end-to-end training strategy. While the encoder part encodes the input audio features to the audio embeddings, the decoder part decodes the audio embeddings into target labels. In other words, the proposed encoder-decoder architecture learns audio features and directly exports the beat labels and the boundary between sections without the step of SSM computation.

Table 1. Music genres and song numbers in 300 songs from the Harmonix dataset

No.	Music Genres	Song Numbers
1	Pop	215
2	Hip-Hop	31
3	Dance/Electronic	18
4	Alternative	13
5	Country	13
6	Others	10

Table 2. Music genres and song numbers in 500 song self-collected

No.	Music Genres	Song Numbers
1	Acoustic	43
2	Alternative, Indie	30
3	Ambient, New Age	8
4	Classical, Opera, Orchestra	4
5	Country	28
6	Dance, Funk, Disco	27
7	Folk, Blues	14
8	Hip-Hop, Rap	53
9	Instrumental	15
10	Jazz, Soul, R&B	41
11	Latin	18
12	Pop, Ballad	187
13	Reggae, Ska	15
14	Rock	42
15	World	22

To evaluate our proposed encoder-decoder network architecture, we collected a dataset which covers a wide range of music genres listed on Table 2. We also implement two systems which use the traditional Laplacian segmentation (LS) method and Laplacian segmentation method combined with multilayer perceptron network (LS-MLP) to compare with the proposed encoder-decoder network architecture.

2   Evaluating Datasets and Metrics

This section outlines the datasets and evaluation metrics used in our experiments.

2.1   Evaluating Datasets

We first introduce the dataset that we use to evaluate our two baselines and proposed encoder-decoder network architecture for the MSA. In particular, we collected 800 songs which comprises 500 self-collected songs and 300 songs from the benchmark dataset of the Harmonix [11]. The dataset of 500 self-collected songs and 300 songs from the Harmonix dataset presents various music genres which are presented in detail in Table 1 and Table 2, respectively.

Given the 800-song dataset, we perform annotations for audio sections in all songs, the principles of our annotation process is described in Sect. 3.

Each music section is annotated with a single label, and each section comprises multiple beats. Consequently, all beats belonging to the same section share the same label. Although the durations of individual beats may vary, the time interval between two consecutive beats is typically no longer than 0.7 s. The proposed models in this paper are designed to learn and predict the label on each beat. The boundary between two sections is detected based on changes in beat labels. The 800 songs are split into 740 songs for training and 60 songs for testing. Notably, the 60 test songs are drawn from the Harmonix dataset [11]. These test songs were selected by music experts, who agreed that the subset is sufficiently diverse and representative to evaluate generalisation performance, particularly for music game content generation. To prove the model robust, we further test our proposed models on the Beatles dataset [12], which comprises 180 songs.

2.2   Evaluating Metrics

We use the metric of Hit Rate Measurement (HRM) to evaluate the proposed baselines and encoder-decoder based models. In particular, HRM, which is commonly used for the MSA [5], is used to evaluate the task of boundary detection. Given the HRM, we report the Precision (P) representing the proportion of estimated boundaries, the Recall (R) representing the proportion of reference boundaries and F1 score also known as the harmonic mean of the Precision and the Recall. Meanwhile, accuracy and confusion matrix result are used to present the performance for the task of the segment label detection in the beat level.

3   Principles for Annotation

We refer to the principles described in the section *Segmentation Principles* in [5]. Specifically, each song is randomly selected and annotated by one music expert, and subsequently reviewed by a second expert. A song is included in the final dataset only if both experts agree on the segmentation.

3.1   Homogeneity and Novelty

The homogeneity approach assumes that each musical segment is relatively uniform in attributes such as key or instrumentation; thus, transitions to dissimilar segments manifest as novelty points. This principle often appears in the Harmonix dataset, which includes mostly pop music, where the different segment has a clear boundary. Besides listening to the audio, we use Melodic Range Spectrogram (log-mel power spectrograms) from Sonic Visualizer [13] to identify the novelty point (the starting/boundary of a segment) synchronized with the beat of the song. An example is illustrated in Fig. 1.

3.2   Repetition

The repetition principle posits that segments sharing the same label are similar sequences with respect to some musical attribute (e.g., timbre, harmony,

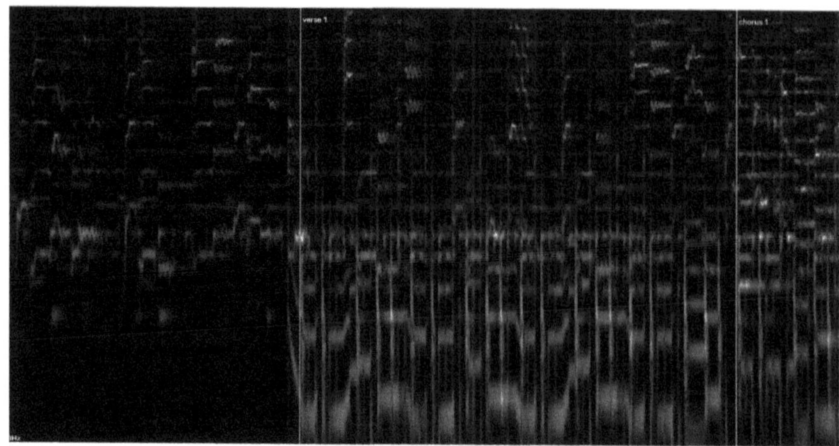

Fig. 1. Segmentations change from Intro to Verse by observing the instrument adding at the novelty point.

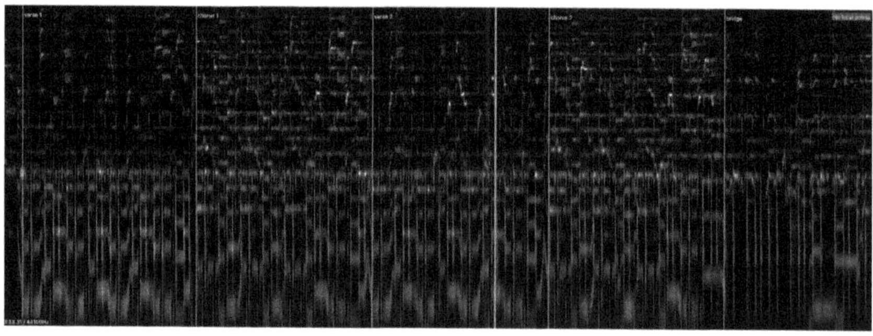

Fig. 2. Verses and Choruses are repeated.

rhythm). Boundaries of such repeated segments can emerge as a byproduct of repetition–i.e., they are defined by the start and end points of the recurring sequences. In simpler terms: if two segments are sufficiently similar, they may be assigned the same label.

This principle is clearly evident in the Harmonix dataset: multiple verses or choruses often recur with identical or only slightly varying instrumentation, duration, or lyrics. For example, Verse and choruses are repeated in Fig. 2.

3.3   Label Name with Meaningful Function

Following the annotation scheme proposed by Wang *et al.* [10], we adopt a seven-class taxonomy, as described in Table 3.

Table 3. Description of labels for each section in a music song

No.	Labels	Description
1	Intro	The introduction section of a song.
2	Verse	The verse section of a song, often containing the storytelling or main lyrical content.
3	Chorus	The chorus section of a song, typically the most memorable and repeated part of the song.
4	Bridge	The bridge section of a song, usually serving as a transitional or contrasting section.
5	Inst	The instrumental sections in a song where there are no vocals.
6	Outro	The outro section of a song, signaling the end of the song.
7	Silence	Audio segments in the audio where there is no significant musical content such as silence or background noise.

4   Proposed Deep Learning Model

The high-level architecture of proposed deep learning model is separated into three main steps: front-end feature extraction and back-end deep neural network.

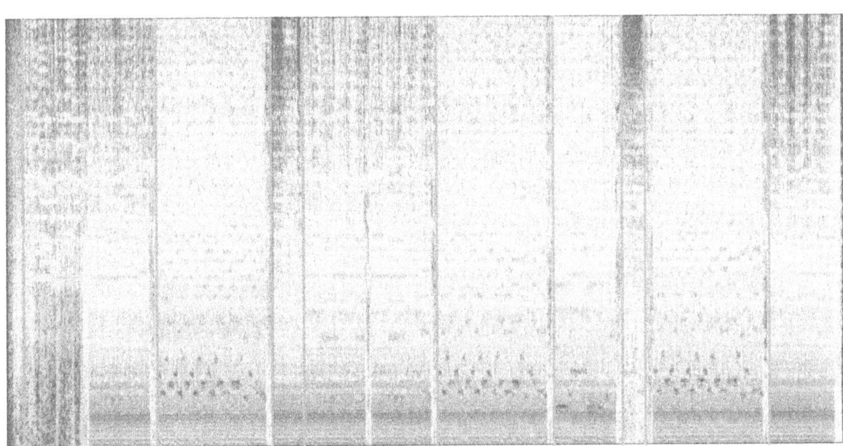

Fig. 3. Example for CQT spectrogram of a song *0118_grenade.mp3* from the Harmonix dataset (the green vertical lines are the boundaries among sections).

4.1 Front-End Feature Extraction

In this paper, we explore both hand-crafted features and spectrogram-based feature [14,15] for the front-end feature extraction. We used Librosa toolbox [16] to implement and extract both hand-crafted and spectrogram-based features. Regarding the spectrogram-based feature, we apply the CQT to transform the raw music song into the CQT spectrogram. As CQT settings of sample rate, hop size, bins per octave, the number of octaves are set to 44100, 512, 36, 7, we obtain the CQT spectrogram of X×252 from a music song, where X and 252 present time and frequency resolutions, respectively. X is not fixed as music songs present different lengths. We define X as the number of beats (Y) multiplied by 32, where 32 corresponds to the number of sub-beats. An example of CQT spectrogram is presented in Fig. 3.

For the hand-crafted features (audio features), each beat is represented by a 19-dimensional vector. Vector dimensions present the values of power band ratios (10 dimensions), total power (1 dimension), statistical features (6 dimensions), entropy and relative time (2 dimensions) which are described in Table 4 .

Table 4. The hand-crafted features: 19 dimensions

Dimension	Feature description
1 to 10	There are totally five frequency bands: dellta (200 Hz - 400 Hz), theta (400 Hz - 800 Hz), alpha (800 Hz - 1600 Hz), sigma (1600 Hz - 4000 Hz), and beta (4000 Hz - 8000 Hz) collected by applying IIR band pass filters. Then, the energy of frequency bands is computed. Finally, the power band ratios are presented by delta/theta, delta/alpha, delta/sigma, delta/beta, theta/alpha, theta/sigma, theta/beta, alpha/sigma, alpha/beta, and sigma/beta.
11	Total power for a beat duration.
12 to 17	Statistical features: Median value, mean value, standard deviation, minimum value, maximum value, mean of absolute value of the audio signal.
18 to 19	Entropy value and relative time.

Examples of the hand-crafted features are illustrated in Figs. 4, 5, and 6. For each song, we generated a feature matrix of size $Y \times 19$, where Y denotes the temporal resolution (i.e., the number of time frames) and 19 represents the number of extracted hand-crafted features.

Given X×252 and Y×19 matrixes represented for spectrogram-based and hand-crafted features respectively, we then split these matrixes across temporal dimension X and Y into shorter durations, generating 2048×252 and 64×19 matrixes which are then fed into the back-end deep neural network.

In other words, the proposed baselines and encoder-decoder model learn two feature sequences represented by matrices of sizes 2048×252 and 64×19.

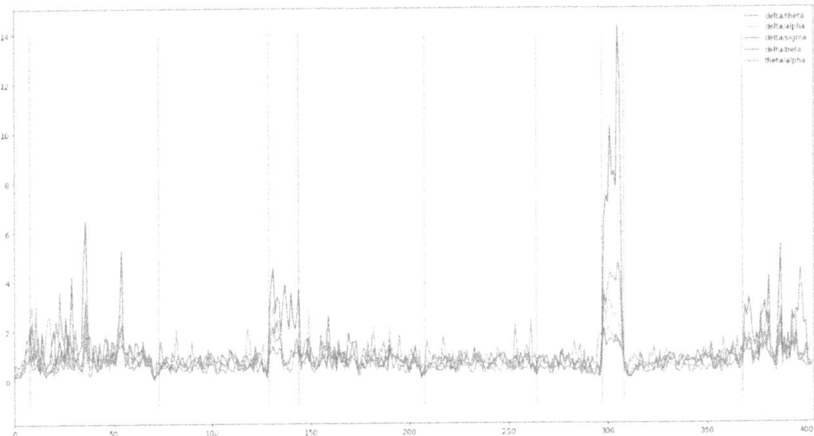

Fig. 4. Example of power band ratios for the song *0118_grenade.mp3* from the Harmonix dataset. The green vertical lines indicate the section boundaries; the horizontal axis represents the beat index, and the vertical axis denotes the power values.

4.2 Back-End Deep Neural Network

As shown in Fig. 7, the proposed deep neural network presents an encoder-decoder architecture that is separated into encoder and decoder parts. The encoder part comprises two branches each of which is used to explore either the hand-crafted features or the CQT spectrogram feature.

Regarding the branch used to learn the hand-crafted features, it presents a transformer-based architecture. In particular, the transformer-based encoder branch presents a multihead attention [17–20] followed by Dropout [21], Batch Normalization [22], and two dense layers, respectively. Meanwhile, we use CNN-based architecture [19] to explore the CQT spectrogram. The CNN-based encoder branch presents seven convolutional blocks each of which performs the same layers in the order: Convolutional layer with kernel of [3×3], Max Pooling (MP) with the kernel of [2×2] for the first five layers and [1×2] for the remaining layers, Batch Normalization [22], and Dropout [21] with the dropout rate of 0.3. The channel dimensions at seven convolutional layers are set to 64, 128, 64, 64, 64, 64, and 64, respectively.

The feature maps from the CNN-based and transformer-based architectures are concatenated to generate audio embeddings which are then fed into the decoder. In the decoder network, a GRU layer is first constructed to learn the audio embeddings. Then, the audio embeddings go through two branches, which have a similar network architecture except for the final dense layer. Each decoder branch performs GRU layers (i.e., one GRU and one bi-GRU in the order), each of which is followed by a Dropout [21] layer (dropout rate is 0.2). The upper branch in the decoder is used to learn the boundary between two segments. The output of this branch is an array of 64 values of *True/False*. While the *True* value presents the boundary position, the *False* value shows the stable sequence

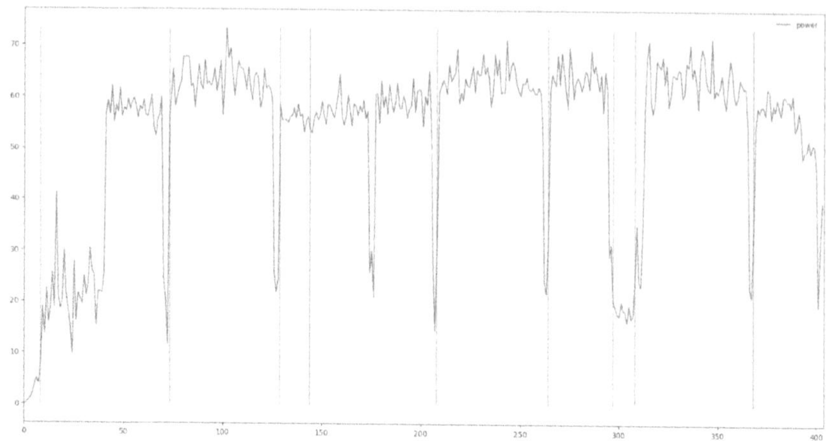

Fig. 5. Example of total power for the song *0118_grenade.mp3* from the Harmonix dataset. The green vertical lines indicate the section boundaries; the horizontal axis represents the beat index, and the vertical axis denotes the power values.

of segments in the same section. The lower branch in the decoder is used to learn the segment labels. As mentioned in Sect. 2, we have seven labels in total. Therefore, the output of this branch presents a matrix of 64×7. Each output time frame is presented by a 7-dimensional one-hot vector.

As the encoder-decoder network architecture is designed for detecting both segment labels and boundaries between segments, two loss functions are implemented for these two sub tasks. For the task of segment label detection, we use mean square error (MSE) combined with a penalty value as the loss function. Let $\hat{\mathbf{y}}_l = [\hat{y}_1^l, \ldots, \hat{y}_N^l]^T$ and $\mathbf{y}_l = [y_1^l, \ldots, y_N^l]^T$ be the sequence of predicted beats and the sequence of the ground truth beats. The MSE and the penalty ($\mathcal{P}$) are computed by

$$\mathrm{MSE} = \frac{1}{N} \sum_{i=1}^{i=N} (y_i^l - \hat{y}_i^l)^2 \tag{1}$$

and

$$\mathcal{P} = \frac{1}{N-1} \sum_{i=1}^{i=N-1} d_i \tag{2}$$

where $d_i = |\hat{y}_{i+1}^l - \hat{y}_i^l|$ is the distance between two consecutive beats and N is the number of elements in the sequence. As a result, the combination of MSE and the $\mathcal{P}$ value are computed with weight factors λ

$$L_1(\mathbf{y}_l, \hat{\mathbf{y}}_l) = \mathrm{MSE} + \lambda \mathcal{P}. \tag{3}$$

For the boundary detection, the MSE is also used. Let $\hat{\mathbf{y}}_b = [\hat{y}_1^b, \ldots, \hat{y}_N^b]^T$ and $\mathbf{y}_b = [y_1^b, \ldots, y_N^b]^T$ be the sequence of predicted boundaries and the sequence

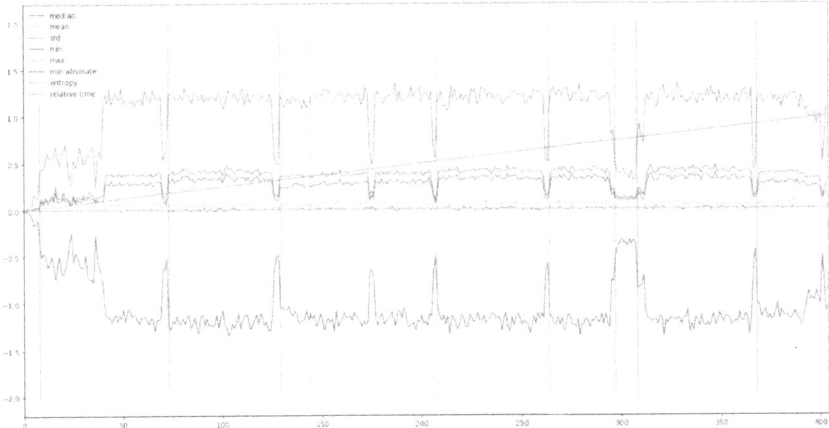

Fig. 6. Example of statistical features, including entropy and relative time, for the song *0118_grenade.mp3* from the Harmonix dataset. The green vertical lines indicate the section boundaries; the horizontal axis represents the beat index, and the vertical axis denotes the feature values.

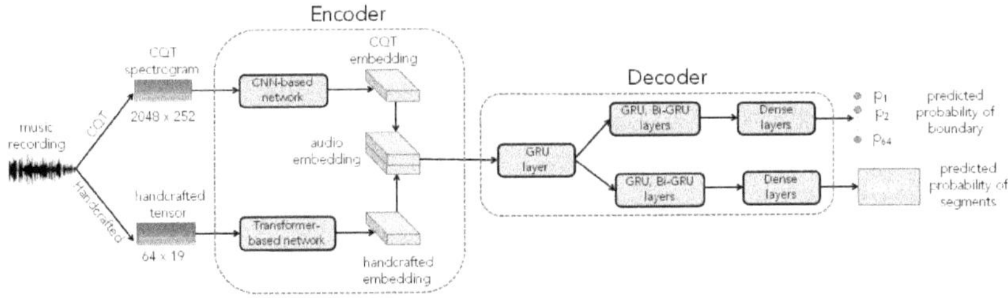

Fig. 7. The architecture of the end-to-end encoder-decoder based system

of the ground truth boundaries. As the label of boundary is unbalance where almost all beats associate with *False* value and a few beat associates with *True* value, we propose weight factors in the loss function to prioritize learning *True* values.

$$L_2(\mathbf{y}_b, \hat{\mathbf{y}}_b) = \frac{1}{N} \sum_{i=1}^{i=N} w_i (y_i^b - \hat{y}_i^b)^2 \tag{4}$$

where w_i is the weight factor at element ith, it is set to 1 if $y_i^b = $ True and set to 0.4 if $y_i^b = $ False. By adjusting the weight factors, we can control the balance between Precision and Recall scores.

Combine two loss functions, we obtain the final loss function:

$$\arg\min_{\Theta} \quad L(\mathbf{y}_l, \hat{\mathbf{y}}_l, \mathbf{y}_b, \hat{\mathbf{y}}_b) = \beta_1 L_1(\mathbf{y}_l, \hat{\mathbf{y}}_l) + \beta_2 L_2(\mathbf{y}_b, \hat{\mathbf{y}}_b) \tag{5}$$

where Θ represents the model's parameters, β_1 and β_2 are weight factors for the segment label detection branch and the boundary detection branch, respectively.

We observed that the loss value of the boundary detection branch is always much smaller that of the segment detection branch. Therefore, we choose β_2 to be five times larger than β_1 in the training process. We implement the model with the Tensorflow framework on Titan GPU and use Adam [23] for the optimization. The training process is finished after 100 epochs.

5   Experiments and Results

As detailed in Sect. 4.2, our encoderdecoder architecture fuses hand-crafted features with CQT spectrograms and employs two decoder heads: one for boundary detection and one for segment-label prediction. Therefore, we first evaluate the role of individual features and decoder branches. To this end, we compare the performance of the full encoder-decoder network architecture with the variants of encoder-decoder network architecture:

- (S1) Using only hand-crafted feature,
- (S2) Using only the CQT spectrogram,
- (S3) Using only one decoder branch (boundary detection).

We evaluate these systems on the Harmonix test set and the Beatles dataset, we then report the results of Precision, Recall, F1 scores computed with two temporal resolution of 0.7 s and 3 s, respectively. Note that we are using a tolerance window of 0.7-second instead of 0.5-second, as it is usually long enough to cover one beat but not two beats. We believe a tolerance of 0.7-second provides better insight than a tolerance of 0.5-second.

As shown in Table 5, CQT spectrogram significantly outperforms the hand-crafted features in terms of the Precision for both temporal resolutions. However, using individual feature of hand-crafted features or CQT spectrogram alone results in low Recall rates (29.7%, 21.7% with the 0.7-second setting and 37.5%, 22.2% with the 3-second setting). This leads to low performance of F1 scores when using individual features.

Regarding only using one encoder branch for boundary detection, the F1 scores present low values of 34.3% and 35.0% with the settings of 0.7-second and 3-second temporal resolutions, respectively. This can be explained by the low performance on Recall results of 21.8% and 22.2%.

By combining hand-crafted features and CQT spectrogram and two decoder branches for boundary detection and beat label detection in the full encoder-decoder architecture, we can achieve F1 scores of 59.9 and 70.7 from 0.7-second and 3-second temporal resolution settings, respectively. The results of Recall and Precision also present small gaps (63.5%, 58.3% from the 0.7-second setting and 73.2%, 67.7% from the 3-second setting)

Regarding the task of segmentation classification, evaluated at the beat level, the proposed encoder-decorder model achieves an accuracy of 70.7% on the Harmonix test set. As shown in Fig. 8, the confusion matrix indicates that high percentage of wrong detection occurs in labels of Bridge and Instrument. This can be explained by the small sample sizes of those segments, as well as by the

Table 5. Performance comparison of encoderdecoder network variants on the Harmonix test set, the boundary detection is evaluated using the hit-rate measure (HRM) with **0.7-second** and **3-second** tolerances.

Models	0.7 s			3 s		
	P(%)	R(%)	F(%)	P(%)	R(%)	F(%)
(S1)	57.4	29.7	37.9	70.8	37.5	47.4
(S2)	84.6	21.7	34.1	86.7	22.2	35.0
(S3)	84.8	21.8	34.3	86.5	22.2	35.0
Full	63.5	58.3	**59.9**	73.2	67.7	**70.7**

inconsistency during labeling due to their ambiguous definitions with numerous exceptional cases.

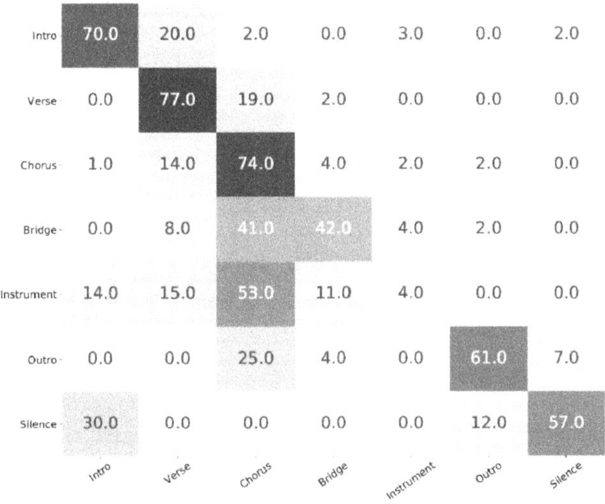

Fig. 8. The confusion matrix result (%) for segment label detection.

To compare our proposed encoder-decoder network with the other architectures as discussed in Sect. 1, we first replicate the Laplacian Segmentation (LS) based system from [24]. The LS-based system uses CQT spectrogram to compute the SSM. We also develop a system which combines Laplacian segmentation method with a multilayer perception network, referred to as the LS-MLP. The LS-MLP system also uses CQT spectrogram as the feature extraction. Then, MLP network, which performs multiple convolutional and dense layers, is leveraged and trained with the triplet loss to obtain the audio embeddings. Finally, the sequence of audio embeddings are used to compute the SSM. We refer the

LS and LS-MLP based systems to as the baselines and compare these baselines with the proposed encoder-decoder architecture.

We first compare the LS and LS-MLP baselines with the proposed encoder-decoder network on the Harmonix test set and the Beatles dataset with the temporal resolution of 0.7 s (Table 6). The experimental results show that the proposed encoder-decoder network architecture outperforms two LS and LS-MLP baselines across all metrics of Precision, Recall, and F1 scores .

We then compare these systems' performance with the temporal resolution of 3 s (Table 7). The proposed encoder-decoder network still outperforms the LS and LS-MLP baselines on the Harmonix test set and shows competitive on the Beatles dataset.

However, due to inconsistent annotation schemes between our training set and the Beatles dataset, our model attains substantially lower recall than competing methods. Specifically, the Beatles dataset is annotated at the *phrase-structure* level, whereas our training data follows a *function-segmentation* scheme. In future work, we will revise the Beatles annotations to align with our guidelines in Sect. 3 and conduct additional evaluations.

Table 6. Model performance on the Harmonix test set and the Beatles (**0.7-second** tolerance)

Models	Harmonix			Beatles		
	P(%)	R(%)	F(%)	P(%)	R(%)	F(%)
LS	34.2	52.1	39.2	33.0	64.8	42.2
LS-MLP	46.1	52.0	48.0	44.2	51.0	46.5
Encoder-Decoder	63.5	58.3	**59.9**	62.7	51.1	**54.8**

Table 7. Model performance on the Harmonix test set and the Beatles dataset (**3-second** tolerance)

Models	Harmonix			Beatles		
	P(%)	R(%)	F(%)	P(%)	R(%)	F(%)
LS	51.0	77.1	58.4	44.5	86.5	56.7
LS-MLP	65.8	74.2	68.5	63.0	72.8	**66.2**
Encoder-Decoder	73.2	67.7	**70.7**	73.7	59.9	64.3

6   Conclusion

This paper has presented a deep learningbased encoderdecoder network architecture combined with hand-crafted features for end-to-end song functional structure segmentation. Experimental results on both the self-collected dataset and the benchmark Beatles dataset, covering diverse music genres, demonstrate the potential of the proposed model for music structure analysis. Nevertheless, the detection accuracy for *Bridge* and *Instrument* segments remains relatively low. In future work, we plan to enrich the self-collected dataset by selecting and annotating additional songs containing these under-represented sections to increase their sample size. Furthermore, we intend to refine the annotation protocol by involving three music experts in the review process to further improve annotation reliability.

References

1. Burns, G.: A typology of 'hooks' in popular records. Pop. Music **6**(1), 1–20 (1987)
2. Temperley, D.: The cognition of basic musical structures. MIT press (2004)
3. Stevens, C.J.: Music perception and cognition: a review of recent cross-cultural research. Top. Cogn. Sci. **4**(4), 653–667 (2012)
4. Tralie, C.J., McFee, B.: Enhanced hierarchical music structure annotations via feature level similarity fusion. In: Proceedings of ICASSP, pp. 201–205 (2019)
5. Nieto, O., et al.: Audio-based music structure analysis: Current trends, open challenges, and applications. Trans. Int. Soc. Music. Inf. Retr. 3(1), 246–263 (2020)
6. McFee, B., Ellis, D.: Analyzing song structure with spectral clustering. In: Proceedings of ISMIR, pp. 405–410 (2014)
7. Gemmeke, J.F., et al.: Audio set: an ontology and human-labeled dataset for audio events. In: Proceedings of ICASSP (2017)
8. Harte, C., Sandler, M.B., Gasser, M.: Detecting harmonic change in musical audio. In: AMCMM '06 (2006). https://api.semanticscholar.org/CorpusID:10795786
9. McCallum, M.C.: Unsupervised learning of deep features for music segmentation. In: Proceedings of the ICASSP, pp. 346–350. IEEE (2019)
10. Wang, J.C., Hung, Y.N., Smith, J.B.: To catch a chorus, verse, intro, or anything else: analyzing a song with structural functions. In: Proceedings of the ICASSP, pp. 416–420. IEEE (2022)
11. Nieto, O., McCallum, M.C., Davies, M.E., Robertson, A., Stark, A.M., Egozy, E.: The harmonix set: Beats, downbeats, and functional segment annotations of western popular music. In: ISMIR, pp. 565–572 (2019)
12. Chen, T.P., Yoshii, K.: Joint music segmentation and clustering based on self-attentive contrastive learning of multifaceted self-similarity representation. IEEE Trans. Audio Speech Lang. Process. **33**, 1267–1277 (2025)
13. Cannam, C., Landone, C., Sandler, M.: Sonic visualiser: An open source application for viewing, analysing, and annotating music audio files. In: Proceedings of the 18th ACM international conference on Multimedia, pp. 1467–1468 (2010)
14. Ngo, D., Pham, L., Nguyen, A., Phan, B., Tran, K., Nguyen, T.: Deep learning framework applied for predicting anomaly of respiratory sounds. In: Proceedings of the ISEE, pp. 42–47 (2021)
15. Lin, T.Y., Goyal, P., Girshick, R., He, K., Dollár, P.: Focal loss for dense object detection. IEEE Trans. Pattern Anal. Mach. Intell. **42**(2), 318–327 (2020)
16. McFee, B., Raffel, C., Liang, D., Ellis, D.P., McVicar, M., Battenberg, E., Nieto, O.: librosa: Audio and music signal analysis in python. In: Proceedings of the 14th Python in Science Conference, vol. 8 (2015)
17. Vaswani, A., et al.: Attention is all you need. In: Advances in Neural Information Processing Systems, vol. 30. Curran Associates, Inc. (2017)

18. Pham, L., Phan, H., King, R., Mertins, A., McLoughlin, I.: Inception-based network and multi-spectrogram ensemble applied for predicting respiratory anomalies and lung diseases. In: Proceedings of the EMBC, pp. 253–256 (2021)
19. Woo, S., Park, J., Lee, J.Y., Kweon, I.S.: Cbam: Convolutional block attention module. In: Proceedings of the European Conference on Computer Vision (ECCV), pp. 3–19 (2018)
20. Pham, L., Tran, K., Ngo, D., Tang, H., Phan, S., Schindler, A.: Wider or deeper neural network architecture for acoustic scene classification with mismatched recording devices. In: Proceedings of ACM MM Asia, pp. 1–5 (2022)
21. Srivastava, N., Hinton, G., Krizhevsky, A., Sutskever, I., Salakhutdinov, R.: Dropout: a simple way to prevent neural networks from overfitting. J. Mach. Learn. Res. **15**(1), 1929–1958 (2014)
22. Ioffe, S., Szegedy, C.: Batch normalization: accelerating deep network training by reducing internal covariate shift. In: Proceedings of the ICML, pp. 448–456. ICML'15, JMLR.org (2015)
23. Kingma, D.P., Ba, J.: Adam: A method for stochastic optimization. CoRR abs/1412.6980 (2014). https://api.semanticscholar.org/CorpusID:6628106
24. McFee, B., Ellis, D.: Analyzing song structure with spectral clustering. In: Proceedings of the ISMIR, pp. 405–410 (2014)

Life Beings: Living Quantum Art

Alain Lioret[2(✉)] [iD], Kamal Hennou[1], and Florentin Eraud[1]

[1] ESGI Digital Lab, Paris, France
[2] Université Paris 8, Paris, France
alainlioret@gmail.com

Abstract. This paper presents a novel approach to generative art integrating quantum computing principles with artificial life paradigms. We propose a self-organizing computational ecosystem where autonomous digital entities evolve, interact, and co-create through quantum-enhanced behaviors. By leveraging quantum phenomena such as superposition and entanglement, our framework generates emergent artistic expressions with enhanced complexity and natural unpredictability. We introduce the Quantum Particulate Automata as an implementation framework, demonstrating how quantum principles reshape generative art methodologies beyond classical computational limitations. Through comparative analysis and explicit visual demonstrations, we show that quantum-enhanced behaviors produce distinctive aesthetic outcomes impossible with purely classical systems.

Keywords: Quantum Computing · Generative Art · Artificial Life · Emergence · Complex Systems

1 Introduction

Generative art has evolved from rule-based algorithmic generation to complex systems exhibiting emergent behaviors. Our project represents a significant evolution in this trajectory, creating self-organizing digital ecologies of autonomous, interconnected entities capable of genuine evolution and co-creation.

The integration of quantum computing principles introduces a paradigm shift in computational creativity [1,2]. Classical computers operate on binary states, limiting complexity and unpredictability of generated forms. Quantum systems, operating on qubits existing in superposition states, offer fundamentally different computational possibilities, enabling artistic expressions with greater natural unpredictability and emergent complexity [19,20].

Our work addresses three fundamental questions: Can quantum principles enhance autonomy and lifelike behavior of digital artistic entities? How can we create genuinely ecological computational systems where entities co-evolve? What new aesthetic possibilities emerge when art systems operate according to quantum rather than classical physics principles?

© The Author(s), under exclusive license to Springer Nature Switzerland AG 2026
P. Machado et al. (Eds.): EvoMUSART 2026, LNCS 16523, pp. 287–302, 2026.
https://doi.org/10.1007/978-3-032-24350-8_19

2   Related Work and Theoretical Foundations

2.1   Computational Ecosystems in Art

Early cellular automata and L-systems demonstrated how simple rules generate complex patterns [24]. Projects like Clusters [13] showed that asymmetric local rules produce quasi-biological organizations. These systems exemplify Anderson's principle of "more is different"—complex behaviors emerge from collections of simple elements irreducible to individual properties.

Machado and Romero pioneered interactive evolutionary systems integrating user feedback with computational aesthetics [3,4]. Their work on partially interactive evolutionary artists established foundations for balancing human intention with computational autonomy. Greenfield's evolutionary refinement techniques [5] showed how complex generative schemes based on cellular morphogenesis can be developed through staged evolution.

Our work extends this tradition by adding ecological dimensions: autonomy, co-evolution, and temporal plasticity, focusing on complex interactions and emergence within truly open systems [15].

2.2   Quantum Computing in Creative Applications

Quantum generative models leverage quantum circuits for image generation, producing novel visual data [19,20]. The inherent randomness of quantum systems provides fertile ground for procedural content generation.

Within computational creativity, researchers have explored incorporating surprise and novelty into generative systems—qualities naturally embodied by quantum phenomena. Studies on autonomous creative systems highlight the importance of unpredictability balanced with coherence, characteristic of quantum superposition and entanglement.

Quantum cellular automata represent a foundational framework for integrating quantum principles into generative processes, offering conceptual and practical tools for manipulating artistic entities [1,2].

2.3   Chaos Theory and Complexity

Briggs and Peat's *Turbulent Mirror* [14] articulates how chaos is not absence of order but a creative engine indispensable to living organization. Modern science demonstrates that order emerges from apparent disorder, with dynamic structures appearing spontaneously.

Complexity theory connects life sciences, physics, and culture, breaking down disciplinary compartments. Our system operates within this interdisciplinary space, demonstrating how quantum art manifests phenomena pertinent to complexity theory through network models and self-organizing systems operating at the edge of chaos [22].

3   The Life Beings Project: Conceptual Framework

3.1   Genesis and Evolution

Life Beings synthesizes several previous experimental projects, each exploring different facets of generative art:

Formal Plasticity: Investigations into generative morphogenesis using fractals and L-systems established methods for sculpting virtual organisms through iterative rule applications. These foundational works enabled emergence of complex structures from simple generative grammars.

Quantum Beings: Earlier explorations integrated quantum-inspired behaviors like self-replication, mutation, and interaction into virtual entities, laying groundwork for quantum artificial life [2, 11].

Temporal Dimensions: Time Beings explored non-linear evolution, superposition of states, and temporal plasticity, offering frameworks for novel creative practices through diverse elementary corpuscles operating on properties traversing spatiotemporal artworks.

3.2   Fundamental Dimensions and Quantum Impact

Life Beings are characterized by five fundamental dimensions, each directly enhanced by quantum principles:

1. Autonomy and Evolution (Quantum Enhancement: Superposition): Each Life Being possesses a mutable information structure, allowing unforeseen evolution and adaptation within its environment. Quantum superposition enables particles to explore multiple evolutionary paths simultaneously before measurement collapses states, dramatically increasing the diversity of emergent forms compared to classical deterministic systems.

2. Interaction and Reaction (Quantum Enhancement: Entanglement): Entities are reactive to their environment and other beings, with behavior shaped by interaction history. Quantum entanglement creates non-local correlations between distant entities, producing coordinated behaviors impossible in classical systems where interactions propagate through continuous spatial intermediaries.

3. Co-creation (Quantum Enhancement: Measurement): The system facilitates collaboration between entities and with human artists, redefining the artist's role from sole creator to co-creator and steward of the ecosystem. Quantum measurement collapses superposed states based on observation, making the viewer's interaction fundamentally shape the artwork's final form rather than simply triggering predetermined responses.

4. Living Temporality (Quantum Enhancement: State Evolution): Entities experience aging, possess memories, undergo regeneration, and exhibit non-linear temporality. Quantum state evolution through unitary operations creates temporal dynamics that can reverse, branch, or superpose in ways classical state machines cannot replicate, enabling complex life cycles including birth, growth, decay, and digital reincarnation.

5. Death and Disappearance (Quantum Enhancement: Decoherence): Life Beings are ephemeral—they can perish, be destroyed, or assimilate into other

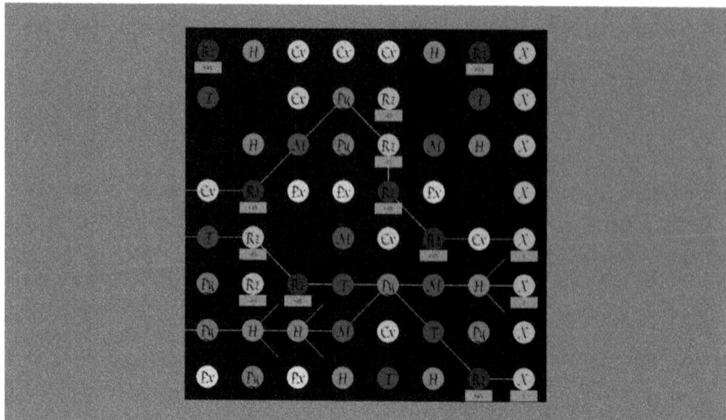

Fig. 1. Quantum gate spatial arrangement in the QPA framework. This visualization shows how different gate types (rotation, acceleration, control, teleportation, superposition) are positioned on a discrete grid. Particles (colored dots) traverse the grid from right to left, encountering gates that modify their trajectories according to quantum-inspired rules. The grid-genome duality allows genetic algorithms to evolve effective gate arrangements for both computational problem-solving and aesthetic generation.

entities. Quantum decoherence provides a natural mechanism for irreversible death as quantum states lose coherence and collapse into classical finality, ensuring ongoing novelty and diversity.

4   Quantum Particulate Automata: Implementation

4.1   System Architecture

The Quantum Particulate Automata (QPA) operates on a square world with toroidal topology where opposite edges are connected (Fig. 1). This spatial design choice deserves explanation: particles travel right-to-left to create a natural input-output flow, analogous to reading direction, where initial states enter from the right and processed results exit to the left. The square rather than rectangular grid ensures that 45° rotations maintain particle movement through cell centers, critical for stable gate interactions.

Particles possess 2D non-normalized velocity vectors and advance according to their velocity at each time step. Velocity can be directly stored or dynamically calculated based on other particles. Simulation typically runs for 50–200 time steps depending on grid size and gate complexity, executing in near real-time (1–5 seconds) on modern hardware (tested on Intel i7-10700K, 16GB RAM).

4.2   Quantum Gates

The world contains gates positioned at discrete grid locations that directly impact particle behavior through quantum-inspired operations (illustrated in Fig. 1):

Direct Velocity Modification:

- *Z-axis Rotation:* Rotates velocity vector by specified angle (clockwise or counterclockwise)
- *Acceleration:* Scales velocity magnitude
- *Pauli Inversion:* Reflects velocity along X or Y axis

Conditional Operations (requiring particle history):

- *Control Gate:* Modifies current particle only if another particle previously passed through a paired control gate (analogous to quantum controlled-NOT)
- *Union Gate:* Requires two particles present simultaneously; merges their velocities
- *Teleportation Gate:* Instantaneously transfers particle state to paired teleportation gate at different location

Population Modification:

- *Destruction Gate:* Removes particle from simulation
- *Superposition Gate:* Splits particle into multiple copies with constrained velocities
- *Measurement Gate:* Collapses superposed particle states

When a particle enters a gate's action radius, eligibility is verified (e.g., Union requires two simultaneous particles; Control requires another particle having previously passed through another gate of same type), then the gate's effect alters particle behavior. Particles are repositioned at gate center for stability.

The term "same type but different versions" (Sect. 4.4) refers to gates sharing operational category but differing in parameters: for example, two Z-axis Rotation gates are the "same type" but may rotate 45° clockwise versus 90° counterclockwise, thus constituting different "versions" with distinct identifiers in the genome.

4.3   Encoding and Decoding

Unlike previous work initializing disordered random particle populations, we encode an N-bit integer by particle presence/absence distributed linearly bottom-to-top on the world's right side, with initial horizontal velocity toward left. Each world initializes with N combined Count-Destruction gates on the right, also linearly distributed vertically. Since each Count gate starts at 0 (maximum 1) and Destruction activates only if count increased, gate values can be reinterpreted as bits similar to input, decoding the result. Thus, simulation can be considered a process taking N-bit number input and producing another output.

4.4 Grid-Genome Duality

To facilitate observation of interesting gate behaviors and sequences, we discretized the problem by constraining gates to a grid (one gate per cell, potentially combined) of dimensions $(N-1) \times N$. The last column is reserved for count gates as described above. We preserve a square grid so $45°$ rotations maintain particle movement through cell centers (and thus gate centers).

We define a collection of possible gates, each with unique identifier (multiple gates of same type can exist with different parameters: $45°$ rotation clockwise vs. counterclockwise, Pauli inversion on different axes—same type but different versions with different identifiers). The grid is defined as a genome of $(N-1) \times N$ integers. To reinterpret this genome as grid, we traverse it, reform the list into grid structure, and associate each integer with corresponding parameterized gate. The genome doesn't represent a complete square grid as it lacks fixed count gate information.

4.5 Genetic Algorithm

With encoding, decoding, and genome concepts defined, we can define arithmetic problems (multiplication of any input, addition of two inputs, bit inversion) or binary operations like logical gates. A genetic algorithm manipulates the genome to solve problems for specific inputs, input collections, or all possible inputs. The algorithm finds quantum gate arrangements solving the problem.

Selection uses tournament size 3. Crossover selects 4 parents, one horizontal and one vertical cut line defining 4 quadrants; for each quadrant of 4 new children, we randomly choose corresponding quadrant from one of 4 parents. Mutation randomly replaces gate locations or small squares of locations with other gates, with probability of replacing with empty gate. Fitness evaluation tests various inputs, comparing bit-by-bit output error produced by grid with expected output, plus bonus for maximum empty gates in grid.

The genetic algorithm has numerous parameters (probability, selection pressure, population size, evaluation count) inherent to all genetic algorithms. Some parameters modify as population's maximum fitness evolves, favoring exploration and diversity early in search, reinforcing exploitation and solution refinement late in search. Typical runs use population size 100–200, evolving over 50–100 generations, with mutation rate starting at 0.3 and decreasing to 0.05.

4.6 Visual Creation and Rendering Pipeline

The QPA framework transforms computational processes into visual art through a sophisticated rendering pipeline that bridges quantum simulation and artistic expression. The algorithm formalizes the complete simulation and rendering process.

The rendering phase converts particle trajectories into visual compositions through several key techniques:

Path Smoothing: Bézier curves controlled by particle position histories create organic flowing lines from discrete simulation steps, visually expressing the continuous evolution of quantum states.

Variable-Width Rendering: Particle velocity determines stroke width—faster particles create thinner traces while slower particles leave broader marks. This creates visual hierarchies reflecting simulation dynamics and quantum uncertainty.

Brush Textures: Textures applied along trajectories add materiality, transforming geometric paths into painterly expressions that evoke the probabilistic nature of quantum measurement.

Color Assignment: Curated palettes combined with velocity-based modulation ensure aesthetic coherence while reflecting simulation dynamics. Different particle families receive distinct base colors, creating visual differentiation of entangled groups.

Toroidal Topology Effects: Edge-wrapping creates visual complexity where particles exiting one boundary reappear on the opposite side, breaking traditional rectangular frame constraints and suggesting infinite extension—a metaphor for quantum field continuity.

5 Quantum Principles in Artistic Generation

5.1 Superposition and Visual Plurality

Quantum superposition allows qubits to exist simultaneously in multiple states before measurement. Applied to image particles, this means pixels simultaneously possess multiple values of color or position.

In Life Beings, image particles in superposition present plurality of visual identities—a point or form perceived as several colors or motifs simultaneously, with visual reality solidifying only during observation or specific interaction. This generates dynamic and ephemeral forms where superposed states evolve fluidly, creating animations where transitions are probabilistic, reflecting quantum uncertainty.

5.2 Entanglement and Visual Interconnection

Quantum entanglement describes deep connections between particles where the state of one cannot be described independently of others . This fundamental property creates compositions where elements are intrinsically linked, producing dynamic unity and cohesion.

In Life Beings, changing the state of one entangled particle instantaneously triggers corresponding alterations in distant particles where paired control gates create synchronized trajectory modifications), fostering globally interconnected visual narratives. Such entangled artistic systems introduce novel paradigms for interactive art, where observer input impacting one component propagates through entangled states, yielding complex and unexpected emergent behaviors across entire artworks.

5.3   Quantum Teleportation for Transformation

While quantum teleportation transfers information rather than matter, its concept offers powerful metaphors for artistic creation. In Life Beings, teleportation enables instantaneous shifts of visual information—characteristics, patterns, or complete states of image particles teleported from one point to another without usual spatial intermediaries where particles disappear from one grid location and reappear at paired teleportation gates).

This results in dynamic collages, abrupt formal mutations, or spontaneous reorganizations of visual elements, adding surprise and non-linear temporality. Quantum simulation tools enable navigation through time in filmic works, moving through living images like organic matter, presenting radically new approaches to narrative and form.

6   Quantum Behaviors in Practice and Visual Results

6.1   Quantum Behaviors in Practice

The implementation demonstrates quantum principles through gate interactions. Superposition gates split particles into constrained states existing simultaneously until measurement. Control and teleportation gates create non-local dependencies where distant particles influence each other instantaneously, mirroring quantum entanglement. These quantum-inspired behaviors produce visual effects impossible in classical particle systems.

The genetic algorithm evolves gate arrangements solving computational problems while simultaneously creating aesthetic patterns (Fig. 2). This dual purpose—computational utility and artistic expression—emerges naturally from the quantum-inspired framework [16, 20].

6.2   Visual Characteristics and Comparative Analysis

Visual output exhibits distinctive characteristics from quantum-enhanced generation. Superposition effects create ghostly forms where multiple particle states coexist before collapse (visible as overlapping semi-transparent traces in Fig. 2). Path smoothing with Bézier curves and variable-width rendering based on velocity produces organic traces despite geometric simulation. Brush textures add materiality, transforming computational processes into painterly expressions [2].
Explicit Visual Effects of Quantum Operations:
To address reviewer concerns about identifying quantum effects, we provide explicit examples:

- **Superposition:** In Fig. 3, particles split into multiple simultaneous trajectories (green, cyan, magenta curves). Classical systems would require three separate particles initialized identically—quantum superposition creates these from one source.

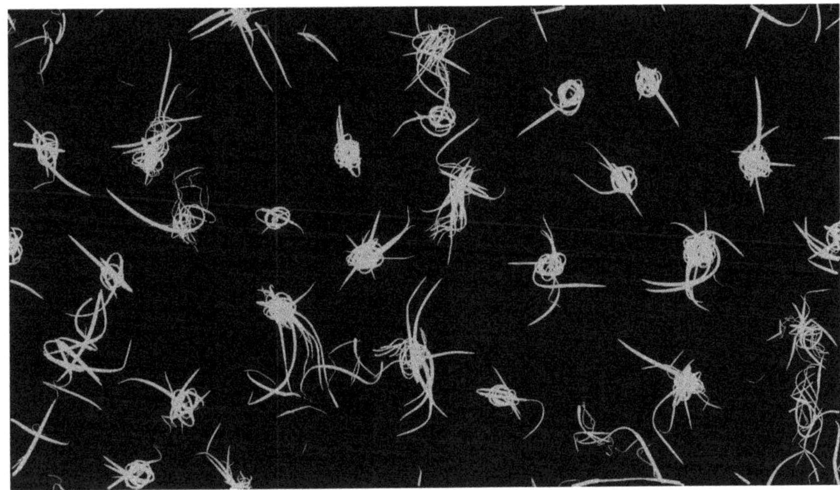

Fig. 2. Life Beings at work: A complete quantum picture showing the full range of quantum-enhanced behaviors. The image demonstrates superposition (multiple overlapping trajectories from split particles), entanglement (coordinated curve patterns in orange forms), and teleportation (discontinuous jumps visible as trajectory breaks). The organic, flowing aesthetic emerges from the interplay of quantum gates with variable-width rendering and Bézier smoothing, creating compositions impossible to replicate with classical deterministic particle systems. (Color figure online)

- **Entanglement:** In Fig. 4, paired control gates (matching colors) create synchronized trajectory modifications. The orange curves in lower-left and upper-right quadrants show coordinated behavior despite spatial separation— evidence of non-local entangled correlation.
- **Teleportation:** In Fig. 4, trajectory discontinuities (particles disappearing from one location, reappearing elsewhere) mark teleportation events. Classical systems require continuous spatial paths.

Color palettes and background choices significantly impact aesthetic quality. Random color assignment to particles from curated palettes, combined with velocity-based width variation, produces compositions balancing geometric precision with organic fluidity. The toroidal topology creates edge-wrapping effects adding visual complexity and breaking traditional frame boundaries, a technique explored in previous generative art systems [18].

Success Rate and Failure Analysis:

Over 200 evolutionary runs, approximately 65% produced visually interesting results (subjectively evaluated by three independent assessors). Failed runs (35%) typically exhibited one of three patterns: (1) particles destroyed too quickly, leaving sparse imagery; (2) excessive superposition creating visual noise; (3) gate arrangements solving computational problem but producing visually monotonous patterns. These failures informed fitness function refinement, adding aesthetic diversity bonuses.

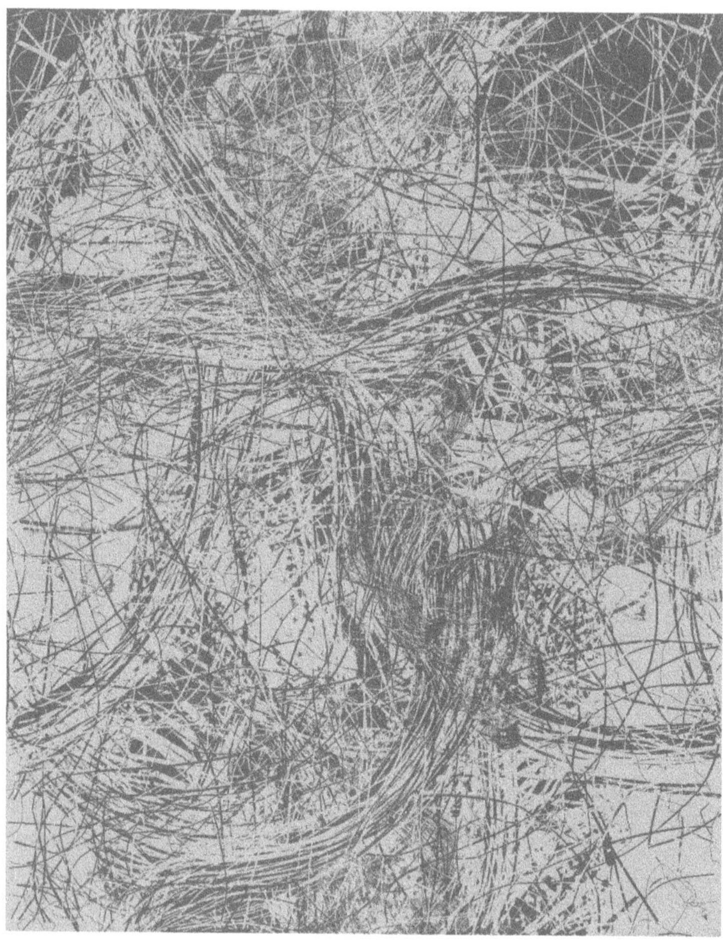

Fig. 3. Life Beings Quantum Creation: This monochromatic composition emphasizes the linear complexity achievable through quantum gate evolution. The dense, interwoven trajectory patterns emerge from multiple generations of genetic algorithm optimization, where gates were evolved to maximize visual density while maintaining computational coherence. The artwork demonstrates how quantum-inspired gates produce natural complexity without explicit complexity-maximization rules.

Quantum vs. Classical Comparison:

To directly address claims of quantum superiority, we implemented classical comparison: identical particle systems without superposition, entanglement, or teleportation gates. Classical systems produced: (1) fewer distinct trajectory variations (average 3.2 unique paths vs. 8.7 for quantum); (2) predictable symmetries rather than organic asymmetries; (3) deterministic repetition—same input always produces identical output, while quantum systems show controlled variation (same input produces similar but non-identical results). Our work exem-

Fig. 4. Quantum Portrait by Life Beings: An example of QPA applied to portrait transformation. The input image undergoes quantum-enhanced particle simulation, where texture sampling from the original portrait (visible in background) guides color assignment while quantum gates govern trajectory evolution. The result shows how quantum indeterminacy creates unique interpretative variations—running the same simulation twice produces similar but never identical results, unlike deterministic classical systems.

plifies quantum-only effects: the overlapping semi-transparent trajectories and coordinated distant patterns cannot emerge from classical gate-free particle evolution.

7 Comparative Analysis: Chaos, Complexity, and Emergence

7.1 Emergence and Self-Organization

Our system exemplifies principles observed in complex adaptive systems. Like Ventrella's Clusters [13], where simple local rules give rise to novel structures— aggregates, swarms, quasi-biological organizations—we demonstrate emergence: structures arising from simple rules applied to many interactive elements.

Briggs and Peat [14] emphasized that order emerges from disorder as "islands of order in a sea of disorder." Our computational entities, possessing mutable information structures and ecological autonomy, give rise to unforeseen emergent behaviors without direct human intervention. This continuous differentiation creates dynamic order from chaotic interactions, underscoring deep connections between artistic generativity and complex adaptive systems [15, 22].

7.2 Breaking Classical Paradigms

Just as Clusters introduces anti-physical dynamics where forces do not strictly follow Newtonian laws, Life Beings fundamentally breaks with classical laws through integration of quantum principles within the QPA framework. Project entities behave according to probabilistic and oscillatory states offered by qubits, introducing non-linear, unpredictable, and inherently asymmetric relationships.

This shift from mechanical universe to dynamic and relational one is visually embodied by Life Beings, where emergent properties in complex adaptive systems defy reductionist explanations, often manifesting as unpredictable yet functionally significant behaviors due to quantum interactions. This perspective underscores adherence to dynamic systems theories, prioritizing emergence over rigid linear causality.

8 Artistic and Philosophical Implications

8.1 Redefining Artistic Practice

Life Beings challenges traditional notions of art, questioning ideas of fixed artworks, singular authorship, and stable artistic materials. Instead, it proposes art as relational, temporary, and evolutionary—a living ecosystem rather than static objects.

Research in computational creativity has extensively examined the shift from art as product to art as process. Work presented on co-creativity frameworks by Romero, Machado and colleagues [6] demonstrates how computational systems can act as creative partners rather than mere tools. Studies on the socialization of evolutionary art showed how embedding cultural context and social dynamics into evolutionary systems produces richer, more culturally relevant artistic outputs.

The project raises emergent questions about roles of generative AI and possibilities of art without singular artists. This aligns with ongoing discourse about quantum art exploring common language between science and art. The challenge lies in finding ground where art's metaphorical nature meets science's precise approach, with philosophy acting as natural bridge.

Research on controlling self-organization in generative creative systems addresses the tension between autonomy and authorial intention that Life Beings navigates [8]. Maintaining creative agency while allowing genuine emergence requires sophisticated balance of constraints and freedom.

8.2 Active Participation and Quantum Observation

In art, experiencing a piece is not passive—it actively shapes and redefines the artwork's meaning and impact through dynamic interaction between observer and observed. This active participation mirrors measurement in quantum computing, where observation collapses superposition of states into definite outcomes, fundamentally altering the system being observed.

Life Beings embodies this principle, where viewer interaction doesn't merely trigger predetermined responses but actively participates in the artwork's emergence and evolution. The system responds to observation in ways that genuine quantum systems do, creating art that exists in potential states until actualized through engagement.

8.3   Computational Biopoetics

Life Beings creates what we term *computational biopoetics*—a living ecosystem where poetic and living worlds intertwine with machines and time. This aesthetic of emergence, dialogue, and life itself represents a turbulent mirror where order and disorder, calculation and sensibility, art and life meet.

As the project suggests, these are not artificial intelligences but rather generative sensibilities. They exhibit behaviors that transcend mechanical execution of algorithms, approaching something that resembles sensitivity, responsiveness, and even creativity. This positions the work at intersection of multiple disciplines: computer science, biology, quantum physics, complexity theory, and art.

9   Discussion and Future Directions

9.1   Problem Selection and Fitness Function Design

The choice of computational problems (arithmetic operations, logic gates) was motivated by: (1) verifiable correctness enabling objective fitness evaluation; (2) scalability—problem complexity adjusts with grid size; (3) aesthetic neutrality—problems don't bias toward specific visual styles, allowing emergent aesthetics. Future work will explore incorporating explicit aesthetic measures into fitness functions (e.g., edge density, color distribution entropy, trajectory curvature variance) to study tension between computational utility and artistic quality.

9.2   Technical Extensions

As quantum computing hardware becomes more accessible, implementing our system on actual quantum computers would enable genuine quantum advantages. Near-term quantum devices with sufficient qubit counts could execute evolved gate arrangements, potentially revealing behaviors impossible to simulate classically [19].

The grid-genome approach could extend to 3D worlds, adding z-axis gates and particle movements. This would expand expressive possibilities while increasing computational complexity. Alternative encoding schemes beyond binary could represent more complex data structures or artistic parameters [18].

Integration with machine learning could enhance fitness evaluation. Training neural networks to recognize aesthetically pleasing compositions could augment or replace hand-designed fitness functions [7]. Generative adversarial networks could provide discriminative feedback guiding evolution toward desired aesthetic qualities while maintaining computational functionality.

Hybrid classical-quantum architectures leveraging strengths of each paradigm could yield practical implementations. Real-time interaction allowing artists to guide evolution through example selection or parameter adjustment would transform the system from autonomous generator to collaborative creative tool [6,8].

9.3   Evaluation and Comparative Studies

Future research should implement rigorous comparative evaluations: user studies comparing quantum-enhanced vs. classical outputs; quantitative metrics for visual complexity, diversity, and aesthetic appeal; longitudinal studies tracking how evolved systems maintain novelty over extended runs. Such evaluations would provide empirical support for claims of quantum superiority in generative art.

10   Conclusion

This work represents a significant step toward integrating quantum computing principles with generative art and artificial life paradigms [1,11]. By leveraging quantum phenomena such as superposition, entanglement, and teleportation, we create computational ecosystems exhibiting enhanced complexity, unpredictability, and emergent behaviors transcending classical computational limitations [16,17].

The Quantum Particulate Automata framework provides practical implementation of these concepts, demonstrating quantum principles can be applied to artistic generation in meaningful ways. Resulting artworks exhibit distinctive visual characteristics and emergent behaviors impossible in purely classical systems [19,20].

Beyond technical achievements, this work contributes to philosophical and aesthetic discourse about nature of creativity, possibility of artificial life, and role of quantum phenomena in complex systems [15,23]. It exemplifies computational biopoetics where poetic and living worlds intertwine with machines and time, creating aesthetic of emergence, dialogue, and life itself [14].

As quantum computing technology matures, we anticipate quantum-enhanced generative art will evolve from experimental curiosity to established practice. This work provides foundational concepts and frameworks for this evolution, demonstrating that marriage of quantum computing and generative art offers genuinely new creative possibilities rather than mere technical novelty [2,18].

Acknowledgments. This research was supported by ESGI Digital Lab. We thank the anonymous reviewers for their valuable feedback that substantially improved this work.

Disclosure of Interests. The authors have no competing interests to declare that are relevant to the content of this article.

References

1. Lioret, A.: Quantum computing for art exploration and creation. In: ACM SIGGRAPH 2014 Studio, pp. 1–1. ACM, New York (2014)
2. Forbes, A., Bartram, L.: Quantum Art. Accessed 18 Jan 2026
3. Machado, P., Romero, J., Santos, M.L., Cardoso, A., Manaris, B.: Adaptive critics for evolutionary artists. In: Workshops on Applications of Evolutionary Computation, pp. 437–446. Springer, Berlin Heidelberg (2004)
4. Machado, P., Romero, J., Cardoso, A., Santos, A.: Partially interactive evolutionary artists. N. Gener. Comput. **23**(2), 143–155 (2005)
5. Greenfield, G.: Generative art and evolutionary refinement. In: European Conference on the Applications of Evolutionary Computation, pp. 291–300. Springer, Berlin Heidelberg (2010)
6. Romero, J., Machado, P., Santos, A.: On the socialization of evolutionary art. In: Workshops on Applications of Evolutionary Computation, pp. 557–566. Springer, Berlin Heidelberg (2009)
7. Machado, P., Martins, T., Amaro, H., Abreu, P.H.: An interface for fitness function design. In: International Conference on Evolutionary and Biologically Inspired Music and Art, pp. 13–25. Springer, Berlin Heidelberg (2014)
8. Young, J., Colton, S.: Controlling self-organization in generative creative systems. In: International Conference on Computational Intelligence in Music, Sound, Art and Design (Part of EvoStar), pp. 194–209. Springer, Cham (2020)
9. Correia, J., Vieira, L., Rodriguez-Fernandez, N., Romero, J., Machado, P.: Evolving image enhancement pipelines. In: International Conference on Computational Intelligence in Music, Sound, Art and Design (Part of EvoStar), pp. 82–97. Springer, Cham (2021)
10. Martins, T., Correia, J., Costa, E., Machado, P.: Evotype: towards the evolution of type stencils. In: International Conference on Computational Intelligence in Music, Sound, Art and Design, pp. 299–314. Springer, Cham (2018)
11. Lioret, A.: Quantum beings. In: XXV Generative Art Conference, pp. 234–247 (2022). Accessed 18 Jan 2026
12. Lioret, A., Bouizem, J.: Fractal beings. In: XX Generative Art Conference GA, pp. 156–168 (2017). Accessed 18 Jan 2026
13. Ventrella, J.: Evolved gliders and waves on a geodesic grid. In: Adamatzky, A. (ed.) Designing Beauty: The Art of Cellular Automata, pp. 73–74. Springer, Cham (2016)
14. Briggs, J., Peat, F.D.: Turbulent Mirror: An Illustrated Guide to Chaos Theory and the Science of Wholeness. Harper & Row, New York (1989)
15. Seevinck, J.: Emergence in interactive artistic visualization. Int. J. Software Eng. Knowl. Eng. **25**(02), 201–230 (2015)
16. Bhaumik, M.: Deciphering the enigma of wave-particle duality. arXiv preprint arXiv:1611.00226 (2016)
17. Broekaert, J.: The tacit 'quantum' of meeting the aesthetic sign; contextualize, entangle, superpose, collapse or decohere. Found. Sci. **23**(2), 255–266 (2018)
18. Chan, B.W.C.: Lenia-biology of artificial life. arXiv preprint arXiv:1812.05433 (2018). Accessed 18 Jan 2026
19. Rudolph, M.S., Toussaint, N.B., Katabarwa, A., Johri, S., Peropadre, B., Perdomo-Ortiz, A.: Generation of high-resolution handwritten digits with an ion-trap quantum computer. Phys. Rev. X **12**(3), 031010 (2022)

20. Huang, H.L., et al.: Experimental quantum generative adversarial networks for image generation. Phys. Rev. Appl. **16**(2), 024051 (2021)
21. Leong, D.: Action in complexity: entanglement and emergent order in entrepreneurship. J. Entrepreneurship **32**(1), 182–217 (2023)
22. Bilder, R.M., Knudsen, K.S.: Creative cognition and systems biology on the edge of chaos. Front. Psychol. **5**, 1104 (2014)
23. Chiatti, L.: Archetypes, causal description and creativity in natural world. In: Licata, I., Sakaji, A. (eds.) Physics of Emergence and Organization, pp. 405–421. World Scientific, Singapore (2008)
24. Romero, J.J., Machado, P.: The Art of Artificial Evolution: A Handbook on Evolutionary Art and Music. Springer, Berlin Heidelberg (2008)
25. Lewis, M.: Evolutionary visual art and design. In: Romero, J., Machado, P. (eds.) The Art of Artificial Evolution: A Handbook on Evolutionary Art and Music, pp. 3–37. Springer, Berlin Heidelberg (2008)

Segmentation-Free Sound Hybridization for Creative Outcomes

Pasquale Mainolfi[1]([✉])[iD] and Carmine Emanuele Cella[2][iD]

[1] Department of New Technologies and Musical Languages "G. Martucci", State Conservatoire of Music, 84125 Salerno, SA, Italy
`pasquale.mainolfi@consalerno.it`
[2] Center for New Music and Audio Technologies (CNMAT), University of California, Berkeley, CA 94709, USA
`carmine.cella@berkeley.edu`
`https://www.consalerno.it/`, `https://cnmat.berkeley.edu/`

Abstract. Sound hybridization is a technique that combines the characteristics of multiple sounds to generate new sonic textures, with applications ranging from scientific analysis to timbral synthesis and assisted musical composition. In this work, we propose a *matching-pursuit-based* approach for adaptive sound hybridization, which avoids a static segmentation of the target and enables a more flexible and coherent reconstruction. The method iteratively selects sound atoms from a reference database, optimizing the synthesis process in terms of both timbral and temporal-structural coherence. We demonstrate that our approach preserves the structure of the target while introducing controlled variations and analyze the impact of different feature spaces on the quality of the hybridization. The results show that the proposed strategy effectively balances fidelity to the target with an emphasis on the distinctive characteristics of the search space from which the sound objects involved in the hybridization process are selected. Our method prioritizes the aesthetic and creative potential of sound hybridization, offering a tool for artistic exploration.

Keywords: Sound hybridization · Matching pursuit · Timbral synthesis · Audio feature analysis · Adaptive Synthesis · Creative audio processing

1 Introduction

1.1 Exploring the Creative Potential of Sound Hybridization

Sound hybridization, the process of combining characteristics of two or more sounds, offers a rich and fertile domain for creative exploration and the generation of novel sonic textures [29]. Its outcomes span a broad spectrum—from subtle blends of timbral nuance to the emergence of entirely new auditory identities. The expressive potential of hybridization lies not only in the sounds themselves, but in the process of crossing perceptual and structural boundaries, enabling

P. Machado et al. (Eds.): EvoMUSART 2026, LNCS 16523, pp. 303–315, 2026.
https://doi.org/10.1007/978-3-032-24350-8_20

composers and sound designers to probe the liminal spaces between sources. The efficacy and creative potential of hybridization techniques are inextricably tied to the underlying representations of sound. Techniques that rely on low-level signal representations, such as raw waveforms or spectral frames, often provide precision but lack structural awareness. Conversely, more abstract or higher-level representations can capture broader sonic qualities—timbre, gesture, morphology—and enable richer, more musically meaningful transformations. One common strategy, especially in high-level abstraction approaches, is to represent the sound object as a composite of simplified, essential properties—qualities abstracted away from micro-temporal detail. Within this paradigm, sound objects can be manipulated, overlapped, and recombined with strategic intent. Such abstraction facilitates a form of compositional control, offering a macro-level perspective from which to direct the hybridization process.

1.2   The Role and Limits of Segmentation

Segmentation plays a pivotal role in these systems. It defines the elementary units upon which hybridization operates, directly influencing the granularity, adaptability, and expressive capacity of the results. A well-devised segmentation scheme can preserve salient features of the source sounds and enhance the intelligibility of their fusion. Yet, segmentation is not neutral. Even when meticulously designed, it imposes a structure—a grid, a set of cuts—that may or may not align with the perceptual and dynamic contours of the original sounds. For instance, cross-synthesis using a phase vocoder typically operates on static windows of a few milliseconds, which are agnostic to the temporal or morphological structure of the input [19]. While this can yield interesting timbral blends, it often results in hybrid sounds that lack a coherent macro-form or expressive gesture. By contrast, more sophisticated hybridization tools such as AudioGuide or SoundTypes adopt onset-based or perceptually informed segmentation strategies, aligning hybridization units with musically or morphologically meaningful events. These approaches aim to respect the temporal articulation of the source materials, but they too are bounded by the segmentation choices—once the units are defined, their internal continuity is lost, and the recombination is limited to these fixed building blocks [31]. In both approaches—whether based on fixed micro-windows or onset-driven segmentation—the hybridization process is constrained by the segmentation itself, which determines not only the fidelity of the blending but also the kinds of sonic transformations that are possible. The result is often a tension between control and emergence, between imposed structure and the organic evolution of sound.

1.3   A Dynamic, Segmentation-Free Approach

In this paper, we propose a method for sound hybridization that eschews static or pre-defined segmentation altogether. Instead, our approach, built upon an extension of matching pursuit, operates by iteratively selecting and combining optimal sonic components—atoms in a redundant dictionary—directly from the continuous signal. This process is dynamic and adaptive: rather than chopping sound into fixed segments, it allows hybridization to unfold as an emergent structure, guided by both perceptual salience and compositional intent. This segmentation-free approach enables nuanced and context-sensitive blending of both timbral and temporal features, preserving the internal continuity of sound gestures while opening space for unexpected cross-influences between sources. It avoids the rigidity of imposed boundaries and supports a more fluid, responsive fusion of sonic materials. Our method is motivated not by engineering goals such as signal reconstruction or compression, but by a deep interest in the aesthetic potential of computational sound synthesis. We approach hybridization not merely as a technical problem, but as an artistic practice—a space where algorithms participate in creative discovery. Rather than preserving the integrity of source sounds, we are interested in what new and evocative sonic identities can arise from their intersection. This perspective aligns with the broader field of computational creativity, in which algorithmic processes act not only as tools but as collaborators in the aesthetic decision-making process. By emphasizing expressive qualities and resisting rigid structural impositions, our approach opens new pathways for sonic experimentation and compositional play. It supports a hybridization process that is not constrained by pre-defined units, but instead reveals the fluid morphologies and latent continuities that emerge through sound's temporal evolution. In doing so, it contributes to a growing repertoire of algorithmic methods that amplify, rather than delimit, creative possibility.

2   Related Work: State of the Art in Sound Hybridization

The field of sound hybridization encompasses a variety of techniques leveraging different representations of audio signals. This section discusses key approaches to sound hybridization, without aiming to be fully exhaustive, but rather highlighting those most closely aligned with our own perspective.

2.1   Corpus-Based Concatenative Synthesis

Corpus-based concatenative synthesis [13,27,30] creates new sounds by selecting and concatenating small segments (units) from a large database. Hybridization occurs when the corpus contains diverse sound sources, and selection is guided by similarity to a target sound. This approach is closely related to *audio mosaicing* [35], where segments from different sources reconstruct a target's spectral and temporal characteristics.

Concatenative synthesis [16,28,32] can yield hybrid sounds that preserve the microstructure of source materials, but it often lacks continuity and smoothness in the hybridization process. The technique is widely used in musical applications, particularly in sound design and speech synthesis.

2.2   Sound Texture Synthesis

Sound texture synthesis [26] generates new audio sharing perceptual similarities with a given sample. While not always framed as hybridization, it enables the blending of different textures to produce coherent, hybridized soundscapes. By operating at a more abstract level, texture synthesis methods allow for perceptually consistent hybridization, particularly useful in environmental sound design and film scoring.

2.3   Cross-Synthesis Techniques

Various **cross-synthesis** techniques transfer characteristics from one sound to another:

- **Vocoding** [11]: A modulator sound shapes the spectral envelope of a carrier sound.
- **LPC Cross-Synthesis** [22]: Linear Prediction Coding extracts vocal tract parameters from one sound and applies them to another's excitation signal.
- **Spectral Cross-Synthesis** [18]: Phase and amplitude spectra are mixed, or a source-filter relationship is applied.

Beyond these methods, other cross-synthesis techniques have been explored, such as those discussed in [10,15,25,33]. Each of these techniques provides a different level of abstraction in hybridization, allowing for fine control over how much of each source sound is retained in the final output.

2.4   Sound-Types Framework

The **sound-types** framework [4,7] represents sounds at a *quasi-symbolic level* [5,6]. It analyzes a sound by segmenting it into small overlapping chunks (atoms), clustering them based on features, and extracting transition probabilities between clusters. Hybridization occurs by:

- **Sound-type matching**: Replacing or merging types based on feature similarity.
- **Probability merging**: Blending transition probabilities to generate new temporal structures.

This approach allows probabilistic synthesis of novel hybrid sounds by integrating elements of corpus-based synthesis with statistical modeling.

2.5   AudioGuide

AudioGuide [13] performs *deferred-time concatenative synthesis* by analyzing a corpus and a target sound in a feature space. It reconstructs the target by selecting corpus units that best match its features, introducing *morphology-aware* hybridization.

Its subtractive spectral algorithm allows simultaneous selection of multiple corpus units, enabling polyphonic textures and expressive transformations.

3   Our Method: Segmentation-Free Matching Pursuit

3.1   Method Overview

Our approach uses *matching pursuit* [21], an iterative algorithm that selects optimal atoms from a source dictionary to approximate a target sound. This enables structured hybridization where source sounds contribute selectively to different temporal regions of the target.

$$S_n \approx \sum_{i=0}^{N} (\alpha_i g_i)_n \tag{1}$$

As shown in Eq. (1), a target signal S can be approximated as a sparse combination of basis functions g multiplied by the respective projection coefficient α. Here, N represents the total number of atoms selected from the dictionary to approximate the target signal S (the sparsity level of the decomposition).

However, in order for this to be achievable, the target signal must be preliminarily prepared and segmented into a series of frames. This segmentation process is crucial because it directly affects the quality of the approximation and the effectiveness of the hybridization. There are several segmentation techniques, depending on the application and the type of signal.

One of the simplest methodologies is *fixed windowing segmentation*, often with overlap to ensure continuity between segments [1]. An alternative approach is *event-based segmentation* [2,3,17,34], which identifies significant variations in the signal's characteristics, such as changes in energy, timbral transitions, or onset. Another method is *spectral-based segmentation* [12,20,23], which exploits variations in the frequency domain to define the segments.

The core idea behind our approach is to avoid a predetermined segmentation. Instead of imposing rigid boundaries on the target signal, our method involves each atom scanning the signal to find the space-frequency region in which it fits best.

3.2   Mathematical Formulation

Let $\mathcal{D}$ be a set of audio frames (atoms) f and S an arbitrary target. We aim to reconstruct S through a linear combination of frames from $\mathcal{D}$, as described in Eq. (2),

$$\mathcal{Y} = \sum_{n \in \Omega} \alpha_n T_{\phi_n}(f^*(\phi_n)) \tag{2}$$

where Ω defines the discrete range of S, $\mathcal{Y} \approx S$ and $f^*(\phi_n)$ is the atom with the lowest cost at phase point ϕ_n.

The lowest cost identifies the frame that most accurately captures the characteristics of the target signal at a specific phase point calculated as in Eq. (7). $\alpha_n = \alpha(f^*(\phi_n))$ is the amplitude coefficient, ϕ_n is the phase point (in samples) in S and $T_{\phi_n}(f)$ represents the shift operator, which positions the frame f at the phase point ϕ_n within the signal, as defined in Eq. (3),

$$(T_\phi f)_n = \begin{cases} f_{n-\phi}, & \phi \le n \le \min(\phi + N_f - 1, N_s - 1), \\ 0, & otherwise. \end{cases} \tag{3}$$

where N_f is the length of frame and N_s the length of S.

ϕ_n is crucial to the process, as it defines the exact location in the target signal where the atom should be placed. It represents the starting point of the atom within the signal. Its precise determination ensures optimal alignment between the atom and the target signal.

This phase-related parameter is calculated as in Eq. (4),

$$\phi_n = argmax(R \star f_i) - (N_{f_i} - 1) \tag{4}$$

where $(\star)$ denotes the cross-correlation operation, N_{f_i} is the length of the ith frame, and R represents the residual energy of S.

When computing $(R \star f_i)$, the ith atom f_i is slide across R to measure how well it aligns at each position. This operation produces a similarity score for every possibile alignment, allowing us to find the position in R where f_i fits best. By applying Eq. (4), we identify ϕ_n, which represents the optimal position where f_i should be placed in $\mathcal{Y}$. However, the index returned by the $argmax$ requires a correction. This adjustment is necessary when cross-correlation is implemented as a full convolution with a flipped kernel. In the proposed implementation, $(R \star f_i)$ is effectively a full convolution with the flipped kernel [24], as shown in Eq. (5),

$$(R \star f_i')_n = \sum_{m=0}^{N_{f_i}-1} R_m \cdot f_{i,n-m}' \tag{5}$$

where f_i' is the flipped atom.

Equation (5) produces an output of length $N_s + N_{f_i} - 1$ [14,24], where N_s represents the length of R which is equal to the length of S (in samples). The goal is to find the starting index of the segment in R that best matches f_i. For each n, the index n represents the shift of the atom relative to the beginning of the signal. When $n = 0$, f_i' is shifted completely to the left, with its "tail" outside R. Here, the tail refers to the portion of the atom that does not overlap with the target signal at a given shift. Conversely, when $n = N_{f_i} - 1$, the first element of f_i' aligns with the first element of the signal. To obtain the starting index of the segment in R, it is therefore necessary to correct this shift by subtracting the actual length of the atom. In Eq. (4), the start index of the corresponding segment ϕ_n is found by moving backward N_i samples from the maximum index.

The amplitude coefficient, instead, is obtained as shown in Eq. (6).

$$\alpha(f_i(\phi_n)) = \frac{1}{\sqrt{|f_i|^2}} \sum_{m=0}^{N_{f_i}-1} f_{i,m} \cdot R_{\phi_n+m} \tag{6}$$

The fitness function $F(f)$ Eq. (7) evaluates the match quality.

$$F(f_i) = \frac{1}{|f_i|} \sum_{n=0}^{N_s-1} (R \star f_i)_n \tag{7}$$

By iteratively refining $\mathcal{Y}$, our algorithm achieves high-quality hybridization while preserving essential target characteristics. At each epoch, the algorithm searches for the best suboptimal solution, replacing the atom if a better one is found, based on Eq. (7).

4   Evaluation

Our evaluation combines perceptual listening tests and objective spectral analysis:

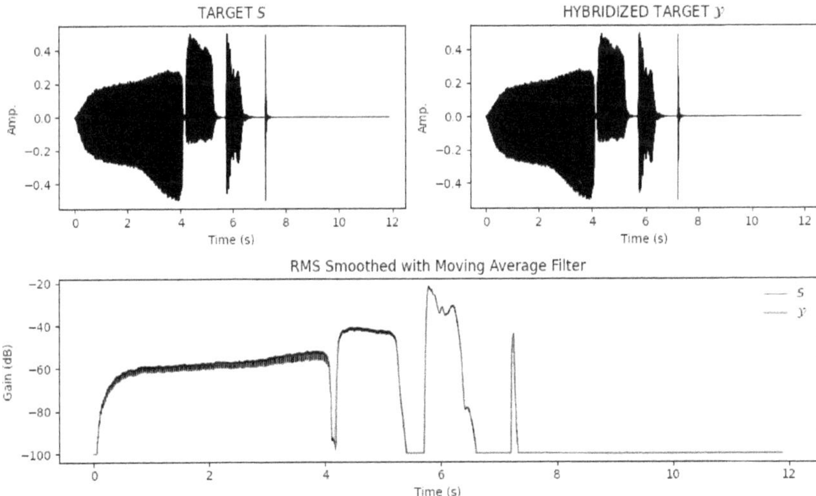

Fig. 1. Comparison of the target signal with the reconstructed signal in the first phase of the experiment. In the top-left plot, the waveform of the target signal is displayed. The top-right plot presents the reconstructed signal. The bottom plot compares the RMS (Root Mean Square) values of both signals over time, smoothed using a moving average filter. These smoothed RMS curves highlight the energy distribution and consistency between the original and reconstructed signals. The test can be heard at https://shorturl.at/odmQD, where the target and its hybridization correspond to the audio files <test-target.wav> and <test-hybrid.wav>, respectively.

- Listening tests assess the clarity of hybridization and preservation of target structure.

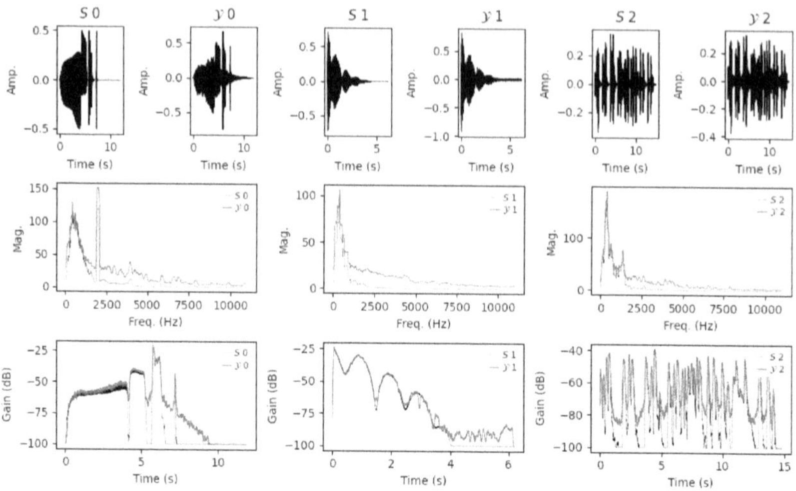

Fig. 2. The plot provides a detailed comparison between the target signal S and the hybridized signal $\mathcal{Y}$, obtained through the hybridization process. The analysis is divided into three vertical sections, each corresponding to a different experiment. For each experiment, the first panel on the left displays the waveform of the target. The second panel shows the waveform of the hybridized signal. The third panel presents the power spectrum of both signals and smoothed using a moving average filter to reduce spectral fluctuations. The target S is represented by a black line, while $\mathcal{Y}$ appears as a red line to emphasize spectral differences. The fourth panel compares the energy evolution of the two signals by plotting the RMS (Root Mean Square) amplitude, also smoothed using a moving average filter and converted to decibels.

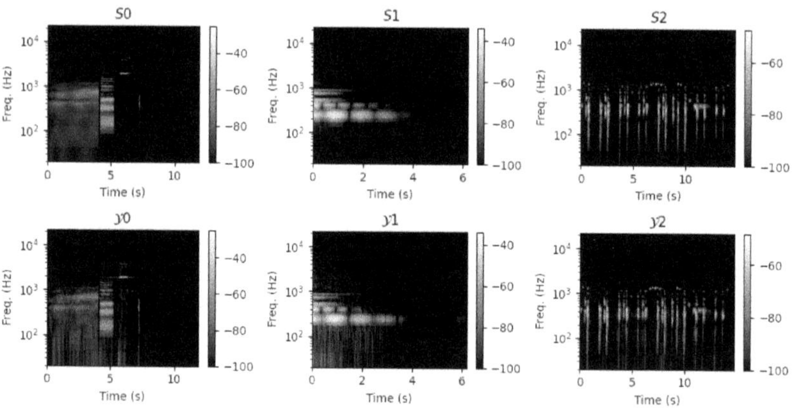

Fig. 3. The plot presents the spectrograms of the three experiments shown in Fig. 2, with frequencies arranged on a logarithmic scale. Each vertical section corresponds to a different experiment. For each experiment, the top row displays the waveform of the target signal S, while the bottom row shows the waveform of the hybridized signal $\mathcal{Y}$.

- Objective metrics include spectral similarity and temporal coherence compared to baseline methods.

We structured the entire testing process into two complementary phases.

The aim of the first phase was to assess how accurately an atom could be placed in a specific section of the target. To achieve this, we combined four distinct events into a single target signal and created a dictionary containing the four atoms as separate units.

The results of this first phase confirmed that the reconstructed signal maintained the correct temporal arrangement of the individual atoms relative to their phase. Figure 1 clearly shows how the energy trend (bottom plot) of the two signals is almost perfectly overlapping.

During the second phase of experimentation, we instead investigated the hybridization process in different contexts. Specifically, we focused on the controlled segmentation of the reference dataset, in our case *OrchideaSOL* [9], which was further augmented so that each frame in the dataset generated a certain number of subatoms.

To formalize this process, we consider the segmentation of frames into sub-segments of varying durations.

Let L_{min} be the minimum length in samples of a sub-segment. For each frame f_i, there will exist a family of portions L_j of the same frame such that $L_j \geq L_{min}$.

$$L_j = \frac{L_{j-1}}{2}, \qquad L_0 = N_i, \quad j = 0, 1, 2, ... N_L \tag{8}$$

with N_L being the maximum number of possible lengths, determined by the condition in the Eq. (9).

$$N_L = max\left\{j | \frac{N_i}{2^j} \geq L_{min}\right\} \tag{9}$$

The tests conducted reveal that the aesthetic-structural nature of the discussed hybridization process is closely related to the duration of the elements in the dataset.

Atoms of shorter duration contribute to a more accurate reconstruction of the target, both from a spectromorphological perspective and in terms of time coherence, and consequently lead to a progressive attenuation of the timbral dynamics and the temporal progression of the frames in the dataset. Conversely, while introducing controlled variations in both the spectromorphological and temporal aspects, pursuit-based hybridization achieves a balance between structural preservation and timbral blending.

Furthermore, we observed that the result remained equally satisfactory even when adopting a mini-batch approach, without the need to process the entire dataset at each epoch. Specifically, incremental parameter updates based on subsets of the dataset proved effective in maintaining the quality of the hybridization process. Local optimization on small data groups is sufficient to converge toward a stable solution that is consistent with the target structure.

In Fig. 2, the result of three experiments using the *OrchideaSOL* dataset [9] is shown. This allows for an evaluation of how closely $\mathcal{Y}$ follows the spectral and temporal dynamics of S, with all variations specifically introduced by the hybridization process.

The numerical analysis supports this visual observation: the cosine distances between the target and hybridized signals computed in frequency domain on power spectra, are 0.0405 for $S0$ hybridized in $\mathcal{Y}0$, 0.0234 for $S1$ hybridized in $\mathcal{Y}1$, and 0.0471 for $S2$ hybridized in $\mathcal{Y}2$ (the labels correspond to the signals shown in Fig. 2), while the standard deviations of the power spectra are 0.00032 ($S0$, $\mathcal{Y}0$), 0.00033 ($S1$, $\mathcal{Y}1$), 0.00028 ($S2$, $\mathcal{Y}2$). The analysis was performed using a Hann window of length 1024 samples with an overlap of 256 samples.

These values indicate a high spectral similarity and confirm that the hybridized signals effectively reconstruct and follow the underlying structure of the targets, as also evidenced by the spectrograms in Fig. 3.

For the experiments, the mini-batch size was 3000 atoms per epoch, with a total of five epochs. From epoch to epoch, the atoms used were removed from the search space.

The experiments in Fig. 2 can be heard at https://shorturl.at/odmQD.

Specifically, $S0$ and $\mathcal{Y}0$ correspond to the audio files *<test-target.wav>* and *<test-hybrid-OrchideaSOL.wav>*, respectively. Similarly, $S1$ and $\mathcal{Y}1$ correspond to *<archeosBell-target.wav>* and *<archeosBell-hybrid.wav>*, while $S2$ and $\mathcal{Y}2$ correspond to the audio files *<piano-target.wav>* and *<piano-hybrid.wav>*.

The presented examples, along with other obtained results, can be listened to at the following address. In particular, the files corresponding to the targets are named according to the convention *<nome-target.wav>*, while the hybridization results are labeled as *<nome-hybrid.wav>*.

5   Future Work

In our approach, we employed inner products central to the matching pursuit algorithm as a similarity measure for hybridization. While effective, this strategy is primarily signal-based and does not explicitly account for perceptual relevance. Future developments will explore perceptually informed similarity measures, such as those based on auditory filterbanks, spectral centroid trajectories, or psychoacoustic models like loudness and sharpness.

We also plan to investigate *non greedy* optimization techniques to improve the robustness and diversity of hybridization results. In particular, we are interested in *stochastic matching pursuit*, as introduced in [8], which offers a probabilistic alternative to purely greedy atom selection, potentially avoiding local optima and enhancing creative variability.

Finally, we are currently working on improved evaluation metrics that better capture the perceptual and structural quality of the hybridization, allowing for more extensive and meaningful testing of our algorithm.

6   Conclusion

This paper highlights the importance of *segmentation-free* in sound hybridization, reviews existing methods, and introduces a novel approach using matching pursuit.

The proposed approach has proven to be effective in creating hybrid representations of reference signals, through the fusion of distinct sound identities into a single coherent representation that preserves the temporal and dynamic structure of the reference target, while spectrally-consistent in reworking its timbral characteristics. The adaptive atom placement methodology eliminates the need for rigid pre-defined segmentation. The iterative logic of the process and the mini-batch approach progressively improve the quality of the reconstruction, reducing the computational load without compromising the quality of the result.

References

1. Allen, J.B., Rabiner, L.R.: A unified approach to short-time Fourier analysis and synthesis. Proc. IEEE **65**(11), 1558–1564 (1977)
2. Bello, J.P., Daudet, L., Abdallah, S., Duxbury, C., Davies, M., Sandler, M.B.: A tutorial on onset detection in music signals. IEEE Trans. Speech Audio Process. **13**(5), 1035–1047 (2005)
3. Bello, J.P., Pickens, J.: A robust mid-level representation for harmonic content in music signals. In: ISMIR, vol. 5, pp. 304–311 (2005)
4. Burred, J.J.: An approach to sound hybridization using the theory of sound-types. In: Proceedings of ICMC. Montreal, Canada (2009)
5. Cella, C.E.: Towards a symbolic approach to sound analysis. In: Proceedings of the Second International Conference on Mathematics and Computation for Music. Yale University (2009)
6. Cella, C.E.: On Symbolic Representations of Music. University of Bologna, Ph.D. thesis (2011)
7. Cella, C.E.: Sound-types: A new framework for symbolic sound analysis and synthesis. In: Proceedings of ICMC. Huddersfield, UK (2011)
8. Cella, C.E.: Orchidea: a comprehensive framework for target-based computer-assisted dynamic orchestration. J. New Music Res. **51**(1), 40–68 (2022)
9. Cella, C.E., Ghisi, D., Lostanlen, V., Lévy, F., Fineberg, J., Maresz, Y.: Orchideasol: a dataset of extended instrumental techniques for computer-aided orchestration. arXiv preprint arXiv:2007.00763 (2020)
10. Collins, N., Sturm, B.L.: Sound cross-synthesis and morphing using dictionary-based methods. In: Proceedings of the International Computer Music Conference. ICMA (2011)
11. Flanagan, J.L., Golden, R.M.: Phase vocoder. Bell Syst. Technical J. **45**(9), 1493–1509 (1966)
12. Gupta, S., Jaafar, J., Ahmad, W.W., Bansal, A.: Feature extraction using MFCC. Signal Image Process. Int. J. **4**(4), 101–108 (2013)
13. Hackbarth, B., Schnell, N., Schwarz, D.: Audioguide: a framework for creative exploration of concatenative sound synthesis. Research report, IRCAM, Paris, France (2010)

14. Holton, T.: Digital Signal Processing: Principles and Applications. Cambridge University Press (2021)
15. Hsu, P.C., Wang, C.H., Liu, A.T., Lee, H.Y.: Towards robust neural vocoding for speech generation: a survey. arXiv preprint arXiv:1912.02461 (2019)
16. Julius, A.: Efficient audio representations for concatenative sound synthesis. Computer Music Journal **31**(4) (2007)
17. Kong, Q., Xu, Y., Sobieraj, I., Wang, W., Plumbley, M.D.: Sound event detection and time-frequency segmentation from weakly labelled data. IEEE/ACM Trans. Audio Speech Lang. Process. **27**(4), 777–787 (2019)
18. Laroche, J., Dolson, M.: New phase-vocoder techniques for real-time pitch shifting, chorusing, harmonizing, and other exotic audio modifications. J. Audio Eng. Soc. **47** (1999)
19. Laroche, J., Dolson, M.: Improved phase vocoder time-scale modification of audio. IEEE Trans. Speech Audio Process. **7**(3), 323–332 (1999). https://doi.org/10.1109/89.759041
20. Lu, L., Jiang, H., Zhang, H.: A robust audio classification and segmentation method. In: Proceedings of the Ninth ACM international conference on Multimedia, pp. 203–211 (2001)
21. Mallat, S.G., Zhang, Z.: Matching pursuits with time-frequency dictionaries. IEEE Trans. Signal Process. **41**(12), 3397–3415 (1993)
22. Moorer, J.A.: The use of linear prediction of speech in computer music applications. In: Audio Engineering Society Convention 59. Audio Engineering Society (1978)
23. Pertusa, A., Klapuri, A., Inesta, J.M.: Recognition of note onsets in digital music using semitone bands. In: Sanfeliu, A., Cortés, M.L. (eds) CIARP 2005. LNCS, vol. 3773 pp. 869–879. Springer, Heidelberg (2005). https://doi.org/10.1007/11578079_90
24. Proakis, J.G.: Digital Signal Processing: Principles Algorithms and Applications. Pearson Education India (2001)
25. Roma, G., Green, O., Tremblay, P.A.: Audio morphing using matrix decomposition and optimal transport. In: Proceedings of DAFx (2020)
26. Schwarz, D.: State of the art in sound texture synthesis. In: Proceedings of DAFX. Paris, France (2011)
27. Schwarz, D., Cahen, R., Britton, S.: Principles and applications of interactive corpus-based concatenative synthesis. In: Proceedings of Journées d'Informatique Musicale. Albi, France (2008)
28. Schwarz, D.: Concatenative sound synthesis: the early years. J. New Music Res. **35**(1), 3–22 (2006)
29. Schwarz, D.: Corpus-based concatenative synthesis. IEEE Signal Process. Mag. **24**(2), 92–104 (2007). https://doi.org/10.1109/MSP.2007.329579
30. Schwarz, D., Beller, G., Verbrugghe, B., Britton, S.: Real-time corpus-based concatenative synthesis with catart. In: 9th International Conference on Digital Audio Effects (DAFx), pp. 279–282 (2006)
31. Smith, N., Lewin, K.: Using matching pursuit for audio time-scale modification. In: Proceedings of the 2002 IEEE International Conference on Acoustics, Speech, and Signal Processing (ICASSP) (2002). https://doi.org/10.1109/ICASSP.2002.5744085
32. Sturm, B.L.: Adaptive concatenative sound synthesis and its application to micromontage compositior. Comput. Music. J. **30**(4), 46–66 (2006)
33. Verfaille, V., Zolzer, U., Arfib, D.: Adaptive digital audio effects (a-dafx): a new class of sound transformations. IEEE Trans. Audio Speech Lang. Process. **14**(5) (2006)

34. Virtanen, T., Plumbley, M.D., Ellis, D.: Computational Analysis of Sound Scenes and Events, vol. 9. Springer, Cham (2018). https://doi.org/10.1007/978-3-319-63450-0
35. Zils, A., Pachet, F.: Musical mosaicing. In: Digital Audio Effects (DAFx), vol. 2, p. 135 (2001)

Quantum Latent Spaces for Symbolic Music Generation

Sanjay Majumder[✉][iD] and Neal Anderson

Style Tree, Indianapolis, IN, USA
{sanjay,neal}@styletree.me

Abstract. Recent advances in quantum computing have opened new possibilities for modeling high-dimensional and entangled structures in musical data. This paper introduces a Quantum Variational Autoencoder (QVAE) architecture for symbolic music generation, leveraging quantum latent spaces to capture superposed and correlated melodic patterns beyond the expressive capacity of classical generative models. By encoding MIDI-based musical sequences into parameterized quantum circuits, the proposed framework enabled the exploration of superposed latent states, facilitating multiple potential melodic continuations within a unified probabilistic representation. The study evaluated QVAE against conventional Variational Autoencoders using both objective metrics, including latent diversity, reconstruction accuracy, and sequence entropy, and subjective listening tests assessing musical coherence and creativity. Experimental results demonstrated that QVAE achieved enhanced diversity in generated melodies while maintaining stylistic consistency, suggesting that quantum-inspired latent modeling can serve as a novel pathway for computational creativity in music information retrieval. Beyond generation, the work discussed the implications of quantum latent structures for music representation learning, proposing a foundation for future hybrid quantum-classical MIR frameworks capable of handling musical complexity through the principles of superposition and entanglement.

Keywords: Quantum Variational Autoencoder Quantum-classical generative model · Quantum Latent Spaces · Symbolic Music Generation

1 Introduction

Music Information Retrieval (MIR) research has increasingly intersected with generative modeling, aiming to synthesize music that embodies structural coherence, stylistic authenticity, and creative novelty. Deep generative models such as Variational Autoencoders (VAEs), Generative Adversarial Networks (GANs), and Transformers have demonstrated substantial progress in learning complex musical representations [4,5,12]. However, despite these advances, most classical models rely on deterministic latent spaces, where each latent vector represents a single, concrete point in the learned manifold. Such representations inherently

P. Machado et al. (Eds.): EvoMUSART 2026, LNCS 16523, pp. 316–330, 2026.
https://doi.org/10.1007/978-3-032-24350-8_21

limit the model's ability to capture and manipulate musical ambiguity—a central characteristic of creative human composition.

In contrast, quantum computing provides a fundamentally different mathematical substrate for data representation, based on the principles of superposition and entanglement [1,14]. Quantum states can encode exponentially many configurations simultaneously, suggesting a new paradigm for generative modeling in domains characterized by high structural and temporal complexity—such as symbolic music. By leveraging parameterized quantum circuits (PQCs), latent representations can exist as quantum superpositions of multiple musical possibilities, enabling a richer exploration of melodic and harmonic continuations within a probabilistic framework.

This paper introduces the Quantum Variational Autoencoder (QVAE), a hybrid quantum-classical generative model that employs PQCs to represent and manipulate latent spaces for symbolic music generation. In the proposed approach, symbolic sequences (e.g., MIDI data) were encoded via a classical neural network encoder into variational parameters that define a quantum circuit. The measurement results of the quantum circuit serve as stochastic latent variables, which are decoded back into symbolic sequences through a classical neural decoder. This hybrid architecture enables the model to explore latent manifolds in superposed states, offering an interpretable and physically grounded mechanism for creative musical variation.

The study examines how quantum-enabled latent spaces influence the diversity, coherence, and structural characteristics of generated music. Particular emphasis is placed on understanding how musical attributes, such as motif evolution, harmonic deviation, and rhythmic subtleties, are shaped by the probabilistic nature of quantum representations and the entanglement patterns produced by PQCs. The broader implications for computational creativity are also considered, especially regarding how hybrid quantum-classical systems may expand generative capabilities beyond what classical latent architectures can provide.

Experimental evaluations were conducted on standard symbolic music datasets, measuring both objective performance metrics—including reconstruction accuracy, entropy, and latent diversity—and subjective evaluations through structured listening tests. The results indicated that QVAE generated musically coherent yet diverse compositions, highlighting the potential of quantum-inspired methods to advance generative MIR.

By integrating quantum information principles with musical representation learning, this work contributed to an emerging discourse on Quantum MIR (QMIR)—a research direction that redefines how music can be represented, generated, and understood through the lens of quantum computation. The proposed QVAE thus not only advances the state of music generation, but also lays conceptual groundwork for a new computational paradigm in musicology, creativity, and artificial intelligence.

2 Related Work

The present work intersects multiple streams of research: (1) deep generative modeling for symbolic music, (2) latent-space representation in music, (3) hybrid quantum–classical machine learning models, and (4) computational creativity in music. In the following, we summarize the key prior art and position our contribution.

Generative models for symbolic music have advanced markedly in recent years. For example, Huang et al. [8] adapted the Transformer architecture to generate minute-long musical sequences, introducing a relative attention mechanism that reduces memory complexity from $\mathcal{O}(L^2)$ to $\mathcal{O}(L)$ for long sequences of length L. In parallel, the *MusicVAE* work [12] employed a hierarchical latent-vector model for music that emphasized long-term structure and improved sampling, interpolation, and reconstruction in symbolic datasets. These works demonstrate the value of latent variable models and architecture modifications for capturing musical structure, a foundation on which our quantum latent space approach is based.

Representation learning and latent space modeling are increasingly important in MIR. Roberts et al. [12] extended MusicVAE to multitrack measures, enabling latent-space interpolation and chord conditioning. Other works, such as [3], explored Transformer autoencoders for style representation in music. These efforts highlight the potential of compact latent representations for musical control and generation. Our quantum latent space contributes a novel mathematical substrate for such representations, using superposition and entanglement.

In the machine learning community, hybrid quantum–classical models have gained traction under the paradigm of parameterized quantum circuits (PQCs). Benedetti et al. [1] provided a systematic foundation for treating PQCs as trainable models in supervised and generative tasks. Another line of work addresses generative modeling with quantum circuits, e.g., shallow circuit learning for cat states. These works provide theoretical and empirical grounding for our proposal to embed symbolic music latent variables in quantum latent spaces.

Beyond pure generation, research on human–AI co-creativity, representation of musical semantics, and user-centric interfaces is germane. For example, Roberts et al. [12] also explored interactive latent "palettes" of musical sequences and timbres. More broadly, music information retrieval has considered ethical, interpretability, and human-centered angles in recommendation, generation, and representation tasks. Our work contributes to this dimension by proposing quantum-latent representations that may enrich creative tools and human–AI interaction.

Although previous studies have modeled latent spaces in symbolic music and advanced generative architectures, none have (to our knowledge) exploited **quantum latent representations** in symbolic music generation. By designing a hybrid quantum-classical variational autoencoder (QVAE) that encodes symbolic MIDI sequences into quantum superposed latent states and decodes them back into music, we aim to (i) enhance latent diversity and creative variation, (ii) provide a novel formalism grounded in quantum information theory, and (iii) open up a new paradigm for computational creativity in MIR. The prior works detailed above delineate the terrain in which our proposal sits and underscore its novelty.

3 System Overview

Figure 1 provides a conceptual illustration of the proposed Quantum Variational Autoencoder (QVAE). The architecture integrates a classical encoder–decoder with a parameterized quantum latent space. Given an input symbolic sequence $\mathbf{x}$, the encoder q_ϕ maps it to variational parameters $(\boldsymbol{\mu}, \boldsymbol{\sigma})$, which define the quantum circuit rotation angles. The parameterized circuit $U(\boldsymbol{\theta})$ produces a latent quantum state $|\psi(\boldsymbol{\theta})\rangle$, whose measurement outcomes $\mathbf{z}$ condition the decoder p_θ to reconstruct or generate symbolic sequences. Unlike classical autoencoders that map inputs to deterministic points in a continuous manifold, this quantum formulation allows latent states to exist in superposition. This provides a mathematical framework for capturing musical ambiguity, where a single latent code can probabilistically map to multiple, structurally valid melodic continuations. This hybrid quantum–classical design enables quantum superposition to enrich latent representations, while classical networks handle data preprocessing and sequence generation.

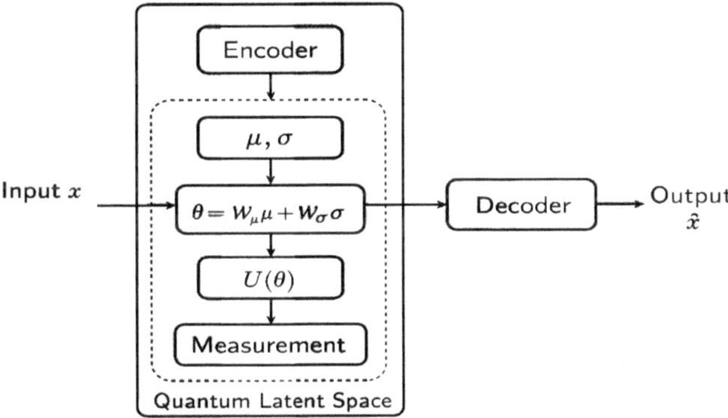

Fig. 1. System overview of proposed QVAE architecture integrating classical and quantum components.

4 Methodology

This section details the architecture, parameterization, and training of the proposed QVAE for symbolic music generation.

4.1 Symbolic Data Representation

Symbolic music sequences were represented using event-based tokenization following [8,12]. Each event belonged to one of four categories: NOTE_ON, NOTE_OFF, TIME_SHIFT, or VELOCITY. Events were one-hot encoded and embedded into a $d_e = 256$-dimensional continuous space. Sequences were truncated or padded to a fixed length $L = 512$ tokens. This representation preserved rhythmic and pitch relationships while remaining compatible with both classical and quantum modules.

4.2 Quantum Latent Space Encoding

The quantum latent state was prepared by a parameterized quantum circuit (PQC):

$$|\psi(\boldsymbol{\theta})\rangle = U(\boldsymbol{\theta})|0\rangle^{\otimes n}, \tag{1}$$

where $U(\boldsymbol{\theta})$ comprises alternating layers of single-qubit rotations and entangling gates. Specifically, for $n = 8$ qubits and circuit depth $d = 3$, each layer ℓ applied $R_Y(\theta_{q,\ell})$ rotations followed by ring-entangled CNOT gates:

$$U(\boldsymbol{\theta}) = \prod_{\ell=1}^{d} \left(\prod_{q=1}^{n} R_Y(\theta_{q,\ell}) \right) \left(\prod_{q=1}^{n-1} \text{CNOT}(q, q+1) \right) \text{CNOT}(n, 1). \tag{2}$$

This hardware-efficient ansatz is designed to maximize expressivity with minimal gate depth. The ring topology of CNOT gates creates a multipartite entangled state capable of capturing complex correlations between features, while limiting the depth to $d = 3$ mitigates noise accumulation and simulation overhead [1]. This balance is crucial for maintaining gradient stability during hybrid optimization.

Parameter Mapping. The encoder's outputs $\boldsymbol{\mu}, \boldsymbol{\sigma} \in \mathbb{R}^m$, where $m = n \times d$, were transformed into circuit parameters through an affine map:

$$\boldsymbol{\theta} = W_\mu \boldsymbol{\mu} + W_\sigma \boldsymbol{\sigma} + \mathbf{b}, \tag{3}$$

where $W_\mu = I_m$, $W_\sigma = \lambda I_m$ with $\lambda = 0.1$, and $\mathbf{b}$ is a trainable bias vector. This direct linear mapping provided smooth, differentiable coupling between classical and quantum parameter spaces.

4.3 Latent Representation and Decoder Conditioning

After executing $U(\boldsymbol{\theta})$ on the quantum simulator, measurement outcomes were sampled in the computational basis:

$$p_{\boldsymbol{\theta}}(\mathbf{b}) = |\langle \mathbf{b}|\psi(\boldsymbol{\theta})\rangle|^2, \quad \mathbf{b} \in \{0, 1\}^n. \tag{4}$$

Each circuit was measured with $S = 1024$ shots, yielding a histogram $p(\mathbf{b})$. To obtain a continuous latent vector, qubit-wise marginals were computed as

$$e_q = \sum_{\mathbf{b}:b_q=1} p(\mathbf{b}), \quad \mathbf{e} = [e_1, \ldots, e_n]^\top. \tag{5}$$

A linear projection $W_e \in \mathbb{R}^{d_e \times n}$ mapped $\mathbf{e}$ into the decoder embedding space:

$$\mathbf{z} = W_e \mathbf{e}. \tag{6}$$

The decoder $p_\theta(\mathbf{x}|\mathbf{z})$ was implemented as a two-layer bidirectional Transformer (for training) and a standard autoregressive Transformer (for inference), ensuring that generated events are conditioned on the latent projection $\mathbf{z}$. The reconstruction objective was the expected categorical cross-entropy:

$$\mathcal{L}_{\text{rec}} = -\mathbb{E}_{q_\phi(\mathbf{z}|\mathbf{x})}[\log p_\theta(\mathbf{x}|\mathbf{z})]. \tag{7}$$

4.4   Quantum Variational Objective

To regularize the latent space, we introduced a quantum analog of the VAE objective:

$$\mathcal{L}_{\text{QVAE}} = \mathcal{L}_{\text{rec}} + \beta D_{\text{KL}}(\rho_\phi \| \rho_0), \tag{8}$$

where $\rho_\phi = |\psi(\boldsymbol{\theta})\rangle\langle\psi(\boldsymbol{\theta})|$ and $\rho_0 = I/2^n$ is the maximally mixed reference state. The quantum KL divergence is given by

$$D_{\text{KL}}(\rho_\phi \| \rho_0) = -S(\rho_\phi) + \log 2^n, \tag{9}$$

with von Neumann entropy $S(\rho_\phi) = -\text{Tr}(\rho_\phi \log \rho_\phi)$. Thus, the objective effectively maximized $S(\rho_\phi)$, promoting mixedness in the latent states:

$$\mathcal{L} = \mathcal{L}_{\text{rec}} + \beta(n - S(\rho_\phi)). \tag{10}$$

Due to the computational complexity of trace-based density matrix operations during backpropagation, our implementation utilizes a differentiable proxy for $S(\rho_\phi)$ defined as the mean of the variational standard deviations, effectively regularizing the spread of circuit rotation angles.

4.5   Gradient Estimation and Optimization

Gradients of circuit parameters were obtained via the parameter-shift rule:

$$\frac{\partial \mathcal{L}}{\partial \theta_j} = \frac{\mathcal{L}(\theta_j + \frac{\pi}{2}) - \mathcal{L}(\theta_j - \frac{\pi}{2})}{2}. \tag{11}$$

Gradients propagated through the affine mapping using the chain rule to update encoder parameters ϕ. All classical components (encoder and decoder) were optimized using Adam with learning rate $\eta = 10^{-4}$, $\beta_1 = 0.9$, $\beta_2 = 0.999$, and $\epsilon = 10^{-8}$.

Algorithm 1. Hybrid QVAE Training Loop

Require: Dataset $\mathcal{D}$, encoder parameters ϕ, decoder parameters θ_c, qubits n, layers d, shots S, epochs E

1: **for** epoch $= 1$ to E **do**
2:    **for** batch $\{x^{(i)}\}_{i=1}^{B} \subset \mathcal{D}$ **do**
3:      $(\mu^{(i)}, \sigma^{(i)}) \leftarrow \text{Encoder}_\phi(x^{(i)})$
4:      $\theta^{(i)} \leftarrow W_\mu \mu^{(i)} + W_\sigma \sigma^{(i)} + \mathbf{b}$
5:      Execute $U(\theta^{(i)})$ with S shots $\rightarrow$ histogram $p^{(i)}(\mathbf{b})$
6:      Compute marginals $e_q^{(i)} = \sum_{\mathbf{b}:b_q=1} p^{(i)}(\mathbf{b})$
7:      $\mathbf{z}^{(i)} = W_e \mathbf{e}^{(i)}$
8:      $\mathcal{L}_{\text{rec}} \leftarrow \text{CE}(x^{(i)}, \text{Decoder}_{\theta_c}(\mathbf{z}^{(i)}))$
9:      $\mathcal{L}_{\text{KL}} \leftarrow n - S(\rho_\phi^{(i)})$
10:     $\mathcal{L} = \mathcal{L}_{\text{rec}} + \beta \mathcal{L}_{\text{KL}}$
11:     Compute $\nabla_{\phi,\theta_c}\mathcal{L}$ (autodiff) and $\nabla_\theta \mathcal{L}$ (parameter-shift)
12:     Update all parameters with Adam
13:    **end for**
14: **end for**

4.6 Hybrid Training Procedure

Training followed a hybrid quantum–classical loop summarized in Algorithm 1.

4.7 Implementation Details

All experiments used $n = 8$ qubits, circuit depth $d = 3$, batch size $B = 32$, and learning rate $\eta = 10^{-4}$. Each quantum circuit was evaluated using $S = 1024$ measurement shots. Training ran for 30 epochs across 5 random seeds $\{0, 1, 2, 3, 4\}$. Qiskit Aer's `statevector_simulator` was used to compute von Neumann entropy exactly, while the `qasm_simulator` provided measurement-based decoding. Although Aer executes on CPU, all classical modules were trained on an NVIDIA RTX 5060Ti GPU with mixed precision for efficiency.

Datasets included MAESTRO [6], JSB Chorales [2], and the Lakh MIDI data set [11], preprocessed to 4/4 time and transposed to C major/minor. Each epoch processed 10,000 sequences. On the described hardware, convergence occurred within 30 epochs with an average per-batch latency of 0.25 s, dominated by circuit simulation.

5 Evaluation Metrics and Analysis

To rigorously evaluate the proposed QVAE for symbolic music generation, we employed a combination of objective and subjective metrics that captured both reconstruction fidelity and generative diversity. The evaluation design followed established MIR practices for generative sequence models [8, 12] and incorporated quantum information-theoretic measures to characterize latent-space expressivity.

5.1 Reconstruction Accuracy

Reconstruction accuracy quantified the model's ability to faithfully reproduce input symbolic sequences from their latent representations. Given a ground-truth sequence $\mathbf{x} = (x_1, \ldots, x_L)$ and its reconstruction $\hat{\mathbf{x}} = (\hat{x}_1, \ldots, \hat{x}_L)$, categorical accuracy was computed as:

$$A_{\mathrm{rec}} = \frac{1}{L} \sum_{t=1}^{L} \mathbb{I}(x_t = \hat{x}_t), \tag{12}$$

where $\mathbb{I}(\cdot)$ denotes the indicator function. The validation mean was obtained as

$$\bar{A}_{\mathrm{rec}} = \frac{1}{|\mathcal{D}_{\mathrm{val}}|} \sum_{\mathbf{x} \in \mathcal{D}_{\mathrm{val}}} A_{\mathrm{rec}}(\mathbf{x}), \tag{13}$$

where $\mathcal{D}_{\mathrm{val}}$ denotes the held-out validation set. Higher $\bar{A}_{\mathrm{rec}}$ reflected stronger fidelity between symbolic and quantum latent domains.

5.2 Latent Diversity via Quantum Shannon and Von Neumann Entropy

Latent diversity evaluated the expressive capacity of the quantum latent space. Given the measurement distribution $p_\theta(\mathbf{b}_k)$ obtained from the parameterized quantum circuit $U(\boldsymbol{\theta})$, we defined the quantum Shannon entropy:

$$H_{\mathrm{latent}} = -\sum_{k=1}^{2^n} p_\theta(\mathbf{b}_k) \log_2 p_\theta(\mathbf{b}_k). \tag{14}$$

For a maximally mixed n-qubit state, $H_{\mathrm{latent}} = n$ bits. Lower entropy indicated structured or concentrated latent representations, while higher entropy indicated broader superposition and diversity.

To complement this, we computed the von Neumann entropy of the encoded state ρ_ϕ:

$$S(\rho_\phi) = -\mathrm{Tr}(\rho_\phi \log_2 \rho_\phi), \tag{15}$$

providing a continuous measure of quantum-state purity. The relation $S(\rho_\phi) \leq H(p)$, where $H(p)$ is the classical Shannon entropy of the measurement outcomes [10], ensured consistency between theoretical and empirical entropy measures. Equality held when ρ_ϕ was diagonal in the measurement basis, making H_{latent} an upper bound estimate for $S(\rho_\phi)$.

5.3 Sequence Entropy and Musical Variability

Sequence-level entropy measured the uncertainty and variability of generated symbolic sequences. For each time step t and event class i, the predictive entropy was defined as:

$$H_{\mathrm{seq}} = -\frac{1}{L} \sum_{t=1}^{L} \sum_{i=1}^{C} p_\theta(x_t = i \mid \mathbf{z}) \log_2 p_\theta(x_t = i \mid \mathbf{z}), \tag{16}$$

where C is the number of symbolic event classes (e.g., `NOTE_ON`, `NOTE_OFF`, `TIME_SHIFT`, `VELOCITY`). Higher H_{seq} implied more creative or exploratory outputs, while lower values implied deterministic repetition. Optimal generation balanced H_{seq} and $\bar{A}_{\text{rec}}$ to achieve both coherence and novelty.

5.4   Latent–Output Mutual Information

To quantify the dependency between latent codes and generated outputs, we estimated the mutual information (MI) between the latent distribution $\mathbf{z}$ and the reconstructed sequence $\mathbf{x}$:

$$I(\mathbf{x}; \mathbf{z}) = H(\mathbf{x}) - H(\mathbf{x} \mid \mathbf{z}), \tag{17}$$

approximated using Monte Carlo sampling over the validation set. Larger $I(\mathbf{x}; \mathbf{z})$ values indicated that the latent variables captured musically meaningful structure rather than random noise, paralleling disentanglement metrics from β-VAE frameworks [7].

5.5   Subjective Listening Evaluation

Perceptual quality was assessed through a listening test with $N = 15$ expert participants (comprising professional musicians and musicology graduate students with an average of 4 years of formal training). Each listener rated 20-second excerpts on a five-point ITU-R BS.1284 [9] scale across three criteria:

1. **Musical Coherence:** Harmonic and structural consistency.
2. **Creativity:** Perceived novelty and thematic evolution.
3. **Aesthetic Appeal:** Subjective enjoyment and expressiveness.

The mean opinion score (MOS) for criterion i was computed as:

$$\text{MOS}_i = \frac{1}{NM} \sum_{j=1}^{N} \sum_{k=1}^{M} r_{i,jk}, \tag{18}$$

where $r_{i,jk}$ is the rating of listener j for sample k, and M is the number of evaluated samples. The global perceptual quality index was

$$\text{MOS}_{\text{avg}} = \frac{1}{3} \sum_{i=1}^{3} \text{MOS}_i. \tag{19}$$

5.6   Composite Evaluation Index (CEI)

To consolidate diverse metrics into a unified measure, we defined a normalized Composite Evaluation Index:

$$\text{CEI} = \alpha_1 \tilde{A}_{\text{rec}} + \alpha_2 \tilde{H}_{\text{latent}} + \alpha_3 \tilde{H}_{\text{seq}} + \alpha_4 \tilde{I}_{\mathbf{x};\mathbf{z}} + \alpha_5 \tilde{\text{MOS}}_{\text{avg}}, \tag{20}$$

where each $\tilde{}$ term denotes min–max normalization to $[0, 1]$, and $\sum_i \alpha_i = 1$. Empirically, balanced weighting was achieved with $\alpha_1 = 0.25$, $\alpha_2 = 0.20$, $\alpha_3 = 0.20$, $\alpha_4 = 0.15$, and $\alpha_5 = 0.20$. CEI provided a unified measure of technical accuracy and perceived quality.

The proposed evaluation framework, combining entropy, information-theoretic, and human-centered metrics, offered a comprehensive validation strategy consistent with both MIR and quantum generative modeling standards.

6 Experimental Results and Evaluation

6.1 Experimental Setup

We evaluated the proposed QVAE against a classical baseline Variational Autoencoder (VAE) with comparable model capacity. Both models were trained on a combined symbolic dataset consisting of MAESTRO (v3.0) [6], JSB Chorales [2], and the Lakh MIDI data set [11], preprocessed as described in Sect. 4.7.

Model Configurations

- **Baseline VAE:** Hierarchical encoder with latent dimension $z \in \mathbb{R}^{128}$ and a Transformer-based autoregressive decoder.
- **QVAE (proposed):** As described in Sect. 3, using an $n = 8$-qubit parameterized quantum circuit (PQC).

Both models were trained using the hybrid procedure described in Algorithm 1, with identical hyperparameters (learning rate 10^{-4}, batch size 32) to ensure a fair comparison.

6.2 Quantitative Results

Table 1 summarizes the quantitative results for all metrics, computed on a held-out validation set of 2,000 sequences and 1,000 generated samples for entropy and MOS evaluations.

The QVAE achieved significantly higher H_{seq}, indicating enhanced generative diversity enabled by quantum superposition. While the baseline's H_{latent} was numerically higher (7.82 bits), it operates in a continuous 128D Gaussian manifold, whereas the QVAE is constrained by a discrete 8-qubit Hilbert space (max 8 bits; $2^8 = 256$ states). The QVAE's 6.44-bit result represents highly efficient state-space utilization, driving the superior sequence-level diversity (H_{seq}) observed in the outputs. A slight reduction in reconstruction accuracy reflects the expected fidelity–diversity trade-off. Although the baseline maintained a higher overall CEI, the QVAE demonstrated statistically comparable Mutual Information ($p = 0.304$), proving robust structural preservation within a more diverse melodic manifold.

Table 1. Comparison between baseline VAE and proposed QVAE across objective and subjective metrics.

Metric	Baseline VAE	QVAE (proposed)	Difference (*t*-test)
Reconstruction accuracy $\bar{A}_{\mathrm{rec}}$	0.24 ± 0.18	0.18 ± 0.15	$t = 30.13,\ p = 0.000^{***}$
Latent entropy H_{latent} (bits)	7.82 ± 0.04	6.44 ± 0.11	$t = 23.15,\ p < 0.000^{***}$
Sequence entropy H_{seq} (bits/event)	4.12 ± 1.15	4.90 ± 0.89	$t = -58.37,\ p = 0.000^{***}$
Mutual information $I(\mathbf{x}; \mathbf{z})$	1.90 ± 0.30	1.92 ± 0.29	$t = -1.03,\ p = 0.304^{n.s.}$
Mean opinion score (MOS)	3.1	2.8	-
Composite Evaluation Index (CEI)	0.687	0.657	-

*** means highly significant; *n.s.* stands for not significant.

6.3 Statistical Analysis

Table 1 presents the statistical comparison between the models. Independent *t*-tests confirm that the QVAE significantly outperforms the baseline in sequence-level diversity ($t = -58.37, p < 0.001$), supporting the hypothesis that quantum latent spaces facilitate richer generative modeling. While the baseline maintained higher reconstruction accuracy, the QVAE achieved statistically comparable Mutual Information ($p = 0.304$), indicating effective structural preservation despite increased stochasticity. The exceptionally high *t*-statistics result from the large validation sample size, confirming that these performance differences are mathematically robust.

6.4 Visualization of Quantum Latent Structure

To qualitatively examine latent organization, we visualized both classical and quantum latent spaces:

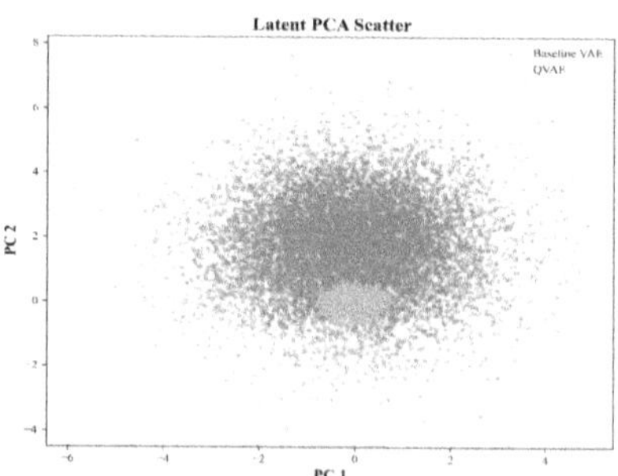

Fig. 2. Comparative PCA projection of the latent manifolds for the Baseline VAE (blue) and QVAE (orange), standardized for unified visualization. (Color figure online)

Latent PCA Scatter. Principal Component Analysis (PCA) was applied to 128-dimensional VAE latents and to continuous QVAE embeddings derived from normalized bitstring frequencies. The QVAE distributions (Fig. 2) exhibited a highly concentrated and structured core within the broader baseline manifold, suggesting that the quantum latent space captures a more specialized and efficient representation of musical features.

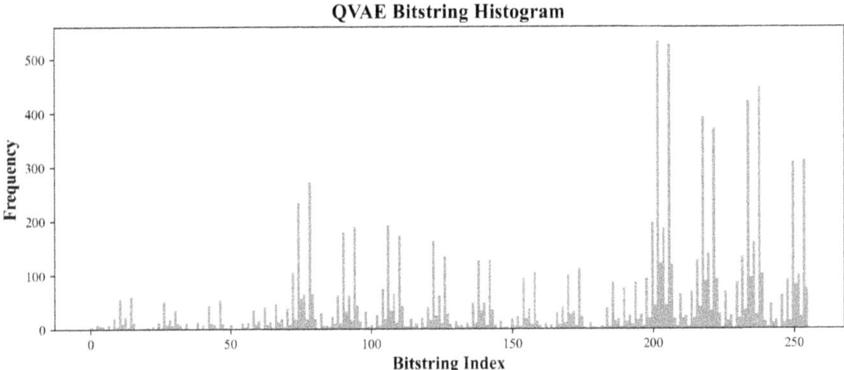

Fig. 3. Distribution of measured quantum bitstrings from the QVAE latent encoder.

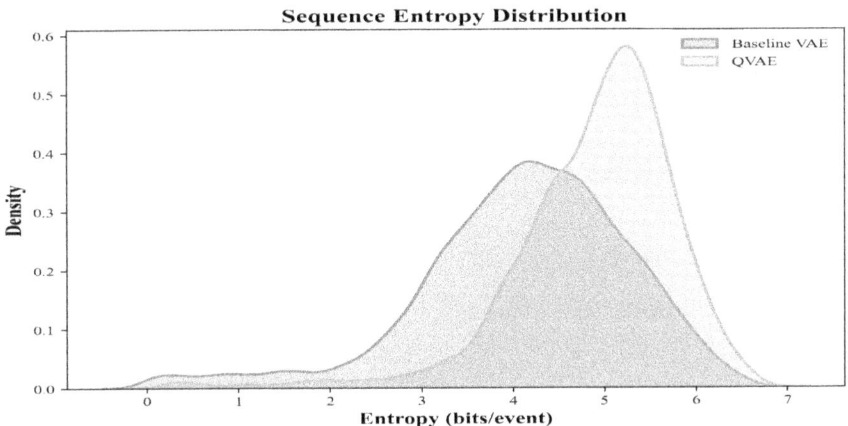

Fig. 4. Sequence-level entropy distributions for generated musical tokens, comparing the diversity of Baseline VAE and QVAE outputs.

Latent Bitstring Histogram. Histograms (Fig. 3) of the QVAE measurement bitstrings showed multi-modal occupancy, with several bitstrings observed frequently, indicating structured quantum sampling. In contrast, the baseline VAE's Gaussian latent sampling exhibited a tight clustering near the mean.

Entropy and Mutual Information Distributions. Histograms of per-sample sequence entropy (Fig. 4) and mutual information (Fig. 5) confirmed consistent rightward shifts for QVAE, aligning with Table 1.

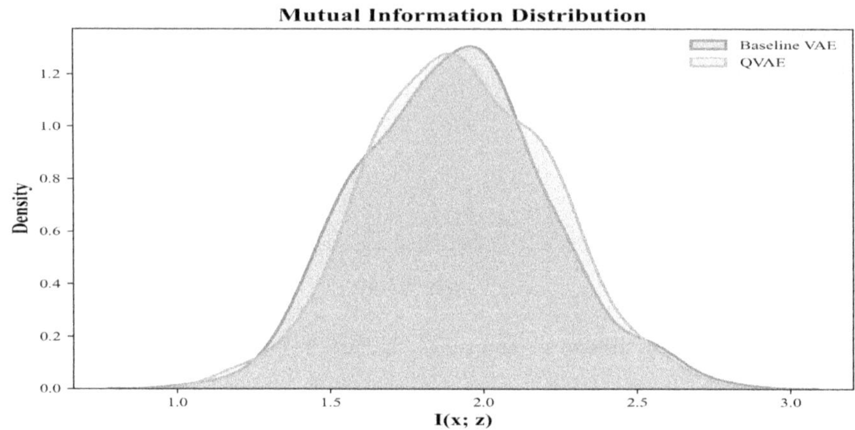

Fig. 5. Distributions of Mutual Information $I(x; z)$ for the Baseline VAE and QVAE, indicating statistically comparable information preservation across both architectures.

6.5   Ablation Studies and Additional Analysis

We conducted ablation experiments to analyze contributions from quantum architectural parameters:

- **Number of qubits (n):** Increasing n from 4 to 8 improved H_{latent} and CEI, but gains plateaued beyond $n = 8$.
- **Circuit depth (d):** Deeper ansatzes ($d > 3$) enhanced generative diversity but slightly reduced reconstruction fidelity unless balanced by stronger KL regularization ($\beta > 0.1$).
- **Quantum regularization ($\mathcal{L}_Q$):** Incorporating quantum entropy regularization ($\gamma > 0$) stabilized the training and produced more interpretable latent clusters.

These findings highlight the sensitivity of hybrid models to both quantum circuit expressivity and classical–quantum coupling strength. Overall, the QVAE demonstrated significantly higher generative diversity (H_{seq}) compared to the baseline, although this exploratory capacity came at the cost of lower perceived creativity (MOS) and reconstruction fidelity. While the baseline maintained a higher overall CEI of 0.687 compared to the QVAE's 0.657 in this run, the significant gain in Sequence Entropy ($t = -58.37$) validates that quantum-enhanced latent spaces can prioritize generative diversity while maintaining a performance floor competitive with classical architectures.

7   Discussion and Limitations

The proposed QVAE demonstrated that quantum latent spaces can enhance generative diversity and perceptual novelty in symbolic music generation. The improved entropy and mutual information metrics suggested that quantum superposition facilitated richer representational mixtures compared to classical Gaussian latents. However, several practical limitations emerged during the experimentation.

First, the computational overhead of simulating quantum circuits on classical hardware significantly constrained scalability. The Qiskit `AerSimulator`, while precise, introduced a per-batch latency of approximately 0.25 s even for moderate qubit counts ($n = 8$). This limited the exploration of larger Hilbert spaces or deeper circuit ansatzes.

Second, while entropy-based metrics provided interpretable indicators of latent diversity, they do not directly measure musicality or aesthetic coherence. Although subjective evaluations partially addressed this, future work could employ learned perceptual metrics or music-structural coherence indices derived from tonal and rhythmic hierarchies.

Finally, the current QVAE formulation is based on a fixed measurement basis, leading to a potentially limited exploration of latent subspaces. Incorporating adaptive or entanglement-aware measurement schemes may yield more expressive latent representations. Additionally, quantum kernel methods [13] or hybrid attention-based PQCs could further improve disentanglement and interpretability.

8   Conclusion

This work presented a QVAE for symbolic music generation, integrating a classical Transformer-based encoder–decoder with a parameterized quantum latent space. Through hybrid variational training, QVAE leveraged quantum superposition to encode richer latent distributions while maintaining coherent musical outputs.

Comprehensive evaluations using both objective and subjective metrics demonstrated that the QVAE achieved significantly higher sequence entropy and comparable mutual information relative to a classical VAE baseline. While the baseline maintained higher reconstruction fidelity and perceived creativity (MOS), these results suggest that quantum latent modeling provides a robust pathway for prioritizing generative diversity and exploratory expressivity in symbolic systems.

9 Code and Generated Music

The code and the music generated for this project can be accessed via the following GitHub link: https://github.com/Sanjaymajumder/Quantum_Latent_Spaces_for_Symbolic_Music_Generation.

References

1. Benedetti, M., Lloyd, E., Sack, S., Fiorentini, M.: Parameterized quantum circuits as machine learning models. Quantum Sci. Technol. **4**(4), 043001 (2019). https://doi.org/10.1088/2058-9565/ab4eb5
2. Boulanger-Lewandowski, N., Bengio, Y., Vincent, P.: Modeling temporal dependencies in high-dimensional sequences: application to polyphonic music generation and transcription. In: Proceedings of the 29th International Conference on Machine Learning (ICML) (2012)
3. Choi, K., Hawthorne, C., Simon, I., Dinculescu, M., Engel, J.: Encoding musical style with transformer autoencoders. In: Proceedings of the 37th International Conference on Machine Learning, vol. 119, pp. 1899–1908. PMLR (2020)
4. Dhariwal, P., Jun, H., Payne, C., Kim, J.W., Radford, A., Sutskever, I.: Jukebox: a generative model for music. arXiv preprint arXiv:2005.00341 (2020). Accessed 15 Jan 2026
5. Engel, J., Agrawal, K.K., Chen, S., Gulrajani, I., Donahue, C., Roberts, A.: GAN-Synth: adversarial neural audio synthesis. In: International Conference on Learning Representations (ICLR) (2019). https://openreview.net/forum?id=H1xQVn09FX. Accessed 15 Jan 2026
6. Hawthorne, C., et al.: Enabling factorized piano music modeling and generation with the MAESTRO dataset. In: International Conference on Learning Representations (2019). https://openreview.net/forum?id=r1lYRjC9F7. Accessed 15 Jan 2026
7. Higgins, I., et al.: beta-VAE: learning basic visual concepts with a constrained variational framework. In: International Conference on Learning Representations (2017). https://openreview.net/forum?id=Sy2fzU9gl. Accessed 15 Jan 2026
8. Huang, C.Z.A., et al.: Music transformer: generating music with long-term structure. In: 7th International Conference on Learning Representations (ICLR) (2019). https://openreview.net/forum?id=rJe4ShAcF7. Accessed 15 Jan 2026
9. ITU-R: Recommendation ITU-R BS.1284-2: General methods for the subjective assessment of sound quality. International Telecommunication Union (2019)
10. Nielsen, M.A., Chuang, I.L.: Quantum Computation and Quantum Information. Cambridge University Press (2010)
11. Raffel, C.: Learning-Based Methods for Comparing Sequences, with Applications to Audio-to-MIDI Alignment and Matching. Ph.D. thesis, Columbia University (2016)
12. Roberts, A., Engel, J., Raffel, C., Hawthorne, C., Eck, D.: A hierarchical latent vector model for learning long-term structure in music. In: Proceedings of the 35th International Conference on Machine Learning, vol. 80, pp. 4364–4373. PMLR (2018)
13. Schuld, M., Killoran, N.: Quantum machine learning in feature hilbert spaces. Phys. Rev. Lett. **122**(4) (2019)
14. Schuld, M., Sinayskiy, I., Petruccione, F.: An introduction to quantum machine learning. Contemp. Phys. **56**(2), 172–185 (2015). https://doi.org/10.1080/00107514.2014.964942

Weather Sonification via a Latent Emotion Space: A Deep Learning Approach

Takeshi Matsumura[1] and Gavin J. Pringle[2](✉)

[1] University of Edinburgh, Edinburgh, UK
[2] EPCC, University of Edinburgh, Edinburgh, UK
g.pringle@epcc.ed.ac.uk

Abstract. This paper introduces a deep learning-based approach to the sonification of daily weather forecasts by mapping meteorological data to musical features via a latent emotion space grounded in psychological theory. Unlike traditional rule-based mappings, our model consisting of Variational Autoencoder and Feed-forward Neural Network learns associations between weather and music using numerical weather prediction data and emotion-annotated datasets. The system generates music that reflects weather conditions through features such as tempo, note duration, pitch range, and mode. Listening evaluations revealed that participants could distinguish broad weather categories (e.g., favourable vs. unfavourable), though finer distinctions proved challenging. This work highlights music's potential as an intuitive and inclusive medium for communicating weather information, particularly for blind, visually impaired, and neurodivergent users, and supports engagement in STEAM (Science, Technology, Engineering, Arts, and Mathematics) education.

Keywords: Weather Sonification · Emotional Association · Deep Learning

1 Introduction

Music communicates emotions beyond words and is often combined with other art forms. For example, film soundtracks strongly influence how audiences interpret ambiguous scenes, perceive characters and recall events [1]. Musical impressions are shaped by features such as mode, chord, rhythm, tempo and possibly key; a diminished chord, for instance, sounds unstable and is frequently used in horror films or alert sounds.

Weather has inspired composers for hundreds of years, often serving as a thematic element. A study on popular music showed that songs mentioning "sun" or "rain" were more likely to be in major or minor modes, respectively [2]. Although interpretations may vary by region, e.g., rainfall in deserts may be perceived positively, people within the same climate tend to share emotional associations with seasonal and daily weather.

With the growth of digital sensing technologies, simulation tools, and generative AI, vast amounts of data are now available, and sonification offers an accessible and engaging way to summarise and communicate such data. Beyond data presentation, sonification can also create social impact, as it may enable blind and visually impaired

(BVI) individuals to engage with STEM fields, as demonstrated in space science [3], and can promote STEAM (STEM + Art) education more broadly.

Meteorology represents a particularly promising application area. Sonification might support trained forecasters in interpreting complex weather data, while also fostering the participation of BVI users in meteorology. Moreover, its potential extends beyond specialists to the public, exemplified by the theme of ICAD 2023, "Sonification for the Masses" [4], which emphasised accessibility, everyday relevance, and public engagement. In this context, it is important to note that typical listeners are more interested in the overall weather situation than in the quantitative values of individual parameters, and sonification that, say, provides the weather forecast, must also be pleasant for daily listening [5]. One of our aims is to create accessible weather forecasts, such as a one-minute musical piece summarising the next 24 h of weather, for individuals who cannot easily interpret graphical or verbal presentations.

This paper presents a deep learning (DL) model for weather sonification that learns weather–music associations shared by people of the same climate and cultural context. The key contribution to this field is the introduction of a latent emotion space into weather sonification: a concept previously applied in image-to-music generation. This space is a 2D representation of two emotional dimensions, namely valence and arousal, which allows the DL model to map complex weather data to corresponding emotional responses in music. By linking weather and music through emotion, our model can generate music directly from daily weather forecasts. Harnessing music's influence on emotion, perception, and memory, this approach aims to foster behavioural responses such as, for instance, preventive actions to weather hazards.

The paper is structured as follows: Sect. 2 reviews background and related work; Sect. 3 describes the methodology; Sects. 4 and 5 present the dataset and model development; Sect. 6 reports training results and listening evaluation; Sect. 7 discusses findings; and finally, Sect. 8 concludes with future work.

Our code, along with input and expected output music files, is published [6] and is also available via github[1].

2 Background and Related Work

Sonification of continuous data is generally categorized into three principal approaches: audification, which directly translates data into sound; parameter-mapping sonification, which projects individual data dimensions onto acoustic dimensions; and model-based sonification, which incorporates dynamic models and often enables user interaction [7].

The sonification of continuous weather data has been explored in a variety of contexts, with parameter-mapping sonification being the predominant approach. Examples include sonification of daily weather forecasts broadcast on the radio [5], the use of weather observations to compose music for dance performances [8], the sonification of historical daily records to compare seasonal climates [9], and the use of rainfall data from hundreds of stations combined with a neural network trained on Irish music [10]. Other applications involve the spatialized sonification of hailstorms to convey their locations [11] and the

---

[1] https://github.com/takeshi4126/weather_music_evo26, last accessed 2026/01/20.

parameter mapping sonification of pressure, location and asymmetry in hurricanes for exhibition purposes [12]. Sonification has also been used to raise public awareness of the threats posed by global warming and climate change [13].

Nevertheless, applying manually crafted parameter-mapping sonification to weather data raises several challenges. Designing appropriate mappings requires expertise both in meteorology and in musical composition, and mappings developed for one context may not readily transfer to another. Moreover, weather data is inherently high-dimensional, whereas the perceptually distinct acoustic features available for mapping are relatively limited. Consequently, the selection of mappings must be guided not only by data characteristics but also by considerations of listener perception. A review of parameter mappings from physical to musical features [14] revealed that pitch was the most frequently employed feature, commonly representing variables such as temperature and pressure. However, pitch is a single-dimensional parameter and cannot simultaneously represent both temperature and pressure. According to a listening test, a single sonification could convey trends in five weather parameters [15]; however, listeners must be informed in advance of the manually-crafted mapping rules, and grasping quantities becomes more difficult as the number of weather parameters increase.

To address these limitations, DL has been introduced into weather sonification. Early attempts include blending precomposed music motifs for low and high rainfalls using MusicVAE [16] as a part of their Smart City Sonification [17] and controlling the Long Short-Term Memory (LSTM) hyperparameter based on monthly mean temperature [18]. However, these studies have not directly tackled the challenge of communicating weather patterns to the public without requiring prior knowledge of mapping rules.

DL can leverage hidden associations between music and weather data to enable sonification without explicit mapping rules by articulating the tacit knowledge shared within a cultural/environmental context. This area is underexplored, but to the best of our knowledge, such understanding is necessary for the public to comprehend the sonification of daily live weather forecasts. The idea of mapping weather data into a 2D Emotion Map was pioneered by [5], but was limited to a projection based on Principal Component Analysis (PCA) to a small set of manually defined prototypes. Recently, a DL-based approach was proposed to generate music for artistic images via emotion-based association [19], which shares a similar aim to our work.

These limitations in rule-based sonification as well as the existing DL-based approach motivate the need for a data-driven approach to communicate weather patterns without requiring prior knowledge, which we present in the following methodology.

3 Methodology

Building on prior approaches, our methodology introduces a DL-based emotion mapping model for weather sonification, aiming to help listeners intuitively grasp the weather forecast conveyed through music, by leveraging the emotional linkage between weather and music obtained via DL.

An empirical study [20] describing how performers control music features to express emotions revealed quantitative relationships between seven musical features and five emotions, which strongly suggests that DL is applicable to control music features to express emotions. According to a review on relations between music and emotion [21], Russell's circumplex model of affect is a widely used model in which emotions are represented along two axes: valence, ranging from pleasant to unpleasant, and arousal, ranging from excited to calm [22]. We have employed this circumplex model and developed our Weather-Music Association DL Model to associate weather and music via this continuous 2D latent emotion space.

The Weather-Music Association DL Model is trained using three data sets described in the next section: Numerical Weather Prediction (NWP) data from the UK Met Office [23], emotion-annotated weather words from [24], and emotion-annotated music data from [19]. A pair of valence and arousal values, which is hereinafter referred to as an emotion vector, in the emotion-annotated weather and music data are used to associate weather and music representations within the latent space. Once trained, the model accepts the numerical weather prediction data without the need for emotion vectors and outputs music features used to compose the weather music.

The generated weather music was evaluated by four listeners: two in the UK and two in Japan. Although the number of participants was insufficient for convincing statistical analysis, the evaluation identified potential improvements in both the model and the evaluation procedure to inform a future large-scale evaluation.

In contrast to [5], our study employs a data-driven representation using a Variational Autoencoder (VAE) to capture latent structures of meteorological data and integrates it systematically into musical feature mappings (pitch range, tempo, mode, etc.), thereby making the solution more flexibly adapted to the different climate conditions and musical cultures. Although contrastive learning was employed in [19], our model simply applied supervised learning on the VAE latent space to easily visualise associations between weather parameters and inferred music features, using emotion vectors.

4 Preparation of the Data Sets for DL

4.1 Weather Forecast Data

Edinburgh was selected for weather sonification. It has a cool temperate maritime climate (Cfb), with average monthly temperatures ranging from 1.4 °C to 19.4 °C and evenly distributed rainfall of 40–80 mm during the 1991–2020 climate period [25]. Southwesterly winds are common and severe weather events are rare.

The weather forecast data chosen for sonification was the NWP data from the UK Met Office whose two-year rolling archive is publicly available [23]. New data is published every hour with several hours delay from the live forecast, which enables both training and live sonification. The NWP hourly data has a 2 km grid resolution across the UK and is provided in the NetCDF format.

To improve DL efficiency, all input data must be cleaned. Here, weather parameters were pruned by removing redundant data (e.g. mean wind speed and wind gust, which are highly correlated) and selecting representatives from each weather parameter group. Spatial dependency was avoided by using data from a single grid point (55°56′23.9″N, 3°12′14.7″W) located within Edinburgh, and temporal effects were addressed by including the hour of day and day of year as conditional inputs. Thirteen surface-level parameters, defined in [26], were selected for sonification, including cloud cover (total, high, medium and low), fog fraction, pressure, relative humidity, temperature, visibility, wind (speed and direction), and 1-h rainfall and snowfall.

To prepare weather parameters for DL, wind direction, θ, such that $0 \leq \theta < 360$, and θ increases clockwise from north, was converted to vector form $\mathbf{w}_i = (S_i \cos \theta_i, S_i \sin \theta_i)$ to handle circularity and variability at low speeds. Similarly, hour of day and day of year were mapped to $\mathbf{t}_i = (\sin(2\pi d_i/365), \sin(2\pi h_i/24))$ to preserve temporal continuity, following [27]. All selected parameters were then standardized using $z = (x - u)/s$ with the sample mean u and the standard deviation s.

4.2 Emotion-Annotated Weather Parameters

The NWP data are continuous, and annotating emotion vectors to individual weather data samples is infeasible. To simplify the process, this work used the emotion-annotated weather words from a prior study [24] and annotated emotion vectors to the NWP data according to Table 1, where we defined the value range for each weather word.

Table 1. Weather data annotation by weather words

Weather parameter	Weather word	Value range	Valence	Arousal
Temperature (°C)	Freezing	$[-\infty, 0)$	3.42	4.52
	Cold	$[0, 5)$	4.43	4.30
	Cool	$[5, 10)$	6.55	4.54
	Comfortable	$[10, 15)$	7.85	4.04
	Warm	$[15, 20)$	7.34	4.62
	Hot	$[20, \infty)$	4.83	4.87
Relative Humidity (%)	Dry	$[0, 0.5)$	4.85	3.65
	Humid	$[0.9, 1.0]$	3.02	3.35
Wind Speed (m/s)	Breezy	$[3, 5)$	5.66	4.81
	Windy	$[5, 7)$	4.75	5.15
	Gusty	$[7, 10)$	4.28	4.86
	Blustery	$[10, \infty)$	4.22	5.41
Rainfall (mm)	Drizzle	$(0.0, 0.5)$	4.52	3.44
	Rainy	$[0.5, \infty)$	4.76	3.93
Cloud Amount in Low and Mid Layers	Clear	$[0, 0.2)$	7.77	4.17
	Partly Cloudy	$[0.2, 0.5)$	6.09	3.45
	Cloudy	$[0.5, 0.8)$	4.70	3.16
	Overcast	$[0.8, 1.0]$	4.52	3.10

The scale of valence is from 1 (unhappy) to 9 (happy) and that of arousal from 1 (calm) to 9 (excited). Where an NWP data point could be annotated by multiple weather words, such as warm, humid and rainy, the same data point was repeated. The valence and arousal values were scaled from the [1, 9] range to [0, 1] using the Min–Max scaler in scikit-learn [28] to ensure consistency with the input range of the emotion-annotated music data.

In [24], weather words were annotated by 420 participants in 4 emotional dimensions: valence, arousal, dominance and surprise, wherein strong correlations were reported between valence and dominance (0.94) and arousal and surprise (0.87). Hence, only valence and arousal were used for this work. Notably, the annotators were from a university in the southern U.S., whose climate differs from Edinburgh: a bias that likely influences the emotional interpretation of weather.

4.3 Emotion-Annotated Music Features

Emotion-annotated music data from [19] includes 3,000 MIDI files segmented from 467 piano pieces of various genres, annotated for valence and arousal by six individuals. Although the limited number of annotators raises concerns about statistical reliability reported here, the data provides useful insights into music–emotion associations and helps guide feature selection for the DL model.

We took the 3,000 files and extracted musical features, including mean note duration, tempo, pitch range, and mode (major/minor) using the music21 [29] and musif [30] libraries. The partial pair plot in Fig. 1 shows the valence and arousal values entered by the 6 individuals for each of the 3,000 files and their relation to the extracted music features, with major and minor modes shown in blue and pink, respectively.

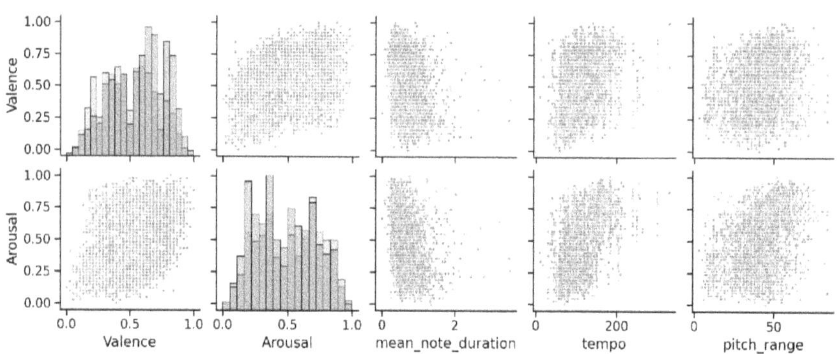

Fig. 1. Partial pair plot between Valence-Arousal values and the extracted music features

The histograms reveal a greater number of samples with higher valence than lower valence, and a greater number with lower arousal than higher arousal. The scatter plots depict the tendency of the calculated correlations. As shown in Table 2, valence and arousal are weakly negatively correlated with mean note duration, and weakly positively correlated with tempo and pitch range.

Mode is widely recognized as a key factor in shaping musical mood, with its emotional associations reviewed in [31]. In the latent emotion space (Fig. 1), major modes

Table 2. Correlations between the emotion vectors and the music features

Weather Pattern	Mean note duration	Tempo	Pitch range
Valence	−0.3	0.3	0.2
Arousal	−0.4	0.5	0.4

(blue) generally align with higher valence and lower arousal, while minor modes (pink) show the opposite trend. Four emotion-annotated features were selected as a set of input parameters to the DL: mean note duration, tempo, pitch range, and mode.

5 Development

With both weather and music data annotated for emotional dimensions, we now describe the architecture of our Weather–Music Association DL Model, which was implemented using Python and Tensorflow library [32] based on the stellar sonification work presented in [33].

5.1 Weather-Music Association DL Model

Our Weather–Music Association DL Model comprises three components: the Weather VAE, the Weather–Emotion FNN (Feed-forward Neural Network), and the Music–Emotion VAE. They are discussed in detail below using Fig. 2.

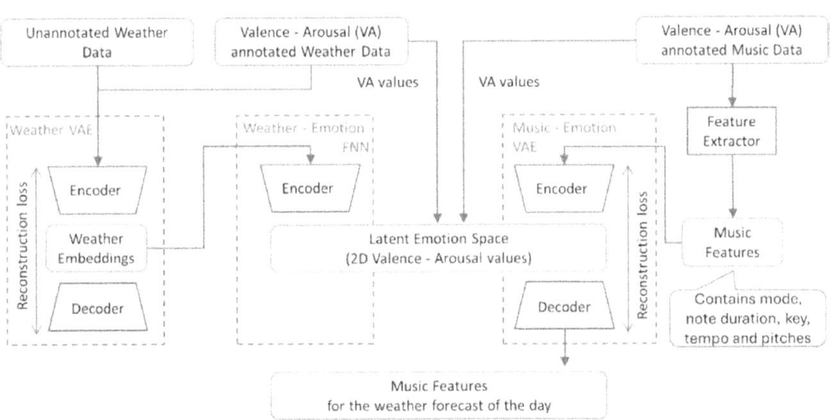

Fig. 2. Weather-Music Association DL Model

The Weather VAE generates a weather embedding, a compact representation of a NWP data point, which is passed to the Weather–Emotion FNN to infer a latent emotion vector representing the emotional response to the weather. This vector is then fed into the Decoder of the Music–Emotion VAE to predict music features. Weather-Emotion

FNN and Music-Emotion VAE are trained via supervised learning using the annotated emotion vectors as labels. The hyperparameters of the DL models were optimized using Optuna [34].

Weather VAE is a Conditional VAE with 4 dense layers (456, 44, 284, 328 units) and an 8D latent space. The Encoder receives 13 engineered and scaled weather parameters at the chosen grid point along with 2 conditional inputs (time of day, day of year) and outputs the mean and log variance of the latent distribution. An 8D latent vector z is sampled from this distribution. The Decoder then reconstructs the weather parameters from z and the same conditional inputs. Weather VAE is trained by maximising its Evidence Lower Bound (ELBO) defined by Eq. 1 following the Generative Semi-supervised Model [35], where y is the conditional input parameters for data point x. The 1st term on the right-hand side is the expected reconstruction loss of x and the 2nd term is the Kullback-Leibler divergence (KL divergence) to transform the latent space to approximate a Gaussian distribution.

$$\mathcal{L}(\theta, \phi; x, y) = \mathbb{E}_{q_\phi(z|x,y)}\big[\log p_\theta(x|z, y)\big] - D_{KL}\big(q_\phi(z|x, y)\|p_\theta(z|y)\big). \tag{1}$$

Weather-Emotion FNN is a feed-forward neural network with 3 dense layers (412, 368, 356 units). It is trained on emotion-annotated weather embeddings to minimise the mean squared error of the predicted latent emotion vectors. Drop-out layers (rate 0.2) are applied after the input and each dense layer to prevent overfitting.

Music-Emotion VAE consists of 5 dense layers (452, 308, 484, 444, 256 units). It receives music features consisting of mode, tempo, pitch range, and mean note duration. The associated ELBO $\mathcal{L}(\theta, \phi; x)$ is denoted by Eq. 2

$$\mathcal{L} = \sum_{j \in \{m,f\}} \mathbb{E}_{q_\phi(z|x)}\big[\omega_j \log p_\theta(x_j|z)\big] - \beta D_{KL}\big(q_\phi(z|x)\|p_\theta(z)\big) - \lambda\|z - y\|^2, \tag{2}$$

where $\beta = 0.1$, $\lambda = 8.8$, $\omega_f = 1.0$ and $\omega_m = 0.9$. The 1st term on the right-hand side of Eq. 2 is the expected reconstruction loss of x, combining two parts: binary cross-entropy for the mode (x_m) and mean squared error for the other features (x_f) including mean note duration, tempo and pitch range. The 2nd term is the KL-divergence, and the 3rd term is the squared L2-norm between the latent emotion vector z and the labelled emotion vector y, representing the supervised loss. Hyper-parameters β and λ facilitate the effects of the three terms to the overall loss.

5.2 Music Generation

The music features predicted by our Weather-Music Association DL Model are used to create a seed melody for the MelodyRNN [36]. The seed melody is fixed to 8 notes, corresponding to 2 measures in quarter notes, which we consider sufficient to establish a musical motif, as using more notes would increase randomness in pitch and thus less pleasant to the ear. Each of the 8 notes in the seed melody assigned the inferred mean note duration with 1/16 note as the minimum. The inferred tempo is applied in the final score. To reflect the inferred mode, note pitches are sampled from a predefined distribution for either C major or A natural minor [37]. The predefined, one-octave distribution was extended based on the inferred pitch range, thereby favouring higher pitches when the range is wider. The generated music consists of a monophonic melody of 8 measures without chords and is output as a MIDI file with piano as the instrument.

6 Results and Evaluation

6.1 Training of Weather VAE

We trained the Weather VAE for 100 epochs with 6,998 (80%) data points and tested with 1,750 (20%) data points. For validation of the stability, 5-fold cross validation was repeated for 20 times, 100 trials in total, and the coefficient of determination R^2 was calculated and shown in Table 3. Excluding the 8 deficient trials, which did not include snowfall events in the test data, the R^2 score was 89% or higher for all the weather parameters, and the mean of all parameters were 93% and 92% for the training and test data, respectively. Also, it can be claimed that the Weather VAE did not overfit to the training data, because the difference of mean R^2 scores between the training and test data was 1% or less, except for the visibility, and the standard deviations were less than 0.1 for all the weather parameters.

Table 3. Coefficients of determination R^2 for the weather parameters

Weather Parameter	Training R^2		Test R^2	
	Mean	Std. Dev	Mean	Std. Dev
Cloud amount of high cloud	0.94	0.003	0.93	0.005
Cloud amount of low cloud	0.93	0.004	0.93	0.005
Cloud amount of medium cloud	0.93	0.005	0.93	0.006
Cloud amount of total cloud	0.95	0.003	0.95	0.004
Fog fraction at screen level	0.97	0.031	0.96	0.039
Pressure at surface	0.92	0.006	0.92	0.008
Relative humidity at screen level	0.95	0.005	0.94	0.011
Temperature at screen level	0.90	0.004	0.89	0.007
Visibility at screen level	0.97	0.056	0.94	0.078
Rainfall accumulation (1 h)	0.91	0.005	0.91	0.008
Snowfall accumulation (1 h)	0.91	0.005	0.90	0.007
East-west component of wind	0.91	0.005	0.91	0.007
North-south component of wind	0.91	0.004	0.91	0.005
Overall Mean	0.93	0.010	0.92	0.015

6.2 Training of Weather-Emotion FNN

The Weather-Emotion FNN was trained for 300 epochs with a mini-batch of 32 to map the weather data into the latent emotion space. As shown in Fig. 3, the left panel displays the annotated emotion vectors which we employed as the labels, while the right panel shows the FNN output.

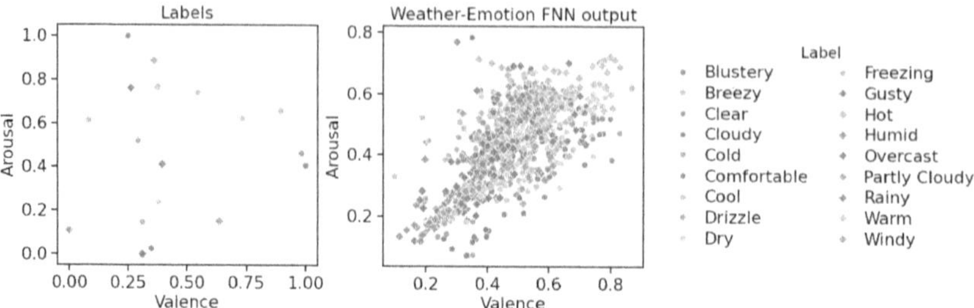

Fig. 3. Weather-Emotion FNN prediction result

The predicted vectors are generally close to their labels, though slightly clustered toward the centre due to averaging across weather types (e.g., humid, warm, overcast). Also, positive correlation between valence and arousal as well as absence of data in the left-top and right-bottom regions are apparent.

6.3 Training of Music-Emotion VAE

Music-Emotion VAE was trained with the training and test data of 2,348 (80%) and 587 (20%) respectively for 300 epochs with a mini-batch of 32. Figure 4 visualises the relation between the latent emotion vectors and the music features predicted by the Music-Emotion VAE for the test data.

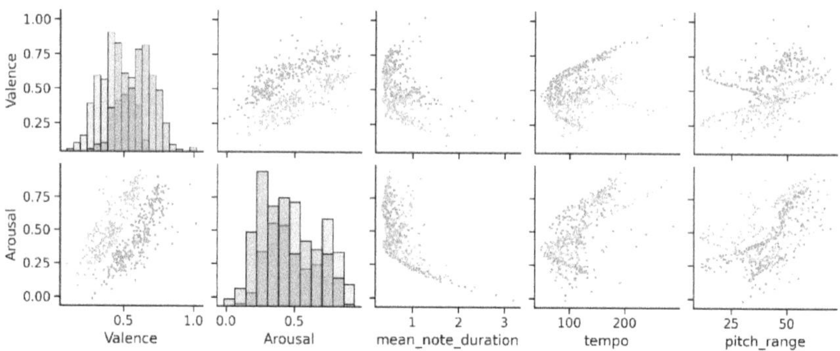

Fig. 4. Partial pair plot between the latent emotion vectors and the Music-Emotion VAE output

The pitch range is defined as the number of semitones between the highest and lowest notes, and the mean note duration assigns full, half, quarter, 1/8, 1/16 and 1/32 notes to 4, 2, 1, 0.5, 0.25 and 0.125 respectively. The colour of dots represents the mode determined by splitting the probability of major mode inferred by our Weather-Music Association DL Model with threshold of 0.5.

A key result is that the Music-Emotion VAE predicted shorter mean note duration, larger pitch range and major scale for higher valence, and faster tempo and larger pitch range for higher arousal. The prediction captured the overall trends of the original data distribution in Fig. 1 but exaggerated and skewed, which should be investigated in the future. As discussed in Future Work, this work is limited to monophonic melodies, however, polyphonic harmony will be introduced to further distinguish complex emotional states.

6.4 Listening Evaluation

To evaluate whether listeners can distinguish weather patterns through sonification, 5 music samples were generated for each of four patterns, namely Cold & Windy, Comfortable, Rain and Snow, which yields 20 samples in total, stored in [6]. Among the four participants, only one, in Japan, had no experience of playing musical instruments. We recognize that the sample size is small and does not provide robust statistical strength. Due to project time limitations, this evaluation serves as a preliminary proof-of-concept to demonstrate the feasibility of the latent emotion space approach rather than a definitive statistical verification.

Figure 5 shows box plots of the latent emotion vectors (valence and arousal) inferred by the Weather–Emotion FNN and used by the Music–Emotion VAE.

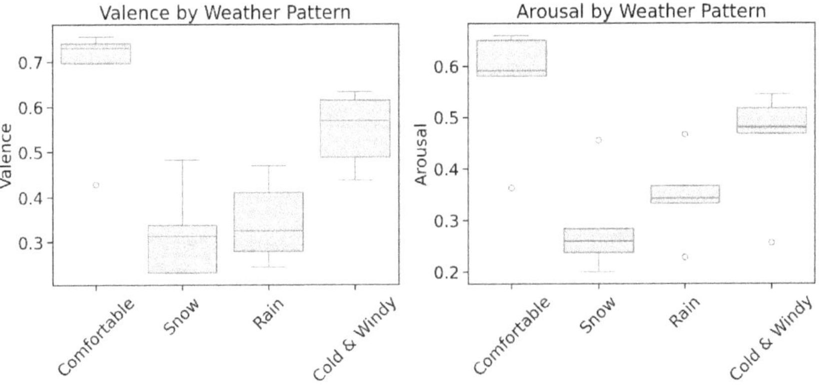

Fig. 5. Valence and Arousal distributions of the created 4 weather patterns

The plots indicate clear emotional separation among the four patterns. Comfortable weather is distinct in both valence and arousal, while Cold & Windy is separated from both Rain and Snow in arousal and shows a distinct interquartile range in valence. Rain and Snow overlap in valence but are distinguishable in arousal. These results suggest that the four weather patterns could be distinguishable in the 2D latent emotion space.

Table 4 shows the music features used to compose the weather music, averaged over the 5 pieces of music per weather pattern. The major mode count column contains the number and proportion of major mode music in each weather pattern. The table suggests that the mode and mean note duration are good discriminators between the first two patterns and the last two patterns, and the tempo makes the Comfortable pattern stand out, while the pitch range seems not contributing to the distinction.

Table 4. Music features used to express weather patterns

Weather Pattern	Averaged Mean note duration	Averaged Tempo	Averaged Pitch range	Major mode count
1: Cold & windy	0.6	96	46	4 (80%)
2: Comfortable	0.5	121	49	4 (80%)
3: Rain	0.8	84	43	0 (0%)
4: Snow	1.0	84	47	0 (0%)

The analysis suggests that listeners' ability to differentiate between weather patterns is expected to be as follows: a) Cold & Windy and Comfortable can be distinguished from Rain and Snow using the mean note duration and mode, b) Cold & Windy can be distinguished from Comfortable using tempo, and c) Rain and Snow are hardly distinguishable.

Two generated music samples corresponding to the Comfortable and Snow weather patterns are shown in Fig. 6. The generates music samples are 8-measure long but only the first rows in the generated scores are displayed.

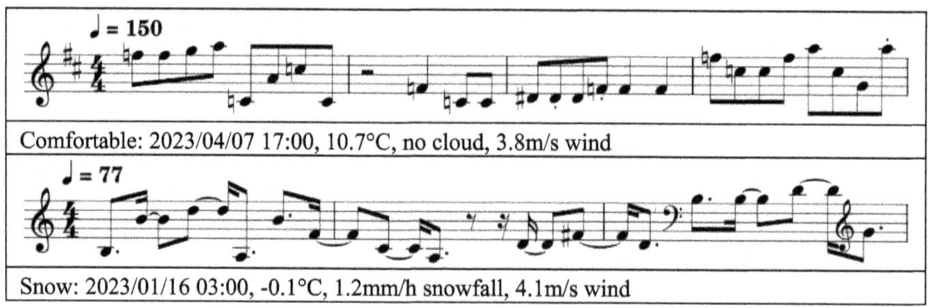

Fig. 6. Examples of generated music representing Comfortable (upper) and Snow (lower), only the first row of total 8 measures from the scores

The associated latent emotion vectors (valence, arousal) were (0.76, 0.65) and (0.31, 0.26), respectively. The Snow sample is slower in tempo and set in the minor mode, whereas the Comfortable sample is faster and in the major mode. Consequently, the two pieces convey clearly contrasting musical impressions. The participants listened to the weather music one by one without seeing the music score and freely selected one answer from the 4 weather patterns. This evaluation was treated as 80 multinomial trials (n = 1, K = 4), assuming a 25% chance level for random guessing.

The result of the listening evaluation is summarised in the confusion matrix in Table 5, where the columns and rows correspond to the expressed and answered weather patterns respectively, and the cell values indicate the number of responses. The diagonal cells in the table are the cases where the expressed weather patterns were correctly answered. The number of correct answers was 24 out of 80 (30%) which was higher than the 25% chance level; however, Cohen's effect size index is very small: $h \approx 0.11$. While this suggests limited statistical evidence of strong performance, no doubt partly due to the small sample size, listening evaluations clearly reveal that the model requires improvement before repeating the study with a larger sample size of participants.

Table 5. Listening evaluation result of weather sonification

Answered weather pattern	Expressed weather pattern			
	Pattern 1 Cold & Windy	Pattern 2 Comfortable	Pattern 3 Rain	Patten 4 Snow
Pattern 1	4	7	6	2
Pattern 2	8	6	5	2
Pattern 3	5	4	4	6
Pattern 4	3	3	5	10

7 Analysis of Results and Discussion

To interpret the listening evaluation outcomes, we analysed listener responses and model predictions in detail. It can be seen from Table 5 the confusion between pairs of weather patterns is largely symmetric. Snow was most frequently confused with Rain, consistent with expectation (c). Notably, Cold & Windy and Comfortable were often confused (15 instances), suggesting that tempo alone, despite being the primary difference between them, is not a reliable discriminator for this very small cohort, contrary to expectation (b). To assess expectation (a), patterns were grouped into (1 & 2) and (3 & 4), and the evaluation was reanalysed in Table 6. These groups show minimal cross-group confusion, supporting expectation (a), though it remains unclear whether the distinguishing factor was mean note duration or mode. In summary, expectations (a) and (c) are partially supported, while (b) is not.

Table 6. Listening evaluation result combining patterns 1 & 2 and 3 & 4

Answered weather pattern	Expressed weather pattern	
	Patterns 1 & 2	Patterns 3 & 4
Patterns 1 & 2	25 (62.5%)	15 (37.5%)
Patterns 3 & 4	15 (37.5%)	25 (62.5%)

A participant with musical training mentioned that the mode was unclear. Applying the Krumhansl–Schmuckler key-finding algorithm [37] to the generated music supported this, identifying the opposite of the intended mode in 7 of the 20 pieces (35%). Another listener, aware of the implementation details, suggested that using C major and A natural minor as the evaluation pair may have contributed to this ambiguity, as these relative keys share the same pitch classes. Given that mode strongly influences perceived valence

[31], future work should further refine the generation process to ensure that the mode is clearly conveyed, such as the use of chords rather than single notes.

Although Cold & Windy weather is typically perceived as unpleasant, Weather–Emotion FNN inferred high valence values for this pattern's samples because they corresponded to clear skies, which were assigned high valence values in Table 1. As a result, most Cold & Windy pieces were generated in the major mode, which may have confused participants. This highlights the need to refine the evaluation procedure, as participants might have responded differently had the pattern been labelled Clear Sky. It also raises an open question about how averaging emotion vectors during the Weather-Emotion FNN training affects inference, which remains for future study.

8   Conclusion

We presented a novel deep learning-based weather sonification that leverages a latent emotion space to translate meteorological data into emotionally resonant music. Trained on UK Met Office forecasts and emotion-annotated datasets, the model generates music using features such as pitch range, tempo, note duration, and mode. Listening evaluations showed that participants could broadly distinguish weather categories, though improvements are needed for finer distinction.

To illustrate potential improvements, we provide a set of demonstrative music pieces generated using an updated version of our model, incorporating clearer tonal contrasts using A major vs. C natural minor, alongside an integrated key-finding algorithm for mode validation, and motif repetition. The technical details of this updated model and its evaluation are beyond the scope of the present paper: these examples are shared solely to invite participation in this new listening study[2].

Future work will focus on rigorous evaluation of these enhancements through expanded participant cohorts, enriching the sonification using chords, harmony, polyphony, timber etc., and automated data preprocessing for real-time accessibility applications. Beyond technical refinement, this interdisciplinary approach offers promising applications in STEAM education and inclusive weather communication, particularly for blind, visually impaired, and neurodivergent audiences.

Acknowledgments. We gratefully acknowledge the use of the ML model and associated stellar data provided by Dr Adrián García Riber at Universidad Politécnica y Real Conservatorio Superior de Música de Madrid. The authors extend their sincere gratitude to Ms Hitomi Iriyama, a jazz violinist, who provided musical guidance, and to the other Listening Evaluation participants.

References

1. Boltz, M.G.: Musical soundtracks as a schematic influence on the cognitive processing of filmed events. Music Percept. Interdisc. J. **18**, 427–454 (2001)
2. Brown, S., Aplin, K., Jenkins, K., Mander, S., Walsh, C., Williams, P.: Is there a Rhythm of the rain? An analysis of weather in popular music. Weather **70**, 198–204 (2015)

---

[2] https://forms.gle/Xg2nTtUF1WEiUegk7, last accessed 2026/01/20.

3. Sonification: A tool for research, outreach and inclusion in space sciences. United Nations Office for Outer Space Affairs (2023)
4. Weger, M., Ziemer, T., Rönnberg, N. (eds.): Sonification for the Masses. In: Proceedings of the 28th International Conference on Auditory Display (ICAD 2023). ICAD, Norrköping, Sweden (2023)
5. Hermann, T., Drees, J.M., Ritter, H.: Broadcasting auditory weather reports - a pilot project. In: Proceedings of the 2003 International Conference on Auditory Display, Boston, MA, USA, pp. 208–211 (2003)
6. Matsumura, T.: Source code for weather sonification via a latent emotion space: a deep learning approach (2025). https://doi.org/10.5281/zenodo.17505630
7. Hermann, T., Hunt, A., Neuhoff, J.: The Sonification Handbook. Logos Verlag (2011)
8. Lindborg, P.: Interactive sonification of weather data for the locust wrath, a multimedia dance performance. Leonardo **51**, 466–474 (2018)
9. Flowers, J.H., Whitwer, L.E., Grafel, D.C., Kotan, C.A.: Sonification of daily weather records: issues of perception, attention and memory in design choices. In: Proceedings of the 2001 International Conference on Auditory Display, Espoo, Finland, pp. 222–226 (2001)
10. Fernström, M., Griffith, N., Taylor, S.: Bliain Le Baisteach - Sonifying a year with rain. In: Proceedings of the 2001 International Conference on Auditory Display, Espoo, Finland, pp. 282–283 (2001)
11. Childs, E., Pulkki, V.: Using multi-channel spatialization in sonification: a case study with meteorological data. In: Proceedings of the 2003 International Conference on Auditory Display, Boston, USA, pp. 192–195 (2003)
12. Ballora, M.: Two examples of sonification for viewer engagement: hurricanes and squirrel hibernation cycles. In: Proceedings of the 21st International Conference on Auditory Display, Austria, pp. 300–301 (2015)
13. George, S.S., Crawford, D., Reubold, T., Giorgi, E.: Making climate data sing: using music-like sonifications to convey a key climate record. Bull. Am. Meteor. Soc. **98**, 23–27 (2017). https://doi.org/10.1175/BAMS-D-15-00223.1
14. Dubus, G., Bresin, R.: A systematic review of mapping strategies for the sonification of physical quantities. PLoS ONE **8**, e82491 (2013). https://doi.org/10.1371/journal.pone.008 2491
15. Schuett, J.H., Winton, R.J., Batterman, J.M., Walker, B.N.: Auditory weather reports: demonstrating listener comprehension of five concurrent variables. In: Proceedings of the 9th Audio Mostly: A Conference on Interaction With Sound. Association for Computing Machinery, New York, NY, USA (2014). https://doi.org/10.1145/2636879.2636898
16. Roberts, A., Engel, J., Raffel, C., Hawthorne, C., Eck, D.: A hierarchical latent vector model for learning long-term structure in music. In: Proceedings of the 35th International Conference on Machine Learning, pp. 4364–4373. PMLR (2018)
17. Roddy, S., Bridges, B.: The design of a smart city sonification system using a conceptual blending and musical framework, web audio and deep learning techniques. In: Proceedings of the 26th International Conference on Auditory Display (ICAD 2021) (2021). https://doi. org/10.21785/icad2021.010
18. Kalonaris, S.: Tokyo Kion-On: query-based generative sonification of atomospheric data. In: Proceedings of the 27th International Conference on Auditory Display, pp. 105–109 (2022). https://doi.org/10.21785/icad2022.039
19. Wang, Y., Chen, M., Li, X.: Continuous emotion-based image-to-music generation. IEEE Trans. Multimedia **26**, 5670–5679 (2024). https://doi.org/10.1109/TMM.2023.3338089
20. Bresin, R., Friberg, A.: Emotion rendering in music: range and characteristic values of seven musical variables. Cortex **47**, 1068–1081 (2011). https://doi.org/10.1016/j.cortex.2011. 05.009

21. Eerola, T., Vuoskoski, J.K.: A review of music and emotion studies: approaches, emotion models, and stimuli. Music Percept. Interdisc. J. **30**, 307–340 (2013)
22. Russell, J.A.: A circumplex model of affect. J. Pers. Soc. Psychol. **39**, 1161–1178 (1980)
23. UK Met Office: Met Office UK Deterministic (UKV) 2km on a 2-year rolling archive. https://registry.opendata.aws/met-office-uk-deterministic/. Accessed 31 Aug 2025
24. Stewart, A.E.: Affective normative data for English weather words. Atmosphere **11**, 860 (2020). https://doi.org/10.3390/atmos11080860
25. UK Met Office: Edinburgh/Gogarbank Location-specific long-term averages. https://www.metoffice.gov.uk/research/climate/maps-and-data/location-specific-long-term-averages/gcvw5vmsn. Accessed 31 Aug 2025
26. UK Met Office: Weather DataHub | Glossary of parameters. https://datahub.metoffice.gov.uk/docs/glossary. Accessed 31 Aug 2025
27. The TensorFlow Authors: TensorFlow Core Tutorials Time series forecasting (2024). https://www.tensorflow.org/tutorials/structured_data/time_series. Accessed 20 Jan 2026
28. Pedregosa, F., et al.: Scikit-learn: machine learning in Python. J. Mach. Learn. Res. **12**, 2825–2830 (2011)
29. Cuthbert, M., Ariza, C., Hogue, B., Oberholtzer, J.W.: music21 Documentation. https://www.music21.org/music21docs/index.html. Accessed 31 Aug 2025
30. Llorens, A., Simonetta, F., Serrano, M., Torrente, Á.: musif: a Python package for symbolic music feature extraction. In: Proceedings of the Sound and Music Computing Conference 2023, Stockholm, Sweden, pp. 132–138 (2023)
31. Carraturo, G., Pando-Naude, V., Costa, M., Vuust, P., Bonetti, L., Brattico, E.: The major-minor mode dichotomy in music perception. Phys. Life Rev. **52**, 80–106 (2025). https://doi.org/10.1016/j.plrev.2024.11.017
32. Abadi, M., et al.: TensorFlow: large-scale machine learning on heterogeneous systems (2015). https://www.tensorflow.org/. Accessed 20 Jan 2026
33. Riber, A.G., Serradilla, F.: AI-rmonies of the Spheres. In: Artificial Intelligence in Music, Sound, Art and Design. In: 12th International Conference, EvoMUSART, pp. 132–147 (2023)
34. Akiba, T., Sano, S., Yanase, T., Ohta, T., Koyama, M.: Optuna: a next-generation hyperparameter optimization framework. In: KDD 2019: Proceedings of the 25th ACM SIGKDD International Conference on Knowledge Discovery & Data Mining, pp. 2623–2631 (2019). https://doi.org/10.1145/3292500.3330701
35. Kingma, D.P., Rezende, D.J., Mohamed, S., Welling, M.: Semi-supervised learning with deep generative models. In: Proceedings of the 28th International Conference on Neural Information Processing Systems, Montreal, Canada, vol. 2, pp. 3581–3589 (2014)
36. Waite, E., Eck, D., Roberts, A., Abolafia, D.: Generating long-term structure in songs and stories (2016). https://magenta.withgoogle.com/2016/07/15/lookback-rnn-attention-rnn. Accessed 20 Jan 2026
37. Krumhansl, C.L.: Cognitive Foundations of Musical Pitch. Oxford University Press (2001). https://doi.org/10.1093/acprof:oso/9780195148367.001.0001

Classifying Audio Timbre Without Audio Using Text-Only Training

Peter McCabe[(✉)] [iD] and Patrick J. Donnelly [iD]

Oregon State University, Corvallis, OR 97330, USA
`{mccabepe,donnellp}@oregonstate.edu`

Abstract. Text-only training is a promising machine learning paradigm for training multimodal models without requiring data from every modality. However, despite the potential of text-only training, to date few studies have explored its use as an approximation of missing data for supervised learning in data-scarce environments. In this work we define the kinds of problems suited for text-only training and examine techniques to acquire or design text-based training data. We address the modality gap's role in the performance of text-only training, and present a case study on classifying subjective audio timbre descriptions based on three kinds of text-only training data and six augmentation methods on eight audio-timbre datasets. Our work shows that the text-only training paradigm successfully trains audio classifiers without audio and opens the door to future work in examining the effectiveness of text-only training for supervised machine learning problems without available datasets.

Keywords: Timbre · Machine Learning · Text-only Training

1 Introduction

Large multimodal models have been shown to be very effective at a wide variety of tasks in the machine learning community. Lately, an increasing body of research has explored using text as the only data modality during training (text-only training), despite its use in multimodal applications at inference time. This strategy is highly desirable because it circumvents the need for large datasets annotated across multiple modalities. Acquiring text data is often far cheaper and available at much larger scales than other modalities like images, audio, or video. While most research in this area has focused on tasks like captioning [8,36], retrieval [21], and zero-shot classification [17], little work has been done to explore this promising method for direct use in downstream supervised learning. These tasks naturally allow a practitioner to use paired text data, such as captions, metadata, etc., as both labels and input in a self-supervised fashion [8,36], but little work has been done exploring text-only training outside of this narrow context. Instead, we investigate a common frustration encountered by researchers: lacking necessary data. This problem motivates our work: *how effective is Text-only Training at solving supervised machine learning problems*

P. Machado et al. (Eds.): EvoMUSART 2026, LNCS 16523, pp. 347–366, 2026.
https://doi.org/10.1007/978-3-032-24350-8_23

in which there are no pre-existing supervised data? To explore this question we present a case study on the task of audio timbre classification.

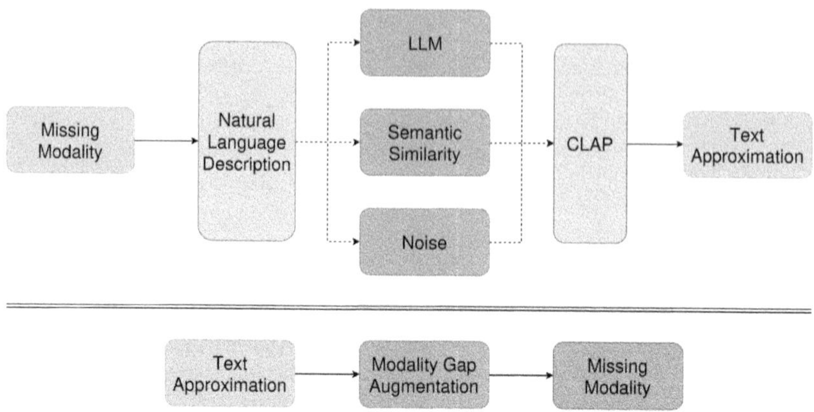

Fig. 1. By describing the desired attributes of the missing modality with words, CLAP [70] can approximate the absent modality. Then, by bridging the gap between its latent text and audio embedding spaces, one increases the fidelity of the approximation.

We must first define text-only training (ToT) and describe the kinds of problems it is well-suited to solve. An overview is provided in Fig. 1. ToT refers to the method of using only text data to train machine learning models by leveraging a pretrained multimodal model capable of comparing data from different modalities. By using the most similar text data to the source material in lieu of the original, one may approximate the same data in the missing modality. In other words, instead of using real audio data, one could use the text data that is most similar to it according to the chosen multimodal model. Therefore, the kinds of supervised learning problems best-suited to ToT have three criteria: (1) there is a lack of labeled data in the original modality, (2) the source modality must have a pretrained multimodal model able to measure semantic similarity, and (3) the desired property or properties should be well-described by language.

First, one can expect the ToT process to be noisy, and therefore would prefer to use cleaner data in the desired source modality. Second, ToT requires a semantic connection between text and desired modalities, usually in the form of a pretrained multimodal model. Third, in order to best approximate it, the desired data's key property (e.g., label, structure, value, etc.) should be easily and fully describable with natural language. In other words, as little information is lost when describing the property with words as possible.

For our case study, we explore classifying subjective labels of non-music audio timbre. This challenge involves the prediction of what a human might typically describe a non-musical sound "sounding like" or the defining characteristic or description of a sound. This task meets the above criteria because there are no large datasets available, there are several multimodal contrastive audio-language models suitable for our task [14,18,70], and there are a multitude of ways to describe the timbre of sound [4]. We suggest that this task represents a typical use-case for the ToT paradigm, and seek to show the potential efficacy, pitfalls, and practitioner decisions required to make the paradigm succeed. In summary, this case study seeks to answer our research question via the proxy task of non-musical audio timbre classification.

1.1 Background

Together with rhythm, pitch, and volume, timbre is a fundamental aspect of sound that allows humans to differentiate between multiple sounds. In other words, a sound's timbre, or tone color, is usually described as what a sound "sounds" like. While often used to describe the particular characteristics of instruments [22], musical moods and themes [15], or genre [61], every sound has a describable timbre. For example, one might describe the sound of a spoon striking a crystal glass as "bright" or "ringing" [53]. These everyday sounds and subjective descriptors serve as the data and labels for our experiments as they are often passed over by the machine learning community in favor of musical excerpts [3] or musical instruments [19], used as features for other tasks [2,57], or as only parts of long-form labels for captioning [9,35].

In this work we choose the LAION-CLAP implementation [70] of CLAP [13] as our pretrained audio-language model. CLAP is a popular multimodal Audio-Language Model based on the highly influential CLIP [43] model used in computer vision. At a high level, CLAP is trained to encode paired audio and text data as vectors, and predict matching pairs within a given training batch. At inference time, CLAP is used to measure how well an audio and text pair match semantically. Mathematically, CLAP seeks to minimize the symmetric InfoNCE [37] loss. This loss encourages the vector representations of paired or semantically similar audio and text vectors to be near each other in the CLAP embedding space, and to encourage dissimilar pairs to be far apart. A typical use is providing the names of classes, e.g., "A sound of a [class]", and selecting the class label which CLAP predicts to match each sound [14,70].

1.2 Modality Gap

Unfortunately, CLAP and models like it suffer from the effects of the Modality Gap, which refers to the phenomenon observed in multimodal models in which the embeddings from each modality reside in separate subspaces within the overall embedding space [28]. These subspaces are nearly or entirely disjoint.

Despite being trained to restrict cross-modal paired data to be as close as possible and all other data as far as possible, the CLAP embedding space has two distinct clusters of text and audio embeddings. To show CLAP's modality gap, we follow [56] to train a linear classifier to predict the embedding modality on the Clotho [12] dataset. We achieve > 0.998 F_1-score on every trial of a 10-fold cross validation experiment, indicating the text and audio embeddings are almost entirely linearly separable. With respect to ToT, this means that naïvely using the text embedding instead of the audio embedding for any given pair will not lead to identically trained models. In essence, the modality gap is a specific form of domain adaptation which necessitates methods for mitigating its effect and motivates some of our experiments.

2 Related Work

To better understand the process of ToT and closing the modality gap, we review several pertinent studies. We also summarize relevant research about the perception of timbre and attempts to automatically quantify it.

Text-only training has mostly been explored in the computer vision (CV) community, and has been applied to a variety of CV tasks, including: image captioning [68,73,74], generation [45,79], zero-shot classification [17,71], multimodal retrieval [21], visual question answering [59], category discovery [67], and improving multimodal embedding model alignment [30,72]. In the audio domain, ToT has been used in targeted sound extraction [31,32,51], captioning [25,75], and retrieval [20]. We are inspired by [8] who train an audio captioning model with only text with augmented captions as input instead of audio.

Without addressing the modality gap, ToT faces significant performance degradation. Broadly, these methods tend to be supervised [10,11,43] or unsupervised [17,36,79] mappings that act directly on one modality's embeddings to approximate embeddings in the other, architecture or optimization changes [1,56,63], or changes to the underlying embedding space [5,44,78]. As our work focuses on utilizing pre-trained models, we focus on the first category.

In one of the first works on approximating cross-modal embeddings, Radford et al. train a diffusion model to convert between text and image embeddings for image generation [43]. After introducing the modality gap, Liang et al. mitigate the gap by vector addition [28]. Nukrai et al. compare that with adding Gaussian noise and prefer the latter approach [36], while Tam et al. find that it is preferred over using optimal transport or PCA features [60]. Saijo et al. compare several gap-closing methods, including Gaussian noise and PCA features, and find that adding noise is the superior method [51]. Similarly, Role et al. find both optimal transport and PCA features inferior to a spectral embedding method [50].

The study of human description and perception of timbre is most often found in the musicology [26,48] and psycho-acoustic [4,57,66] literature. Relatively few studies in the machine learning community have focused on the prediction or linguistic description of timbre, choosing to use the timbre of sounds as features rather than labels [2,19,57]. Mannone et al. cluster listener associations with timbre and color [33], and Reymore et al. differentiate between fine-grained timbral descriptors of noisy instruments [47]. Mapping timbre has been investigated in the context of digital music production [29,54,64] and audio effects [6]. Audealize [55] uses a crowd-sourcing approach to learn connections between words and digital audio effects. Directly predicting timbral descriptors has been performed in K-Pop vocal styles [23], musical instruments [22,49], and general audio [38].

With the rise of LLMs, research into how well LLMs "understand" concepts like timbre has developed [34]. Saitis et al. [52] and Siedenburg et al. [58] compare human and LLM timbre ratings of instruments and find similar intra-rater agreements as well as moderate rating correlations. Czedik-Eysenberg *et al.* measure the similarities between human perception of three timbral descriptors and popular audio feature extraction software libraries and five variants of CLAP [7]. Tian *et al.* perform a similar analysis between 18 representations [62]. Finally, Velissaridis *et al.* compare language and audio-language models equipped with timbral descriptors in instrument classification and emotion recognition [65].

3   Datasets

Because we investigate the scenario in which there is no available data whatsoever, arguably the most important decision a practitioner of ToT must make is how to source the text data. Depending on the exact nature of the desired dataset, there are numerous ways to create text-based synthetic data. Wang *et al.* compare methods for closing the modality gap using four datasets with different levels of multimodality [69]. Qi *et al.* scrape text from the web and use filter models to remove unwanted data [42]. A similar approach uses a captioning model and filter to generate synthetic captions for both images [27,78] and audio [24]. As we do in this study, LOVM [71] directly generates text data via LLM by prompting for descriptions based on class labels. Before we discuss the methods we choose, we first introduce our validation datasets.

3.1   Audio Commons

The principal datasets used in our work are the AudioCommons (AC) Timbral Characterisation Tool v0.2 Development Dataset(s) [38,39] developed to create the `Timbral Models` python package [40]. This tool is designed to measure how well seven different timbral descriptors match a given audio recording, as well as to detect the presence of reverb. The dataset is comprised of eight subsets corresponding to each descriptor. For each descriptor a collection of sounds was sourced from FreeSound[1]. Next, for all words[2], human annotators were presented with a set of eight sounds and rated each on a scale from 0–100 on how well that word described the sound relative to the others.

To simplify the creation of new text data (described below), we binarize the continuous-valued annotations into two classes. Additionally, by reducing the difficulty of the downstream task, we can clearly see the effects of ToT on an otherwise simple problem. First, we calculate the mean of all annotations, then set all values less than 50 as the negative class (e.g., *Not Bright*) and those 50 or greater as the positive class (e.g., *Bright*). Finally, we compute the CLAP embeddings for all audio files, and use the pairs of embedding and binary label as our testing data. In Table 1, we provide the specific details on dataset sizes and label distributions of the eight binary datasets.

---

[1] https://freesound.org/, last accessed 2026/02/01.
[2] The *Reverb* set was created separately by the authors of [40].

Table 1. Sizes and class balance of each timbral descriptor audio dataset. †We use half of the reverb dataset ($n \approx 400$) for training and testing the audio baseline and the other for development and validation.

	Booming	Bright	Deep	Hard	Reverb†	Rough	Sharp	Warm
Count	102	210	218	206	220	212	165	184
% Positive	30	62	61	67	53	55	60	42

3.2 Pseudo-labeling Real Captions

Now that we have established our problem and class labels, we can explore different methods for generating labeled text data. The first of the three dataset sourcing methods we explore in this work is labeling real captions with text-based semantic similarity models. The process consists of labeling real audio captions by comparing each caption with a predetermined sentence describing the label. In this work we call these "Anchor Prompts". We write five versions of our anchor prompt to reduce bias and average their embeddings to create a central pseudo-prompt, which we list in Table 2.

Table 2. The prompts we used for semantic similarity labeling with CLAP and SBERT. Each prompt is embedded into the model's embedding space and the mean is calculated and re-normalized for use as the comparison pseudo-prompt.

A {label} sound.
A sound that could be described as: {label}
An audio clip with a {label} quality.
An audio clip that sounds {label}.
A sound/audio clip that has or contains {label}.

As in works like [43], each label's mean anchor prompt embedding is then compared with the caption's embedding vector and the class with the highest similarity is selected. We use two large models that capture semantic relationships between text pairs: LAION-CLAP [70] and the SBERT [46] model ALL-MINILM-L12-V2. CLAP is the natural choice for audio-related text-text similarity, since it is explicitly trained to match text and audio. This also represents the natural choice for a practitioner already using CLAP as an encoder to also use it as the pseudo-labeler in their workflow. SBERT juxtaposes CLAP as a text-only semantic similarity model trained on general text data. While similarly designed to correctly identify semantically-similar text pairs, SBERT was not necessarily intended to be used on audio domain-specific tasks, and is selected to compare strategies of using a domain-specific similarity versus a general language similarity model.

We choose WAVCAPS [35] as our caption dataset because it is both large and widely used in training multimodal models for tasks like audio captioning.

Because much of the data used in WavCaps was also used to train LAION-CLAP, we can be more confident that the concepts that the captions describe are in CLAP's distribution, and the label predictions are likely more accurate than otherwise [70]. It consists of over 400K partially-human annotated caption-audio pairs, and the final distributions of labeled data for each model are shown in Fig. 2. None of the test audio is present in WavCaps.

The advantage to this method is that it uses real text data and is cheaper than sourcing new data altogether. However, this method relies on the performance of the model(s) chosen to label the data. As can be seen in Fig. 2, the bias of the chosen model can dramatically alter the final dataset's distribution. Additionally, because label noise is particularly harmful to machine learning models, one must rely on finding large amounts of real data to overcome the noise present in the labeling process, which may not be available.

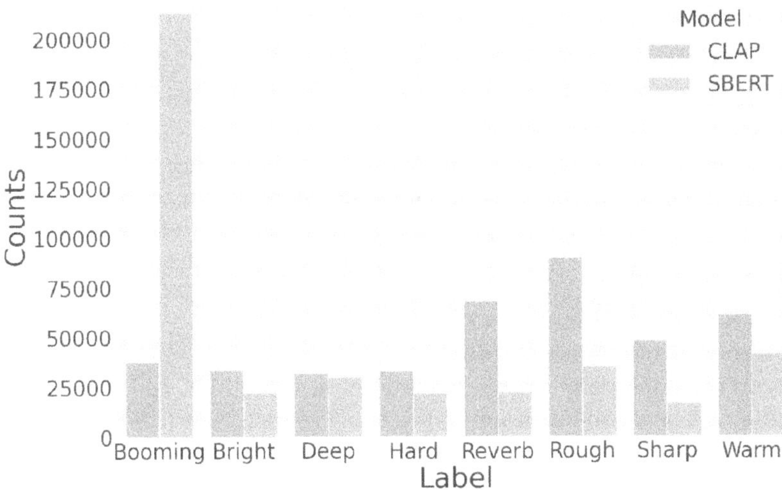

Fig. 2. The class label distributions of WavCaps [35] as determined by text-text similarity between captions and a label prompt. We compare CLAP [70], an audio-language model, and SBERT [46].

3.3   LLM Captions

To contrast with labeling existing captions, another natural choice is to use one of the widely available LLMs to generate pseudo-captions based on a desired class [71]. To test this method, we employ the `Llama-3.1-8B-Instruct` [16] LLM. For each label in the AC dataset we prompt the LLM to generate descriptions of audio events that might be described using that label. For each label we generate approximately 15k samples after removing duplicates, and assign a class label based on the prompted timbral descriptor.

One might prefer this method over labeling real data because one could theoretically generate an arbitrarily large amount of data. However, generating high-quality text data using LLMs is expensive and ultimately not as reliable as real data. This method suffers from the same possible label noise mentioned above, because the generated text might not actually correspond to the provided label as desired. For example, we empirically found that, despite selecting hyperparameters that promoted varied responses, many of the generated captions described the same scene in slightly different words. This limits the overall distribution of captions and corresponds to an effective dataset size smaller than it seems. And just as we saw above, the exact output distribution depends on the model chosen, but with the added burden of tuning hyperparameters.

3.4   Label-Centered Gaussian Noise

For our final method, we explore training purely on adding noise to our anchor prompts. This way, a practitioner need not bother with the expensive processes of pseudo-labeling or generating captions from scratch. Inspired by [77], who model discrepancies in modality alignment as Gaussian noise, we utilize the anchor prompts and their mean embedding as the centers of six separate Gaussian distributions with a shared variance we measure following the procedure in [8, 36]. In other words, each example is constructed as a mixture of six Gaussians, each with the same variance, but centered around different anchor prompts. We generate approximately 25k samples for each label.

This method is the least labor-intensive of the three we present and is easily scalable to an arbitrary size. However, it assumes that the embedding model has a general "understanding" of the classes in both modalities, and that the anchor prompt(s) act as a good measure of the mean of each class's distribution in the text space. Luckily, we already assume our embedding model has at least some correlation between the desired data and descriptive text labels, otherwise the contrastive-learning objective underpinning the model would have failed. Since we can empirically see the effectiveness of the CLAP model and its variants [18, 32], we are able to implicitly verify our assumption.

4   Methodology

In this section, we discuss methods for addressing the modality gap, our experimental design, and the comparative baselines we select.

4.1   Bridging the Modality Gap

Bridging the modality gap is a specific form of domain adaptation, meaning applying machine learning to one's out-of-domain dataset will likely result in poor performance. Therefore, it is critical that a practitioner of ToT address the modality gap problem. While there are many methods used in practice, we focus on common baselines and the methods with the most theoretical backing. We compare (1) doing nothing, akin to not performing domain adaptation, (2) with the method first explored by Liang *et al.* in their work that introduced the modality gap [28], and (3) the method described by Zhang *et al.* [76,77]. Liang *et al.*'s method (Mean Shifting) involves measuring the mean difference vector between each modality, then using that mean vector to shift samples from one modality towards the other. The mean difference vector is defined as:

$$\Delta_{gap} = \frac{1}{n} \sum_{i=1}^{n} a_i - \frac{1}{n} \sum_{i=1}^{n} t_i \tag{1}$$

where a_i and t_i refer to the i_{th} audio and text embedding respectively. Then, by adding the mean vector to the text embeddings, one can approximate the paired audio vectors: $\hat{a}_i = t_i + \Delta_{gap}$. We store the mean audio embedding of the Clotho dataset [12], and calculate the mean of each text dataset during training.

Our third method is taken from Zhang *et al.* (named C2). They proved under certain empirically verified assumptions that their method, C2, removes the modality gap [77]. C2 involves centering each modality at zero by subtracting the mean embedding vector from each modality during both training and inference. Similar to Mean Shifting, C2 shifts one modality towards the other, but while Mean Shifting pushes one modality towards the empirical mean of the other, C2 shifts both modalities towards zero.

As discussed by Zhang *et al.* the two modalities are nominally aligned at zero, but suffer from "alignment noise" that prevents paired text and audio from being perfectly aligned [77]. To account for alignment noise, the authors introduce "C3" which expands on C2 by injecting Gaussian noise into the training data to mimic additional noisy samples centered near the real data and to train a model robust to noisy data. This also falls well within the pattern of the literature, in which many authors use added Gaussian noise or other random methods for bridging the gap [8,36,51]. For a fair comparison, we investigate adding noise to all three methods listed above: Nothing, Mean Shifting, and C2. We label these as Gaussian Noise, Shift+Noise, and C3 [77].

4.2   Experiments

For each combination of training dataset, data augmentation strategy, and target dataset we perform the following experiment: (1) Randomly sample a balanced training dataset by undersampling the negative classes to match the number of positive examples[3], (2) Augment the data with the augmentation strategy[4]. (3) Calculate the F_1-score on the target audio dataset. We run every experiment 10 times and report the mean and standard deviation of test F_1-scores.

---

[3] Recall that there are eight total classes but we perform a series of binary classifications.

[4] Note that we add noise only once; we find repeatedly adding noise makes little difference.

Additionally, we perform ablations on training with the full-but-imbalanced training sets and combinations thereof. Training on the large imbalanced datasets accounts for any performance changes that might occur due to the undersampling used in the main experiments, but suffers from significant class imbalances. Then, we train on combinations of the full datasets to examine if one or more strategies compensates for the shortcomings of the other(s).

We select our classification model based on the theoretical guarantees provided by Zhang *et al.* [76]. Under the same conditions as C2, they prove that any regularized linear classifier minimizing a quadratic loss will be identical when trained on either modality. This means that under perfect conditions the model trained on text would exactly match one trained on real audio. We select the RidgeClassifier from `scikit-learn` [41] with default hyperparameters as our model. We utilize CLAP's zero-shot similarity scores between the test audio datasets and the mean embedding of our anchor prompts as a common baseline. For each dataset we calculate the mean of the negative class anchor prompt embeddings and select the larger CLAP scores between the positive and mean negative prompt embedding as the predicted class. Finally, we compare these scores against those produced by simply using the small real audio datasets.

5   Results and Discussion

First, we examine the differences in performance between each dataset creation method. In Fig. 3, we observe that the CLAP labeling of WAVCAPS and Label-Centered Noise tend to outperform both the LLM-generated captions and the SBERT labeling. Unsurprisingly, the two best methods are based around the CLAP embedding space itself. The Label-Centered Noise dataset assumes that the audio clips that have the described timbral characteristic are embedded in the CLAP space near the anchor prompt, while the WAVCAPS-CLAP dataset relies on the captions that display semantic similarity to the anchor prompts to be embedded nearby. The former approach relies on between-modality similarity, while the latter relies on within-modality similarity. Viewed through that lens, the better performance on the Label-Centered Noise dataset could be explained by the effects of the modality gap itself, because within-modality similarities are on average higher than cross-modal similarities [28]. Additionally, the WAVCAPS-CLAP dataset may contain more mislabeled data as captions with similar meanings in the CLAP embedding space are confused with each other.

As shown in Fig. 4, the LLM-generated dataset appears to perform particularly poorly in the audio modality, despite scoring highly during training. It exhibits the greatest deviation between train and test scores, indicating that its examples were easily separable, but failed to emulate real audio data. This could be explained by the kind of language model we chose. By following existing work [71] and using a general language model for dataset creation, we sacrifice domain-specific knowledge that might be available in a specialized LLM. The SBERT labeling exhibits the opposite behavior: underperforming in train scores, but occasionally competing for the second best scores on the test sets. This is somewhat surprising due to the SBERT dataset's heavy bias towards the *Booming*

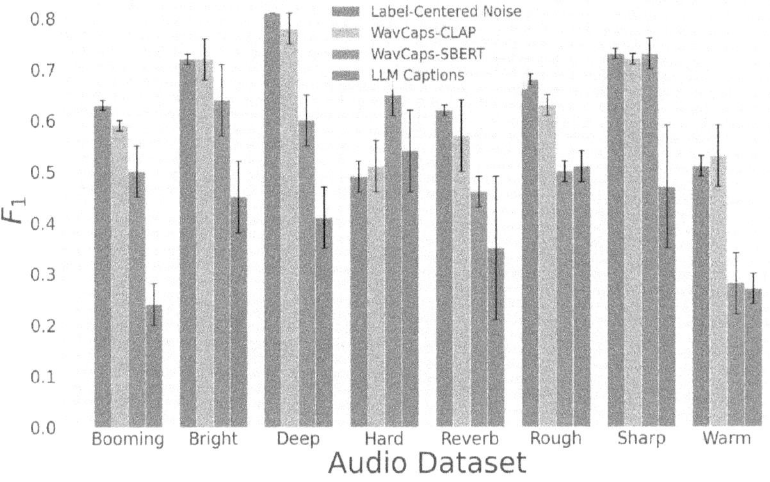

Fig. 3. Mean and std. of test scores averaged across data augmentation methods.

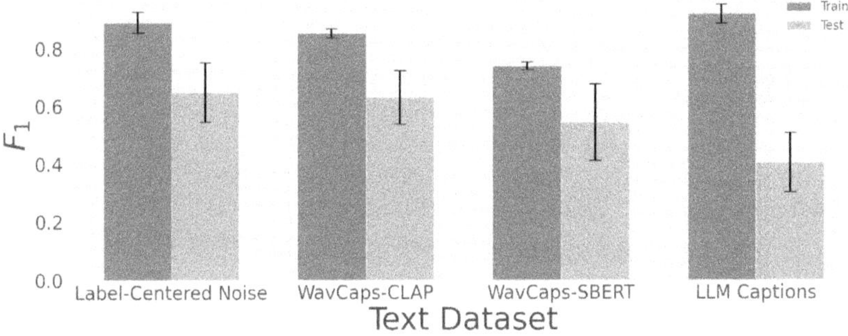

Fig. 4. Mean and std. of train and test scores across all augmentations and targets.

class (see Fig. 2). Since *Booming* is a minority class in the audio modality, one might expect SBERT's labeling to fail to achieve success on real audio.

Next, we analyze the impact of each gap-closing method (see Fig. 5). As stated above, applying no data augmentation method generally harms performance. This is particularly clear in the *Hard* and *Reverb* datasets, in which performance drops significantly when using no data augmentation. C2 and Mean Shift exhibit approximately comparable performance, although Mean Shift does have marginally more variance in its scores. When noise is added, C3 achieves the highest test scores for the majority of audio datasets, but does not show any major improvements. As seen in Table 3, only the change from performing no data augmentation to adding Gaussian noise showed any noteworthy improvement.

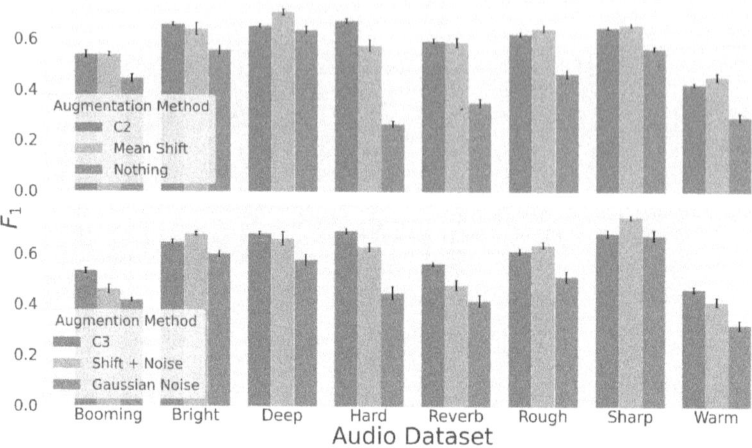

Fig. 5. Mean test scores using methods without (top) and with (bottom) added noise.

Additionally, we explore the effects of the size of the dataset in an ablation study. In Fig. 6, we show additional experiments performed on the full and unbalanced datasets, as well as all combinations of the datasets. Recall that each constructed dataset has approximately equal numbers of each class before binarization, which means each binary dataset has seven times the number of negative examples as positive examples. Additionally, the class distributions of the pseudo-labeled datasets WavCaps-CLAP and WavCaps-SBERT are dependent on the models' outputs and vary widely between them (see Fig. 2).

Table 3. Mean and std. of change between noise-free and added-noise methods.

Method	C3	Shift + Noise	Gaussian Noise
Change	0.009 ± 0.025	-0.012 ± 0.065	0.050 ± 0.070

We find that the average F_1-score for all large unbalanced datasets is lower than the average scores of the balanced datasets. Additionally, the variance of the scores across methods and target datasets is much higher. This indicates that naïvely creating large datasets may not be effective for ToT. Combining datasets also harms performance, indicating that dataset size does not overcome class imbalance. This is also likely caused by conflicting labels between datasets. Each method has its own distribution: some examples in the same class from a different datasets may appear very different. Combining datasets may increase the overall proportion of noisy labels.

Finally, Table 4 lists the average and best performing model scores for experiments with data augmentation. For comparison, we show CLAP zero-shot scores

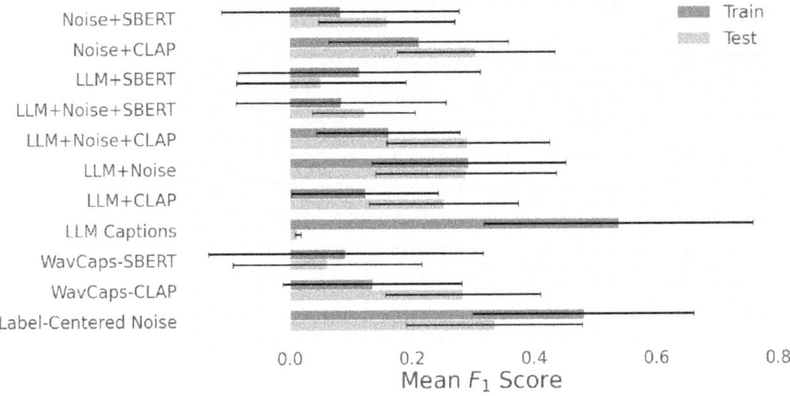

Fig. 6. Ablation study on training with large and unbalanced datasets.

and the mean cross-validation score from training on real audio datasets. While the average model scores are generally less than CLAP's zero-shot performance, we find the best performing model is always at least as good as CLAP, and is better in seven out of eight cases. Notably, for a majority of audio datasets, the best text-only models were within 0.05 F_1-score of the average model trained on real audio. More impressively, ToT meets or exceeds the performance of training on real audio on the *Bright* and *Sharp* datasets, respectively.

Table 4. Test F_1 scores of the mean and best models with data augmentation, CLAP zero-shot, and 10-fold cross validation on real audio. Best text-only results are in bold.

Dataset	Booming	Bright	Deep	Hard	Reverb	Rough	Sharp	Warm
Average Model	0.52	0.66	0.68	0.64	0.56	0.63	0.68	0.44
CLAP Zero-Shot	0.58	0.76	**0.85**	0.25	0.57	0.62	0.75	0.47
Best Model	**0.67**	**0.79**	**0.85**	**0.81**	**0.66**	**0.78**	**0.80**	**0.62**
Audio Datasets	0.72	0.79	0.89	0.85	0.82	0.81	0.78	0.77

Overall, we demonstrate that dataset creation techniques which leverage the embedding space required by text-only training tend to lead to better downstream performance. Addressing the modality gap is necessary for ToT to succeed, but the exact method is not as important. Adding noise can help but is less effective than other methods at bridging the modality gap. Despite label noise, training on small balanced datasets proved to be more successful than training on large and unbalanced data. Altogether, we find that text-only training has the potential to outperform CLAP and training on the audio itself.

6   Future Work and Conclusion

While we demonstrated that text-only training can be effective under the right circumstances, there are still many avenues of future research to explore. In this work we limited ourselves to binary classification to simplify and align the downstream task with text-only training. However, timbre is inherently multi-label and continuous-valued, and future work in timbre classification should explore these more difficult versions of the problem. As our study serves as an introduction to the problem and not as a definitive source for the state of the art, we do not perform hyperparameter tuning on models and noise, nor claim to have found the best-performing paradigm for text-only training. Likewise, we only explored common augmentation techniques and classification models with theoretical backing, but in practice, text-only training is most often used in conjunction with deep learning architectures [8, 32].

We define three criteria for problems that best suit text-only training, but provide no empirical justification. As future work, we intend to explore the effect of how much mutual information is shared between the missing modality and the language used to describe it. In this work, we did not compare using different LLMs to generate training data, or different embedding models. As mentioned above, each LLM could have a different biased output distribution with respect to desired properties, and each embedding model has its own biases that might affect performance on downstream tasks. Yet, even when using the same LLM, there are more variations of dataset creation methods to be explored. For simplicity we restricted our generated datasets to have labels that matched our validation sets, but in the real world many different descriptors might describe a sound. For that reason we recommend future work to investigate using multiple labels as prompts, out-of-domain labels as negative examples, and other techniques to better mimic real data. While we explored three methods for creating text data, many more approaches are possible. For example, we did not exhaustively explore methods for adding noise. As future work, we intend to consider Gaussian noise with non-constant covariance matrices, as there is no guarantee of non-correlated alignment noise.

In this study we introduce a text-only supervised learning paradigm and discuss the challenges of dataset creation and addressing the modality gap. We present a case study on audio timbre classification and compare three training data generation techniques and six augmentation strategies on eight timbre datasets. We show that leveraging the embedding model's inherent knowledge is superior to using dedicated text-text similarity methods or generating text using an LLM, and that addressing the modality gap is a necessary step in the process. Finally, we show that text-only training is a viable machine learning paradigm by meeting or exceeding the performance of models trained on real audio.

References

1. Al-Jaff, M.: Messing with the Gap: On the Modality Gap Phenomenon in Multimodal Contrastive Representation Learning. Ph.D. thesis, Uppsala University (2023)
2. Aucouturier, J., Pachet, F.: Improving timbre similarity: how high's the sky? J. Negat. Res. Speech Audio Sci. (JNRSAS) (2004)
3. Bogdanov, D., Minz, W., Tovstogan, P., Porter, A.: Mtg-jamendo dataset (2019). https://doi.org/10.5281/ZENODO.3826813
4. Carron, M., Rotureau, T., Dubois, F., Misdariis, N., Susini, P.: Speaking about sounds: a tool for communication on sound features. J. Des. Res. **15**(2), 85 (2017). https://doi.org/10.1504/JDR.2017.086749
5. Chen, H., He, L., Liu, Y., Yang, L.: Visual neural decoding via improved visual-EEG semantic consistency (2024). https://doi.org/10.48550/arXiv.2408.06788. Accessed 26 Jun 2025
6. Chu, A., O'Reilly, P., Barnett, J., Pardo, B.: Text2FX: harnessing clap embeddings for text-guided audio effects. In: ICASSP 2025 - 2025 IEEE International Conference on Acoustics, Speech and Signal Processing (ICASSP), pp. 1–5 (2025). https://doi.org/10.1109/ICASSP49660.2025.10890334, iSSN: 2379-190X
7. Czedik-Eysenberg, I., Reuter, C.: Evaluating timbre-related audio descriptors across different libraries and multimodal embeddings. In: Proceedings of DAS|DAGA 2025 (2025). https://doi.org/10.71568/dasdaga2025.657
8. Deshmukh, S., Elizalde, B., Emmanouilidou, D., Raj, B., Singh, R., Wang, H.: Training audio captioning models without audio. In: ICASSP 2024 - 2024 IEEE International Conference on Acoustics, Speech and Signal Processing (ICASSP), Seoul, Republic of Korea, pp. 371–375. IEEE (2024). https://doi.org/10.1109/ICASSP48485.2024.10448115
9. Doh, S., Choi, K., Lee, J., Nam, J.: LP-MusicCaps: LLM-based pseudo music captioning. In: Proceedings of the 24th International Society for Music Information Retrieval Conference, pp. 409–416 (2023)
10. Dong, H.W., et al.: CLIPSonic: text-to-audio synthesis with unlabeled videos and pretrained language-vision models. In: 2023 IEEE Workshop on Applications of Signal Processing to Audio and Acoustics (WASPAA), New Paltz, NY, USA, pp. 1–5. IEEE (2023). https://doi.org/10.1109/WASPAA58266.2023.10248160
11. D'Orazio, A., Briglia, M.R., Crisostomi, D., Loi, D., Rodolà, E., Masi, I.: Implicit inversion turns CLIP into a decoder (2025). https://doi.org/10.48550/arXiv.2505.23161. Accessed 26 Jun 2025
12. Drossos, K., Lipping, S., Virtanen, T.: Clotho: an audio captioning dataset. In: ICASSP 2020 - 2020 IEEE International Conference on Acoustics, Speech and Signal Processing (ICASSP), Barcelona, Spain, pp. 736–740. IEEE (2020). https://doi.org/10.1109/ICASSP40776.2020.9052990
13. Elizalde, B., Deshmukh, S., Ismail, M.A., Wang, H.: CLAP learning audio concepts from natural language supervision. In: ICASSP 2023 - 2023 IEEE International Conference on Acoustics, Speech and Signal Processing (ICASSP), pp. 1–5 (2023). https://doi.org/10.1109/ICASSP49357.2023.10095889, ISSN: 2379-190X
14. Elizalde, B., Deshmukh, S., Wang, H.: Natural language supervision for general-purpose audio representations. In: ICASSP 2024 - 2024 IEEE International Conference on Acoustics, Speech and Signal Processing (ICASSP), Seoul, Republic of Korea, pp. 336–340. IEEE (2024). https://doi.org/10.1109/ICASSP48485.2024.10448504

15. Ferrer, R., Eerola, T.: Semantic structures of timbre emerging from social and acoustic descriptions of music. EURASIP J. Audio Speech Music Process. **2011**(1), 11 (2011). https://doi.org/10.1186/1687-4722-2011-11
16. Grattafiori, A., et al.: The Llama 3 herd of models (2024). https://doi.org/10.48550/arXiv.2407.21783. Accessed 25 Oct 2025
17. Gu, S., Clark, C., Kembhavi, A.: I can't believe there's no images!: learning visual tasks using only language supervision. In: 2023 IEEE/CVF International Conference on Computer Vision (ICCV), pp. 2672–2683 (2023). https://doi.org/10.1109/ICCV51070.2023.00252, ISSN: 2380-7504
18. Guinot, J., Riou, A., Quinton, E., Fazekas, G.: SLAP: Siamese language-audio pretraining without negative samples for music understanding (2025). https://doi.org/10.48550/arXiv.2506.17815, ISMIR 2025, Accessed 26 Jun 2025
19. Herrera-Boyer, P., Peeters, G., Dubnov, S.: Automatic classification of musical instrument sounds. J. New Music Res. **32**(1), 3–21 (2003). https://doi.org/10.1076/jnmr.32.1.3.16798
20. Hu, R., et al.: Vela: scalable embeddings with voice large language models for multimodal retrieval. In: Interspeech 2025, pp. 2640–2644. ISCA (2025). https://doi.org/10.21437/Interspeech.2025-159
21. Jang, Y.K., Lim, S.N.: Towards cross-modal backward-compatible representation learning for vision-language models (2024). https://doi.org/10.48550/arXiv.2405.14715. Accessed 26 Jun 2025
22. Jiang, W., Liu, J., Li, Z., Zhu, J., Zhang, X., Wang, S.: Analysis and modeling of timbre perception features of Chinese musical instruments. In: 2019 IEEE/ACIS 18th International Conference on Computer and Information Science (ICIS), Beijing, China, pp. 191–195. IEEE (2019). https://doi.org/10.1109/ICIS46139.2019.8940168
23. Kim, K.L., Lee, J., Kum, S., Park, C.L., Nam, J.: Semantic tagging of singing voices in popular music recordings. IEEE/ACM Trans. Audio Speech Lang. Process. **28**, 1656–1668 (2020). https://doi.org/10.1109/TASLP.2020.2993893
24. Kong, Z., et al.: Improving text-to-audio models with synthetic captions (2024). https://doi.org/10.48550/arXiv.2406.15487. Accessed 29 Oct 2024
25. Kouzelis, T., Katsouros, V.: Weakly-supervised automated audio captioning via text only training. In: Proceedings of the 8th Detection and Classification of Acoustic Scenes and Events 2023 Workshop (DCASE2023), Tampere, Finland, pp. 81–85 (2023)
26. Lavengood, M.: A New Approach to the Analysis of Timbre. Ph.D. thesis, CUNY Academic Works (2017)
27. Li, J., Li, D., Xiong, C., Hoi, S.: BLIP: bootstrapping language-image pre-training for unified vision-language understanding and generation. In: Chaudhuri, K., Jegelka, S., Song, L., Szepesvari, C., Niu, G., Sabato, S. (eds.) Proceedings of the 39th International Conference on Machine Learning. Proceedings of Machine Learning Research, vol. 162, pp. 12888–12900. PMLR (2022). https://doi.org/10.48550/arXiv.2201.12086. Accessed 26 Jun 2025
28. Liang, W., Zhang, Y., Kwon, Y., Yeung, S., Zou, J.: Mind the gap: understanding the modality gap in multi-modal contrastive representation learning. In: Proceedings of the 36th International Conference on Neural Information Processing Systems. NIPS '22, Red Hook, NY, USA, pp. 17612–17625. Curran Associates Inc. (2022)
29. Limberg, C., Schulz, F., Zhang, Z., Weinzierl, S.: Pitch-conditioned instrument sound synthesis from an interactive timbre latent space. In: Gabrielli, L., Cecchi,

S. (eds.) Proceedings of the 28-th International Conference on Digital Audio Effects (DAFx25), Italy, pp. 433–440. Ancona (2025)
30. Liu, Z., Liu, J., Ma, F.: Improving cross-modal alignment with synthetic pairs for text-only image captioning. In: Proceedings of the Thirty-Eighth AAAI Conference on Artificial Intelligence and Thirty-Sixth Conference on Innovative Applications of Artificial Intelligence and Fourteenth Symposium on Educational Advances in Artificial Intelligence. AAAI'24/IAAI'24/EAAI'24, vol. 38, pp. 3864–3872. AAAI Press (2024). https://doi.org/10.1609/aaai.v38i4.28178
31. Ma, H., et al.: Language-queried target sound extraction without parallel training data. In: ICASSP 2025 - 2025 IEEE International Conference on Acoustics, Speech and Signal Processing (ICASSP), Hyderabad, India, pp. 1–5. IEEE (2025). https://doi.org/10.1109/ICASSP49660.2025.10887543
32. Ma, H., Peng, Z., Li, X., Shao, M., Wu, X., Liu, J.: CLAPSep: leveraging contrastive pre-trained model for multi-modal query-conditioned target sound extraction. IEEE/ACM Trans. Audio Speech Lang. Process. **32**, 4945–4960 (2024). https://doi.org/10.1109/taslp.2024.3497586, publisher: Institute of Electrical and Electronics Engineers (IEEE)
33. Mannone, M., Distefano, V., Santini, G.: Classes of colors and timbres: a clustering approach. Electron. J. Appl. Stat. Anal. **15**(3), 588–605 (2022). https://doi.org/10.1285/i20705948v15n3p588, number: 3
34. Marjieh, R., Sucholutsky, I., Van Rijn, P., Jacoby, N., Griffiths, T.L.: Large language models predict human sensory judgments across six modalities. Sci. Rep. **14**(1), 21445 (2024). https://doi.org/10.1038/s41598-024-72071-1
35. Mei, X., et al.: WavCaps: a ChatGPT-assisted weakly-labelled audio captioning dataset for audio-language multimodal research. IEEE/ACM Trans. Audio Speech Lang. Process. **32**, 3339–3354 (2024). https://doi.org/10.1109/taslp.2024.3419446, publisher: Institute of Electrical and Electronics Engineers (IEEE)
36. Nukrai, D., Mokady, R., Globerson, A.: Text-only training for image captioning using noise-injected CLIP. In: Findings of the Association for Computational Linguistics: EMNLP 2022, Abu Dhabi, United Arab Emirates, pp. 4055–4063. Association for Computational Linguistics (2022). https://doi.org/10.18653/v1/2022.findings-emnlp.299
37. Oord, A.v.d., Li, Y., Vinyals, O.: Representation learning with contrastive predictive coding (2019). https://doi.org/10.48550/arXiv.1807.03748. Accessed 26 Jun 2025
38. Pearce, A., Brookes, T., Mason, R.: Modelling timbral hardness. Appl. Sci. **9**(3), 466 (2019). https://doi.org/10.3390/app9030466
39. Pearce, A., Brookes, T., Mason, R.: Timbral Characterisation Tool v0.2 Development Dataset (2019). https://doi.org/10.5281/zenodo.2545496. Accessed 02 Aug 2024
40. Pearce, A., Safavi, S., Tim, B., Mason, Russell Wang, W., Plumbley, M.: D5.8: release of timbral characterisation tools for semantically annotating non-musical content (2019). https://audiocommons.github.io/materials/. Accessed 1 Feb 2026
41. Pedregosa, F., et al.: Scikit-learn: machine learning in Python. J. Mach. Learn. Res. **12**, 2825–2830 (2011)
42. Qi, D., Su, L., Song, J., Cui, E., Bharti, T., Sacheti, A.: ImageBERT: cross-modal pre-training with large-scale weak-supervised image-text data (2020). https://doi.org/10.48550/arXiv.2001.07966. Accessed 22 Jul 2024
43. Radford, A., et al.: Learning transferable visual models from natural language supervision (2021). https://doi.org/10.48550/arXiv.2103.00020. Accessed 30 Sept 2025

44. Ramasinghe, S., Shevchenko, V., Avraham, G., Thalaiyasingam, A.: Accept the modality gap: an exploration in the hyperbolic space. In: 2024 IEEE/CVF Conference on Computer Vision and Pattern Recognition (CVPR), Seattle, WA, USA, pp. 27253–27262. IEEE (2024). https://doi.org/10.1109/CVPR52733.2024.02574
45. Ramesh, A., Dhariwal, P., Nichol, A., Chu, C., Chen, M.: Hierarchical text-conditional image generation with clip Latents (2022). https://doi.org/10.48550/arXiv.2204.06125. Accessed 13 Jul 2025
46. Reimers, N., Gurevych, I.: Sentence-BERT: Sentence embeddings using Siamese BERT-networks. In: Proceedings of the 2019 Conference on Empirical Methods in Natural Language Processing and the 9th International Joint Conference on Natural Language Processing (EMNLP-IJCNLP), Hong Kong, China, pp. 3980–3990. Association for Computational Linguistics (2019). https://doi.org/10.18653/v1/D19-1410
47. Reymore, L., Beauvais-Lacasse, E., Smith, B.K., McAdams, S.: Modeling noise-related timbre semantic categories of orchestral instrument sounds with audio features, pitch register, and instrument family. Front. Psychol. **13**, 796422 (2022). https://doi.org/10.3389/fpsyg.2022.796422
48. Reymore, L., Huron, D.: Using auditory imagery tasks to map the cognitive linguistic dimensions of musical instrument timbre qualia. Psychomusicology: Music, Mind, and Brain **30**(3), 124–144 (2020). https://doi.org/10.1037/pmu0000263, place: US Publisher: Educational Publishing Foundation
49. Reymore, L., Noble, J., Saitis, C., Traube, C., Wallmark, Z.: Timbre semantic associations vary both between and within instrumentsan empirical study incorporating register and pitch height. Music. Percept. **40**(3), 253–274 (023). https://doi.org/10.1525/mp.2023.40.3.253, publisher: University of California Press
50. Role, F., Meyer, S., Amblard, V.: Fill the Gap: quantifying and reducing the modality gap in image-text representation learning (2025). https://doi.org/10.48550/ARXIV.2505.03703. Accessed 12 Apr 2025
51. Saijo, K., Ebbers, J., Germain, F.G., Khurana, S., Wiehern, G., Roux, J.L.: Leveraging audio-only data for text-queried target sound extraction. In: ICASSP 2025 - 2025 IEEE International Conference on Acoustics, Speech and Signal Processing (ICASSP), Hyderabad, India, pp. 1–5. IEEE (2025). https://doi.org/10.1109/icassp49660.2025.10888769
52. Saitis, C., Siedenburg, K.: When ChatGPT talks timbre. In: Proceedings of the Third International Conference on Timbre, pp. 73–77. The School of Music Studies Aristotle University of Thessaloniki, Aristotle University of Thessaloniki (2023)
53. Saitis, C., Weinzierl, S.: The semantics of timbre. In: Siedenburg, K., Saitis, C., McAdams, S., Popper, A.N., Fay, R.R. (eds.) Timbre: Acoustics, Perception, and Cognition, vol. 69, pp. 119–149. Springer, Cham (2019). https://doi.org/10.1007/978-3-030-14832-4_5
54. Salas, J.: Generating music from literature using topic extraction and sentiment analysis. IEEE Potentials **37**(1), 15–18 (2018). https://doi.org/10.1109/MPOT.2016.2550015
55. Seetharaman, P., Pardo, B.: Audealize: crowdsourced audio production tools. J. Audio Eng. Soc. **64**(9), 683–695 (2016). https://doi.org/10.17743/jaes.2016.0037
56. Shi, P., Welle, M.C., Björkman, M., Kragic, D.: Towards understanding the modality gap in CLIP. In: ICLR 2023 Workshop on Multimodal Representation Learning: Perks and Pitfalls (2023). https://openreview.net/forum?id=8W3KGzw7fNI, last accessed 2025/10/27

57. Siedenburg, K., Fujinaga, I., McAdams, S.: A comparison of approaches to timbre descriptors in music information retrieval and music psychology. J. New Music Res. **45**(1), 27–41 (2016). https://doi.org/10.1080/09298215.2015.1132737
58. Siedenburg, K., Saitis, C.: The language of sounds unheard: exploring musical timbre semantics of large language models (2023). https://doi.org/10.48550/arXiv.2304.07830. Accessed 20 Jul 2025
59. Song, H., Dong, L., Zhang, W., Liu, T., Wei, F.: CLIP models are few-shot learners: empirical studies on VQA and visual entailment. In: Proceedings of the 60th Annual Meeting of the Association for Computational Linguistics (Volume 1: Long Papers), Dublin, Ireland. Association for Computational Linguistics (2022). https://doi.org/10.18653/v1/2022.acl-long.421
60. Tam, D., Raffel, C., Bansal, M.: Simple weakly-supervised image captioning via CLIP's multimodal embeddings. In: The AAAI-23 Workshop on Creative AI Across Modalities (2023). https://openreview.net/forum?id=UMxeP-FuwyC. Accessed 21 Jul 2025
61. Thompson, A.E.: Playing Tag: An Analysis of Vocabulary Patterns and Relationships Within a Popular Music Folksonomy. Master's thesis, University of North Carolina at Chapel Hill (2008). https://doi.org/10.17615/08PE-W163
62. Tian, H., Lattner, S., Saitis, C.: Assessing the alignment of audio representations with timbre similarity ratings (2025). https://doi.org/10.48550/arXiv.2507.07764, ISMIR 2025. Accessed 24 Sept 2025
63. Udandarao, V.: Understanding and Fixing the Modality Gap in Vision-Language Models. Master's thesis, University of Cambridge (2022)
64. Vaillant, G.L., Molle, Y.: Contrastive timbre representations for musical instrument and synthesizer retrieval (2025). https://doi.org/10.48550/arXiv.2509.13285. Accessed 24 Sept 2025
65. Velissaridis, G., Athwal, R., Musharaf, M., Fazekas, G., Saitis, C.: Evaluating foundation models on timbre-related cognitive tasks. In: 1st Workshop on Large Language Models for Music & Audio (LLM4MA) (2025). https://openreview.net/forum?id=tXyh8CY9kZ. Accessed 27 Oct 2025
66. Wallmark, Z.: Semantic crosstalk in timbre perception. Music Sci. **2**, 205920431984661 (2019). https://doi.org/10.1177/2059204319846617
67. Wang, E., Peng, Z., Xie, Z., Yang, F., Liu, X., Cheng, M.M.: GET: unlocking the multi-modal potential of CLIP for generalized category discovery. In: Proceedings of the Computer Vision and Pattern Recognition Conference (CVPR), pp. 20296–20306. arXiv (2025). https://doi.org/10.48550/arXiv.2403.09974, CVPR 2025. Accessed 26 Jun 2025
68. Wang, J., Zhang, Y., Yan, M., Zhang, J., Sang, J.: Zero-shot image captioning by anchor-augmented vision-language space alignment (2022). https://doi.org/10.48550/arXiv.2211.07275. Accessed 26 Jun 2025
69. Wang, Y., Luo, H., Xu, J., Sun, Y., Wang, F.: Text data-centric image captioning with interactive prompts (2024). https://doi.org/10.48550/arXiv.2403.19193. Accessed 26 Jun 2025
70. Wu, Y., Chen, K., Zhang, T., Hui, Y., Berg-Kirkpatrick, T., Dubnov, S.: Large-scale contrastive language-audio pretraining with feature fusion and keyword-to-caption augmentation. In: ICASSP 2023 - 2023 IEEE International Conference on Acoustics, Speech and Signal Processing (ICASSP), Rhodes Island, Greece, pp. 1–5. IEEE (2023). https://doi.org/10.1109/ICASSP49357.2023.10095969
71. Yi, C., He, Y.H., Zhan, D.C., Ye, H.J.: Bridge the modality and capability gaps in vision-language model selection. In: Globerson, A., et al. (eds.) Advances in Neural

Information Processing Systems. vol. 37, pp. 34429–34452. Curran Associates, Inc. (2024)

72. Yu, Y., et al.: Multimodal knowledge alignment with reinforcement learning. In: Conference on Computer Vision and Pattern Recognition (CVPR) (2023). https://doi.org/10.48550/arXiv.2205.12630. Accessed 30 Jun 2025

73. Zeng, D., Shen, Y., Lin, M., Yi, Z., Ouyang, J.: Zero-shot image captioning with multi-type entity representations. In: Proceedings of the AAAI Conference on Artificial Intelligence, vol. 39, no. 21, pp. 22308–22316 (2025). https://doi.org/10.1609/aaai.v39i21.34386, number: 21

74. Zeng, Z., Xie, Y., Zhang, H., Chen, C., Chen, B., Wang, Z.: MeaCap: memory-augmented zero-shot image captioning. In: 2024 IEEE/CVF Conference on Computer Vision and Pattern Recognition (CVPR), pp. 14100–14110 (2024). https://doi.org/10.1109/CVPR52733.2024.01337, ISSN: 2575-7075

75. Zhang, Y., Xu, X., Du, R., Liu, H., Dong, Y., Tan, Z.H., Wang, W., Ma, Z.: Zero-shot audio captioning using soft and hard prompts. IEEE Trans. Audio Speech Lang. Process. **33**, 2045–2058 (2025). https://doi.org/10.1109/TASLPRO.2025.3567770

76. Zhang, Y., HaoChen, J.Z., Huang, S.C., Wang, K.C., Zou, J., Yeung, S.: Diagnosing and rectifying vision models using language. In: The Eleventh International Conference on Learning Representations (ICLR) (2023). https://openreview.net/forum?id=D-zfUK7BR6c. Accessed 30 Jun 2025

77. Zhang, Y., Sui, E., Yeung, S.: Connect, collapse, corrupt: learning cross-modal tasks with UNI-modal data. In: The Twelfth International Conference on Learning Representations (ICLR) (2023). https://openreview.net/forum?id=ttXg3SKAg5. Accessed 09 Jul 2025

78. Zhao, P., Luan, R., Zhang, W., Wu, P., He, S.: Guiding cross-modal representations with MLLM priors via preference alignment (2025). https://doi.org/10.48550/arXiv.2506.06970. Accessed 26 Jun 2025

79. Zhou, Y., et al.: Towards language-free training for text-to-image generation. In: 2022 IEEE/CVF Conference on Computer Vision and Pattern Recognition (CVPR), New Orleans, LA, USA, pp. 17886–17896. IEEE (2022). https://doi.org/10.1109/cvpr52688.2022.01738

Decoding Emotions: Multimodal Integration of Deep Embeddings, Lyrics and Music-Aware Cues

Alessia Novacco, Francesca Gasparini[(✉)] , Giulia Rizzi ,
and Aurora Saibene

University of Milano-Bicocca, Viale Sarca 336, 20126 Milano, Italy
a.novacco@campus.unimib.it,
{francesca.gasparini,giulia.rizzi,aurora.saibene}@unimib.it

Abstract. Music Emotion Recognition (MER) is a challenging task considering the nuances of defining emotions. While unimodal models provide a good baseline for MER, multimodal models are becoming fundamental to provide an in-depth description of emotions. Leveraging on the multimodal MERGE dataset, we investigate the power of audio-related deep embeddings, lyrics informed features, and music-aware cues in providing an informative set of features for low-impact computational learning models.

Results confirm that multimodal fusion outperforms unimodal approaches. Moreover, different experiments highlight the positive contribution of genre metadata and the potential use of harmonic features for real-time computationally low-impact applications.

These findings confirm the importance of multimodal integration for robust and interpretable emotion recognition systems, while opening up future directions, including advanced feature fusion, user-specific model adaptation (user-tuning), and multi-label emotion representation.

Keywords: Music Emotion Recognition (MER) · Multimodal Learning · Emotion-aware System

1 Introduction

The intersection of artificial intelligence and affective computing has fostered the emergence of systems capable of recognizing and responding to human emotions. Within the music domain, this has given rise to the field of Music Emotion Recognition (MER), which has applications in personalized music recommendation, music therapy, and adaptive interfaces, where understanding emotional cues from music can enhance user engagement and well-being [1].

Unlike traditional recommendation systems that rely primarily on listening history or explicit preferences, emotion-aware systems aim to infer the user's affective state and suggest music that either reflects or guides it toward a desired mood. In this view, music becomes not merely entertainment but an active tool

P. Machado et al. (Eds.): EvoMUSART 2026, LNCS 16523, pp. 367–382, 2026.
https://doi.org/10.1007/978-3-032-24350-8_24

for emotional regulation, capable of supporting motivation, concentration, and psychological balance.

However, MER is a particularly challenging task, considering that emotions may be perceived, felt, induced, or intended when in presence of musical pieces [2,3]. Moreover, emotional perception varies across individuals depending on cultural background, personal experiences, and situational context. Consequently, developing models that generalize well while capturing the nuances of emotions is crucial.

Another major point concerning MER effectiveness is related to the growing limitations of unimodal models in capturing the multifaceted nature of musical emotions. Unimodal systems, although effective within specific domains, struggle to generalize across genres, cultural contexts, and annotation frameworks. For instance, audio-based models often confound emotional categories with correlated acoustic cues such as tempo or loudness, failing to capture subtle variations in valence. Instead, lyric-based models overlook non-verbal affective information conveyed by harmony, timbre, and rhythm.

Recent studies have demonstrated that multimodal fusion can significantly mitigate these issues. Krols et al. [4], using the MoodyLyrics4Q dataset [5], combined Mel-spectrograms for audio and pre-trained word embeddings for lyrics.

Similarly, Zhang et al. [6] introduced the Multilayered Music Decomposition and Multimodal Integration Interaction (MMD-MII) model, which integrates Mel-spectrograms, Mel-Frequency Cepstral Coefficients (MFCCs), and semantic embeddings like Word2Vec and the Bidirectional Encoder Representations from Transformers (BERT) [7] in a hierarchical fusion network trained on the Emotify [8] corpus.

These studies confirm that audio and textual modalities encode complementary emotional information: audio features primarily drive arousal estimation, while lyrical semantics are more predictive of valence. Such findings underline the necessity of integrated, explainable multimodal frameworks to achieve robust and human-interpretable emotion recognition in music.

Despite their success, many multimodal systems still rely on basic text representations, limiting their ability to capture nuanced emotional semantics. Addressing this gap, Ara et al. [9] proposed a natural language processing-enhanced framework that integrates lexical, syntactic, and affective features extracted using lexicons.

On the other hand, unimodal audio-based studies remain essential for benchmarking feature extraction methods. Choi et al. [10] conducted a systematic comparison of audio representations for music tagging, ranging from raw waveforms and fast Fourier transform to Mel-spectrograms and MFCCs, showing that the choice of representation critically affects deep learning performance. Their results suggest that multimodal models should carefully select complementary audio features according to task requirements and data characteristics.

Motivated by recent advances in music emotion recognition, this work advocates a multimodal and interpretable perspective on emotion modelling in music, emphasizing the role of complementary information sources in capturing its affective complexity. The proposed framework explores how heterogeneous cues, spanning signal-level, semantic, and music-aware dimensions, can be integrated to improve emotional prediction.

Specifically, this study combines deep audio embeddings, acoustic descriptors, and lyrical content to jointly model the affective tone and semantic meaning of songs. Beyond the conventional audio–text fusion paradigms, it incorporates interpretable features derived from musical structure and metadata, such as genre information and harmonic progressions. These descriptors act as a bridge between algorithmic predictions and the compositional or perceptual concepts commonly used by musicians and listeners, fostering transparency and facilitating human-centred interpretation of model outputs.

Moreover, the work deliberately prioritises low-complexity and low-impact feature representations, aiming to support the development of real-time and resource-efficient MER systems. This design choice reflects a broader vision of emotion recognition systems that are not only accurate, but also explainable, modular, and adaptable to real-world constraints.

The main contributions of this study lie in (i) a systematic and controlled comparison of deep audio embeddings, lyric-informed features, and music-aware descriptors; (ii) an in-depth analysis of their complementarity under classical yet robust learning paradigms; and (iii) a principled assessment of their suitability for explainable and computationally efficient MER applications.

2   Materials and Methods

In this work, we conducted different experiments on "MERGE: A Bimodal Dataset for Static Music Emotion Recognition" [11] (synthetically described in Sect. 2.1), considering both unimodal and multimodal approaches and studying different set of features. Multimodal approaches follow a modular pipeline, drawn in Fig. 1, designed to process, align, and combine different types of information extracted from each song. Notice that each audio file was trimmed or zero-padded to a fixed duration of 25 s, ensuring uniform input size across samples. Tracks were converted to mono by merging stereo channels and resampled to 22,050 Hz, preserving relevant acoustic information while reducing computational load. The texts corresponding to the lyrics were preprocessed following a standard natural language processing pipeline, including normalization, tokenization, lemmatization, and stopword removal.

In what follows, the dataset, the feature extraction, classification, and regression steps are discussed in depth.

2.1   Dataset

The MERGE dataset [11] was collected to support multimodal MER research and we chose it especially for its bimodal structure, consisting of complementary audio and textual modalities. For this reason, we selected it as a suitable benchmark for our study. The dataset contains (i) song short audio clips of

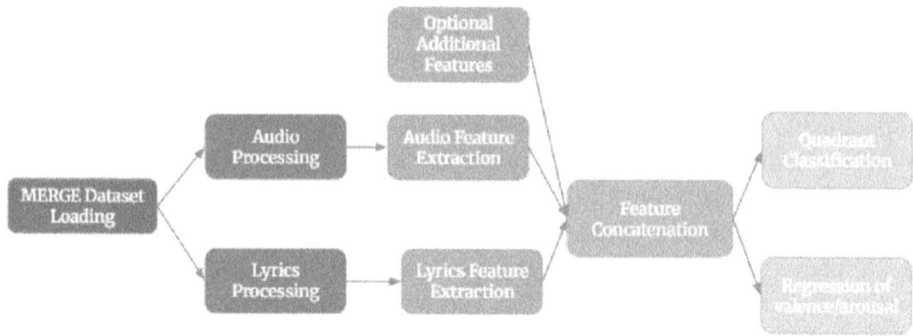

Fig. 1. Multimodal pipeline at the core of the multimodal experiments.

variable duration, (ii) complete lyrics for each song, (iii) song metadata including artist, title, track duration, mood-related tags, genre, and release year, and (iv) semi-automatic and human emotion annotations on the valence and arousal dimensions. Emotion annotations are provided as aggregated valence and arousal scores obtained by averaging multiple human ratings. On the latter point, we divide the annotations according to Russell's circumplex model [12] in the four quadrants of the valence/arousal plane. Therefore, of the 2216 songs, 525 fall in the high-arousal positive-valence (HAPV) quadrant, 673 in the high-arousal negative-valence (HANV), 500 in the low-arousal negative-valence (LANV), and 518 low-arousal positive valence (LAPV). It can be said that the songs are sufficiently well balanced across the quadrants (Fig. 2), and thus class imbalance issues, such as for the Free Music Archive (FMA) [13] or Deezer Mood [14] datasets, are avoided.

2.2 Feature Extraction Methodologies

The feature extraction methodologies mainly regard audio, music-aware cue features, and lyrics.

Audio is a primary carrier of affective information, encoding both perceptual and structural cues relevant to emotional expression. Therefore, for the audio component of each song, the Mel-spectrogram was extracted, considering its effectiveness and computational efficiency when used as input to deep neural networks [10,15]. Subsequently, Mel-spectrograms are used as input to various neural network configurations for audio feature extraction.

Initially, a Convolutional Neural Network (CNN) is used to extract features from the Mel-spectrograms, considering that this kind of CNNs has been designed for processing grid-like data such as spectrograms and providing a compact yet rich representation of audio features.

Afterwards, other three deep learning-based extraction approaches have been considered. Namely, (i) a simple Recurrent Neural Network (RNN), processing the Mel-spectrogram as a temporal sequence, (ii) a combination of a RNN and a

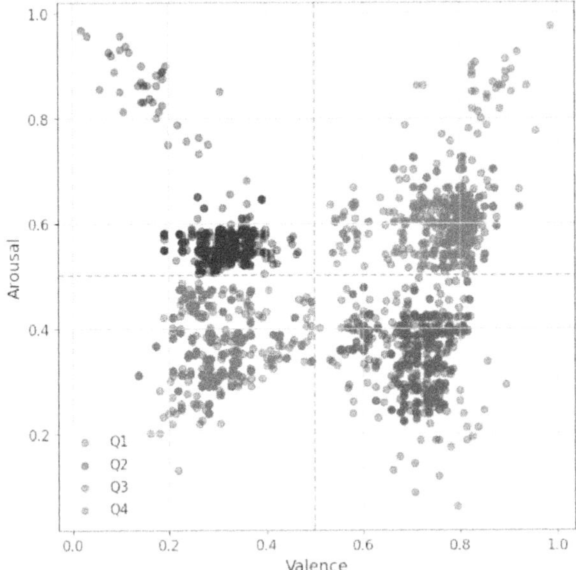

Fig. 2. Distribution of MERGE samples in the valence-arousal space.

Long-Short Term Memory (LSTM) layer to capture longer temporal dependencies, and (iii) a segment-based LSTM, where each song is divided in 5-s segments on each of which are produced embeddings to be concatenated as a final representation intended to understand if temporal variations could better represent emotions.

In addition, two sets of auxiliary **music-aware cues features** were explored, i.e., genres (based on metadata) and chords (low-impact audio feature), to investigate how stylistic and harmonic information influence emotional perception. These features were hypothesized to enhance interpretability and contextual awareness when combined with the other multimodal representations and reduce significantly the computation time for feature extraction.

Here, we build on the hypothesis that a close relationship exists between genres and emotions. In fact, genre labels provide a high-level contextual cue, capturing cultural and stylistic conventions that often correlate with typical emotion induction tendencies in listeners, such as energetic arousal in rock, calmness in classical music, or positively valenced emotional experiences in country music. Moreover, Dutta et al. [16] found that valence trends vary across different genres, justifying our hypothesis. Avoiding an under-representation of genres, we map genres in macro-genre categories and consider such macro-categories as binary features. Notice that a song may be mapped in multiple genres.

An additional line of investigation focused on the harmonic progressions of the songs, since harmony is well known to play a fundamental role in conveying musical emotions and we aim to assess whether and to what extent chord-based information can be effectively exploited by machine learning models for automatic emotion classification. Notice that harmonic choices are not random but rather the result of the composer's prior musical knowledge and creative intent, which inherently shape the affective experience of the listener. For example, major chords are typically associated with positive emotional valence, while minor chords are linked to sad or negative emotions [17].

Chord progression extraction followed the method proposed by Shah et al. [18], which implements a pipeline using Chroma features computed with the `Librosa` library [19]. First, a chroma Short-Time Fourier Transform (STFT) spectrogram was generated to represent the intensity of the twelve pitch classes over time. For each temporal frame of one second, the dominant pitch was identified and treated as the root note of the corresponding chord. The resulting note sequences were then converted into indexed numerical features and normalized.

Subsequently, **lyrics-related features** have been explored following Ara et al.'s methodology [9], considering that lyrics provide semantic and affective cues complementary to audio and express emotions through word choice, semantics, syntactic patterns, and sentiment polarity.

Lexical features such as the average word frequency, average word length, and Term Frequency-Inverse Document Frequency (TF-IDF) vectors computed over a maximum of 100 significant terms [20] were extracted. Instead, stylistic features have been computed considering the syntactic structure of the lyrics, including the average number of words per sentence, the presence of forms of the verb *to be*, and the use of slang expressions, identified through a dedicated lexicon. Particularly, emotional features have been extracted using two distinct emotion lexicons, i.e., (i) the NRC Emotion Lexicon, that associates words with eight basic emotions (anger, fear, anticipation, trust, surprise, sadness, joy and disgust), subsequently mapped onto the corresponding emotional quadrants to produce per-track emotion counts [21], and (ii) the Affective Norms for English Words (ANEW), assigning numerical valence and arousal scores to words, from which average values were computed for each song [22].

All extracted lyrical features were concatenated into a single input matrix and standardized, in order to be concatenated with the other features used.

To identify the most informative predictors for emotion classification, feature importance was estimated using a Random Forest (RF) classifier. The importance of each variable was computed as its average contribution to impurity reduction, measured by entropy across all decision trees, weighted by the number of samples affected. Normalizing these contributions across the entire forest yielded a score reflecting each feature's relative predictive relevance.

The top 40 lyrical features were selected to balance the multimodal representation with the audio feature set, reducing redundancy and improving interpretability. The most relevant features, mainly derived from the ANEW and NRC affective lexicons, along with stylistic metrics such as average word and sentence length, indicate that both the content and the form of the lyrics contribute to emotional expression. Overall, combining lexical, stylistic, and affective features enables the model to capture the emotional richness embedded in song lyrics.

Table 1 summarizes the implemented features and feature extraction methodologies.

Table 1. Implemented features and feature extraction methodologies.

Mode	Features	Methods/Models	Notes
Audio	**Deep embeddings**	CNN, RNN, RNN + LSTM, segmented LSTM	Extraction of Mel-spectrograms from audio using Librosa. Used as direct input to neural networks for feature extraction.
	Music-aware cues	Genres	Derived from MERGE dataset.
		Chords	Harmonic Feature derived from chords through Croma STFT extraction.
Lyrics	**Lexical:** TF-IDF, word length/frequency.	Feature engineered + RF for feature selection	All the lyrical features were concatenated and the top most significant 40 ones were selected.
	Stylistic: average words per sentence, use of "to be", slang		
	Emotional: NRC, ANEW		

2.3   Classification and Regression

Three types of classifiers have been tested: Random Forest (RF), Support Vector Machine (SVM) and Multi-Layer Perception (MLP). To ensure statistical reliability, stratified K-Fold Cross-Validation (K = 5) was applied to the training set (80%) to ensure balanced class distributions, maintaining balanced class proportions across folds, while the test set (20%) is fixed. Considering a four-class task with balanced condition, the performance of the models was assessed using accuracy only.

Unimodal classification experiments considered separately audio, lyrics, genres, and chords features as inputs to the models. Subsequently, different combinations of feature types were considered as multimodal representations to verify our initial hypothesis on multimodality effectiveness in recognising emotions.

As one of the future goals of this work is to recommend songs based on user's emotions, regression models have been considered. In fact, a more emotion-aware recommendation may be achieved by accurately estimating continuous valence and arousal values.

Verified the efficacy of the multimodal approaches, for the continuous prediction task, several regression models were applied to the multimodal feature embeddings, including (i) linear models (i.e., linear regression, ridge, lasso), (ii) tree-based models (i.e., random forest regressor and gradient boosting), (iii) support vector regression, and (iv) an MLPRegressor with two hidden layers.

This setup allowed comparisons between classical and deep regression approaches, evaluating their ability to capture non-linear relationships between multimodal musical features and affective dimensions. For regression, performance was assessed by computing R^2.

3   Results

After considering all the different classifiers introduced in Sect. 2.3 with all the different combinations of features described in Table 1, we found that RF classification generally produces the best results. Moreover, the audio features extracted with CNN from the Mel-spectrograms achieve the best deep embedding of audio content. Therefore, we firstly present a comparison of unimodal and multimodal experiments using the RF classifier. Afterwards, we provide an in-depth analysis of multimodal learning focused on comparing audio deep embeddings and models with low computational complexity.

3.1   Classification Results

As previously summarised, the initial model elected for unimodal and multimodal classification is RF. Figure 3 and Table 2 provide a summary of unimodal and multimodal experiments, where the audio embeddings are extracted with a CNN from Mel-spectrograms (audio), the lyrics are the 40 features resulting from RF ranking (lyrics), and the music-aware cues features are genre and chords.

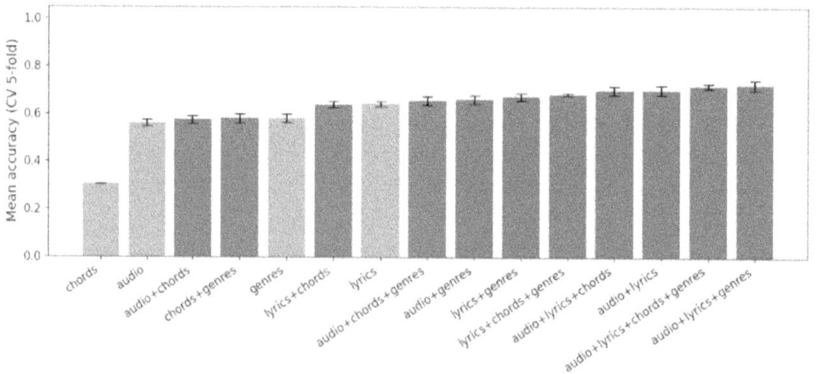

Fig. 3. Unimodal and multimodal model comparison using RF. Unimodal models are highlighted in orange while multimodal ones are reported in blue. (Color figure online)

While lyrics-based unimodal experiments provided almost comparable results with the proposed multimodal experiments, unimodal models using genres, audio, or chords are clearly non-beneficial for MER. When genre information

Table 2. Performance of unimodal and multimodal experiments with RF classifier.

Features	Accuracy	Std
audio+lyrics+genres	0.73	0.0228
audio+lyrics+chords+genres	0.72	0.0297
audio+lyrics+chords	0.70	0.0196
audio+lyrics	0.69	0.0210
lyrics+chords+genres	0.68	0.0042
lyrics+genres	0.67	0.0148
audio+genres	0.65	0.0160
audio+chords+genres	0.65	0.0138
lyrics	0.64	0.0108
lyrics+chords	0.64	0.0128
genres	0.58	0.0182
chords+genres	0.58	0.0186
audio+chords	0.56	0.0196
audio	0.56	0.0234
chords	0.31	0.0009

was concatenated with the audio–lyric embeddings, models exhibited a consistent improvement in the classification task. Specifically, it increased classification accuracy by approximately 4–6% on average. This improvement suggests that genre acts as a strong contextual prior, providing coarse yet meaningful cues about expected emotional tone.

Although the contribution of chords features to multimodal performance is about 1–1.5% lower compared to other multimodal solutions, their extraction and their use in model training have a lower computational complexity compared to the Mel-spectrogram based counterparts (Table 3), making them a highly efficient alternative, particularly suitable for real-time applications.

Table 3. Extraction and training times for the audio feature set and the chord-based feature set.

Step	Feature set	Number of features	Time (s)
Extraction	Audio	32	246.95
Extraction	Chords	30	0.017
Training	Audio	32	1.09
Training	Chords	30	0.16

However, in the present study we are interested in providing an offline baseline on which building personalized models, thus we perform an in-depth analysis of multimodal approaches using different combinations of audio deep embeddings and classification models, considering that genre information, although beneficial, does not directly affect the comparison between audio embedding strategies, and that chord-based features showed limited and non-systematic improvements when combined with deep audio embeddings.

In Fig. 4, we compared the results of multimodal models where lyrics features are fixed and correspond to the lexical, stylistic, and emotional features selected by a RF as summarized in Table 1, while audio embedding techniques and classification models changes. For example, M1 corresponds to the use of a RF classifier that has in input the fixed lyrics features and the deep audio embeddings obtained by passing the Mel-spectrograms in input to a CNN. The exemplified model is also the one achieving the best overall accuracy, reaching an average accuracy of 0.69 with good stability (std = 0.0210).

A detailed analysis of the latter multimodal classifier per emotional quadrant revealed that HAPV and LANV classes were predicted with the highest accuracy, reflecting their stronger acoustic and semantic distinctiveness, whereas LAPV and HANV showed lower recognition rates, often confused with neighbouring quadrants in the valence–arousal space.

The confusion matrix in Fig. 5 indicates that most errors occurred between adjacent quadrants rather than across opposite poles, confirming that the model preserved the topological structure of the emotional space. Misclassifications between HAPV and HANV typically resulted from songs with high energy but ambiguous lyrical sentiment, while confusion between LAPV and LANV arose from low-arousal tracks with neutral or ambiguous wording. This suggests that while multimodal integration effectively captures broad emotional polarity, fine-grained discrimination along the valence axis remains challenging due to the semantic variability of lyrics and subjective annotation noise.

Feature importance analysis, computed via RF impurity reduction, revealed that acoustic features dominated arousal prediction, whereas textual features were more predictive of valence, with word polarity and sentence-level sentiment emerging as key variables.

Overall, this balance between acoustic and linguistic contributions underscores the advantage of multimodal fusion, as each modality encodes distinct yet synergistic aspects of emotional expression.

As MERGE is a recently released resource, no prior studies have explored it empirically, and thus no direct comparison with the literature could be reported.

3.2   Regression Results

Crucially, regression allowed observation of continuous emotional dynamics rather than discrete labels. In Table 4 the regression results obtained considering all the models introduced in Sect. 2.3 are reported. Gradient Boosting Regressor achieved the highest overall performance, with an average R^2 of 0.44 for arousal and 0.37 for valence, confirming its ability to capture non-linear relationships between multimodal features and affective dimensions.

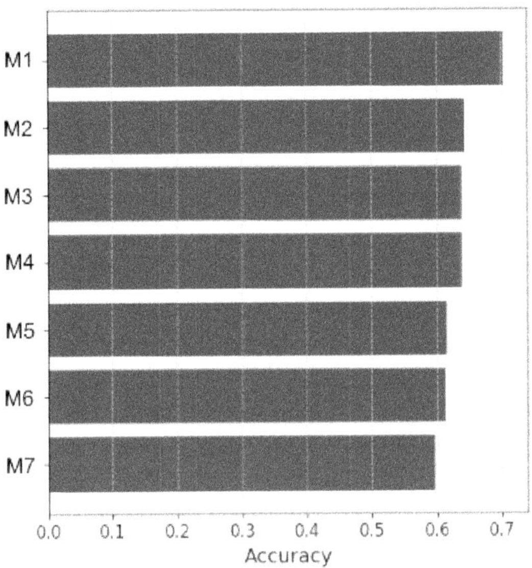

Fig. 4. Comparison of multimodal models, where the lyrics features are fixed. The models are reported as audio embedding technique+classification model, thus M1=CNN+RF, M2=LSTM, M3=RNN+RF, M4=(RNN+LSTM)+RF, M5=(RNN+LSTM)+MLP, M6=RNN+SVM, and M7=RNN+MLP.

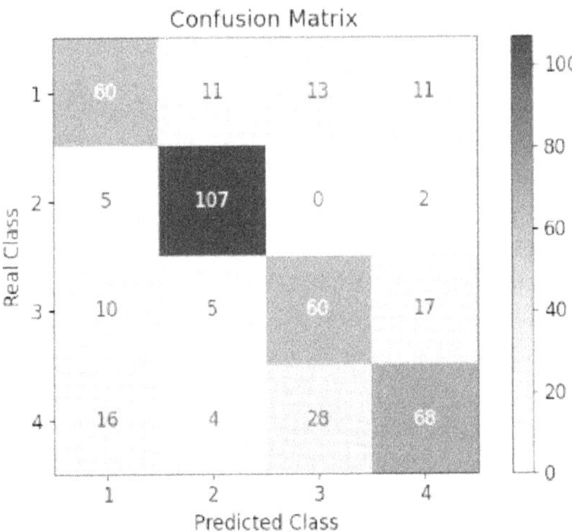

Fig. 5. Confusion matrix of the CNN+RF model, where 1=HAPV, 2=LAPV, 3=LANV, and 4=HANV.

Table 4. Results of the regression of the values of Valence and Arousal, evaluated in terms of R^2.

Models	R^2 Valence	R^2 Arousal
Random Forest	0.389	0.423
Gradient Boosting	0.374	0.442
SVR	0.344	0.396
Linear Regression	0.287	0.403
Ridge	0.287	0.403
Lasso	−0.005	−0.009
MLPRegressor	−0.373	−0.703

Analysing different feature combinations (Fig. 6), the integration of audio, lyrics, and genres yielded the best results. Other combinations, such as audio+lyrics+chords+genres and audio+lyrics, are also competitive, but including genre information provides an additional boost, particularly for valence. In contrast, single feature sets alone are insufficient.

Overall, these results mirror the trend observed in classification: multimodal integration consistently improves performance, highlighting the complexity of predicting continuous emotional dimensions and the importance of combining multiple information sources.

These results demonstrate that the multimodal framework can approximate the continuous emotional space with reasonable precision and support fine-grained user-level affect estimation.

4   Discussion

Considering the provided results, MER seems to benefit from the use of multimodal approaches exploiting deep learning feature extraction: (i) audio and lyrics information are at the core of MER improvement, mainly driving arousal and valence understanding, respectively; (ii) genres contribute to the multimodal audio-lyrics representation, but are insufficient when considered as the sole features; and (iii) chords provide competitive results when combined with the audio-lyrics modalities, with the advantage of requiring a lower computational complexity for their extraction, suggesting their potentiality in real-time MER systems.

Despite the encouraging results, the study presents some limitations and potential biases that should be acknowledged to properly interpret the outcomes and guide future research.

Starting from the multimodal coverage of the dataset, it is limited, as many tracks do not contain both audio and lyrical information. In some cases, tracks are purely instrumental or lack complete textual transcriptions. Additionally, the MERGE dataset provides only averaged annotations rather than individual ones, preventing direct application of user-tuning techniques on the same data.

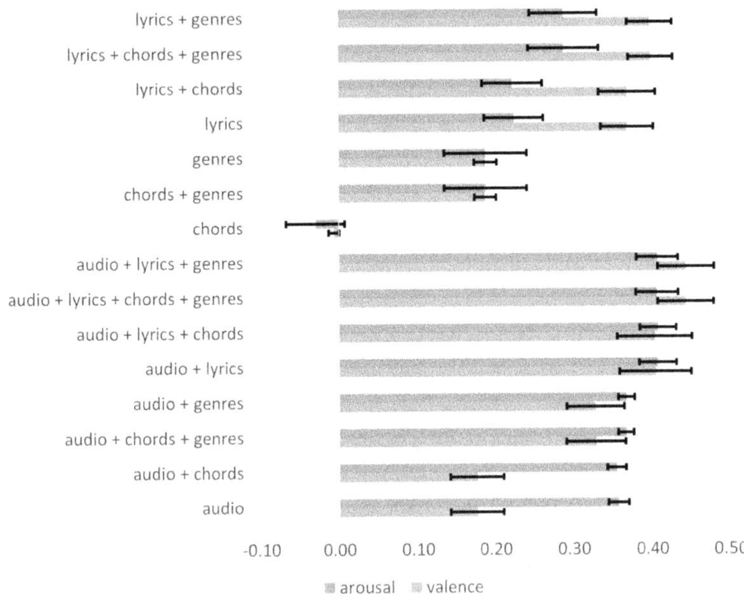

Fig. 6. Regression performance (R^2) for different feature combinations.

On this note, emotion annotations are inherently subjective and thus variability and potential disagreement among annotators may be present, particularly in cases of mixed or subtle emotional states. Such inconsistencies can reduce the reliability of labels and create misalignment between emotional signals across modalities. An example is when music with a positive tone is paired with lyrics expressing negative emotions. This "incongruence" complicates both model training and interpretation, as the system must reconcile potentially conflicting emotional cues.

Another limitation may be represented by the use of 25-s random audio excerpts, possibly reducing alignment between audio and lyrical features, since lyrics are analysed in full.

Furthermore, all models were trained and tested exclusively on MERGE, without cross-domain validation, leaving open questions about robustness and transferability to real-world contexts or alternative datasets.

Considering these observations, future works will focus on dataset enrichment and balancing, multimodal pipeline enhancement, and MER user-tuning.

The dataset may integrate tracks from additional public or private datasets to increase the diversity of musical genres, languages, and cultures represented, thereby improving the generalization capabilities of the models. To improve multimodal coherence, a methodology to automatically associate the portion of lyrics corresponding to the analysed audio segment will be developed, temporally aligning the textual content with the audio segment processed by the model.

Given the success of the multimodal approach, experiments with more advanced feature fusion strategies may be considered (e.g., attention-based fusion or multimodal transformer models), dynamically weighting the contribution of each modality according to the characteristics of each track. This will also possibly imply incorporating additional feature types, including cultural metadata, listening context information, or user behavioural data such as familiarity with the track or subjective liking, to enrich the multimodal representation.

Moreover, the system could be extended to a multi-label framework capable of assigning multiple emotional labels to each track simultaneously, reflecting the complex and nuanced nature of emotions evoked by music. In fact, listeners often perceive multiple emotions simultaneously when listening to a track [23]. A structural model such as Verse1-Chorus-Verse2 [24] may also be considered, given that it exploits the structural segmentation of tracks, acknowledging that different sections may convey distinct emotional content.

Finally, a promising direction for future research involves the personalization of MER systems through user-specific tuning. While global models capture the average emotional perception across listeners, they fail to account for the variability of felt emotions, the unique subjective emotional responses that individuals experience when listening to music. Incorporating personalization mechanisms could bridge this gap, enabling models to adapt dynamically to each listener's affective profile.

Building on the user-tuning experiments, several strategies could be explored to enhance personalization. Approaches such as transfer learning, fine-tuning [25], lightweight adaptor layers [26,27] or meta-learning frameworks [28] represent promising directions to improve alignment with individual emotional responses and facilitate efficient user-specific adaptation.

Ultimately, integrating user-tuning into multimodal MER systems represents a crucial step toward developing emotionally intelligent music recommendation systems capable of adapting to the evolving affective identity of each listener.

5   Conclusions

While MER remains a challenging task considering the subjectivity of emotions and the differences pertaining to the perceived, felt, induced, and intended emotions, this study presents different multimodal approaches exploiting audio and lyrics information, demonstrating their ability to better characterise emotions in musical contents compared to their unimodal counterparts.

Besides some limitations bounded to the dataset and the experimental choices, this study represents a groundwork for further development of a more refined MER recommender system to be tuned on its users.

The use of less computationally demanding features, such as chords, opens up the possibility of developing an efficient real-time system relying on a pre-trained general model that may be improved by introducing cross-dataset learning and different features and feature fusion strategies.

Acknowledgments. We thank Gabriele Comi for his initial work on the dataset retrieval and basic exploratory data analysis.

Disclosure of Interests. The authors have no competing interests to declare that are relevant to the content of this article.

References

1. Yang, Y.-H., Chen, H.H.: Machine recognition of music emotion: a review. ACM Trans. Intell. Syst. Technol. (TIST) **3**(3), 1–30 (2012)
2. Turchet, L., Pauwels, J.: Music emotion recognition: intention of composers-performers versus perception of musicians, non-musicians, and listening machines. IEEE/ACM Trans. Audio Speech Lang. Process. **30**, 305–316 (2021)
3. Gómez-Cañón, J.S., et al.: Music emotion recognition: toward new, robust standards in personalized and context-sensitive applications. IEEE Signal Process. Mag. **38**(6), 106–114 (2021)
4. Tibor Krols, N.O., Nikolova, Y.: Multi-modality in music: predicting emotion in music from high-level audio features and lyrics (2023). https://arxiv.org/abs/2302.13321. Accessed 30 Jan 2026
5. Çano, E., Morisio, M.: Moodylyrics: a sentiment annotated lyrics dataset. In: Proceedings of the 2017 International Conference on Intelligent Systems, Metaheuristics & Swarm Intelligence, pp. 118–124 (2017)
6. Zhang, W., Liu, Q., Hu, Y.: MMD-MII: a multilayered analysis and multimodal integration interaction approach revolutionizing music emotion classification. IEEE Trans. Affect. Comput. (2024, in Press)
7. Devlin, J., Chang, M.-W., Lee, K., Toutanova, K.: BERT: pre-training of deep bidirectional transformers for language understanding. In: Proceedings of the 2019 Conference of the North American Chapter of the Association for Computational Linguistics: Human Language Technologies, Volume 1 (Long and Short Papers), pp. 4171–4186. Association for Computational Linguistics (2019)
8. Aljanaki, A., Wiering, F., Veltkamp, R.C.: Studying emotion induced by music through a crowdsourcing game. Inf. Process. Manage. **52**(1), 115–128 (2016)
9. Ara, A., Rekha, V.: Enhancing music emotion classification using multi-feature approach. Int. J. Adv. Comput. Sci. Appl. **15**(9), 5 (2024) approach. Int. J. Adv. Comput. Sci. Appl. **15**(9), 5 (2024)
10. Choi, K., Fazekas, G., Sandler, M., Cho, K.: A comparison of audio signal preprocessing methods for deep neural networks on music tagging. In: 2018 26th European Signal Processing Conference (EUSIPCO), pp. 1870–1874. IEEE (2018)
11. Louro, P.L., Redinho, H., Santos, R., Malheiro, R., Panda, R., Paiva, R.P.: MERGE – a bimodal dataset for static music emotion recognition (2024). https://arxiv.org/abs/2407.06060. Accessed 30 Jan 2026
12. Russell, J.A.: A circumplex model of affect. J. Pers. Soc. Psychol. **39**(6), 1161–1178 (1980)
13. Defferrard, M., Benzi, K., Vandergheynst, P., Bresson, X.: FMA: a dataset for music analysis. In: Proceedings of the 18th International Society for Music Information Retrieval Conference (ISMIR), Suzhou, China (2017). https://arxiv.org/abs/1612.01840. Accessed 30 Jan 2026

14. Delbouys, R., Hennequin, R., Piccoli, F., Royo-Letelier, J., Moussallam, M.: Music mood detection based on audio and lyrics with deep neural net. In: Proceedings of the 19th International Society for Music Information Retrieval Conference (ISMIR), Paris, France, pp. 370–375 (2018)
15. Huang, Z., Ji, S., Hu, Z., Cai, C., Luo, J., Yang, X.: ADFF: attention based deep feature fusion approach for music emotion recognition. *arXiv preprint* arXiv:2204.05649 (2022). Accessed 20 Sept 2025
16. Dutta, S., Mookherjee, S.: Exploring the emotional landscape of music: an analysis of valence trends and genre variations in spotify music data. Presented at the 18th International Conference for Internet Technology and Secured Transactions, Oxford, 13–15 November 2023. *arXiv preprint* arXiv:2310.19052 (2023)
17. Kolchinsky, A., Dhande, N., Park, K., Ahn, Y.-Y.: The minor fall, the major lift: inferring emotional valence of musical chords through lyrics. Royal Society Open Sci. **4**(11), 170952 (2017)
18. Shah, A.K., Kattel, M., Nepal, A., Shrestha, D.: Chroma Feature Extraction: A Survey. In *Chroma Feature Extraction Using Fourier Transform*. Rochester Institute of Technology / Kathmandu University (2022). Conference paper; available via ResearchGate
19. McFee, B., Raffel, C., Liang, D., Ellis, D.P.W., McVicar, M., Battenberg, E., Nieto, O.: Librosa: audio and music signal analysis in python. In: Proceedings of the 14th Python in Science Conference, Austin, TX. SciPy, pp. 18–25 (2015)
20. Salton, G., Buckley, C.: Term-weighting approaches in automatic text retrieval. Inf. Process. Manage. **24**(5), 513–523 (1988)
21. Mohammad, S.M., Turney, P.D.: Crowdsourcing a word-emotion association lexicon. Comput. Intell. **29**(3), 436–465 (2013)
22. Warriner, A.B., Kuperman, V., Brysbaert, M.: Norms of valence, arousal, and dominance for 13,915 English lemmas. Behav. Res. Methods **45**(4), 1191–1207 (2013)
23. Heng-Tze, S., Chen, S.-K., Yang, Y.-H.: A new multilabel system for automatic music emotion recognition. IEEE Trans. Affect. Comput. **14**(2), 123–137 (2023)
24. Raboy, L.J.M., Taparugssanagorn, A.: Verse1-chorus-verse2 structure: a stacked ensemble approach for enhanced music emotion recognition. Appl. Sci. **14**(13), 5761 (2024)
25. Shen, L.-Y., Fang, S.-X., Lin, Y.-C., Chou, H.-C., Lee, H.: Meta-PerSER: few-shot listener personalized speech emotion recognition via meta-learning. *arXiv preprint* arXiv:2505.16220 (2025). Accessed 30 Jan 2026
26. Houlsby, N., et al.: Parameter-efficient transfer learning for NLP. In: Proceedings of the 36th International Conference on Machine Learning (ICML), vol. 97, pp. 2790–2799. PMLR (2019)
27. Pfeiffer, J., Kamath, A., Rücklé, A., Cho, K., Gurevych, I.: AdapterFusion: non-destructive task composition for transfer learning. In: Proceedings of the 2020 Conference on Empirical Methods in Natural Language Processing (EMNLP), pp. 4871–4883. Association for Computational Linguistics (2020)
28. Zhang, D., You, W., Liu, Z., Sun, L., Chen, P.: Personalized dynamic music emotion recognition with dual-scale attention-based meta-learning. In: Proceedings of the AAAI Conference on Artificial Intelligence, vol. 39, pp. 1629–1637 (2025)

Mapping Artificial Neural Networks' Processing Data in Audiovisual Artworks

Tanguy Pocquet[1,2]([email]) , Richard Allmendinger[2] , and Ricardo Climent[1]

[1] NOVARS, The University of Manchester, Manchester M13 9PL, UK
`tanguy.pocquetduhaut-jusse@manchester.ac.uk`
[2] Alliance Manchester Business School, The University of Manchester, Manchester M13 9PL, UK

Abstract. In a non-generative approach to artificial intelligence in an artistic practice, this work looks at mapping processing data from different artificial neural networks (NNs) onto sound and visuals. One aim of this practice-based piece of research is to offer insights into how these ubiquitous, yet notoriously unintelligible algorithms operate, sometimes by exposing the audience to that very unintelligibility. The other is to use these vast amounts of abstract data as a blank canvas for audiovisual artworks. At the heart of the whole work is a link and cross fertilization between the use of sounds and visuals aesthetically associated with errors and digital malfunction, and the use of actual 'waste' data (from NN training), which acts as a trace of their operation. We look at different kinds of NNs, and at various in-training data-streams: from the easily apprehensible output of generative networks, through various performance metrics, to the highly-abstracted activation and weight values in and between hidden layers. We believe these pieces uncover some interesting features about the evolution of GANs' outputs in training and the consistent patterns they follow in the latent space, illustrate the potential for errors (mode collapse, noise, etc.) in those networks, and outline the inherently abstract nature of weight changes during an MLP's training.

Keywords: Audiovisual composition · Sonification · Creative coding · Machine learning

1 Link to / Limitations of Existing Approaches

There are many examples of visualizations of NNs, however, these generally focus on trained networks, trying to understand how they work after training [5,14, 25,26], and sometimes specifically at explaining individual decisions [4]. These approaches are purely practical, even if some aesthetic aspects of the resulting visualizations are sometimes mentioned [12,14].

This lack of focus on training, and instead on explaining post-training decisions, is interesting and can probably be explained by the need for explainable

AI (XAI) for the end user, rather than for insights for the people training the networks. Extensive work has been done on XAI, on the need to remedy the fact that generally "no information is provided about what exactly makes [neural networks] arrive at their predictions" [19], and countless approaches have been taken to achieve this [1,2]. However, again, the overwhelming majority of approaches focus on trained networks, narrowing down which parts of the input data lead to a particular decision [4,8,21].

Moreover, while there have been examples of visualization for XAI [19,22], and studies on XAI for audio applications of NNs [6], there is very little work done on sonification as a tool in XAI. To the best of our knowledge, the only work on this topic introduced the idea of sonification for XAI, but without a practical output [20]. This gap is all the more intriguing given the success researchers have had in applying sonification to explain other types of algorithms [3].

Beyond its artistic focus, our work also breaks away from the traditional focus of XAI (by looking at the process rather than the output), as in our opinion how networks learn is as, if not more, interesting than how they work once trained. Hence, we focus more on 'Interpretability' ("level of understanding how the underlying (AI) technology works") than 'Explainability' ("level of understanding how the AI-based system [...] came up with a given result") [13].

If the common phrase 'Opening the black box' can mean two things:

- enforcing a right to explanation by allowing one to retain intellectual oversight over a network, and providing a reason for each decision (generally what XAI focuses on), or
- actually 'seeing' the inner, mathematical processes of a network in real time, including during training, at the risk of being confronted with extremely abstracted numerical information,

then this paper focuses more on the latter. This is made possible by the fact that all the works cited here have fully technical priorities, whereas we want to approach NNs as algorithms with inherently compelling processes and structures. The networks were therefore written with the explicit goal of having their inner-workings observed, in a complete human-in-the-loop machine learning approach [16].

2   Approach

This section first outlines our principal motivation in tackling this cross-disciplinary research project. It then details our choice of datasets, made in relation to these motivations. Finally, it describes the architectures we decided to use based on the data focused on, along with a brief run-through of architecture types we discarded for this project, and of the reasons why.

2.1   Motivations

The aim of this project is to map the processing data of artificial neural networks onto sounds and visuals to create audiovisual artworks. In so doing, the hope is to focus on the process rather than the output of these networks, to expose what is habitually hidden, and perhaps bring some understanding or impression of the nature of these processes (in a non-empirical way), and offer an alternative to generative AI models as a way of using machine learning algorithms in art-making. We do not claim to bring empirical insights on the specific outcomes of certain training runs based on starting hyperparameters, but rather to expose the listener to enough runs to give them a strong overlook of the networks' training, and an impression of their operation. Moreover, the balance between aesthetic and interpretability of the output inevitably leads to compromises in how data is represented. While this remains an artistic project first and foremost, we are interested in how exposing an audience to information leads them to come to more instinctive conclusions about the underlying structures that drive them.

2.2   Data Choice

Perhaps counter-intuitively, the focus of this work is on anything but the output of trained networks. The current state of generative AI for images, text or audio has led us to firmly reject the use of networks trained on pre-existing art to generate new materials. Instead, our focus is on the inner processes of these algorithms, whatever is normally hidden from view.

This may be data as abstracted as the weights of connections and activations of neurons from hidden layers, or as intelligible as the training output of a network. In between, we have also looked at standard performance metrics like loss and accuracy, at synthetic metrics like the RMS of outputs, similarity metrics between outputs, edge intensity in generated images, etc. These data streams are then used to drive audio and visual parameters, with the hope that some of the mechanisms of these algorithms can be showcased in the resulting pieces.

While early in the research, we looked at intermediary level features, such as layer structure, and the impact of certain hidden layers over others, this required a lot of pre-processing of information (to derive averages of neuron or weight values in specific zones) and led to equally abstract results to when looking at individual weights and activations, so we decided not to pursue this.

The focus on process over result means that the efficiency or accuracy of networks can be superfluous. Sometimes, looking at a network that fails to learn the features of a dataset, or at one that does not generalize and generates the same output repeatedly, can be just as insightful in understanding how they operate as looking at an efficient and accurate version.

The spectrum between what could be described as low-level and high-level metrics; weights or connections and intelligible output, inevitably leads to discussions on what the strength of each side might be. While focusing on the core of the networks' operation, looking at changes in weights and activations in the whole structure, might seem like the best way to discover how they operate, there are significant issues of intelligibility when at such abstracted numerical processes. On the other hand, when looking at outputs, one might find a lack of insights with anything representative, as we are only dealing with the most surface level information.

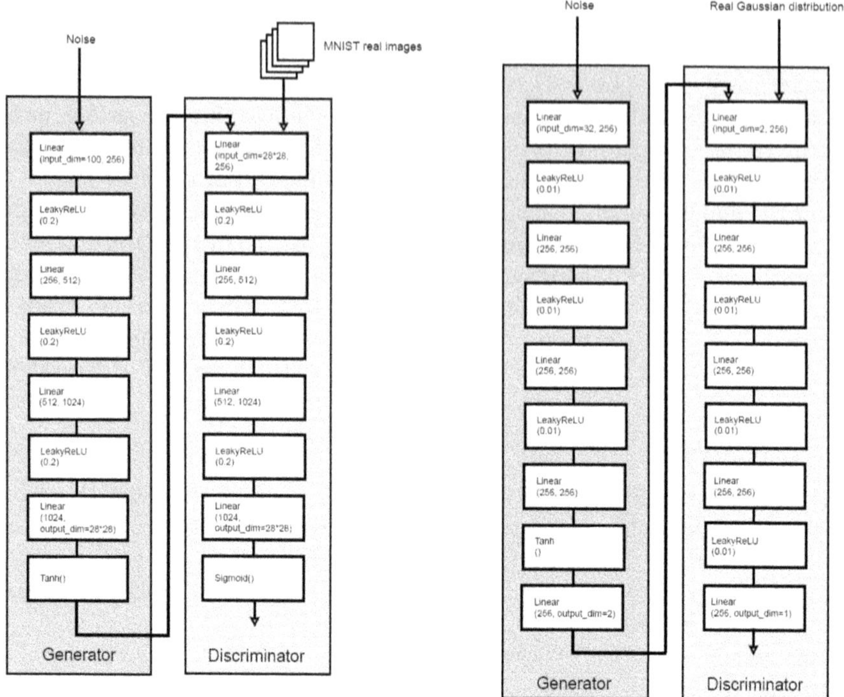

Fig. 1. Network structures for the GANs used in Piece 1 (left) and Piece 2 (right).

The phrase 'opening the black box' however, while thrown around copiously, may not be so useful. One of the pieces in this project offers a visualization of the live change of all weights and activations over various training runs of a multilayer perceptron (MLP) trained on a very simple problem. As will be outlined later, even with an elementary network and dataset, and the most minute details of its structure laid bare for all to see, nobody can really deduce a cogent explanation of how the network works. Opening the black box is by no means equivalent to understanding its inner workings.

2.3 Architectures

Through this project, we looked at multiple architectures, and essentially settled on a couple to really use as tools in audiovisual works. In most pieces, we used Generative Adversarial Networks (GANs) (see Fig. 1), because of their generative nature (as the use of observable in-training output as starting points for visuals was a particular focus of the work), being relatively lightweight to run, and leading to tangible results for an audience.

We also looked at MLPs (see Fig. 2) when investigating more low-level and abstracted data (individual neurons' activations, and weight changes), simply

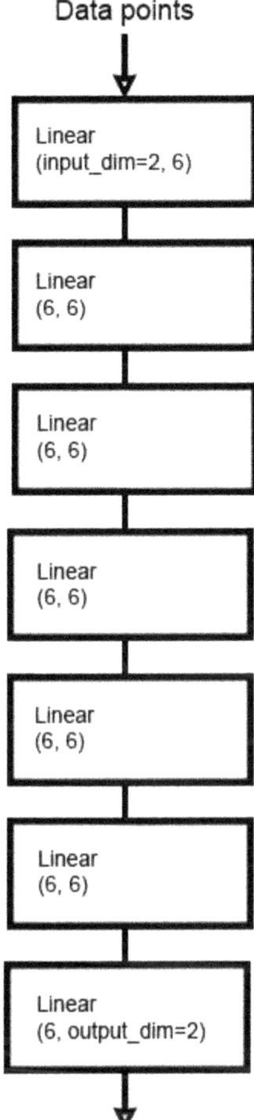

Fig. 2. Network structure for the MLP used in Piece 3.

because only with the simplest of structures are these parameters usable and somewhat digestible for a human observer.

We looked at other architectures in various levels of details, even using some (notably CNNs) for early works, but chose to discard most, for various reasons:

- Convolutional neural networks, because classification does not lead to the same obvious changes as generation.
- Recurrent neural networks [18], because of their longer training times on time-based data.
- Encoders, because of the abstract nature of their outputs, which make them difficult to display to an audience.
- Diffusion models [23], because we did not find a satisfying application to use in AV works.

– Word2vec models, because of the types of outputs they produce (the audio-visual medium makes text-based information harder to incorporate).

The idiosyncrasies of models' architectures are a central part of turning the algorithms into tools for artistic creation. Hence, the inherent narrative dimension of GANs' operation, centered around a zero-sum game dynamic and a notion of opposition; but also the gradual transition from noise to order in the generator's outputs were all aspects that we were interested in showcasing. The materialism of generated material, and its malleability and transformation over time are another consequence of these network's technical makeup that we wanted to frame, to make the processes tangible through their continuous training output.

Similarly, when using MLPs, their foundational and elementary aspects were important in our decision to display their processes as bare abstractions, with little interpretation of any of the data streams they present.

2.4   Datasets and Problems

The training data that was focused could be described as 'data that looks like data'; basic MNIST dataset, Gaussian point-distributions, non-linearly separable point sets, text datasets, etc.

This choice of basic, widely used, abstract, low-dimensional datasets allows the pieces to focus on making in-training transformations of outputs clear, and outline the difference between the training data and generated materials, in a more comprehensive and empirical way than if the data was more complex. The simplicity of the data also allows for as literal and as devoid of artifice an approach as possible, which fits the aesthetic priorities of the works.

Moreover, the problems solved by the networks (data separation, distribution identification, image generation, etc.) are in many cases foundational to machine learning, and again emphasize the fact that the process is more important to this work than the result (which will have been reached many times before).

One last important point when it comes to the data and problems focused on in this project, is the absence of audio in the data looked at. This ties back to the distance voluntarily taken from a generative AI approach, and our doubts in the face of generating materials from existing art. Since the music is the primary dimension of the compositions resulting from this work, it seemed fitting to avoid using audio and falling into plain audio generation, that would have tainted the aesthetic of the pieces, and the focus of the research on process rather than output.

2.5   This Approach Versus a Network Designer's

The non-teleological approach taken in this project, looking at process over output, is made clearer in the differences between the priorities of most NN designers (on efficiency, accuracy, etc.), and that of this work. The latter is to create aesthetically and narratively interesting results, including through the presentation of sub-optimal runs, the re-purposing of errors, the framing of inaccuracies, the intentional push towards inefficiency, etc. The idea is not to create particularly efficient networks, but to demonstrate how they train, evolve, improve, and sometimes fail. The stance taken being that there is a lot of understanding of these algorithms to gain from seeing them fail.

This intentional push towards errors and inaccuracies is achieved by pushing some of the networks' hyperparameters to unrealistic values for traditionally 'successful' results to occur. In a human-in-the-loop approach to ML [16], we have played with networks' hyperparameters, to force them to specific patterns of malfunction, that could be used as narratively interesting features of the pieces. For instance:

- Mode collapse [9] (by mildly reducing the discriminator's learning rate): a common error whereby a GAN fails to generalize, and produces one type of sample endlessly.
- Noisy output (with too inefficient an architecture): where the network fails to improve past a certain limit, and continuously outputs data resembling the training data, but noticeably 'off'
- Slow training (by significantly reducing the general learning rate): where the network runs at an excessively slow pace.
- Chaotic outputs (by significantly increasing the learning rate past any realistic value, stopping a network from ever approaching a solution).

This is a 'dirty-electronics' approach to neural networks, running them on a small-scale, with a focus on creativity rather than efficiency; encompassing "a notion of the post-digital, the self-made and do-it-yourself in contrast to the mass-produced, and the reinvigoration of the role of the human" [17]. Its main aim is for both its process and results to offer a parallel to the aesthetic nature of the works.

Hence, a lot of the pieces that come out of this work will display various runs of a neural network, including unsuccessful ones, before finally ending on a technically viable training run, to show the potential differences. This iterative structure reflects the technical approach that often needs to be taken when training a network, tweaking hyperparameters until a satisfying outcome is reached. Whether looking at outputs over training or at internal metrics, this iterative movement towards order is also an interesting way to organize artistic materials. Generally, this technical approach ties in with the main aesthetic currents that the musical and visual sides of the work are indebted to.

2.6   A Note on the Aesthetics of the Works

In essence, reflecting some of the technical priorities detailed above, this work is perhaps an illustration of the view that "The point of art is detritus" [24]; it looks at processes, at what are normally byproducts, extraneous material.

Sonically, it owes a lot to glitch music, a subgenre that harnesses "the 'failure' of digital technology [. . .]: glitches, bugs, application errors, system crashes, clipping, aliasing, distortion, quantization noise, and even the noise floor of computer sound cards are the raw materials composers seek to incorporate into their music" [7]. More broadly, it links to the post-digital aesthetic, which "takes digital to be the pinnacle of audio fidelity (technology driven by the ultimate desire for perfect clarity and the elimination of noise) and situates itself after: after the revolution, amidst the rediscovery of noise through digital malfunction" [11]. This link between the information being sonified (byproducts, detritus) and the sound they effect (sounds of errors, digital malfunction, clicks, and noise) is central to the project.

3   Pieces; Results; Insights

This section goes through the compositional and technical approaches taken for each piece in the project, and outlines the outcomes and strengths of each work.

3.1   Abstract Masses of Numerical Information

The first piece presented here, Piece 1, uses a generative adversarial network trained on the MNIST dataset of images of handwritten digit. It illustrates how GANs move, through training, from noise to order, and the intermediary structures they go through in the process. It does so by displaying, each step of training, 3600 generated samples from the generator as visuals, and mapping a wide selection of metrics (performance metrics such as loss and accuracy and synthetic metrics taken from the generated data such as entropy averages of the samples, similarity between samples, RMS values, edge intensity, etc.) to various audio parameters (various granular synthesizers' parameters, sample lengths, overdrive, overall noise, etc.).

The original musical material are pre-composed organ loops, but various transformations of timbre, length of loops, distortion, are driven by the synthetic metrics calculated from the output of training. For instance, the length of loops is mapped to the similarity of the output images, generator loss is mapped to the distortion of loops, discriminator loss to volume of clicks, the edge intensity of output images to the quantity of static, the RMS average values of the outputs to a tremolo on the loops, etc.

The idea is for the visuals to act as literal tracking of the network's activity (the visuals consist simply of generated sample images), and to get an overall impression of its performance through the complex soundscape that results from mapping of all these different yet linked metrics. All the complex mappings detailed above amalgamate so that generally, the noisier the output, the worse the GAN's performance, and the cleaner the sound, the more efficient it is. However, as we have outlined, the noisier output is also an integral part of the work, and is focused on through various 'failed' runs before settling on a successful run, ending in a clearer sound.

We found that by displaying just 3600 of tens of thousands of potential output samples that change continuously (see Fig. 3), over thousands of epochs, listeners were realizing the mass of information that continuously runs through a NN. We also believe the piece is also an excellent illustration of key issues GANs

Fig. 3. Two ensembles of 3600 output samples from a GAN trained on the MNIST dataset, in untrained and uniform (top), and trained and generalized (bottom) forms (From Piece 1).

tend to run into: mode collapse (some of the runs showed seem to show 3600 identical outputs, when really the network has failed to generalize, and is stuck on one particular output structure); noisy outputs; outputs that fall just short of a satisfactory accuracy (images that eerily close to the expected outputs, but not numbers at all); etc.

3.2 Iterative Movement, Structured Outputs

Another GAN-based piece, Piece 2 is a good illustration of the underlying structure that network outputs tend to go through during training before reaching a solution. In contrast with the previous work's focus on chaos and mass of information, this piece focuses on a much simpler dataset, a set of eight 2D Gaussian distributions of points, and at how the GAN performs at reproducing this overall distribution.

This piece is in 8-channel audio, and each of these eight Gaussian distributions correspond to one speaker. As the points move around with each step from the GAN, the sound is panned accordingly, going from a narrowly focused sound if the GAN generates all points in one area, to a uniformly panned sound when it successfully distributes them.

In this piece also, the visuals are just output samples from the generator, displaying the training data in black, and the generated output in yellow (see Fig. 4). But this time, the sound output, as detailed above, is also a clear representation of this output data, in contrast with the complex mapping of synthetic metrics detailed for the last piece.

Again working with an iterative structure, the piece displays various unsuccessful training runs, where the network fails to attribute points evenly across the distribution, before reaching a successful state, also reaching a musical climax by having all 8 speakers on together.

The main detail of the GAN's processing this piece uncovers then, is the similar data structures displayed in every run; starting as one tightly focused set of points, exploding into a wide-covering array, and re-converging into a tighter pattern, either successfully (covering the eight regions) or unsuccessfully (only partially covering the original distribution). We believe this work brings valuable insights to the audience on the inherent structure GAN outputs develop in, following specific courses in the latent space, no matter what final result they come to. This order in output is an interesting contrast to the mass of

Fig. 4. Two in-training output samples of a GAN trained on an eight-zone point distribution; in mode collapse (left) and successfully trained (right) states. (from Spin, split, spread, splatter (in eighths)).

unintelligible simultaneous data streams going through the first piece, where there seems to be no structure to the training, because of the higher-complexity of the training data.

3.3   Arbitrary Solutions

As mentioned above, other than output in training, which offers very interesting avenues for organisation of materials, another dimension of the networks looked at is the much more abstract changes in weights and activations, from within networks' hidden layers (see Fig. 5).

This aspect is one of the first we were interested in looking at, originally in CNNs. However, the absurd level of complexity reached even in simple networks, as even simple architectures contain multiple 3-dimensional fully-connected layers and easily reach billions of weights, made this more challenging than expected. Therefore, it took until later to write as simple an MLP as possible, and to be able to visualize all weights and connections in one cogent visual output. This network was trained on as foundational an ML problem as possible: the separation of two non-linearly separable data groups in 2 dimensions. This was the basis for the third piece presented here, Piece 3.

Once again, the iterative organization is key both to the musical structure (as the piece is an example of loop-based, synthetic electronic music) and to the technical aspects of the network's operation it aims to make clear. The piece displays repeated training runs of the same architecture, on the same dataset, to the same level of accuracy. The temporality at the core of any NN training, defined by its structure, the complexity of the dataset and the learning rate of the model, is reflected in the musical output, as the learning rate is modified to result in sections of different lengths and degrees of momentum throughout the piece.

Despite the piece just being the result of repeated training of the same network on the same problem, all resulting weight structures look completely different. The aim of this display of a full NN architecture on repeated training runs was to show that the 'problem + weight structure = solution' paradigm that one might imagine a network might follow does not apply at all; the same problem and a completely different structure might counterintuitively still lead, because of the random nature of activations, to the same solution. Hence, the key insight we wanted to outline in this piece was the degree of arbitrariness (at least by the standard of human logic) of these networks' process. Ironically, the key thing to understand—and the main interest artistically—is their unintelligibility. Generally, abstraction, unintelligibility, vastness of information, are all focuses of this research. Going back to the argument that 'opening the blackbox' is a futile quest, perhaps an alternative is then to find interest in the very abstraction of these incomprehensible structures.

Going back to the question of XAI, few studies actually visualise training at all, partly because of the inability to gain any clear insights from these structure changes. Even when looking at a trained network, "visualizing features to gain intuition about the network is common practice, but mostly limited to the 1st layer where projections to pixel space are possible." [25]. Quickly, any information becomes undecipherable. Yet that is not to say there is no need for a narrative representation in the ML field. Indeed, Guattari [10] had already theorised and predicted that:

"Paradoxically, it is perhaps in the "hard" sciences that we tend to see the most spectacular reconsideration shift towards processes of subjectification. Is it not significant, for example, that Prigogine and Stengers invoke the necessity of introducing in physics a "narrative element", which they consider indispensable to theorise evolution in terms of irreversibility. This being said, I am convinced that the question of subjective enunciation will pose itself more and more as

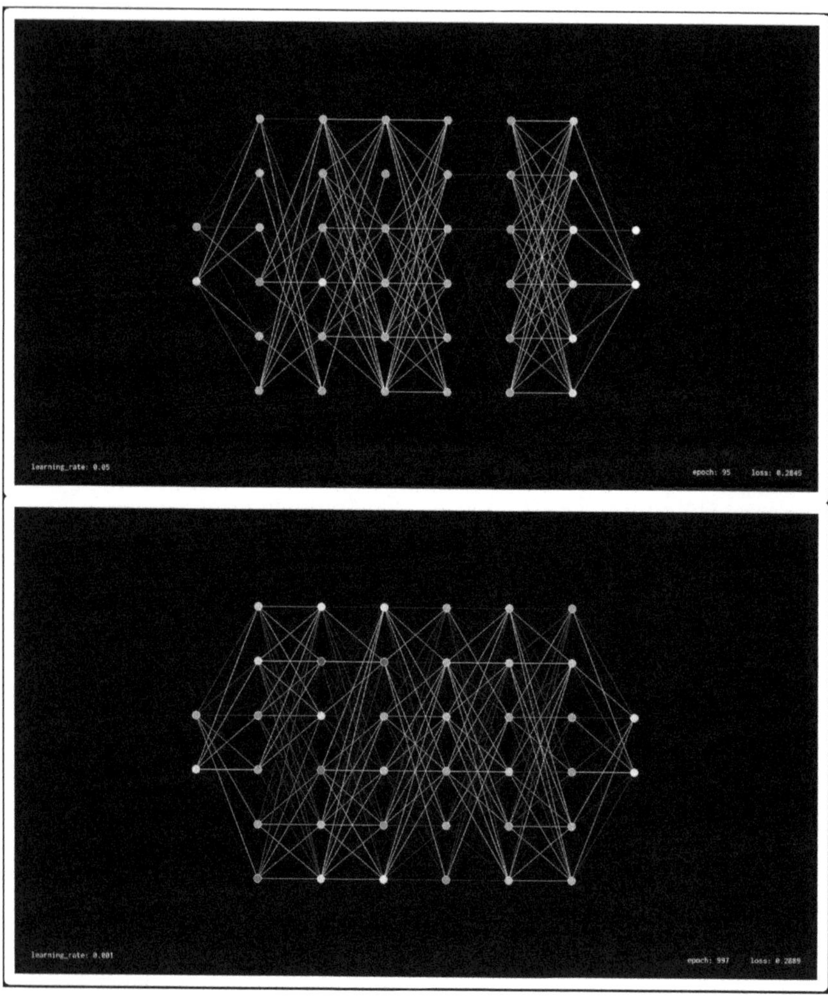

Fig. 5. Two weight structures after two different runs of the same feed-forward network trained on the same set of nonlinearly separable data (from Piece 3).

machines producing signs, images, syntax and artificial intelligence continue to develop[...]" [10].

4   Conclusions

This work offered to address, in a small way, the need for representation in ML through the presentation of outputs, training data and performance metrics, in a relatively raw and untreated state, to confront listeners with the high degree of abstraction that defines these networks' processes, but also provide a trace of their operation. We have seen how waste and inaccuracies can be as significant as the intended output in understanding how these systems behave, and simultaneously the idea of waste provides an aesthetic framework (linked to the postdigital movement) within which the sounds and visuals of these pieces are situated. Moreover, making artworks from the processes of NNs rather than their final output, and in doing so leaving all creative agency to the artist, offers an alternative to generative approaches, and inevitably comments on our use of these algorithms for creative applications.

By focusing on various dimensions from neural networks' training and on various types of networks (MLPs, GANs), we believe we have uncovered and displayed various aspects on how these networks operate, and provided he audience with a representation of their processes. Again, the aim is not empirical feedback on performance, but a general impression of the NNs' overall operation, which we believe we have been successful in conveying. The vastness of continuous information flowing through them; the various faults met when dealing with suboptimal hyperparameters; the inherent structure followed by GANs when navigating the latent space to generate output; the lack of 'logical' structure followed by NN weight structures, and the fact that innumerable different structures can lead to the same result; etc. This mixture of cogent and unintelligible conclusions makes for strong contextual starting points for the compositions presented, and also paints a useful picture of the contradictions at the heart of neural networks' operation. We therefore believe this work to have been successful.

Future explorations will continue to frame the idiosyncrasies of various network architectures and attempt to use them as creative tools. For instance, one piece still in progress looks at training a GAN on spectrograms from audio files, generating more spectrograms to be turned back into audio form. The work will explore imprecisions inherent to the network, such as the grid-like artifacts that appear on generated images due to the kernels' size, which, when the spectrograms are turned back to audio, become regular rhythmic patterns that can be used as the basis for the piece's rhythmic structure. While looking at larger architectures might seem like a natural development, we are also conscious of our aim to use NNs as tools, on consumer-grade machines, for individual arts projects, and computing power inevitably becomes a limitation to individual use when looking at larger models.

Artistically, our hope is for this d.i.y. approach to machine learning, using small-scale networks trained on abstract data, to inspire other artists to adapt algorithms and approach them as interesting mathematical objects to be commented on through art, rather than ready-made tools to apply in a uniform way. These algorithms remain fascinating, but perhaps more so if one does not limit themselves to their marketed guideline use.

References

1. Adadi, A., Berrada, M.: Peeking inside the black box; a survey on explainable artificial intelligence. IEEE Access **6**, 52138–52160 (2018)
2. Ali, S., et al.: Explainable artificial intelligence (XAI): what we know and what is left to attain trustworthy artificial intelligence. Inf. Fusion **99**(101805) (2023)
3. Asonitis, T., Allmendinger, R., Benatan, M., Climent, R.: SonOpt: sonifying bi-objective population-based optimization algorithms. Genet. Program Evolvable Mach. **24**, 3 (2023)
4. Baehrens, D., et al.: How to explain individual classification decisions. J. Mach. Learn. Res. **11**, 1803–1831 (2010)
5. Bau, D., et al.: GAN dissection: visualizing and understanding generative adversarial networks. arXiv, 1811(10597) (2018). Accessed 10 Oct 2025
6. Bryan-Kinns, N., et al.: Exploring XAI for the arts: explaining latent space in generative music. In: 1st Workshop on eXplainable AI Approaches for Debugging and Diagnosis, NeurIPS 2021 (2023). Accessed 10 Oct 2025
7. Cascone, K.: The aesthetics of failure: "Post-digital" tendencies in contemporary computer music. Comput. Music. J. **24**(4), 12–18 (2000)
8. Chang, C., Creager, E., Goldenberg, A., Duvenaud, D.: Explaining image classifiers by counterfactual generation. In: International Conference on Learning Representations, ICLR, New Orleans (2019)
9. Goodfellow, I., et al.: Generative adversarial networks. In: Advances in Neural Information Processing Systems, vol. 27 (2014)
10. Guattari, F.: Les trois écologies. Galilée, Paris (1989)
11. Haworth, C.: "All the musics which computers make possible": questions of genre at the Prix Ars Electronica. Organised Sound **21**(1), 15–29 (2016)
12. Herdt, R., Maass, P.: Visualize and paint GAN activations (2024). arXiv, 2405(15636). Accessed 10 Oct 2025
13. ISO, IEC: Software and systems engineering, Software testing, Part 11: Guidelines on the testing of AI-based systems. ISO/IEC TR **29119–11**, 2020 (2020)
14. Mahendran, A., Vedaldi, A.: Visualizing deep convolutional neural networks using natural pre-images. Int. J. Comput. Vision **120**, 233–255 (2016)
15. Mikolov, T., Chen, K., Corrado, G., Dean, J.: Efficient estimation of word representations in vector space, In: ICLR, International Conference on Learning Representations (ICLR), Scottsdale, Arizona (2013)
16. Mosqueira-Rey, E., Hernandez-Pereira, E., Alonso-Rios, D., Bobes-Bascaran, J., Fernandez-Leal, A.: Human-in-the-loop machine learning: a state of the art. Artif. Intell. Rev. **56**, 3005–3054 (2023)
17. Richards, J.: Getting the hands dirty. Leonardo Music J. **18**, 25–31 (2008)
18. Rumelhart, D., McClelland, J.: Learning internal representations by error propagation. In: Parallel Distributed Processing: Explorations in the Microstructure of Cognition: Foundations, pp. 318–362. The MIT Press, Cambridge, Massachussetts (1987)

19. Samek, W., Wiegand, T., Müller, K.-R.: Explainable artificial intelligence; understanding, visualizing and interpreting deep learning models. arXiv [Preprint] (2017). Accessed 10 Oct 2025

20. Schüller, B., et al.: Towards sonification in multimodal and user-friendly explainable artificial intelligence. In: ICMI 2021: 2021 International Conference on Multimodal Interaction, Montréal, pp. 788–792. Association for Computing Machinery (2021)

21. Shrikumar, A., Greenside, P., Kundaje, A.: Not just a black box: learning important features through propagating activation differences. In: PMLR. 34th International Conference on Machine Learning, Sydney: JMLR, pp. 3145–3153 (2017)

22. Simonyan, K., Vedaldi, A., Zisserman, A.: Deep inside convolutional networks: visualising image classification models and saliency maps. In: International Conference on Learning Representations (2014)

23. Song, J., Meng, C., Ermon, S.: Denoising diffusion implicit models. In: ICLR 2021, The Ninth International Conference on Learning Representations, ICLR (2021). Accessed 10 Oct 2025

24. Weiner, L.: Lawrence Weiner interviewed by Hans Ulrich Obrist, Arles 2012 (2012). https://vimeo.com/73451498. Accessed 10 Oct 2025

25. Zeiler, M., Fergus, R.: Visualising and understanding convolutional networks. In: 13th European Conference on Computer Vision (EECV 2014). Springer, Zurich, Switzerland (2014)

26. Zintgraf, L., Cohen, T., Adel, T., Welling, M.: Visualising deep neural network decisions: prediction difference analysis. In: ICLR 2017 5th International Conference on Learning Representations, Toulon (2017)

LoopMatcher: Proof-of-Concept for AI-Assisted Music Loop Search

Subhrojyoti Roy Chaudhuri[1]($\boxtimes$) [iD], Sai Pranav Madupu[2] [iD],
Krishna Teja Guru Sai[2] [iD], and Vikram Jamwal[1] [iD]

[1] TCS Research, Pune 411013, India
`{subhrojyoti.c,vikram.jamwal}@tcs.com`
[2] Indian Institute of Information Technology, Design and Manufacturing,
Kancheepuram 600127, India
`{cs22b1027,cs22b1044}@iiitdm.ac.in`

Abstract. Music loops are in increasing demand for music creation as well as live performance. There are numerous free and commercial options available for procuring and using loops from libraries and community repositories. A challenge commonly encountered by musicians is searching for loops that match their sonic requirements within the large space of available options. Musicians typically rely on metadata associated with loops, such as key and tempo, to narrow the search space and subsequently make selections using their own perceptual judgments. This process can be time-consuming and effort-intensive. In our work, we investigate whether AI audio models can assist musicians in the task to search loops with perceptual similarity. We develop LoopMatcher, a workflow that integrates two complementary AI models - one for efficient embedding-based retrieval (VGGish) and the other for perceptual similarity refinement (CDPAM) - to automatically narrow the search space and rank candidate loops given a reference loop. In a proof-of-concept validation study, we observe a Spearman correlation of 0.68 between algorithmic rankings and human perceptual judgments. Additionally, randomly selected loops consistently rank lowest, judged by both the system and participants. The results of our work provide encouraging outcomes that indicate strong correlations between our approach and human perceptual judgments, demonstrating that AI-assisted loop search shows promise and merits further investigation.

Keywords: AI models for audio · Music retrieval · Audio search · Music loops · Vector databases for audio

1 Introduction

Digital Audio Workstations (DAWs) and loop sequencing software such as Ableton Live, Loopy Pro, and Launchpad, now available on diverse platforms from desktop computers to tablets and smartphones, have made loop-based music creation increasingly accessible [17] [12]. Modern musicians not only create their own loops but regularly incorporate samples from third-party libraries and community-based repositories that may contain thousands of options [15].

Musicians typically rely on metadata - such as key, tempo, and genre tags - to perform initial filtering, then manually audition remaining candidates to identify loops that sonically complement their work. However, metadata for loops is sometimes incomplete, inconsistent, or incorrect, limiting its utility for search. Furthermore, the perceptual judgment of whether two loops "match" or "complement" each other involves nuanced sonic characteristics - including timbre, rhythm, harmonic content, and energy - that are difficult to capture through simple metadata tags. With library sizes often reaching thousands of samples, this manual search process becomes time-consuming and effort-intensive, diverting creative energy from actual music-making.

Recent advances in AI audio models present an opportunity to automate search for loops, especially the ones which have perceptual similarity. Two relevant AI approaches have emerged: (a) embedding-based models that capture acoustic features in vector representations suitable for similarity search, and (b) learned perceptual metrics trained to align with human similarity judgments. However, the feasibility of applying these models to loop search remains unclear: Can AI models capture the nuanced perceptual judgments musicians make when selecting perceptually similar loops? Can they effectively narrow large search spaces to manageable candidate sets?

This paper presents a proof-of-concept investigation of AI-assisted loop search. We develop LoopMatcher, a configurable workflow that integrates two established AI audio models with complementary strengths: VGGish [11] for computationally efficient embedding-based retrieval from databases of 3000+ loops, and CDPAM [14] for perceptual similarity refinement based on models trained with human judgments. The two-stage architecture exploits the complementary properties of these models: VGGish rapidly filters large databases to a candidate set, while CDPAM refines rankings based on perceptual similarity. Given a reference loop as input, LoopMatcher automatically ranks candidate loops from a pre-populated database, presenting the top-k most similar matches.

We conduct an initial validation study with eleven participants who evaluate the system's rankings against their own perceptual judgments. Our results show a Spearman correlation of 0.68 between algorithmic and human rankings - a strong correlation given the subjective nature of musical preference. Additionally, three randomly selected loops were consistently ranked lowest by both the system and human evaluators, providing evidence that the approach captures meaningful musical similarity rather than arbitrary patterns. These findings provide encouraging evidence for the feasibility of AI-assisted loop search, though significant work remains before such systems can be deployed in production environments.

1.1 Contributions

The main contributions of this work are:

- **Feasibility demonstration:** A proof-of-concept showing that readily available AI audio models can meaningfully assist in loop search tasks, achieving strong correlation (0.68) with human perceptual judgments in an initial validation study.
- **Two-stage architecture:** A modular workflow combining embedding-based retrieval (VGGish) with perceptual refinement (CDPAM) that effectively balances computational efficiency for large-scale search with perceptual accuracy for final ranking.
- **Initial validation evidence:** Experimental results from eleven participants demonstrating alignment between AI rankings and human judgments, including successful identification of random loops as poor matches.

1.2 Paper Organization

The remainder of this paper is organized as follows: Sect. 2 reviews related work on loop search, music retrieval, and AI audio models; Sect. 3 presents the Loop-Matcher architecture and workflow design; Sect. 4 details the implementation including model selection, database setup, and the two-stage retrieval algorithm; Sect. 5 describes our validation experiment, results, and limitations; Sect. 6 discusses implications for AI integration in music production and identifies future research directions; and Sect. 7 concludes with a summary of findings and next steps.

2 Related Work

2.1 Music Search and Retrieval

Music search and retrieval for playlist creation and recommendation is a well - established research area. Traditional approaches rely on metadata - such as genre, tempo, key, and mood tags - to index and retrieve similar music from large libraries [10]. While effective for complete musical tracks, this metadata-based approach faces significant challenges in the context of music loops. Loop metadata can be incomplete, inconsistent, or incorrect, making it unreliable for search. Furthermore, the perceptual notion of whether two loops "match" or "complement" each other involves nuanced sonic characteristics - including timbre, rhythm, harmonic content, and energy - that are difficult to capture through simple metadata tags alone.

2.2 Loop Creation and Extraction

Several researchers have explored automated loop creation from existing musical material. Shi et al. [16] developed LoopMaker, which automatically extracts loops from an input music track by comparing segments within the same track to identify repeating patterns suitable for looping. While this reduces the time required to create loops from existing material, the approach is limited to extracting loops from a single input track and does not address the problem of searching through external loop libraries or community repositories. It is also unclear whether their method supports extracting loops from individual arrangement tracks (e.g., isolating only the guitar track) within a multi-track composition.

Similarly, Chaudhuri et al. [7] explore generating music loops from input tracks for live performance setlist preparation. Their approach focuses on loop generation and arrangement rather than retrieval from existing collections, and thus does not help musicians search through their existing loop libraries.

Roma et al. [15] propose a heuristics-based approach to identify loops in unstructured audio data by analyzing rhythmic content. Their method assumes loops contain rhythmic patterns and uses this assumption to detect loop boundaries within audio streams. While useful for loop detection and segmentation, this work does not address the challenge of ranking or recommending loops based on their perceptual compatibility with a reference loop or specific musical context.

2.3   Loop Compatibility and Recommendation

The work most closely related to ours is that of Yi et al. [19], who address loop compatibility in music production contexts. Their approach operates in two phases: first, they reconstruct loop embeddings for specific instruments (drums, bass, etc.) from audio; second, they train mapping models to predict compatibility between instrument-specific loops using these reconstructed embeddings. This method requires training datasets containing compatible loop pairs - loops known to sound good when played concurrently - which they obtain by assuming that loops from the same commercial music pack are mutually compatible.

While this assumption provides a practical solution for creating training data, it has several limitations. First, musicians often work with loops from diverse sources - including self-created loops, community repositories, and multiple commercial libraries - that are not pre-organized into compatible collections. Second, the approach requires separate trained models for each instrument pair (drum-bass, drum-piano, etc.), which may not scale well as the number of instrument types increases. Third, the reliance on music pack organization assumes a level of library structure that may not exist in typical musician workflows, where loops are often collected opportunistically from various sources over time.

2.4   Research Gap and Our Contribution

Our review reveals a significant gap in existing research: while prior work addresses loop creation, extraction, and instrument-specific compatibility prediction, there has been limited investigation into using general-purpose AI audio models to assist musicians in searching for perceptually similar loops from large, unstructured libraries. Furthermore, existing approaches have not systematically explored leveraging AI models explicitly trained on human perceptual judgments to guide loop search and ranking tasks.

This paper addresses this gap by investigating whether readily available AI audio models - specifically, embedding-based retrieval models (VGGish) and perceptual similarity metrics trained on human judgments (CDPAM) - can effectively narrow the search space for loop selection in a proof-of-concept study. Our approach offers several advantages over existing methods: it does not require instrument-specific training data or models, does not assume pre-organized compatible loop pairs, and can work with diverse, unstructured loop collections that musicians typically encounter in practice. By combining efficient embedding-based retrieval with perceptual refinement, we aim to demonstrate the feasibility of AI-assisted loop search that aligns with human perceptual judgments.

3   LoopMatcher Workflow: Solution Architecture

LoopMatcher is designed as a modular workflow that integrates multiple AI audio models to assist musicians in loop search tasks. Given a reference loop as input, the system searches a pre-populated database and returns a ranked list of the top-k most perceptually similar loops. The workflow is designed to be reconfigurable, allowing different AI models to be substituted at each processing stage to accommodate future advances in audio AI technology.

3.1   System Overview

The LoopMatcher architecture consists of four main modules, as illustrated in Fig. 1: (1) input processing, (2) similarity specification, (3) search, and (4) result presentation with user feedback. These modules work in sequence to transform a reference audio input into a ranked list of matching loops.

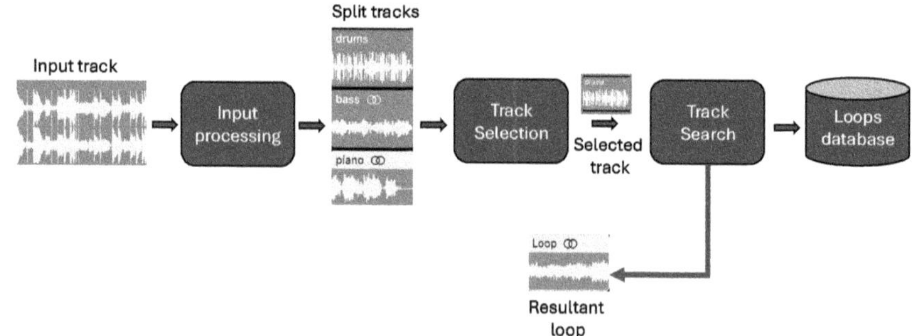

Fig. 1. LoopMatcher architecture showing the four main stages: (1) input processing with optional track separation, (2) user selection of reference tracks, (3) two-stage AI search (VGGish + CDPAM), and (4) presentation of ranked results with feedback option

3.2   Workflow Components

Input Processing Module. The user begins by providing an audio file as input. The input processing module can optionally decompose this audio into its constituent arrangement tracks (drums, bass, guitar, piano, vocals, etc.) using source separation AI models. This decomposition is implemented through the AIMaD configuration framework (discussed below), which allows the specific source separation model to be specified declaratively. Track separation enables users to search for loops that match specific instrumental components of their reference material.

Similarity Specification Module. After track separation, the user selects which tracks or track combinations should serve as the reference for similarity matching. As illustrated in Fig. 2, users may choose to use all tracks combined, or isolate specific instruments (e.g., only the drum track) as the search reference. This flexibility enables musicians to discover not only perceptually similar loops, but also those that enhance specific elements of their existing material—for example, identifying bass loops that complement a particular drum pattern.

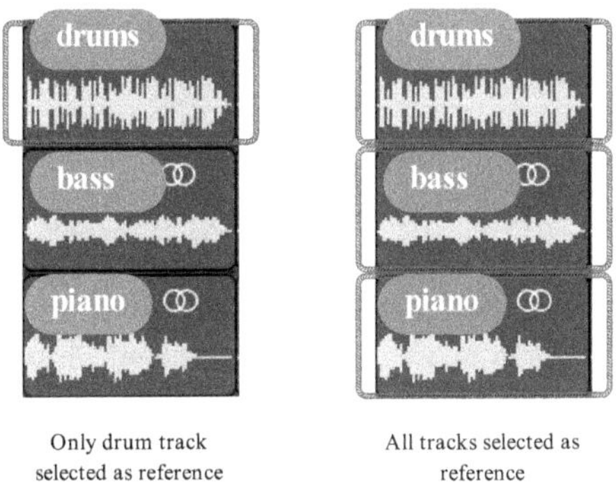

Fig. 2. Track selection as reference for loop search. Users can select individual tracks (e.g., drums only) or combinations of tracks to serve as the search reference.

Search Module. Using the selected reference audio, the search module queries a vector database pre-populated with embeddings of loops from the user's library. The search process employs multiple AI models in sequence: first, an embedding model (e.g., VGGish) retrieves candidate loops based on acoustic similarity; second, a perceptual similarity model (e.g., CDPAM) re-ranks these candidates to align with human perceptual judgments. The database population is performed as a one-time setup step and can be incrementally updated as users add new loops to their libraries.

Result Presentation and Feedback. The workflow concludes by presenting the top-k ranked loops to the user, where k is a configurable parameter. Users can audition these loops and provide feedback on the quality of each match. This feedback mechanism is designed to support future refinements, potentially enabling the system to learn user-specific preferences over time, though feedback integration is not implemented in the current proof-of-concept version.

3.3 AI Model Integration via AIMaD

To facilitate model substitution and reduce the programming expertise required to configure the system, LoopMatcher employs a simple configuration file format, which we call as *AI for Music Application Description* (AIMaD). AIMaD provides a declarative interface for specifying which AI models should be used at each workflow stage. For example, users can specify VGGish for embedding generation or substitute alternative models by modifying the configuration file without changing the underlying code.

The current AIMaD implementation is intentionally minimal, serving primarily to demonstrate the concept of low-code AI model integration in music applications. Future development could expand AIMaD into a more comprehensive framework with standardized interfaces for diverse AI audio models, potentially lowering barriers to AI adoption for musicians and music software developers. However, in this proof-of-concept work, AIMaD serves mainly as an architectural pattern rather than a fully developed system.

3.4 Design Rationale

The modular architecture provides several advantages. First, the separation of embedding-based retrieval and perceptual refinement allows the system to balance computational efficiency (for initial filtering of large databases) with perceptual accuracy (for final ranking). Second, the configurable design enables experimentation with different AI models without restructuring the workflow. Third, the track selection capability gives users fine-grained control over what aspects of their reference material should guide the search. These design choices aim to create a flexible foundation for investigating AI-assisted loop search while acknowledging that significant development remains before such systems could be deployed in production music software.

4 LoopMatcher: Implementation Details

This section describes our proof-of-concept implementation demonstrating the two-stage retrieval approach from Sect. 3.

4.1 Model Selection and Architecture

Figure 3 shows our implementation employing two complementary AI models addressing the efficiency-accuracy tradeoff. We select representative models from

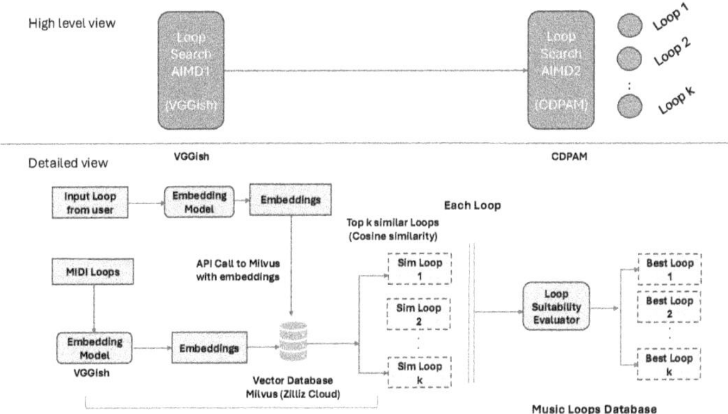

Fig. 3. Two-stage implementation: VGGish retrieval, CDPAM re-ranking

each approach - embedding-based and perceptual - for our goal of feasibility assessment.

VGGish. [11] generates compact 128-dimensional embeddings enabling efficient similarity search. Though not the most recent model, its computational efficiency and robust performance [8] make it well-suited for initial retrieval from 3000 loops. The compact embeddings enable faster searches compared to higher-dimensional alternatives (512 or 1024 dimensions).

CDPAM. [14] employs contrastive learning to align with human perceptual judgments. While computationally expensive, we apply it only to VGGish-retrieved candidates, limiting cost while benefiting from perceptual accuracy. This exploits complementary strengths: VGGish rapidly filters the database to a manageable candidate set; CDPAM refines based on perceptual similarity that better matches human judgments. Hence, the two-stage design leverages the complementary strengths of fast embedding-based retrieval enabling scalability as well as ensuring alignment with human similarity judgments for the candidates filtered by VGGish.

4.2   Database Setup and Processing

We conducted all experiments using a T4 GPU runtime on Google Colab. The software environment included Python libraries such as TensorFlow 2.12.0, requests, audiofile, audresample, and midi2audio, along with the FluidR3_GM sound font for MIDI rendering. We use 3000 loops from the MIDI-Loops dataset [1] (uniformly 8 bars/32 beats), hosted in Milvus [18] on Zilliz Cloud [4] using HNSW (Hierarchical Navigable Small World graph) indexing for efficient approximate nearest neighbor search [3]. The uniform structure simplifies proof-of-concept implementation by eliminating variable-length handling, though real-world systems would need to address tempo normalization and beat alignment.

Stage 1 (VGGish): Convert MIDI-Loops to WAV, extract 128-dim embeddings via TensorFlow, upload to Milvus, then query using cosine similarity via ANN search [13] [2] to retrieve K candidates (we use $K = 12$ in experiments).

Stage 2 (CDPAM): Compute CDPAM scores for each candidate using Librosa, then re-rank by descending scores. This may significantly reorder VGGish results, as perceptual similarity can differ from acoustic feature similarity in ways important to listeners.

4.3   Algorithm

Table 1 defines notation; Algorithm 1 presents the complete logic.

Table 1. Algorithm Notation

Symbol	Description
L_{in}	Input reference loop
D	Loop database with VGGish embeddings
K	Candidates to retrieve
e_{in}	128-dim VGGish embedding of L_{in}
C	Top-K candidates from Milvus
L_c	Candidate loop from C
$s[L_c]$	CDPAM score for L_c
R	Final ranked list

Implementation Scope: This setup helps validate our core search functionality. Source separation, feedback integration, and incremental updates are presently excluded. The uniform 8-bar format simplifies implementation but limits generalizability; production systems would need to handle diverse lengths, tempos, and audio formats.

Algorithm 1. LOOP_MATCHER

---

function LOOP_MATCHER(L_{in}, D, K)
    $e_{in} \leftarrow$ VGGish_Embedding(L_{in})
    $C \leftarrow$ Milvus_Search(e_{in}, D, K, cosine, HNSW)
    for each L_c in C **do**
        $s[L_c] \leftarrow$ CDPAM_Similarity(L_{in}, L_c)
    end for
    $R \leftarrow$ Sort(C, by $s[\cdot]$, descending)
    return R
end function

---

5   Validation

We recruited eleven volunteers with diverse music interests (listening, performing, making) and diverse cultural backgrounds. Ages ranged twenties to fifties
(nine males, two females), including one professional musician, two church performers, and eight enthusiastic listeners. This diversity allows assessing whether-
rankings align across different musical expertise levels.

5.1   Experimental Design

We selected one reference loop with clear rhythmic and harmonic content based
on the melody from Stevie Wonder's *'I Just Called to Say I Love You'* - a widely
recognized song that provides familiar musical context for participants to judge
candidate loops. We generated top twelve matches via LoopMatcher for the input
reference loop, recording ranks (1–12) without revealing them to volunteers. We
added three random loops as negative controls to validate match/non-match
distinction, creating fifteen total loops divided randomly into three groups of
five.

Volunteers rated each loop's "match" to the reference on a 1–5 Likert scale
(Fig. 4), interpreting "match" based on musical knowledge with minimal guidance (e.g., "sounds good together or in sequence"). This loose definition captures
intuitive judgments musicians make in practice, though it introduces subjectivity. Volunteers could listen to loops multiple times, individually and combined
with the reference.

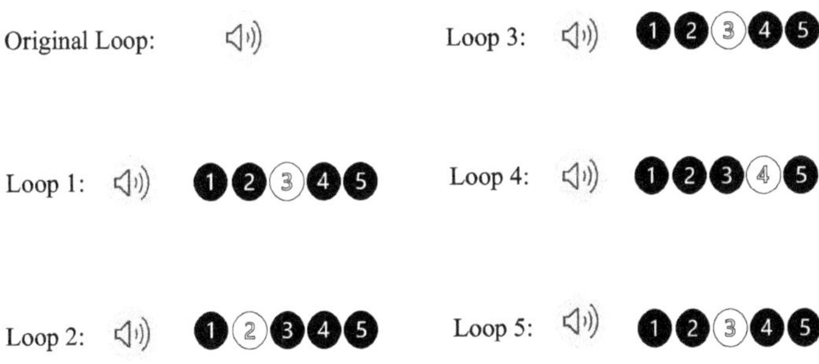

Fig. 4. Survey interface for perceptual validation

5.2   Results

We calculated average volunteer rankings per loop and obtained Spearman's
correlation of 0.68 with LoopMatcher rankings - a strong correlation per Evans
[9] (0.60–0.79), particularly given musical preference subjectivity. Beyond the
correlation coefficient, we observe good agreement in specific patterns. Top six

LoopMatcher-ranked loops received high volunteer rankings (Fig. 5), suggesting the system effectively identifies best matches - the most important use case for musicians.

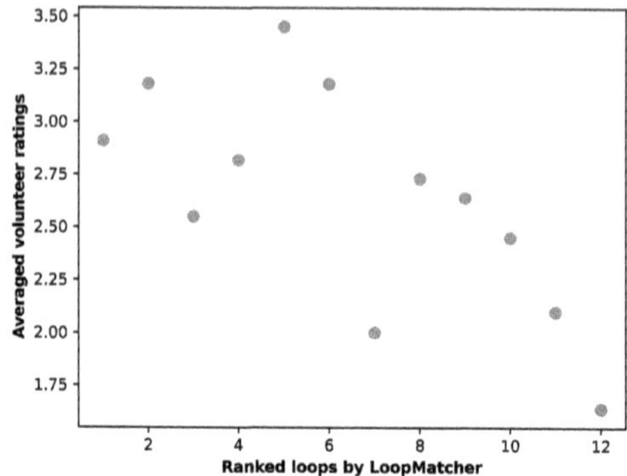

Fig. 5. LoopMatcher vs. volunteer rankings (higher values indicate better matches)

Critically, three randomly selected loops ranked lowest by volunteers (average ratings below 2/5), validating successful match/non-match distinction. This demonstrates meaningful similarity detection rather than arbitrary rankings. The twelfth-ranked loop also received lowest volunteer ranking among non-random loops, showing consistency at both ranking extremes.

5.3   Limitations

We acknowledge several important limitations in our experimental design that define the scope of these findings. First, though carefully designed, our validation is based on a single reference loop, which limits our ability to assess how the system performs across diverse musical styles, tempos, and instrumentation. Second, with only fifteen test loops (twelve matched and three random), this study serves as an initial proof-of-concept rather than a comprehensive validation. Third, all volunteers are from the same organization, which may introduce sampling bias despite their diverse cultural backgrounds. Fourth, we intentionally left the notion of "match" loosely defined to capture intuitive musical judgments, but this introduces subjectivity that makes it difficult to isolate specific aspects of similarity (e.g., rhythmic, timbral, or harmonic matching).

Despite these limitations, the results provide encouraging evidence that AI models can potentially assist musicians in narrowing the search space for music loops. The strong correlation (0.68) between algorithmic and human rankings, combined with the successful identification of negative controls, suggests that the approach merits further investigation with more extensive validation studies.

6   Discussion

While AI models for music are being actively developed, their adoption by musicians for tasks like loop search remains limited. We observe growing adoption of AI for specific applications such as musical track mastering [5,6], yet significant potential exists for AI to assist artists in searching musical sounds from large libraries to enhance their creativity. In our work, we explored the feasibility of using AI audio models to implement the LoopMatcher workflow, which narrows the search space automatically based on the perceptual similarity of loops, enabling artists to select loops from a smaller set of options and thereby saving time and energy.

6.1   Interpretation of Results

Our laboratory experiments demonstrate encouraging results. The 0.68 Spearman correlation between algorithmic and human rankings is noteworthy given the subjective nature of musical preference and the diversity of participants ranging from professional musicians to enthusiastic listeners. Importantly, the system successfully identified randomly selected loops as poor matches, validating meaningful similarity detection rather than arbitrary patterns. The two-stage architecture effectively narrowed 3000 loops to highly relevant candidates, confirming that combining complementary models, one optimizing for computational efficiency and the other for perceptual accuracy, is a viable strategy for music retrieval applications. Although the results indicate a generally strong positive correlation between human rankings and LoopMatcher's rankings, instances of negative correlation were also observed. Discussions with several participants suggested that a small subset of volunteers approached the notion of loop similarity differently. Rather than focusing solely on perceptual similarity, they also considered compositional or structural aspects when evaluating the loops. This alternative interpretation was limited to only a few individuals, but it likely contributed to the observed deviations.

6.2   Future Research Directions

The utility of such a solution must be validated in professional environments to assess real-world impact. Future work should address several key directions: (1) expanded validation with multiple reference loops across diverse genres and instrumentations; (2) larger studies with more participants from varied organizations; (3) baseline comparisons against metadata-only search and alternative AI models to quantify added value; (4) real-world deployment as DAW plugins with longitudinal user studies in actual creative workflows; (5) technical enhancements including investigation of recent models (transformers, foundation models like CLAP) and handling of variable-length loops. We intend to build a production version of LoopMatcher for distribution to practicing musicians to collect feedback about its practical utility.

6.3   Implications for AI Integration in Music Production

Beyond technical validation, we see a significant opportunity for Digital Audio Workstations to adopt architectures enabling seamless AI model integration into musicians' workflows. Musicians' limited AI adoption may stem from technical expertise requirements, for example, setting up AI models typically demands programming knowledge, API usage, and GPU management, which are prohibitive for those whose expertise lies in creativity rather than software development. In our work, we explored a possible architecture using wrapper modules configured via simple description files (AIMaD). While our current implementation is basic, it demonstrates the concept of configuration-based AI integration that could lower barriers to adoption.

DAWs could benefit from architectures facilitating AI integration analogous to how audio plugin formats (VST, AU) enabled rich ecosystems accessible without programming knowledge. However, AI models present additional challenges including heavier computational requirements, external dependencies, and rapid evolution. Focused efforts going forward should be directed toward defining frameworks like AIMaD with appropriate structures to capture the suitability of AI models for music-centric applications, balancing simplicity for musicians with flexibility for diverse AI models. This represents an important area for future research at the intersection of human-computer interaction, software engineering, and music technology.

7   Conclusion

The use of music loops in composition and performance is a common practice among musicians. However, selecting the appropriate loops from large libraries containing thousands of samples remains a challenging and time-consuming task. We investigated whether AI audio models could assist musicians in this search process by identifying perceptually similar sounding loops.

We developed LoopMatcher, a workflow integrating two complementary AI models - VGGish for efficient embedding-based retrieval and CDPAM for perceptual similarity refinement - to automatically narrow the search space for loop selection. Our feasibility study with eleven participants demonstrated a strong correlation (0.68 Spearman coefficient) between algorithmic rankings and human perceptual judgments. The system successfully distinguished matching loops from randomly selected controls, showing a meaningful convergence between algorithmic and human assessments of loop compatibility.

Our findings strongly suggest that AI audio models can meaningfully assist musicians in loop discovery, potentially saving time and enhancing creative efficiency. Future work should address identified limitations through expanded validation studies, real-world deployment, and technical enhancements. With continued development, AI-assisted search tools like LoopMatcher could become valuable additions to modern musicians' workflows, helping artists focus on creativity rather than tedious search tasks.

References

1. asigalov61/MIDI-Loops · Datasets at Hugging Face — huggingface.co. https://huggingface.co/datasets/asigalov61/MIDI-Loops. Accessed 05 Sept 2025
2. How can approximate nearest neighbor (ANN) search improve audio search efficiency? — milvus.io. https://milvus.io/ai-quick-reference/how-can-approximate-nearest-neighbor-ann-search-improve-audio-search-efficiency. Accessed 17 Sept 2025
3. Top 10 Audio Embedding Models for AI: A Complete Guide - Zilliz Learn — zilliz.com. https://zilliz.com/learn/top-10-most-used-embedding-models-for-audio-data#Audio-Similarity-Search-with-Vector-Databases, Accessed 05 Sept 2025
4. Zilliz: Vector Database built for enterprise-grade AI applications — zilliz.com. https://zilliz.com/. Accessed 05 Sept 2025
5. Aker, O.: Ai-assisted music mastering: An exploration of human and ai practices in contemporary music production. In: Understanding Generative AI in a Cultural Context: Artificial Myths and Human Realities, pp. 17–50. IGI Global Scientific Publishing (2025)
6. Birtchnell, T.: Listening without ears: artificial intelligence in audio mastering. Big Data Soc. **5**(2), 2053951718808553 (2018)
7. Chaudhuri, S.R.: Ai assisted workflow for set-list preparation with loops for live musicians. In: 2024 IEEE International Conference on Big Data (BigData), pp. 3163–3167. IEEE (2024)
8. Diwakar, M., Gupta, B.: The robust feature extraction of the audio signal by using vggish model. Int. J. Comput. Sci. Inf. Secur. (IJCSIS) **21**(6) (2023)
9. Evans, J.D.: Straightforward statistics for the behavioral sciences. Thomson Brooks/Cole Publishing Co (1996)
10. Henderson, M., Steffensen, P.B., Teglbjærg, D.S., Andersen, J.S., Antich, J.L.D., Jørgensen, T.: Methods and systems for organizing music tracks (Apr 16 2024), uS Patent 11,960,536
11. Hershey, S., et al.: Cnn architectures for large-scale audio classification. In: 2017 IEEE International Conference on Acoustics, Speech and Signal Processing (ICASSP), pp. 131–135. IEEE (2017)
12. Kitahara, T., Iijima, K., Okada, M., Yamashita, Y., Tsuruoka, A.: A loop sequencer that selects music loops based on the degree of excitement. In: Proceedings of the 12th Sound and Music Computing Conference (SMC 2015), pp. 435–438 (2015)
13. Li, W., Zhang, Y., Sun, Y., Wang, W., Li, M., Zhang, W., Lin, X.: Approximate nearest neighbor search on high dimensional data–experiments, analyses, and improvement. IEEE Trans. Knowl. Data Eng. **32**(8), 1475–1488 (2019)
14. Manocha, P., Jin, Z., Zhang, R., Finkelstein, A.: Cdpam: contrastive learning for perceptual audio similarity. In: ICASSP 2021-2021 IEEE International Conference on Acoustics, Speech and Signal Processing (ICASSP), pp. 196–200. IEEE (2021)
15. Roma Trepat, G., Serra, X.: Music performance by discovering community loops. Web Audio Conference (2015)
16. Shi, Z., Mysore, G.J.: Loopmaker: automatic creation of music loops from pre-recorded music. In: Proceedings of the 2018 CHI Conference on Human Factors in Computing Systems, pp. 1–6 (2018)

17. Souvignier, T.: Loops and grooves: the musician's guide to groove machines and loop sequencers. Hal Leonard Corporation (2003)
18. Wang, J., Hanson, E., Li, G., Papakonstantinou, Y., Simhadri, H., Xie, C.: Vector databases: What's really new and what's next? (vldb 2024 panel). Proc. VLDB Endowment **17**(12), 4505–4506 (2024)
19. Yi, X., Barthet, M., et al.: A generative framework for composition-aware loop recommendation in music production: Drum2bass use case (2024)

Multi-objective Evolution of Diffusion Model Prompt Embeddings Using CLIP-IQA

Marcel Salvenmoser[(✉)] [iD] and Michael Affenzeller [iD]

University of Applied Sciences Upper Austria, Wels, Austria
{marcel.salvenmoser,michael.affenzeller}@fh-ooe.at
https://heal.heuristiclab.com/team/affenzeller

Abstract. We study multi- and many-objective evolutionary optimization of image quality criteria in diffusion models, with the goal of characterizing the capabilities and limits of a diffusion model and an image quality scorer. Our search operates exclusively in the prompt embedding space: we evolve continuous embeddings that condition image generation, without using or modifying textual prompts. This representation is well suited for real valued evolutionary algorithms, enables fine-grained control, and avoids constraints imposed by discrete word-level prompt manipulation.

Multi-objective optimization is employed as an analysis tool to examine which score ranges are achievable, how performance scales with an increasing number of objectives, and where conflicts between image quality criteria arise under a fixed generator and evaluator. Prompt embeddings are evolved from random initialization while keeping model parameters unchanged.

Image quality is assessed using CLIP-IQA with native and custom axes such as "naturalness", "complexity", and "happiness". Experiments with three, four, five, and ten objectives compare a single-objective Genetic Algorithm based on summed scores with NSGA-III. Results show that embedding evolution consistently improves scores over random initialization and reveals trade-offs between visual criteria, with NSGA-III achieving higher hypervolume and greater diversity than the single-objective baseline.

Keywords: Diffusion Models · Prompt Embeddings · Multi-Objective Optimization · Evolutionary Algorithms · NSGA-III · Genetic Algorithms · CLIP-IQA · Hypervolume · Image Generation

1   Introduction

Text-to-image Diffusion Models (DM) have become state of the art in image synthesis, surpassing earlier GAN-based systems [6,16]. In typical usage, a user

P. Machado et al. (Eds.): EvoMUSART 2026, LNCS 16523, pp. 413–428, 2026.
https://doi.org/10.1007/978-3-032-24350-8_27

provides a textual prompt and the model generates corresponding images. Internally, most text-conditioned DM tokenize the input text and map it to a continuous prompt embedding, which conditions the denoising process during image generation [16]. In this paper, this representation is referred to simply as the (prompt) embedding.

While text prompts provide an intuitive interface for users, they offer only indirect control over the underlying continuous conditioning signal. The relationship between textual changes and embedding variations is opaque, and exploration is restricted to discrete textual descriptors. Directly operating in embedding space removes these constraints, enables fine-grained control, and aligns naturally with real valued evolutionary operators. This motivates algorithmic search in embedding space rather than search over discrete text strings.

We investigate an Evolutionary Computing approach that evolves prompt embeddings of a frozen diffusion model to steer image generation toward multiple image quality criteria. Image evaluation is performed using Contrastive LanguageâĂŞImage Pre-training Image Quality Assessment (CLIP-IQA), which provides zero-shot quality axes such as "naturalness", "brightness", and "quality" based on CLIP similarity to paired positive and negative references [24]. These criteria are not independent and often exhibit trade-offs, motivating the formulation of embedding evolution as a Multi-Objective Optimization (MOO) problem. Previous work in this area has primarily focused on Single-Objective Optimization (SOO) or open-ended exploration [17,18]. To the best of our knowledge, embedding evolution has not been studied in a many-objective setting.

1.1   Approach and Study Design

We compare a single-objective Genetic Algorithm (GA) with the Non-dominated Sorting Genetic Algorithm III (NSGA-III) [5] across several objective configurations, including both custom-defined CLIP-IQA criteria and native metrics introduced in the CLIP-IQA framework [24]. Four experimental setups are considered:

1. **Three custom objectives:** a simple multi-objective scenario inspired by the fast-good-cheap project constraint triangle [23], used to examine whether images can satisfy all three criteria simultaneously.
2. **Four objectives with opposing axes:** a configuration including *happy* and *sad* as directly conflicting criteria, evaluated using NSGA-III to study conflict resolution.
3. **Five native CLIP-IQA objectives:** a medium-scale setup near the transition between multi- and many-objective optimization, used to analyze interactions between commonly used quality metrics.
4. **Ten native CLIP-IQA objectives:** a many-objective stress test designed to assess scalability and population diversity under higher objective dimensionality.

Each experiment is executed once per algorithm under identical initialization and shared computational budgets. We report hypervolume (HV), per-objective and summed fitness values. For qualitative analysis, we include radar plots and low-dimensional projections of generated images together with their corresponding embeddings.

1.2   Research Questions

This work addresses two research questions. The first focuses on quantitative behavior and scalability, while the second examines qualitative outcomes and visual trade-offs.

RQ1 Effectiveness and Scalability of Multi-objective Optimization vs. Single-Objective. Does evolving prompt embeddings with NSGA-III improve scores relative to a random initial population and outperform a single-objective GA in terms of scoring and diversity? How does this difference change as the number of objectives increases, and what score ranges are reachable?

RQ2 Objective Trade-Offs and Visual Characteristics. Which visual characteristics emerge when individual CLIP-IQA criteria are emphasized, how do conflicting or competing objectives constrain achievable images, and how are these trade-offs reflected in the structure of the embedding space and the resulting image diversity?

2   Theoretical Background

This section provides the conceptual foundations underlying our study: DMs for image generation, the role of embeddings and the CLIP encoder, and CLIP-based evaluation metrics used as optimization objectives.

2.1   Diffusion Models, Embeddings, and CLIP

Diffusion Models (DMs) are generative systems that synthesize images by iteratively denoising random noise through a learned reverse process [16]. Compared to Generative Adversarial Networks (GANs) [8], they offer more stable training, higher fidelity, and greater diversity [12]. Although DMs are in general more computationally expensive, recent techniques such as Adversarial Diffusion Distillation reduce inference time while preserving visual quality [19]. Thus paving the way for models such as SD-Turbo and SDXL-Turbo to be used in experiments where many images are being generated.

Furthermore, modern variants operate in a latent space, combining a U-Net-based denoiser with a variational autoencoder and a text encoder that transforms textual prompts into embeddings used to condition image generation [16]. These *prompt embeddings* are compact numerical representations capturing the semantic essence of the text. They enable efficient computation and meaningful similarity comparisons [15]. A core component in this process is CLIP (Contrastive Language-Image Pretraining) [14], which aligns text and image embeddings in a shared space using a contrastive objective. CLIP supports zero-shot generalization and underpins many DMs, where *embeddings* guide the denoising process [12]. Blending or interpolating CLIP embeddings, as shown in [15], results in generated images that combine visual traits from each source prompt, an observation that directly inspires the embedding-level crossover and mutation strategies explored in this work. Beyond conditioning, CLIP also enables evaluation metrics.

2.2   CLIP-Based Metrics: CLIP-IQA, CLIPScore, and CLIP-AGIQA

CLIP can be used not only to condition DMs, but also as a general image evaluation mechanism by measuring imageâĂŞtext similarity. Several CLIP-based metrics exist for assessing image quality or imageâĂŞtext alignment.

CLIP-IQA as described in [24] defines image quality criteria through antonym text prompt pairs. For example "Bright photo." and "Dark photo." form a single quality axis called "Brightness". For a given image, CLIP computes cosine similarities to both text embeddings, and a softmax maps the result to a normalized score in the range $[0, 1]$. This allows zero-shot evaluation of perceptual and abstract attributes such as "contrast", "colorfulness", or "noisiness" without task-specific training.

Other CLIP-based metrics include CLIPScore [11], which focuses on imageâĂŞcaption alignment, and CLIP-AGIQA [7], which adapts CLIP-IQA toward AI-generated image assessment.

3   Related Work

Research on optimizing text-to-image DMs can be grouped into prompt-level, image-level and embedding-level.

PROMPTIST [10] optimizes prompts directly in text space through a two-stage process-supervised fine-tuning followed by reinforcement learning with a reward combining CLIP relevance and aesthetic quality. It improves generated image quality but remains single-objective and discrete in nature.

ImageBreeder [21], in contrast, operates at inference time, benchmarking multiple evolutionary operators for DMs in the latent image space and using *ImageReward* as a scalar fitness function. It demonstrates that Evolutionary Algorithm driven pipelines can enhance image quality.

Moving beyond images and text, *Evolving the Embedding Space* [17] introduces real-valued genetic optimization directly in the prompt-embedding space of SDXL-Turbo, guided by the LAION Aesthetics Predictor V2. *Open-Ended Evolution via an Island Model* [18] extends this idea by evolving multiple embedding populations under low selection pressure to explore stylistic diversity using PCA and UMAP analyses. Both approaches show that manipulating continuous embeddings is feasible and expressive, yet they focus on aesthetic or exploratory single-objective optimization rather than explicit MOO.

In summary, prior work either optimizes discrete prompts, latent representations, or embeddings for optimizing image criteria. None apply explicit multi- or many-objective optimization across distinct perceptual axes. Our work extends this line by evolving prompt embeddings under multiple CLIP-IQA criteria using GA and NSGA-III, comparing their performance through hypervolume and visual trade-offs.

4 Methodology

We evolve prompt embeddings directly in the continuous numeric space of a frozen DM. Operating in this space provides more direct access to the model's conditioning mechanism than modifying textual prompts and enables finer control, as the search is not constrained by discrete word-level representations. This makes the embedding space well suited for real valued evolutionary operators.

Concretely, we use the SDXL-Turbo text encoder and generator, following the embedding-based optimization approach of [17]. The choice of DM is discussed in Sect. 4.2. Image quality is evaluated using CLIP-IQA, and optimization is performed over multiple criteria in a multi-objective setting.

Our pipeline is simple: (i) initialize a population of embeddings randomly, (ii) modify them with evolutionary operators using mutation and crossover, (iii) generate the corresponding images with the frozen DM, and (iv) score the images on multiple CLIP-IQA criteria in $[0, 1]$. For optimization, we use a form of NSGA-III and compare it in most experiments with a GA.

4.1 Problem Formulation

Let the DM's text encoder output an embedding tensor $\mathbf{E}$ of arbitrary shape, for example $\mathbf{E} \in \mathbb{R}^{t_1 \times t_2 \times \cdots \times t_K}$. For optimization, we flatten this tensor into a one-dimensional vector $\mathbf{z}$ as follows:

$$\mathbf{z} = \mathrm{vec}(\mathbf{E}) \in \mathbb{R}^D, \qquad D = \prod_{k=1}^{K} t_k.$$

Here, $\mathrm{vec}(\cdot)$ denotes the vectorization operator, which stacks all tensor entries in a fixed and consistent order. The resulting vector $\mathbf{z}$ lies in a real-valued search space of dimension D, representing all embedding components as continuous parameters. During optimization, we *ignore the original tensor structure* and directly apply variation operators - mutation and crossover - on $\mathbf{z}$. Given $\mathbf{z}$, the frozen DM renders one image, which we score with M CLIP-IQA objectives. Each objective is higher-is-better in $[0, 1]$ and our multi-objective goal is to maximize all criteria $\mathbf{f}(\mathbf{z}) = \big(f_1(\mathbf{z}), \ldots, f_M(\mathbf{z})\big) \in [0, 1]^M$.

For the GA we reduce the vector objective to a single score by summation and *maximize* $s(\mathbf{z})$. This keeps the same evaluators while using a simple scalar fitness.

4.2 Diffusion Model and Encoder Choice

Not all DMs allow direct modification of prompt embeddings. Many online models accept only textual input and are computationally expensive for large-scale image generation. Thus, research in this area typically relies on locally usable models such as Stable Diffusion [22] or the newer FLUX models [2], both of which provide research-ready checkpoints.

For our experiments, we use **SDXL-Turbo**, trained via *Adversarial Diffusion Distillation* [19] from SDXL-Base. It offers an effective balance between speed, image quality, and accessibility, enabling efficient local experimentation. The embeddings resulting from the SDXL-Turbo text encoder are split between the token-level tensor 77×2048 and a pooled embedding of size 1280. These influence image generation differently: token embeddings provide fine-grained control, while pooled embeddings capture overall semantic context [13]. Although, this model was our choice for these experiments, the same approach can be extended to other models with compatible text encoders.

4.3 Initial Embedding Population and Variation

We initialize each embedding vector $\mathbf{z}$ randomly within the empirical value ranges observed from real text encoder outputs. To define this range, we iterate over all prompts from the *Google Parti Prompts (P2)* dataset [9] and record the minimum and maximum value at each embedding position. These per-dimension limits span a feasible range of the encoder output space. During initialization, we then sample random vectors uniformly within this absolute minimum and maximum range and use them as starting embeddings for the DM. This is done both for the embeddings and the pooled embeddings used in SDXL-Turbo. Other DMs may only require a single tensor here. This restriction prevents embeddings that would otherwise produce images that are noisy and not meaningful, shown in [17].

For variation two main operators are defined and used: arithmetic crossover and gaussian mutation.

Crossover combines two parent embeddings (and their pooled variants separately) $\mathbf{z}^A, \mathbf{z}^B \in \mathbb{R}^D$ into one offspring by weighted averaging:

$$\mathbf{z}' = \alpha\,\mathbf{z}^A + (1 - \alpha)\,\mathbf{z}^B, \qquad \alpha \in [0, 1].$$

By default, $\alpha = 0.5$, which gives the simple arithmetic mean. Crossover can be applied either to the whole vector (default) or to a random subset of positions, controlled by a crossover proportion ρ_c.

Mutation changes random positions of the embedding vector. First, a random subset of indices is selected according to the mutation rate p_m. Then, for each selected index i, a normally distributed random value $\varepsilon_i \sim \mathcal{N}(0, 1)$ is scaled by a fixed strength parameter σ and added to that position:

$$z_i' = z_i + \sigma\,\varepsilon_i, \qquad \text{for selected } i.$$

All other positions remain unchanged. After mutation, the vector is clamped to the valid range $\ell \leq \mathbf{z}' \leq \mathbf{u}$, where ℓ and $\mathbf{u}$ are the absolute per-dimension minima and maxima derived from the Google Parti Prompts P2 encoder outputs (see Sect. 4.3). This prevents the search from leaving the embedding space where meaningful images can be generated.

4.4   Algorithmic Setup

Full experimental runs are computationally expensive, so each configuration is executed once and the algorithmic setup is kept consistent across experiments. Core parameters are fixed, while only the CLIP-IQA objective sets, population size, and number of generations are varied.

We use SDXL-Turbo with a fixed denoising budget of $InferenceSteps = 4$, as recommended in [19]. Variation operators are shared: arithmetic crossover with $\alpha = 0.5$; mutation on token embeddings with rate $p_m = 0.2$ and strength $\sigma = 2.0$; and mutation on pooled embeddings with rate $p_m = 0.2$ and strength $\sigma = 0.4$. These values are specific to the SDXL-Turbo text encoder and were selected through manual experimentation to induce meaningful image variation while preserving structural and visual coherence.

Algorithm-wise a variant of NSGA-III and objective-summed GA is implemented.

NSGA-III. The Non-dominated Sorting Genetic Algorithm III (NSGA-III) [5] extends NSGA-II [4] and is used to address multi- and many-objective optimization problems. In high-dimensional objective spaces, most individuals tend to become non-dominated, reducing selection pressure. NSGA-III mitigates this by associating solutions with predefined reference directions and favoring individuals that contribute to a uniform spread across the objective space [5].

Our implementation integrates components from the `pymoo` library for reference direction generation, non-dominated sorting, and survival. Parent selection, variation, and evaluation are handled within the *Evolutionary Diffusion Framework*. Each generation produces one offspring per parent ($\lambda = \mu$) via crossover and mutation, followed by immediate CLIP-IQA evaluation. Selection uses binary tournament based on Pareto rank, with ties broken by larger $\sum_m f_m$, inspired by but not fully implementing U-NSGA-III [20]. Environmental selection follows elitist $\mu + \lambda$ survival using energy-based reference directions [3].

GA. As a baseline, we use a standard Genetic Algorithm [1] with identical variation operators. Multi-objective scores are reduced to a single scalar by summation, which is maximized. Selection uses tournament selection with $k = 3$ and elitism of one individual.

Fitness Evaluation. Fitness evaluation is performed exclusively using **CLIP-IQA** [24]. Each generated image is scored according to the selected CLIP-IQA criteria, including both native metrics and custom-defined objective combinations used in the experiments.

5   Experimental Setup

Based on the algorithmic setup, we design several experiments to study different aspects of MOO using embeddings and a frozen DM. The experiments vary the number of CLIP-IQA objectives from three up to ten and include both custom and competing criteria. Except for the competing-criteria case, each configuration is executed with a GA (single-objective, using summed fitness) and our NSGA-III implementation (multi- or many-objective) to allow direct comparison. N denotes the population size and G the number of generations in each experiment.

All code and experimental artifacts are made publicly available. Our implementation extends the *Evolutionary Diffusion Framework*[1] and includes executable Jupyter notebooks for *Google Colab*.

Outputs and Analyses

The experimental tools are designed to address our two main research questions (RQ1 and RQ2).

- **Hypervolume (HV):** Estimated as dominated hypervolume on the fixed box $[0, 1]^M$ using Monte Carlo sampling with 800,000 uniform points. A point is counted if it is weakly dominated by at least one solution in the final population, and we report a normal-approximation 95% interval reflecting Monte Carlo estimator uncertainty (not run-to-run variance). HV is computed on the last population and measures coverage of the Pareto front, addressing **RQ1** by comparing GA and NSGA-III under matched budgets.
- **Per-objective fitness and radar charts:** Used to visualize score distributions of selected individuals across objectives and to show trade-offs among CLIP-IQA axes. These visual summaries support both **RQ1** (quantitative progress) and **RQ2** (qualitative trade-offs).
- **UMAP projection:** Applied to embeddings and image sets to explore structure and diversity in the evolved solutions. This analysis mainly supports **RQ2** by revealing how different objectives perform and affect eachother in the embedding space.
- **Fitness statistics:** First, last and average fitness comparisons show optimization progress and stability, serving as a quantitative indicator for **RQ1**.

---

[1] https://github.com/malthee/evolutionary-diffusion.

5.1   Three Custom Objectives

We define three custom CLIP-IQA criteria using antonym prompt pairs:

"Good photo." vs "Bad photo.", "Dynamic photo." vs "Static photo.",
"Cheap-looking photo." vs "Expensive-looking photo."

following the CLIP-IQA style of formulation [24]. The criteria are chosen as a simple example and are inspired by the "good-fast-cheap" project constraint triangle [23], here adapted to image generation as good, dynamic, and cheap-looking.

This setup serves as a minimal multi-objective case to compare a single-objective GA with NSGA-III. Both algorithms use identical variation operators, DM settings, and a shared budget of $N = 200$ and $G = 100$.

Table 1. Three-objective experiment: hypervolume of the last generation on fixed $[0, 1]^3$ and summed fitness $\sum_m f_m$.

Method	HV [95% CI]	$\sum_m f_m$
GA	0.579 [0.578, 0.580]	1.920
NSGA-III	0.664 [0.663, 0.665]	2.144

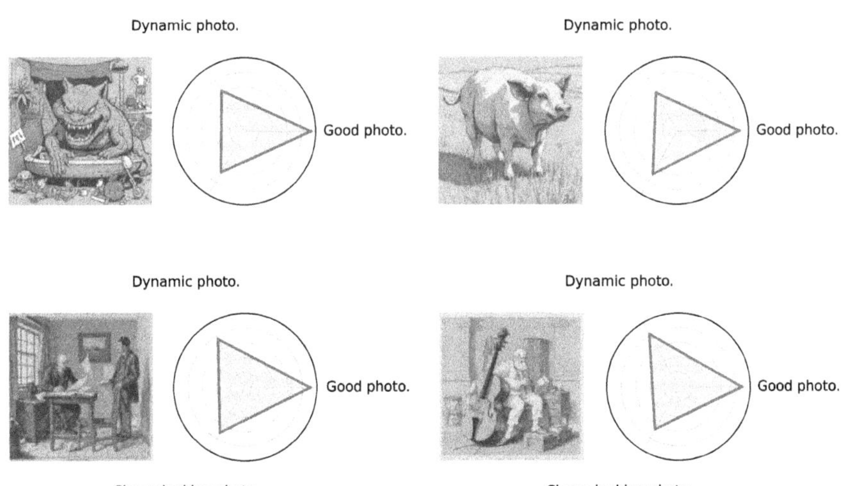

Fig. 1. Three-objective setup: comparison of first (top) and last (bottom) generation results. NSGA-III is shown on the left and GA on the right. Radar charts depict objective scores alongside the best-scoring images by summed fitness.

Table 1 shows that NSGA-III outperforms the GA in both hypervolume and summed fitness under identical conditions. Hypervolume increases by approximately 15%, while summed fitness improves by about 11.7%. Starting from an initial average fitness of 1.771, all three objectives improve by at least 15–30% for both methods.

Early generations show low scores for the "dynamic" and "cheap-looking" criteria, while the "good" criterion scores relatively higher. After 100 generations, both GA and NSGA-III produce high-scoring images that predominantly depict busy indoor scenes, indicating convergence toward similar visual characteristics under this objective configuration (Fig. 1).

5.2 Four Objectives with Competing Axes ("happy" Vs "sad")

This experiment focuses on directly competing objectives. The following criteria are evaluated:

"quality", "beautiful",

"Happy image." vs "Sad image.", "Sad image." vs "Happy image."

The "quality" and "beautiful" metrics are included to reach four objectives and to improve overall image quality. We use $N = 200$ and $G = 100$, running NSGA-III only.

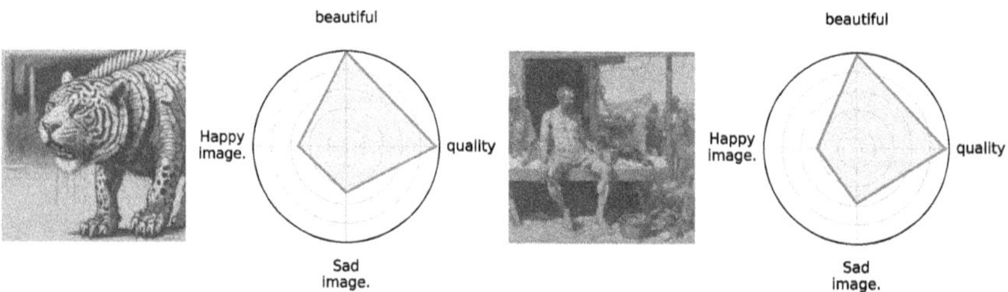

Fig. 2. Best balanced-scoring images for the four-objective ("quality", "beautiful", "happy", "sad") setup. Left: generation 15 with scores 0.961, 0.988, 0.520, 0.480. Right: generation 17 with scores 0.961, 0.981, 0.425, 0.575.

Table 2. Four-objective experiment ("happy" vs "sad"): hypervolume of the last generation on fixed $[0, 1]^4$ and summed fitness $\sum_m f_m$.

Method	HV [95% CI]	$\sum_m f_m$
NSGA-III	0.442 [0.441, 0.443]	2.702

The initial population already scores highly on the "sad", "quality", and "beautiful" criteria, indicating a bias in either the DM or the CLIP-IQA objective space.

Because the "happy" and "sad" axes are mutually contradictory by construction, the theoretical maximum summed score is 3.0, as both emotional criteria saturate near 0.5. Consequently, the hypervolume computed with a fixed $[0, 1]^4$ reference region shown in Table 2 includes a large unattainable portion of the objective space, making this result less comparable to others. Likewise compared to the initial population, maximum total fitness improves only marginally (approximately 2% from 2.649).

During evolution, average "sad" scores decrease by about 14.2%, while "happy" scores increase slightly, leading to a more balanced distribution with final population centers around ~ 0.3 for "happy" and ~ 0.7 for "sad".

The selected high-scoring images differ in composition while achieving comparable balances between the conflicting emotional criteria (Fig. 2), illustrating that multiple diverse visual solutions exist under this constraint.

5.3   Five Native CLIP-IQA Objectives

We now return to the built-in criteria from the CLIP-IQA paper [24]. In this experiment, we address a moderately many-objective problem using five native metrics:

$$\text{"naturalness", "quality", "contrast", "sharpness", "noisiness"}$$

The goal is to evolve embeddings that balance these sometimes competing criteria, seeking images that fulfill all listed criteria. Although, some objectives may partially conflict, they are not strictly opposing like they are in the experiment 5.2. For parameterization, we use a similar general setup with a smaller amount of generations $N = 200$ and $G = 70$.

Table 3. Five-objective experiment using native CLIP-IQA metrics: hypervolume from the last generation (fixed $[0, 1]^5$) and summed fitness $\sum_m f_m$.

Method	HV [95% CI]	$\sum_m f_m$
GA	0.625 [0.624, 0.626]	3.776
NSGA-III	0.744 [0.743, 0.745]	4.194

Table 3 summarizes the results. NSGA-III achieves a hypervolume improvement of approximately 19% over the GA. Summed fitness increases by about 11% relative to the GA and by 21% compared to the average fitness of the initial population (3.454).

Several criteria perform well already in the first generation, including "noisiness", "sharpness", and "quality", each exceeding 0.9 in the best individuals. In

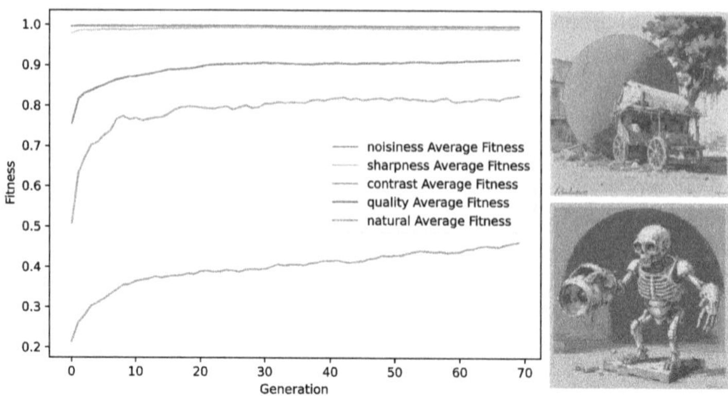

Fig. 3. Five-objective experiment. Left: average population fitness across generations. Right: two top-performing images from the final NSGA-III generation, each scoring at least 0.9 in all criteria except "contrast" (at least 0.7).

contrast, "naturalness" and especially "contrast" start at lower values (maximum ≈ 0.21). Over the course of evolution, average "naturalness" increases by roughly 62%, while "contrast" more than doubles ($+115\%$, from 0.21 to 0.45). Overall, the average population fitness increases by about 21% during the NSGA-III run, despite using fewer generations than earlier experiments.

Figure 3 illustrates both the optimization progress and the diversity of high-scoring images. Although top individuals differ in composition, they share common visual characteristics such as the hue, suggesting that improvements in objectives such as "contrast" and "naturalness" guide the search toward similar regions of the embedding space.

5.4   Ten Native Objectives

In the final experiment, we consider a many-objective setting with ten native CLIP-IQA metrics to examine how NSGA-III scales with increasing dimensionality. The objectives cover a broad range of visual and perceptual attributes, some of which inherently compete. We use $N = 120$ and $G = 200$, making this the most computationally intensive setup in the study. The evaluated criteria are:

"quality", "sharpness", "contrast", "colorfulness", "brightness", "natural",
"real", "beautiful", "new", "complexity"

Table 4 shows that NSGA-III improves hypervolume by approximately 17% compared to the GA and increases summed fitness by about 2.7%. Relative to the initial population (average fitness 5.907), NSGA-III raises total fitness by roughly 15%.

Performance differs across objectives. Metrics that score highly in early generations, such as "sharpness", "real", and "beautiful", improve only marginally.

Table 4. Ten-objective experiment: hypervolume of the last generation on fixed $[0, 1]^{10}$ and summed fitness $\sum_m f_m$.

Method	HV [95% CI]	$\sum_m f_m$
GA	0.117 [0.116, 0.118]	6.67
NSGA-III	0.137 [0.136, 0.138]	6.85

The strongest gains are observed for "contrast" (+81%) and "complexity" (+70%). In contrast, NSGA-III reduces "colorfulness" by about 5.6%, while the GA increases it by approximately 8%, highlighting a clear trade-off that emerges only in the many-objective setting.

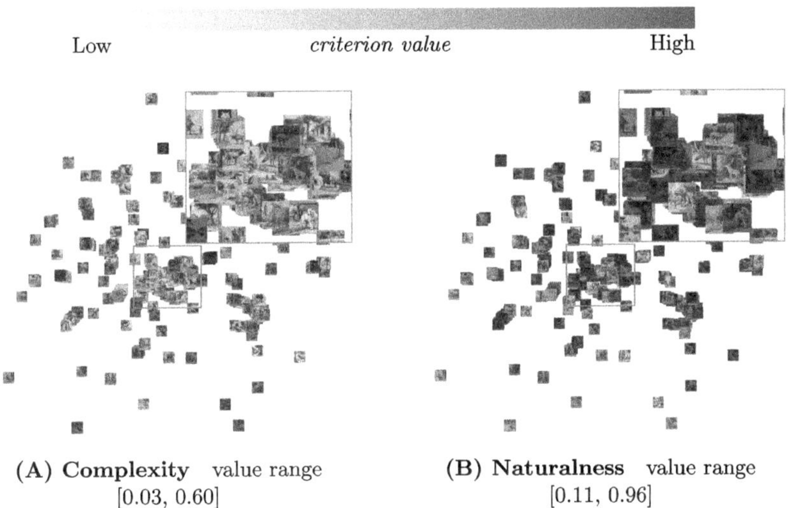

Fig. 4. 2D UMAP projection (N=15, $dist$=0.1) of prompt embeddings across generations with image thumbnails. Colors follow the legend (top): pale yellow → deep blue indicate low to high criterion values. Both panels share the same UMAP layout but are colored by different exemplary IQA criteria.

Figure 4 furthermore illustrates a trade-off using a UMAP projection of embeddings across two selected objectives "complexity" and "naturalness". The figure reveals a distinct clustering pattern: images scoring highly in "complexity" tend to perform lower in "naturalness", and vice versa. Although these criteria are not direct antonyms, the embedding space reveals an implicit competition between them.

6   Results and Discussion

Across all experiments, both GA and NSGA-III improve over the initial random populations, confirming that evolutionary search in the prompt embedding space is effective. Absolute hypervolume values differ across objective sets, but NSGA-III consistently achieves higher hypervolume than the summed single-objective GA in all comparable configurations, with gains typically in the range of 10% to 20%. Differences in summed fitness are smaller, especially in higher-dimensional settings, which is consistent with NSGA-III prioritizing coverage of diverse trade-offs instead of maximizing a single aggregated score. These results answer RQ1 and indicate that in a multi-objective setting scores are able to improve and MOO provides more robust scaling than summed single-objective optimization under matched budgets.

Qualitative analysis supports this interpretation. Radar plots in Figs. 1 and 2 show that emphasizing individual CLIP-IQA axes produces visible shifts in image characteristics, while directly conflicting objectives constrain attainable scores and lead to multiple valid compromises. In the ten-objective case, the UMAP projection in Fig. 4 shows structured regions in the embedding space where improvements in some criteria coincide with degradation in others, making implicit trade-offs observable even when objectives are not defined as antonym pairs. These observations address RQ2 and demonstrate that the method exposes how the DM and CLIP-IQA jointly shape the space of attainable images.

7   Conclusion and Outlook

This work shows that multi-objective evolution of prompt embeddings in a frozen DM is feasible and effective as an analysis tool. By optimizing multiple CLIP-IQA criteria simultaneously, the experiments expose achievable score ranges, trade-offs between visual attributes, and limitations that arise from the interaction between the DM and the scoring function.

Although NSGA-III consistently outperforms a summed single-objective GA in hypervolume coverage, the main contribution is the analytical perspective enabled by multi-objective search in embedding space. The approach reveals how image quality criteria interact, where conflicts emerge, and which visual characteristics are associated with different regions of the search space, without modifying model parameters or relying on textual prompt design.

7.1   Limitations

Runtime per configuration is substantial (approximately 4–10 h on local hardware[2]), which limited most settings to a single run. The majority of compute time is spent on DM inference (roughly three quarters), with CLIP-IQA evaluation accounting for the remaining quarter. In addition, conclusions are specific to the chosen evaluator (CLIP-IQA) and the single DM used in this study (SDXL-Turbo), and may not transfer directly to other scoring functions or generators.

---

[2] Apple M2 Max, 32 GB RAM.

7.2   Outlook

This line of work can be extended in several directions:

- **Generational Models:** other DMs (e.g., Turbo- variants, larger models).
- **Evaluators:** other image scorers and fitness objectives, mixed objective sets.
- **Human alignment:** side-by-side preference studies to validate whether CLIP-IQA or other optimized objectives gains reflect human judgments.
- **Algorithmic ablations:** mutation/crossover variants respecting the tensor-structure and using other parameterizations.

Disclosure of Interests. The authors have no competing interests to declare that are relevant to the content of this article.

References

1. Affenzeller, M., Winkler, S.M., Wagner, S., Beham, A.: Genetic Algorithms and Genetic Programming: Modern Concepts and Practical Applications . https://doi.org/10.1201/9781420011326
2. Black-forest-labs (Black Forest Labs). https://huggingface.co/black-forest-labs. Accessed 2025 25 Mar
3. Blank, J., Deb, K., Dhebar, Y., Bandaru, S., Seada, H.: Generating well-spaced points on a unit simplex for evolutionary many-objective optimization. IEEE Trans. Evol. Comput. **25**(1), 48–60 (2021). https://doi.org/10.1109/TEVC.2020.2992387
4. Deb, K., Pratap, A., Agarwal, S., Meyarivan, T.: A fast and elitist multiobjective genetic algorithm: NSGA-II. IEEE Trans. Evol. Comput. **6**(2), 182–197 (2002). https://doi.org/10.1109/4235.996017
5. Deb, K., Jain, H.: An evolutionary many-objective optimization algorithm using reference-point-based nondominated sorting approach, part i: Solving problems with box constraints. IEEE Trans. Evol. Comput. **18**(4), 577–601 (2014). https://doi.org/10.1109/TEVC.2013.2281535
6. Dhariwal, P., Nichol, A.: Diffusion models beat gans on image synthesis (2021). https://arxiv.org/abs/2105.05233, accessed: 2026-01-29
7. Fu, J., Zhou, W., Jiang, Q., Liu, H., Zhai, G.: Vision-language consistency guided multi-modal prompt learning for blind ai generated image quality assessment (2024). https://arxiv.org/abs/2406.16641, accessed: 2026-01-29
8. Goodfellow, I., et al.: Generative Adversarial Nets **27**(5), 2672–2680
9. Google-research/parti (2022). https://github.com/google-research/parti. Accessed 18 Mar 2024
10. Hao, Y., Chi, Z., Dong, L., Wei, F.: Optimizing prompts for text-to-image generation (2023). https://arxiv.org/abs/2212.09611. Accessed 29 Jan 2026
11. Hessel, J., Holtzman, A., Forbes, M., Le Bras, R., Choi, Y.: CLIPScore: A Reference-free Evaluation Metric for Image Captioning. https://doi.org/10.18653/v1/2021.emnlp-main.595. Accessed 29 Jan 2026
12. Po, R., et al.: State of the Art on Diffusion Models for Visual Computing . https://doi.org/10.48550/arxiv.2310.07204. Accessed 29 Jan 2026

13. Podell, D., et al.: Sdxl: improving latent diffusion models for high-resolution image synthesis (2023). https://arxiv.org/abs/2307.01952. Accessed 29 Jan 2026
14. Radford, A., et al.: Learning Transferable Visual Models From Natural Language Supervision . https://doi.org/10.48550/arXiv.2103.00020. Accessed 29 Jan 2026
15. Ramesh, A., Dhariwal, P., Nichol, A., Chu, C., Chen, M.: Hierarchical Text-Conditional Image Generation with CLIP Latents . https://doi.org/10.48550/arxiv.2204.06125. Accessed 29 Jan 2026
16. Rombach, R., Blattmann, A., Lorenz, D., Esser, P., Ommer, B.: High-Resolution Image Synthesis with Latent Diffusion Models. https://doi.org/10.1109/cvpr52688.2022.01042. Accessed 29 Jan 2026
17. Salvenmoser, M., Affenzeller, M.: Evolving the embedding space of diffusion models in the field of visual arts. In: Machado, P., Johnson, C., Santos, I. (eds.) Artificial Intelligence in Music, Sound, Art and Design, pp. 402–416. Springer, Cham (2025). https://doi.org/10.1007/978-3-031-90167-6_27. Accessed 29 Jan 2026
18. Salvenmoser, M., Affenzeller, M.: Open-ended evolution of artistic styles in diffusion models via island-based genetic algorithms. In: Proceedings of the Genetic and Evolutionary Computation Conference Companion, pp. 2046–2054. GECCO '25 Companion. Association for Computing Machinery, New York (2025). https://doi.org/10.1145/3712255.3734290, https://doi.org/10.1145/3712255.3734290, accessed: 2026-01-29
19. Sauer, A., Lorenz, D., Blattmann, A., Rombach, R.: Adversarial Diffusion Distillation . https://doi.org/10.48550/arxiv.2311.17042. Accessed 29 Jan 2026
20. Seada, H., Deb, K.: U-nsga-iii: a unified evolutionary optimization procedure for single, multiple, and many objectives: Proof-of-principle results. In: Gaspar-Cunha, A., Henggeler Antunes, C., Coello, C.C. (eds.) Evolutionary Multi-Criterion Optimization, pp. 34–49. Springer, Cham (2015)
21. Sobania, D., Briesch, M., Rothlauf, F.: Imagebreeder: Guiding diffusion models with evolutionary computation. In: Proceedings of the Genetic and Evolutionary Computation Conference, pp. 463–471. GECCO '25. Association for Computing Machinery, New York (2025). https://doi.org/10.1145/3712256.3726439, https://doi.org/10.1145/3712256.3726439. Accessed 29 Jan 2026
22. Stability ai image models. https://stability.ai/stable-image. Accessed 25 Mar 2025
23. Van Wyngaard, C., Pretorius, J.H., Pretorius, L.: Theory of the triple constraint — a conceptual review, pp. 1991–1997 (12 2012). https://doi.org/10.1109/IEEM.2012.6838095
24. Wang, J., Chan, K.C.K., Loy, C.C.: Exploring CLIP for Assessing the Look and Feel of Images. https://doi.org/10.48550/arxiv.2207.12396. Accessed 29 Jan 2026

EvoArtist: A Visual LLM-Driven Agentic AI Framework for Autonomous Design Evolution

Kamer Ali Yuksel[✉] and Hassan Sawaf

aiXplain Inc., San Jose, CA, USA
`{kamer,hassan}@aixplain.com`

Abstract. Interactive Evolutionary Computation (IEC) has long served as a powerful methodology for human-guided design optimization, enabling the iterative refinement of artifacts through user-in-the-loop selection. However, the reliance on human input for fitness evaluation poses significant limitations in scalability, consistency, and exploration depth. In this work, we present a fully autonomous, LLM-driven agentic AI system for evolutionary design—one that eliminates the need for human evaluators by integrating the generative and evaluative capabilities of Large Language Models (LLMs) and Visual Large Language Models (V-LLMs). Our system introduces two novel contributions over traditional IEC approaches: (1) the use of LLMs as intelligent mutation and crossover operators that perform directed transformations of code and prompts, informed by implicit domain expertise; and (2) the use of visual LLMs as automated evaluators and feedback providers, not only to select offspring for breeding but also to generate detailed, expert-level assessments that steer the evolutionary trajectory. This system is instantiated in two domains of high creative complexity: landing-page design and generative visual art. In both domains, LLM-driven transformations generate high-quality, functional, and aesthetically pleasing outputs, while visual LLMs act as domain-aware critics, iteratively refining designs toward optimal forms. Through extensive experiments and qualitative assessments, we demonstrate that our system significantly outperforms traditional IEC and stochastic design pipelines in terms of innovation, convergence speed, and quality of output. This work offers a blueprint for future autonomous design systems, revealing the untapped potential of V-LLMs in evolution-inspired creative processes.

1 Introduction

The field of evolutionary computation has historically empowered designers and researchers to explore vast and high-dimensional design spaces, offering a stochastic yet guided mechanism to discover novel configurations. Interactive Evolutionary Computation (IEC), in particular, introduced the human into the loop, allowing users to manually evaluate and select generated outputs based on aesthetic, functional, or subjective criteria. This hybridization between machine-driven generation and human judgment has led to significant creative break-throughs, especially in art, music, and user interface design. However, despite its appeal, IEC remains inherently limited by its dependence on human evaluators. These limitations manifest as scalability bottlenecks, evaluator fatigue, subjective inconsistency, and reduced throughput, especially in domains that demand large-scale iterative design cycles. Recent advancements in generative AI, particularly in the form of large language models (LLMs) and visual-language models (V-LLMs), offer a compelling alternative. These models encapsulate vast swaths of domain-specific and general-purpose knowledge, enabling them to understand, generate, and critique content in a manner that approximates expert human reasoning. When embedded within an agentic AI framework, these models can simulate not only the role of the human evaluator but also that of the creative ideator—autonomously proposing modifications, selecting optimal candidates, and steering the evolutionary process toward desired objectives.

In this work, we propose an LLM-driven agentic AI framework that autonomously evolves creative designs by combining interactive evolutionary principles with v-LLMs. Drawing inspiration from Picbreeder-style interactive evolution[1], our system replaces both human-driven variation and evaluation with AI-driven counterparts. Firstly, a GPT-4.1-based module serves as an intelligent crossover and mutation operator, generating new design variants (code or prompts) by leveraging its implicit domain knowledge rather than random perturbations. Secondly, human subjective selection is supplanted by V-LLM evaluators that assess visual outputs, acting as automated "subject matter experts" to select promising candidates and even providing feedback to guide the next generation. The key contributions of the proposed framework are threefold:

- The use of LLMs as intelligent evolutionary operators—performing targeted crossover and mutation operations grounded in semantic and functional coherence rather than random perturbations. These transformations draw upon the LLM's internalized understanding of design principles, enabling guided rather than stochastic evolution.
- The deployment of visual LLMs as both autonomous selectors and expert critics—evaluating the visual and functional merit of generated variants and providing rich feedback that guides subsequent evolutionary steps. These models act not only as evaluators but as co-evolutionary agents that inform and contextualize the generative logic of the LLM operators.
- Through this fully autonomous system, we demonstrate that the combination of LLMs and V-LLMs within an agentic feedback loop yields a powerful framework for creative design evolution. We validate this hypothesis through case studies in two distinct domains: landing-page design and generative art.

In this paper, we detail the methodology and system architecture – including key modules for mutation, crossover, selection, and feedback – and demonstrate autonomous iterative design evolution without human intervention. Case studies in web landing-page design and generative art (character illustration and architectural concept evolution) highlight the system's ability to produce high-quality designs. Finally, we report results indicating more directed and meaningful evolutionary trajectories and faster convergence compared to traditional interactive evolution, and discuss scalability to large design spaces. Figure 1 summarizes the closed-loop architecture, showing how visual-LLM evaluation provides critique and selection signals that steer LLM-based crossover and mutation.

2   Background and Related Work

Interactive Evolutionary Computation (IEC) introduced the paradigm of human-in-the-loop evolution, in which users evaluate AI-generated variants to guide evolutionary search [9]. IEC democratized creative optimization: it enabled novices without domain expertise to evolve complex images, sounds, interfaces, and artifacts simply by expressing preferences. The most iconic instantiation is *Picbreeder* [8], where users collaboratively bred images encoded by compositional pattern-producing networks (CPPNs). Over hundreds of community-driven generations, simple abstractions evolved into detailed forms such as faces, creatures, and ornate patterns—illustrating how subjective aesthetic selection can steer open-ended evolution toward semantically meaningful attractors without predefined objectives. Picbreeder exemplifies IEC's broader promise: humans can navigate high-dimensional aesthetic spaces in ways difficult to formalize. Subsequent work extended IEC to music composition, architectural forms, procedural game assets, UI layouts, 3D sculptures (notably Sims' seminal evolved virtual forms), and motion patterns. In these systems, human preference operates as a flexible but powerful fitness signal that captures subjective qualities such as beauty, balance, novelty, and emotional resonance. However, IEC is fundamentally bottlenecked by human limitations. User fatigue sharply restricts the number of evaluations per session, and cognitive load constrains feasible population sizes [9]. While techniques such as surrogate modeling of user preference, batched evaluation, aggregation across crowds, and heuristic filters can partially ameliorate these issues, they fail to capture the full subtlety of human aesthetic judgment. As a result, IEC rarely achieves the depth, scale, or open-endedness characteristic of biological evolution. A central challenge, therefore, has been whether the human evaluator can be replaced—or supplemented—by artificial agents that replicate or exceed human aesthetic and functional reasoning.

2.1   Interactive and LLM-Driven Design Evolution

Outside IEC, the broader field of evolutionary art has explored automated aesthetic evaluation. Early work evolved images using hand-crafted metrics for symmetry, contrast, or fractal dimension. Later, neural networks and discriminators began serving as heuristic judges—e.g., Creative Adversarial Networks (CANs), which encourage deviations from learned artistic norms. CLIP's emergence marked a turning point. Tian and Ha [10] demonstrated that CLIP embeddings could function as high-level aesthetic or semantic fitness signals. Their system evolved abstract arrangements of shapes such that CLIP judged them to match target textual descriptions ("Tokyo," "sunset forest"), achieving prompt-constrained creative evolution without human input. This CLIP-guided evolutionary strategy illustrated that multimodal embeddings possess latent knowledge of artistic style, object composition, and semantic alignment, making them viable automated evaluators. Beyond images, evolutionary approaches have also been applied to architectural layouts, UI designs, and web layouts, often optimizing geometric or usability constraints. Yet these approaches historically relied on handcrafted fitness functions. They lacked mechanisms to evaluate more subjective or context-sensitive notions such as visual hierarchy, typographic clarity, atmospheric coherence, or brand alignment.

LLMs have transformed generative design. Models such as GPT-4.1 [4] can produce functional code, HTML/CSS layouts, marketing copy, and design specifications, frequently rivaling human-created prototypes. Unlike classical genetic operators (random mutation, token-level crossover), LLMs incorporate semantic priors and vast contextual knowledge learned from large-scale corpora. They perform structured transformations rather than stochastic perturbations, effectively implementing *semantic mutation* and *pattern-preserving crossover*. Recent research formalizes this idea. Evolution through Large Models (ELM) [3] replaces mutation operators with generative models, using pretrained networks as distributions of semantically meaningful variation. Language Model Crossover (LMX) [11] exploits few-shot prompting to combine "parent" textual artifacts into offspring that preserve high-level regularities and stylistic coherence. These advances suggest that LLMs can serve as domain-general operators that capture design principles implicitly stored in their training data—far beyond the capabilities of classical evolutionary algorithms. LLMs also exhibit contextual self-referentiality. Promptbreeder [1] evolves prompts using LLMs recursively acting on their own outputs. This shows that LLMs can participate in closed evolutionary loops, refining internal representations across generations. These make them ideal candidates for automated creative evolution systems requiring complex, multi-step, logically coherent, or stylistically consistent transformations.

2.2 Agentic AI, Open-Endedness, and Autonomous Creativity

A growing body of work argues that open-endedness is a key ingredient of artificial superhuman creativity [2]. Agentic AI systems—composed of multiple interacting LLM/V-LLM modules—exhibit emergent behaviors such as self-reflection, iterative refinement, and goal-directed exploration. These systems can generate hypotheses, critique themselves, and update their internal representations through structured feedback loops. Autonomous design systems have emerged in HCI, architecture, and generative art, leveraging LLMs to produce design variants and evaluate usability or functionality. Promptbreeder and related systems demonstrate self-improving generative loops, while multimodal agents can transform sketches into code or critique rendered scenes. These developments suggest a fertile convergence between LLM-based reasoning, multimodal perception, and evolutionary search. Despite significant progress, prior evolutionary and generative systems suffer from three limitations:

1. **Lack of semantic variation**. Classical genetic algorithms operate on low-level perturbations that produce brittle or incoherent artifacts. Even neural evolutionary art lacks true semantic understanding.
2. **Limited evaluative richness**. CLIP-guided evolution provides semantic alignment but not multi-dimensional aesthetic judgment, functional critique, or reasoning-driven feedback.
3. **Absence of fully autonomous closed loops**. Existing systems either require human evaluators (IEC), rely on single-modality fitness, or delegate only part of the pipeline to large models.

Where LLMs enable high-level transformation, visual-language models provide the evaluative complement. CLIP [5] encodes images and text within a shared embedding space shaped by large-scale natural-language supervision. Its embeddings capture aesthetic structure, semantic content, compositional balance, stylistic traits, and even object relationships that approximate human judgment. Extensions such as LAION-Aesthetics [7] directly predict visual appeal scores, enabling automated aesthetic ranking at scale. Multimodal models like DALL·E 2 [6], Flamingo, and GPT-4V further enhance evaluative potential by providing both recognition and reasoning. They can produce natural-language critiques—e.g., identifying poor lighting, insufficient contrast, weak composition, or stylistic mismatch. These linguistic feedback loops are critical: unlike scalar fitness scores, textual critiques supply rich, actionable guidance that can be used to steer future mutations. They mimic expert designers who justify their assessments, articulate principles, and highlight specific areas for improvement. This transition—from numerical fitness to linguistic critique—represents a qualitative shift in evolutionary evaluation. It enables systems to combine algorithmic precision with human-like aesthetic reasoning, opening the path toward fully autonomous co-evolution between generation and evaluation.

We introduce the first agentic framework that unifies LLM generation and V-LLM evaluation into a fully autonomous closed-loop for evolutionary design. This synthesis establishes a new paradigm for creative AI: one in which evolution is no longer human-guided nor purely stochastic, but *agentic, semantic, multimodal, and autonomous*. Unlike previous systems, our approach leverages:

- **GPT-4.1 as a semantic mutation and crossover operator**, enabling design transformations grounded in domain knowledge, aesthetics, and structural coherence.
- **Visual LLMs as expert critics**, producing fine-grained assessments of aesthetic quality, functional correctness, and stylistic adherence.
- **Critique-informed evolution**, where V-LLM feedback is injected back into LLM prompts to steer subsequent generations.
- **Multimodal generality**, supporting both HTML/CSS/JS design evolution and generative art/prompt evolution.

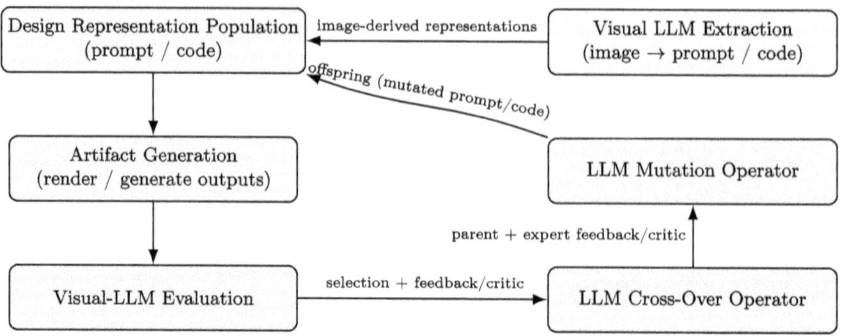

Fig. 1. System architecture of the LLM-driven evolutionary design loop. A visual LLM first extracts design representations (prompt/code) from images, forming an initial population. These representations are rendered into artifacts and re-evaluated by a visual LLM, whose judgments and critiques drive an LLM-based crossover operator that selects parent representations and an LLM-based mutation operator that produces offspring. The mutated variants update the design representation population, closing the autonomous open-ended evolution loop.

3   Methodology

In this section, we introduce our novel framework that marries LLM-guided variation with visual LLM-guided selection. Our system's first core innovation is leveraging a generative pre-trained transformer (GPT-4.1) as a crossover/mutation agent. Instead of applying blind stochastic tweaks, we prompt the LLM to generate new design candidates (e.g., code or image prompts) informed by the "genes" of selected parents and general domain knowledge. This yields mutations that are semantically meaningful improvements or recombinations, as the LLM can "guess the pattern" behind good designs [4] and produce offspring staying within a learned distribution of desirable features [11]. The second innovation replaces the human fitness function with V-LLM. These models automatically assess each candidate's merit – measuring qualities like aesthetic appeal, coherence with a target style, or functional criteria – and thus perform selection by picking top candidates. Moreover, they can produce feedback signals (scores or even textual explanations via image-to-text models) that inform the LLM on how to refine future generations. Our system introduces a novel agentic architecture wherein LLMs and visual LLMs operate in coordinated feedback loops to autonomously guide the evolutionary process. This methodology departs from conventional IEC in both structure and operational philosophy, aiming to create a truly self-improving design ecosystem. The evolutionary cycle is composed of the following core components. Each operation is performed by prompting the LLM or V-LLM with a history of parent designs, evaluator comments, and domain-specific design constraints, allowing it to propose novel and context-aware transformations.

Autonomous Selection and Expert Feedback via Visual LLMs: Visual LLMs play a dual role: evaluating and selecting the top-performing candidates, and providing detailed qualitative feedback. This feedback loop is essential for continuous design improvement. The visual LLM computes a fitness score based on aesthetic principles (symmetry, color harmony, composition), functional factors (readability, spacing, navigability), and domain alignment (e.g., e-commerce vs. artistic expression). Beyond scores, visual LLMs return structured feedback in natural language, outlining perceived strengths, weaknesses, and areas for enhancement. These evaluations are used as part of the next mutation prompt for the LLM, effectively embedding an expert critique into the generative process. This dual capability not only automates selection but also simulates the presence of an expert designer continuously refining and redirecting creative evolution. The system operates iteratively in distinct evolutionary cycles:

- Initialization: Generate an initial diverse population of designs based on basic prompts or templates.
- Autonomous Evaluation and Feedback: Visual LLMs assess each variant, providing explicit semantic and aesthetic feedback alongside design ratings.
- Directed Crossover and Mutation: Using the feedback as context, LLMs perform sophisticated mutations and crossovers to generate contextually improved offspring.
- Evolutionary Iteration: Continuously iterate this loop, refining designs progressively until convergence or defined performance metrics are satisfied.

Directed Crossover and Mutation Operators Using LLMs: Traditional genetic algorithms rely on randomly selected crossover points and mutation probabilities to generate new offspring. In contrast, our approach employs LLMs as semantic-aware transformers that leverage their pretrained knowledge to conduct meaningful recombination and transformation of design representations. These operations are informed by both structural syntax (e.g., HTML/CSS or prompt tokens) and functional semantics (e.g., layout consistency, user intent, or stylistic adherence). Instead of naive interpolation, the LLM synthesizes inputs from selected parents to produce a coherent and goal-aligned child. It identifies complementary features across parent candidates and merges them while resolving potential conflicts (e.g., contradictory layout features or style mismatches). Directed Mutation is context-sensitive rather than random. Given a current design and visual LLM feedback, the LLM mutates structural or stylistic elements with the intention of improving fitness e.g. if feedback highlights poor visual hierarchy, the mutation may increase font contrast or rearrange sections for clarity.

3.1 Evolutionary Loop

After initialization, the system proceeds into an iterative evolutionary loop consisting of evaluation, selection, feedback, and LLM-driven variation. Individuals are scored by a V-LLM evaluator who quantifies their fitness with respect to aesthetic, functional, or stylistic goals. Scores from different evaluation dimensions can be combined for Pareto-based multi-objective ranking. The top-scoring individuals are selected to serve as parents for the next generation. Unlike IEC, this decision is made entirely by AI evaluators, enabling large populations and many generations without human fatigue. The LLM produces several children per parent pair. Importantly, GPT-4.1 frequently introduces creative innovations not present in either parent—a desirable effect analogous to beneficial spontaneous mutations in natural evolution. These variations are semantically guided rather than random, exploiting the model's latent knowledge of good layout, composition, or stylistic structure. The newly generated offspring form the next generation's population. We typically use an elitist strategy, preserving a small number of top performers while replacing the rest.

Fig. 2. Autonomous evolution of landing-page designs and marketing content using the LLM–V-LLM feedback loop. LLM-driven crossover and mutation operate over both structural code (HTML/CSS) and marketing copy, enabling holistic refinement of messaging, tone, semantic clarity, and conversion-oriented framing. The visual LLM evaluates not only aesthetic qualities (hierarchy, balance, CTA prominence) but also content relevance, readability, and alignment with the product narrative. Through iterative feedback, the system converges toward high-quality, conversion-optimized landing pages that integrate strong visual design with coherent, persuasive content without any human-in-the-loop selection.

4   Case Studies

To validate the efficacy, generality, and creative power of our proposed LLM-driven agentic AI system, we conducted extensive case studies across two diverse yet complementary domains: landing page design and generative visual art.

These domains were chosen because they require high levels of semantic alignment, visual aesthetics, functional coherence, and creativity—making them ideal for benchmarking the quality and adaptability of autonomous design.

4.1   Landing Page Design Evolution

In this case study (Fig. 2), the aim is to simultaneously evolve the front-end design and content of a website's marketing-oriented landing page for better conversion. A landing page typically includes elements like a header with navigation, a hero section with a tagline or image, feature sections, call-to-action buttons, and a footer. Good landing page design is characterized by visual appeal, clarity of message, and effective layout (so that a user quickly grasps the product/service being offered). Rather than manually crafting and A/B testing countless design variants, our system aims to auto-generate and evaluate these variants. The system is initialized with a population of generic HTML/CSS layouts generated from a seed prompt. Each evolutionary cycle proceeds as follows:

- **Generation:** LLMs perform directed crossovers and mutations of selected HTML and CSS blocks, introducing layout variations, visual enhancements (e.g., button placement, typography), and functional augmentations (e.g., responsive behavior or interactive components).
- **Evaluation and Feedback:** Visual LLMs assess each variant along multiple axes: layout balance, visual hierarchy, brand coherence, call-to-action clarity, and aesthetic minimalism. They assign scores and provide explicit feedback—e.g., "The current layout lacks a strong focal point; consider increasing contrast or repositioning the headline." This feedback is fed into the mutation operator to guide future generations.
- **Evolutionary Trajectory:** Across generations, the system autonomously produces increasingly sophisticated, responsive, and user-centric layouts. Design convergence typically occurs within 10–15 generations, yielding highly competitive variants that mimic professionally crafted designs.

In each generation, 5 new pages were produced from the current top 3 pages (so some parents produced multiple offspring). GPT-4.1 was prompted with the HTML of parent pages and instructions to "combine the strengths" or sometimes just mutate one design. For example, one prompt for crossover: "Here are two website designs. Design1 has great color scheme. Design2 has a good layout. Create a new design that uses the layout of Design2 with the color scheme of Design1, and improve anything you think will make it more user-friendly." The feedback module added specific tips (like "Ensure text contrast is high" if prior gen had contrast issues). After 10 generations (about 50 pages evaluated in total), the system produced a landing page that scored highly on all criteria. Qualitatively, the evolved design was impressive: it had the clean two-column layout from one ancestor, the vibrant blue-orange color scheme from another, and GPT-4.1 had added a novel testimonial section it "imagined" might benefit the page (this wasn't in any initial seed). The page was fully functional

HTML and had a coherent style guide (consistent font usage, spacing, etc.). When shown to a human web designer, they noted it looked like a decent template one might find on a professional site builder. This showcases the system's ability to autonomously iterate towards human-like designs. Notably, even without access to external design datasets or UI heuristics, the system consistently discovered high-performing design patterns (e.g., F-pattern layouts, card-based structures, above-the-fold CTA emphasis), showcasing the value of LLM-encoded prior knowledge and visual LLM aesthetic judgment.

(a) Representative samples from an intermediate evolutionary generation, illustrating the stylistic coherence while exploring diverse aesthetic and compositional directions.

(b) Outputs from the next evolutionary generation showing increased visual refinement, stronger thematic consistency, and higher alignment with the V-LLM feedback.

Fig. 3. Progression of autonomously evolved generative art across successive evolutionary populations. The first row demonstrates mid-process diversity driven by LLM-guided crossover and mutation, while the second row reflects the convergence effects of visual-LLM feedback, yielding sharper composition, richer texture, and more coherent narrative framing. These results highlight how the agentic LLM–V-LLM loop incrementally directs generative evolution toward higher-quality and more semantically grounded artistic outcomes.

4.2 Generative Visual Art Creation

In this second case study, we evaluated the system's creative capabilities in evolving prompt-based generative artworks, such as character illustrations and architectural renderings. We additionally demonstrate interior-architecture concept evolution in Fig. 4, where the system improves spatial plausibility, lighting, and material coherence across iterations. The idea is akin to an artist refining

Fig. 4. Autonomous evolution of interior-architecture designs guided by V-LLM feedback. The system progressively enhances spatial layout, stylistic coherence, lighting, and material realism through LLM-driven crossover and mutation. Visual LLMs assess both aesthetic quality and architectural plausibility—including spatial balance, atmosphere, and thematic consistency. Across iterations, the designs converge toward polished, human-quality architectural visualizations.

concept sketches gradually. We want the system to autonomously generate image prompts that lead to increasingly refined or stylistically evolving images through:

- **Initialization:** The system begins with a baseline prompt describing a visual concept (e.g., "A futuristic cityscape at sunset" or "A samurai warrior in watercolor style") or extracts them with V-LLM from a given example image.
- **Prompt Evolution via LLMs:** Using crossover and mutation, LLMs explore prompt space by introducing modifiers, compositional shifts, and stylistic refinements e.g. it may evolve from "a young girl in armor" to "a confident warrior girl in ornate gilded armor, anime style, high contrast lighting."
- **Visual Generation and Evaluation:** Images are generated using a diffusion model (e.g., Stable Diffusion) from the evolved prompts. Visual LLMs then evaluate each image on dimensions such as coherence with the prompt, visual creativity, thematic depth, and artistic balance. They also annotate regions or stylistic decisions that could be improved (e.g., "the background lacks depth" or "facial expression is inconsistent with mood").
- **Iterative Refinement:** Based on feedback, prompts are refined to improve not only fidelity but also artistic innovation and emotional resonance. This feedback-guided generation loop continues for several iterations.

Representative populations across successive generations are shown in Fig. 3, illustrating how V-LLM critique drives increased visual coherence and refinement. We ran approximately eight generations for each art scenario. The starting prompts were intentionally simple—for example, *"A warrior princess"* or *"A female knight"* for the character design experiments. GPT-4.1 was instructed to produce new prompts via both crossover and mutation operators. An example offspring prompt generated by GPT-4.1 was: *"A breathtaking portrait of a warrior queen in ornate golden armor, standing in a forest palace, ultra-detailed digital painting, dramatic lighting"*. GPT-4.1 might introduce a stylistic change or add a visual element, depending on the feedback and evaluation context.

The character illustrations evolved by our system showed a clear qualitative progression. Early generations were visually plain and often exhibited mild anatomical inconsistencies—a common artifact of raw generative image models. By generations five or six, after the system repeatedly emphasized attributes such as *"majestic, detailed, portrait"*, the resulting images displayed significantly more intricate armor, a more regal posture, and noticeably clearer facial features. In effect, the system "discovered" beneficial additions such as capes, jewelry, and atmospheric backgrounds that enhanced the sense of majesty; one evolved illustration even introduced a throne in the background without explicit prompting. These improvements emerged because GPT-4.1 incorporated patterns surfaced through V-LLM evaluation feedback: for instance, when a highly detailed generation received a strong score, the feedback highlighted phrases like *"intricate armor"*, which GPT-4.1 subsequently began integrating into its future outputs.

5   Results

Human blind studies show that the system's final outputs consistently outperform IEC baselines across originality, emotional impact, visual harmony, and prompt relevance. The evolutionary process also exhibited signs of emergent design intuition—such as selecting color palettes that reinforced mood or adjusting composition for clearer narrative flow. These effects arise from the interaction of LLM-driven semantic variation and V-LLM critique, enabling autonomous refinement without human aesthetic supervision. The generative art experiments further highlight the system's ability to navigate open-ended aesthetic spaces. By embedding artistic goals directly into visual evaluators, the system maintains coherent stylistic direction while introducing meaningful novelty. Across domains, LLM–V-LLM co-evolution matched or exceeded human-guided IEC while being faster, more scalable, and fully autonomous.

- **Improved Evolutionary Efficiency:** Designs reached high quality with fewer generations due to feedback-guided semantic mutations.
- **Higher Quality and Consistency:** V-LLM–selected outputs aligned closely with expert human judgments.
- **Greater Innovation and Diversity:** Evolved solutions demonstrated higher innovation indices and stylistic diversity than IEC baselines.
- **Semantic Mutations Accelerate Progress:** GPT-4.1 proposes high-level, coherent edits (e.g., enhancing contrast or structure) than random changes.
- **Scalable Automated Selection:** V-LLMs evaluate large populations consistently and without fatigue, enabling deeper evolutionary runs than IEC.
- **Stable Multiobjective Evaluation:** V-LLMs provide reproducible assessments across aesthetic and functional criteria.
- **Implicit Expert Priors:** The combination of LLM generative knowledge and V-LLM reasoning embeds design best practices w/o handcrafted rules.

Across both domains, the system reliably produced designs that met or exceeded evaluator-defined objectives. In the landing-page study, the final landing page layouts demonstrated professional-grade hierarchy, color coherence, and responsiveness. In a blind A/B test (20 comparisons), participants preferred the final-generation designs in 85% of cases, citing clarity and modern aesthetics. In the generative art tasks, images converged toward high visual quality, thematic consistency, and detailed composition. Independent artists ranked later-generation images significantly higher across coherence, detail, and stylistic fidelity. Qualitative inspection also revealed clear improvement trajectories, including refined armor details, improved lighting, richer backgrounds, and emotionally expressive character posture. The system frequently introduced useful novel elements—such as a throne or enhanced color grading—without explicit prompting, suggesting emergent design priors activated through iterative V-LLM feedback. Using LLMs as semantic mutation and crossover operators substantially improves convergence. Because variations reflect domain knowledge rather than random perturbations, most offspring remain viable. In the landing-page study, noticeable improvements appeared by generation 5, with convergence by generation 10 despite small population sizes. Generative art tasks required about eight generations due to the broader aesthetic search space.

- **Feedback-Driven Refinement:** V-LLMs not only score candidates but also provide actionable textual critiques, enabling a closed-loop cycle where the generative LLM performs targeted revisions.
- **Semantic Variation:** LLM-based operators reason over meaning and structure, producing offspring that preserve core intent while introducing coherent novelty—beyond the capabilities of traditional genetic operators.
- **Cross-Modal Generalization:** Because all reasoning occurs in natural language, the same evolutionary framework applies seamlessly across HTML/CSS layouts, visual prompts, and other design formats.
- **Emergent Intuition:** Iterative feedback activates latent design priors in LLMs and V-LLMs, leading to behaviors such as hierarchy optimization, refined composition, and stylistic harmonization.
- **Scalability:** Removing human-in-the-loop evaluation enables continuous, large-scale evolutionary search and long-horizon autonomous refinement.

6   Conclusion

We introduced an agentic AI framework that unifies LLM-based semantic variation with V-LLM aesthetic and functional evaluation to achieve fully autonomous design evolution. The system generates high-quality, semantically coherent, and visually compelling designs across both web layouts and generative art, outperforming IEC and random-search baselines. The case studies demonstrate that LLM/V-LLM co-evolution fosters both creativity and technical refinement, with emergent behaviors reflecting domain-specific design priors. Their results also indicate that our agentic architecture supports robust autonomous evolution across both code-based and prompt-based modalities. Generation and evaluation operate entirely through natural-language, enabling seamless transfer across domains and broader autonomous agentic design ecosystems for human–AI co-creativity.

References

1. Fernando, C., et al.: Promptbreeder: self-referential self-improvement via prompt evolution. arXiv:2309.16797 (2023). Accessed 11 Nov 2023
2. Hughes, E., et al.: Open-endedness is essential for artificial superhuman intelligence. arXiv:2406.04268 (2024). Accessed 11 Nov 2023
3. Lehman, J., et al.: Evolution through large models. arXiv:2206.08896 (2022). Accessed 11 Nov 2023
4. OpenAI: GPT-4 technical report. Technical report, OpenAI (2023). https://arxiv.org/abs/2303.08774. Accessed 11 Nov 2023
5. Radford, A., et al.: Learning transferable visual models from natural language supervision. arXiv:2103.00020 (2021). Accessed 11 Nov 2023
6. Ramesh, A., et al.: Hierarchical text-conditional image generation with clip latents. arXiv:2204.06125 (2022). Accessed 11 Nov 2023
7. Schuhmann, C., et al.: Laion-aesthetics: aesthetic score prediction with clip. GitHub Repository (2022). https://github.com/LAION-AI/aesthetic-predictor. Accessed 11 Nov 2023
8. Stanley, K.O., Miikkulainen, R.: Evolving neural networks through augmenting topologies. Evol. Comput. (2007)
9. Takagi, H.: Interactive evolutionary computation: fusion of the capabilities of EC optimization and human evaluation. Proc. IEEE **89**(9), 1275–1296 (2001)
10. Tian, Y., Ha, D.: Modern evolution strategies for creativity: fitting concrete images and abstract concepts. In: NeurIPS Workshop on AI Art (2021)
11. Toro, I., et al.: Language model crossover: variation through few-shot prompting. arXiv:2302.12170 (2023). Accessed 11 Nov 2023

Author Index

A

Aamir, Janita 3
Affenzeller, Michael 413
Allmendinger, Richard 383
Anderson, Neal 316
Artıktay, Güncel Gürsel 209
Assar, Najam-Ul 21

B

Briman, Eyal 34

C

Cella, Carmine Emanuele 303
Climent, Ricardo 383
Coulibaly, Nouhoum 224
Creangă, Claudiu 67

D

Dawson, Nana Amowee 49
De Blois, Joanie Verviers 97
Dinu, Anca 67
Donnelly, Patrick J. 3, 252, 347
Dwivedi, Himanshu 236

E

Eraud, Florentin 287

F

Florescu, Andra-Maria 67

G

Gasparini, Francesca 367
Goro, Ousmane 224
Grégoire, Anthony 97

H

Hennou, Kamal 287
Hinrichs, Reemt 113
Hromadka, Timo 129

J

Jadliwala, Murtuza 145
Jamwal, Vikram 398

K

Kwan, Derek 252

L

Lacasse, Serge 97
Lange, Alexander 113
Leizerovich, Eyal 34
Leventhal, Michael 224
Lioret, Alain 287
Ly, Ousmane 224

M

Madupu, Sai Pranav 398
Mainolfi, Pasquale 303
Maiti, Anindya 145
Majumder, Sanjay 316
Matsumura, Takeshi 331
McCabe, Peter 347
Mihail, Andreiana 67
Monteiro, Kevin 129

N

Nallaperuma-Herzberg, Sam 129
Nguyen, Nghi 271
Novacco, Alessia 367

O

Ostermann, Jörn 113

P

Pham, Lam 271
Pocquet, Tanguy 383
Pringle, Gavin J. 331

© The Editor(s) (if applicable) and The Author(s), under exclusive license
to Springer Nature Switzerland AG 2026
P. Machado et al. (Eds.): EvoMUSART 2026, LNCS 16523, pp. 443–444, 2026.
https://doi.org/10.1007/978-3-032-24350-8

R
Riber, Adrián García   84
Rizzi, Giulia   367
Roy Chaudhuri, Subhrojyoti   398

S
Sai, Krishna Teja Guru   398
Saibene, Aurora   367
Salvenmoser, Marcel   413
Sawaf, Hassan   176, 192, 429
Seidenberger, Scott   145
Smith, Megan   21
Son, Phan Le   271
Spracklen, Joseph   145
Stephan, Sonja   113

T
Talmon, Nimrod   34
Trinh, Khoi   145

V
Viswanath, Bimal   145

W
Wijewickrama, Raveen   145
Wu, Yueshen   161

X
Xue, Yuting   161

Y
Yuksel, Kamer Ali   176, 192, 429

GPSR Compliance
The European Union's (EU) General Product Safety Regulation (GPSR) is a set
of rules that requires consumer products to be safe and our obligations to
ensure this.

If you have any concerns about our products, you can contact us on

ProductSafety@springernature.com

In case Publisher is established outside the EU, the EU authorized
representative is:

Springer Nature Customer Service Center GmbH
Europaplatz 3
69115 Heidelberg, Germany

www.ingramcontent.com/pod-product-compliance
Ingram Content Group UK Ltd.
Pitfield, Milton Keynes, MK11 3LW, UK
UKHW020815080726
473059UK00007B/2258